USED BOOK

MAY 1 7 2008

Sold to Aztec Shops Ltd.

ANALYSIS AND PERFORMANCE OF FIBER COMPOSITES

THIRD EDITION

Bhagwan D. Agarwal
Consultant
Lombard, Illinois, USA

Lawrence J. Broutman
Consultant
Chicago, Illinois, USA

K. Chandrashekhara
University of Missouri–Rolla
Rolla, Missouri, USA

John Wiley & Sons, Inc.

This book is printed on acid-free paper. ∞

Copyright © 2006 by John Wiley & Sons, Inc. All rights reserved

Published by John Wiley & Sons, Inc., Hoboken, New Jersey
Published simultaneously in Canada

No part of this publication may be reproduced, stored in a retrieval system, or transmitted in any form or by any means, electronic, mechanical, photocopying, recording, scanning, or otherwise, except as permitted under Section 107 or 108 of the 1976 United States Copyright Act, without either the prior written permission of the Publisher, or authorization through payment of the appropriate per-copy fee to the Copyright Clearance Center, Inc., 222 Rosewood Drive, Danvers, MA 01923, (978) 750-8400, fax (978) 750-4470, or on the web at www.copyright.com. Requests to the Publisher for permission should be addressed to the Permissions Department, John Wiley & Sons, Inc., 111 River Street, Hoboken, NJ 07030, (201) 748-6011, fax (201) 748-6008, e-mail: permcoordinator@wiley.com.

Limit of Liability/Disclaimer of Warranty: While the publisher and the author have used their best efforts in preparing this book, they make no representations or warranties with respect to the accuracy or completeness of the contents of this book and specifically disclaim any implied warranties of merchantability or fitness for a particular purpose. No warranty may be created or extended by sales representatives or written sales materials. The advice and strategies contained herein may not be suitable for your situation. You should consult with a professional where appropriate. Neither the publisher nor author shall be liable for any loss of profit or any other commercial damages, including but not limited to special, incidental, consequential, or other damages.

For general information on our other products and services or for technical support, please contact our Customer Care Department within the United States at (800) 762-2974, outside the United States at (317) 572-3993 or fax (317) 572-4002.

Wiley also publishes its books in a variety of electronic formats. Some content that appears in print may not be available in electronic books. For more information about Wiley products, visit our web site at www.wiley.com.

Library of Congress Cataloging-in-Publication Data:
Agarwal, Bhagwan D.
 Analysis and performance of fiber composites / Bhagwan D. Agarwal, Lawrence J. Broutman, K. Chandrashekhara.—3rd ed.
  p. cm.
 Includes bibliographical references and index.
 ISBN-13: 978-0-471-26891-8 (cloth)
 ISBN-10: 0-471-26891-7 (cloth)
  1. Fibrous composites. 2. Reinforced plastics. I. Broutman, Lawrence J. II. Chandrashekhara, K. III. Title.
TA418.9.C6A34 2006
620.1'18—dc22

2005044699

Printed in the United States of America

10 9 8 7 6 5 4 3 2

CONTENTS

Preface xiii

1 Introduction 1

    1.1 Definition / 1

    1.2 Characteristics / 2

    1.3 Classification / 3

    1.4 Particulate Composites / 5

    1.5 Fiber-Reinforced Composites / 7

    1.6 Applications of Fiber Composites / 10

    Exercise Problems / 14

    References / 15

2 Fibers, Matrices, and Fabrication of Composites 16

    2.1 Advanced Fibers / 16
        2.1.1 Glass Fibers / 16
            2.1.1.1 Production of Glass Fibers / 17
            2.1.1.2 Glass Composition and Properties / 18
            2.1.1.3 Surface Treatment of Fibers: Sizes and Coupling Agents / 18
            2.1.1.4 Forms of Glass Fiber / 21
        2.1.2 Carbon and Graphite Fibers / 23
        2.1.3 Aramid Fibers / 26
        2.1.4 Boron Fibers / 27
        2.1.5 Other Fibers / 28

    2.2 Matrix Materials / 30
        2.2.1 Polymers / 30

                2.2.1.1 Thermosetting and Thermoplastic Polymers / 31
                2.2.1.2 Polymer Properties of Importance to the Composite / 31
                2.2.1.3 Common Polymeric Matrix Materials / 34
                2.2.1.4 Fillers / 39
        2.2.2   Metals / 39
    2.3 Fabrication of Composites / 41
        2.3.1 Fabrication of Thermosetting Resin Matrix Composites / 42
            2.3.1.1 Hand Lay-up Technique / 43
            2.3.1.2 Bag Molding Processes / 46
            2.3.1.3 Resin Transfer Molding / 49
            2.3.1.4 Filament Winding / 49
            2.3.1.5 Pultrusion / 51
            2.3.1.6 Preformed Molding Compounds / 53
        2.3.2 Fabrication of Thermoplastic–Resin Matrix Composites (Short-Fiber Composites) / 55
        2.3.3 Fabrication of Metal Matrix Composites / 58
        2.3.4 Fabrication of Ceramic Matrix Composites / 59
    Suggested Reading / 60

3 Behavior of Unidirectional Composites — 62

    3.1 Introduction / 62
        3.1.1 Nomenclature / 62
        3.1.2 Volume and Weight Fractions / 64
    3.2 Longitudinal Behavior of Unidirectional Composites / 67
        3.2.1 Initial Stiffness / 68
        3.2.2 Load Sharing / 71
        3.2.3 Behavior beyond Initial Deformation / 73
        3.2.4 Failure Mechanism and Strength / 74
        3.2.5 Factors Influencing Longitudinal Strength and Stiffness / 76
    3.3 Transverse Stiffness and Strength / 80
        3.3.1 Constant-Stress Model / 80
        3.3.2 Elasticity Methods of Stiffness Prediction / 83
        3.3.3 Halpin–Tsai Equations for Transverse Modulus / 85

    3.3.4 Transverse Strength / 87
        3.3.4.1 Micromechanics of Transverse Failure / 88
        3.3.4.2 Prediction of Transverse Strength / 90
3.4 Prediction of Shear Modulus / 91
3.5 Prediction of Poisson's Ratio / 95
3.6 Failure Modes / 96
    3.6.1 Failure under Longitudinal Tensile Loads / 100
    3.6.2 Failure under Longitudinal Compressive Loads / 102
    3.6.3 Failure under Transverse Tensile Loads / 106
    3.6.4 Failure under Transverse Compressive Loads / 107
    3.6.5 Failure under In-Plane Shear Loads / 107
3.7 Expansion Coefficients and Transport Properties / 108
    3.7.1 Thermal Expansion Coefficients / 108
    3.7.2 Moisture Expansion Coefficients / 114
    3.7.3 Transport Properties / 114
    3.7.4 Mass Diffusion / 117
3.8 Typical Unidirectional Fiber Composite Properties / 123
Exercise Problems / 124
References / 129

4 Short-Fiber Composites — 132

4.1 Introduction / 132
4.2 Theories of Stress Transfer / 133
    4.2.1 Approximate Analysis of Stress Transfer / 133
    4.2.2 Stress Distributions from Finite-Element Analysis / 137
    4.2.3 Average Fiber Stress / 139
4.3 Modulus and Strength of Short-Fiber Composites / 140
    4.3.1 Prediction of Modulus / 141
    4.3.2 Prediction of Strength / 145
    4.3.3 Effect of Matrix Ductility / 150
4.4 Ribbon-Reinforced Composites / 152
Exercise Problems / 155
References / 156

5 Analysis of an Orthotropic Lamina — 158

- 5.1 Introduction / 158
  - 5.1.1 Orthotropic Materials / 158
- 5.2 Stress–Strain Relations and Engineering Constants / 160
  - 5.2.1 Stress–Strain Relations for Specially Orthotropic Lamina / 161
  - 5.2.2 Stress–Strain Relations for Generally Orthotropic Lamina / 164
  - 5.2.3 Transformation of Engineering Constants / 166
- 5.3 Hooke's Law and Stiffness and Compliance Matrices / 174
  - 5.3.1 General Anisotropic Material / 174
  - 5.3.2 Specially Orthotropic Material / 177
  - 5.3.3 Transversely Isotropic Material / 180
  - 5.3.4 Isotropic Material / 181
  - 5.3.5 Specially Orthotropic Material under Plane Stress / 182
  - 5.3.6 Compliance Tensor and Compliance Matrix / 184
  - 5.3.7 Relations between Engineering Constants and Elements of Stiffness and Compliance Matrices / 185
  - 5.3.8 Restrictions on Elastic Constants / 187
  - 5.3.9 Transformation of Stiffness and Compliance Matrices / 189
  - 5.3.10 Invariant Forms of Stiffness and Compliance Matrices / 194
- 5.4 Strengths of an Orthotropic Lamina / 196
  - 5.4.1 Maximum-Stress Theory / 197
  - 5.4.2 Maximum-Strain Theory / 200
  - 5.4.3 Maximum-Work Theory / 203
  - 5.4.4 Importance of Sign of Shear Stress on Strength of Composites / 205

Exercise Problems / 209

References / 212

6 Analysis of Laminated Composites — 213

- 6.1 Introduction / 213
- 6.2 Laminate Strains / 213
- 6.3 Variation of Stresses in a Laminate / 216

6.4 Resultant Forces and Moments: Synthesis of Stiffness Matrix / 218

6.5 Laminate Description System / 225

6.6 Construction and Properties of Special Laminates / 226
 6.6.1 Symmetric Laminates / 227
 6.6.2 Unidirectional, Cross-Ply, and Angle-Ply Laminates / 228
 6.6.3 Quasi-isotropic Laminates / 229

6.7 Determination of Laminae Stresses and Strains / 238

6.8 Analysis of Laminates after Initial Failure / 247

6.9 Hygrothermal Stresses in Laminates / 263
 6.9.1 Concepts of Thermal Stresses / 263
 6.9.2 Hygrothermal Stress Calculations / 264

6.10 Laminate Analysis Through Computers / 272

Exercise Problems / 277

References / 281

7 Analysis of Laminated Plates and Beams — 282

7.1 Introduction / 282

7.2 Governing Equations for Plates / 283
 7.2.1 Equilibrium Equations / 283
 7.2.2 Equilibrium Equations in Terms of Displacements / 286

7.3 Application of Plate Theory / 288
 7.3.1 Bending / 288
  7.3.1.1 Bending of General Laminates / 294
 7.3.2 Buckling / 295
 7.3.3 Free Vibrations / 301

7.4 Deformations Due to Transverse Shear / 306
 7.4.1 First-Order Shear Deformation Theory / 306
  7.4.1.1 Transverse Shear Deformation Effects in Bending of a Simply Supported Rectangular Specially Orthotropic Plate / 309
 7.4.2 Higher-Order Shear Deformation Theory / 311

7.5 Analysis of Laminated Beams / 314

7.5.1 Governing Equations for Laminated Beams / 314
7.5.2 Application of Beam Theory / 315
- 7.5.2.1 Bending / 315
- 7.5.2.2 Buckling / 318
- 7.5.2.3 Free Vibrations / 319

Exercise Problems / 320

References / 322

8 Advanced Topics in Fiber Composites — 324

8.1 Interlaminar Stresses and Free-Edge Effects / 324
- 8.1.1 Concepts of Interlaminar Stresses / 324
- 8.1.2 Determination of Interlaminar Stresses / 326
- 8.1.3 Effect of Stacking Sequence on Interlaminar Stresses / 328
- 8.1.4 Approximate Solutions for Interlaminar Stresses / 330
- 8.1.5 Summary / 334

8.2 Fracture Mechanics of Fiber Composites / 335
- 8.2.1 Introduction / 335
  - 8.2.1.1 Microscopic Failure Initiation / 335
  - 8.2.1.2 Fracture Process in Composites / 336
- 8.2.2 Fracture Mechanics Concepts and Measures of Fracture Toughness / 338
  - 8.2.2.1 Strain-Energy Release Rate (G) / 339
  - 8.2.2.2 Stress-Intensity Factor (K) / 341
  - 8.2.2.3 J-Integral / 345
- 8.2.3 Fracture Toughness of Composite Laminates / 346
- 8.2.4 Whitney–Nuismer Failure Criteria for Notched Composites / 349

8.3 Joints for Composite Structures / 355
- 8.3.1 Adhesively Bonded Joints / 355
  - 8.3.1.1 Bonding Mechanisms / 355
  - 8.3.1.2 Joint Configurations / 356
  - 8.3.1.3 Joint Failure Modes / 357
  - 8.3.1.4 Stresses in Joints / 358
  - 8.3.1.5 Advantages and Disadvantages of Adhesively Bonded Joints / 359

8.3.2 Mechanically Fastened Joints / 360
 8.3.2.1 *Failure Modes of Mechanically Fastened Joints / 360*
 8.3.2.2 *Advantages and Disadvantages of Mechanically Fastened Joints / 361*
8.3.3 Bonded-Fastened Joints / 361

Exercise Problems / 362
References / 363

9 Performance of Fiber Composites: Fatigue, Impact, and Environmental Effects   368

9.1 Fatigue / 368
 9.1.1 Introduction / 368
 9.1.2 Fatigue Damage / 370
  9.1.2.1 *Damage/Crack Initiation / 370*
  9.1.2.2 *Crack Arrest and Crack Branching / 370*
  9.1.2.3 *Final Fracture / 373*
  9.1.2.4 *Schematic Representation / 373*
  9.1.2.5 *Damage Characterization / 374*
  9.1.2.6 *Influence of Damage on Properties / 375*
 9.1.3 Factors Influencing Fatigue Behavior of Composites / 378
 9.1.4 Empirical Relations for Fatigue Damage and Fatigue Life / 385
 9.1.5 Fatigue of High-Modulus Fiber-Reinforced Composites / 386
 9.1.6 Fatigue of Short-Fiber Composites / 390
9.2 Impact / 395
 9.2.1 Introduction and Fracture Process / 395
 9.2.2 Energy-Absorbing Mechanisms and Failure Models / 396
  9.2.2.1 *Fiber Breakage / 396*
  9.2.2.2 *Matrix Deformation and Cracking / 398*
  9.2.2.3 *Fiber Debonding / 399*
  9.2.2.4 *Fiber Pullout / 399*
  9.2.2.5 *Delamination Cracks / 401*
 9.2.3 Effect of Materials and Testing Variables on Impact Properties / 401

9.2.4 Hybrid Composites and Their Impact Strength / 407
9.2.5 Damage Due to Low-Velocity Impact / 411
9.3 Environmental-Interaction Effects / 416
9.3.1 Fiber Strength / 416
9.3.1.1 Features of Stress Corrosion / 416
9.3.1.2 Static Fatigue and Stress–Rupture of Fibers / 417
9.3.1.3 Stress Corrosion of Glass Fibers and GRP / 419
9.3.2 Matrix Effects / 422
9.3.2.1 Effect of Temperature and Moisture / 422
9.3.2.2 Degradation at Elevated Temperatures / 426
9.3.2.3 Stress–Rupture Characteristics at Modest Temperatures / 429

Exercise Problems / 431

References / 431

10 Experimental Characterization of Composites   439

10.1 Introduction / 439
10.2 Measurement of Physical Properties / 440
10.2.1 Density / 440
10.2.2 Constituent Weight and Volume Fractions / 441
10.2.3 Void Volume Fraction / 442
10.2.4 Thermal Expansion Coefficients / 442
10.2.5 Moisture Absorption and Diffusivity / 443
10.2.6 Moisture Expansion Coefficients / 444
10.3 Measurement of Mechanical Properties / 445
10.3.1 Properties in Tension / 445
10.3.2 Properties in Compression / 449
10.3.3 In-Place Shear Properties / 452
10.3.3.1 Torsion Tube Test / 452
10.3.3.2 Iosipescu Shear Test / 453
10.3.3.3 [±45]$_s$ Coupon Test / 455
10.3.3.4 Off-Axis Coupon Test / 456
10.3.3.5 Other Tests / 458
10.3.4 Flexural Properties / 459

10.3.5 Measures of In-Plane Fracture Toughness / 463
  10.3.5.1 Critical Strain-Energy Release Rate (G_c) / 463
  10.3.5.2 Critical Stress-Intensity Factor or Crack Growth Resistance (K_R) / 464
  10.3.5.3 Critical J-Intergral (J_c) / 470
10.3.6 Interlaminar Shear Strength and Fracture Toughness / 471
10.3.7 Impact Properties / 475

10.4 Damage Identification Using Nondestructive Evaluation Techniques / 481
  10.4.1 Ultrasonics / 481
  10.4.2 Acoustic Emission / 483
  10.4.3 x-Radiography / 485
  10.4.4 Thermography / 486
  10.4.5 Laser Shearography / 488

10.5 General Remarks on Characterization / 488

Exercise Problems / 490

References / 491

11 Emerging Composite Materials   496

11.1 Nanocomposites / 496

11.2 Carbon–Carbon Composites / 498

11.3 Biocomposites / 498
  11.3.1 Biofibers / 498
  11.3.2 Wood–Plastic Composites (WPCs) / 501
  11.3.3 Biopolymers / 502

11.4 Composites in "Smart" Structures / 503

Suggested Reading / 504

Appendix 1   Matrices and Tensors   507

Appendix 2   Equations of Theory of Elasticity   530

Appendix 3   Laminate Orientation Code   542

Appendix 4 Properties of Fiber Composites   **548**

Appendix 5 Computer Programs for Laminate Analysis   **553**

Index   **555**

PREFACE

The importance of fiber-reinforced composites can be gauged from the fact that the U.S. fiber reinforced polymer matrix composites industry has grown at an average rate of 6.5% since 1960, which is far greater than that of the conventional metallic materials aluminum and steel, and is approximately twice the growth rate of the U.S. economy (see Chapter 1). This growth is due to the outstanding mechanical properties, unique flexibility in design capabilities, and ease of fabrication offered by the composites. Additional advantages of these composites include lightweight, corrosion resistance, impact resistance and excellent fatigue strength. New applications of composites are being developed continuously and the development of new composites have resulted in the sustained growth of the composites industry.

The first and second editions of this book have been widely accepted as a textbook for university-level composite materials courses in several countries. They also served as a useful reference source for practicing engineers and scientists wishing to continue their education. The third edition has been prepared based on the authors continuing experiences in teaching university-level courses and conducting seminars for industry.

This book provides a complete treatment of the subject, covering mechanics, materials, analysis, fabrication, characterization, performance and other topics of practical importance. Basic consepts are explained in simple language, and the subject is developed gradually, maintaining a balance between mechanics and materials aspects. A basic knowledge of the strength of materials is sufficient to pursue most of the topics in this book. Example problems and numerous illustrations throughout the book help to develop a better understanding of the subject. The exercise problems at the end of each chapter are provided for practice on application of the principles.

The revisions for the third edition were aimed at making the text more self-sufficient by providing greater coverage of the composites technology. A new chapter on analysis of laminated plates and beams (Chapter 7) greatly enhances the analysis capabilities of the book. Another new chapter on emerging composite materials (Chapter 11) provides brief coverage of this topic. Significant additions have been made to Chapter 2 (discussion on fillers and resin transfer molding) and Chapter 10 (new section on measurement of physical properties, and discussion on two new damage identification techniques). Chapter 5 on analysis of an orthotropic lamina has been completely rewritten and reorganized to improve presentation and readability. Addition of several example problems and explanations in this chapter should enhance understanding of the subject. Several sections have been rewritten in many other chapters, and new material added, most notably, the concepts of thermal stresses (Chapter 6), free-edge effects and joints (Chapter 8), and fracture mechanics (Chapter 9). The survey of commercially available computer packages in Appendix A.5 has been updated.

With these additions and modifications, this book will serve the needs of undergraduate and graduate courses, as well as the needs of the practicing engineers and scientists. The entire book will be difficult to cover in a one-semester course. Depending on the background of the students and the level of the course, appropriate topics may be selected. In the first course of composites Chapters 7–9 may be omitted. Chapters 1 through 4 should be adequate to introduce composites in a general course on materials. Chapter 7, which deals with the analysis of laminated plates and beams, and Chapters 8 and 9, which deal with advanced topics and performance of composites, may be covered in an advanced course. The contents and references of Chapter 8, 9 and 11 should be helpful to those using or engaged in research studies dealing with composite materials. Exercise problems in Chapter 7, and some (marked with an asterisk) in Chapter 6 require lengthy calculations and should be assigned only selectively to the students who have competence with personal computers and the appropriate software (see Appendix A.5). A solution manual for all the exercise problems is available and may be requested from the authors or the publisher.

The authors would like to acknowledge the help of Saikrishna Sundararaman, a graduate student at the University of Missouri–Rolla, in solving new example and exercise problems. He also typed most of the new material in the book.

BHAGWAN D. AGARWAL
LAWRENCE J. BROUTMAN
K. CHANDRASHEKHARA

1

INTRODUCTION

1.1 DEFINITION

The word "composite" means "consisting of two or more distinct parts." Thus a material having two or more distinct constituent materials or phases may be considered a composite material. However, we recognize materials as composites only when the constituent phases have significantly different physical properties, and thus the composite properties are noticeably different from the constituent properties. For example, common metals almost always contain unwanted impurities or alloying elements; plastics generally contain small quantities of fillers, lubricants, ultraviolet absorbers, and other materials for commercial reasons such as economy and ease of processing, yet these generally are not classified as composites. In the case of metals, the constituent phases often have nearly identical properties (e.g., modulus of elasticity), the phases are not generally fibrous in character, and one of the phases usually is present in small-volume fractions. Thus the modulus of elasticity of a steel alloy is insensitive to the amount of the carbide present, and metallurgists generally have not considered metal alloys as composites, particularly from the point of view of analysis. Nevertheless, two-phase metal alloys are good examples of particulate composites in terms of structure. Although plastics—which are filled for cost purposes and contain small amounts of additives—are composites, they need not be considered as such if their physical properties are not greatly affected by the additives. Thus classification of certain materials as composites often is based on cases where significant property changes occur as a result of the combination of constituents, and these property changes generally will be most obvious when one of the phases is in platelet or fibrous form, when the volume fraction is greater than 10%, and when the property of one constituent is much greater (≥ 5 times) than the other.

Within this wide range of composite materials, a definition may be adopted to suit one's requirements. For the purpose of discussion in this book, composites can be considered to be materials consisting of two or more chemically distinct constituents, on a macroscale, having a distinct interface separating them. This definition encompasses the fiber composite materials of primary interest in this text. This definition also encompasses many other types of composites that are not treated specifically in this book.

1.2 CHARACTERISTICS

Composites consist of one or more discontinuous phases embedded in a continuous phase. The discontinuous phase is usually harder and stronger than the continuous phase and is called the *reinforcement* or *reinforcing material,* whereas the continuous phase is termed the *matrix.* The most notable exception to this rule is the class of materials known as *rubber-modified polymers,* consisting of a rigid polymer matrix filled with rubber particles.

Properties of composites are strongly influenced by the properties of their constituent materials, their distribution, and the interaction among them. The composite properties may be the volume-fraction sum of the properties of the constituents, or the constituents may interact in a synergistic way so as to provide properties in the composite that are not accounted for by a simple volume-fraction sum of the properties of the constituents. Thus, in describing a composite material as a system, besides specifying the constituent materials and their properties, one needs to specify the geometry of the reinforcement with reference to the system. The geometry of the reinforcement may be described by the shape, size, and size distribution. However, systems containing reinforcements with identical geometry may differ from each other in many ways; for example, the reinforcement in the systems may differ in concentration, concentration distribution, and orientation. Therefore, all these factors may be important in determining the properties of the composites, but seldom are all accounted for in the development of theoretical descriptions of composites.

The shape of the discrete units of the discontinuous phase often may be approximated by spheres or cylinders. There are some natural materials such as mica and the clay minerals and some man-made materials such as glass flakes that can best be described by rectangular cross-sectioned prisms or platelets. The size and size distribution control the texture of the material. Together with volume fraction, they also determine the interfacial area, which plays an important role in determining the extent of the interaction between the reinforcement and the matrix.

Concentration is usually measured in terms of volume or weight fraction. The contribution of a single constituent to the overall properties of the composite is determined by this parameter. The concentration generally is regarded as the single most important parameter influencing the composite properties. Also, it is an easily controllable manufacturing variable used to alter the properties of the composite. The concentration distribution is a measure of homogeneity or uniformity of the system. The homogeneity is an important characteristic that determines the extent to which a representative volume of material may differ in physical and mechanical properties from the average properties of the material. Nonuniformity of the system should be avoided because it reduces those properties that are governed by the weakest link in the material. For example, failure in a nonuniform material will initiate in an area of lowest strength, thus adversely affecting the overall strength of the material.

The orientation of the reinforcement affects the isotropy of the system. When the reinforcement is in the form of particles, with all their dimensions approximately equal (equiaxed), the composite behaves essentially as an isotropic material whose properties are independent of direction. When the dimensions of the representative reinforcement particles are unequal, the composite may behave as an isotropic material provided that the particles are randomly oriented, such as in the randomly oriented short-fiber-reinforced composites. In other cases the manufacturing process (e.g., molding of a short-fiber composite) may induce orientation of the reinforcement and hence induce some anisotropy. In continuous-fiber-reinforced composites, such as unidirectional or cross-ply composites, anisotropy may be desirable. Moreover, the primary advantage of these composites is the ability to control anisotropy by design and fabrication.

The *concentration distribution* of the particles refers to their spatial relations to each other. Particles may by uniformly dispersed in a composite and placed at regular spacings so that no two particles touch each other. On the other hand, it is possible to imagine a dispersion of particles so arranged that they form a network such that a continuous path connects all particles. This happens at a much lower concentration than that at which the close packing of particles becomes possible. Such network-forming dispersions may have a significant influence on the electrical properties of the composites. An interesting example of this is the dispersion of carbon black in rubber. Above a volume concentration of about 10%, the electrical conductivity of the mixture increases markedly. This has been attributed to the network formation of carbon-black particles.

1.3 CLASSIFICATION

Most composite materials developed thus far have been fabricated to improve mechanical properties such as strength, stiffness, toughness, and high-temperature performance. It is natural to study together the composites that have a common strengthening mechanism. The strengthening mechanism strongly depends on the geometry of the reinforcement. Therefore, it is quite convenient to classify composite materials on the basis of the geometry of a representative unit of reinforcement. Figure 1-1 represents a commonly accepted classification scheme for composite materials.

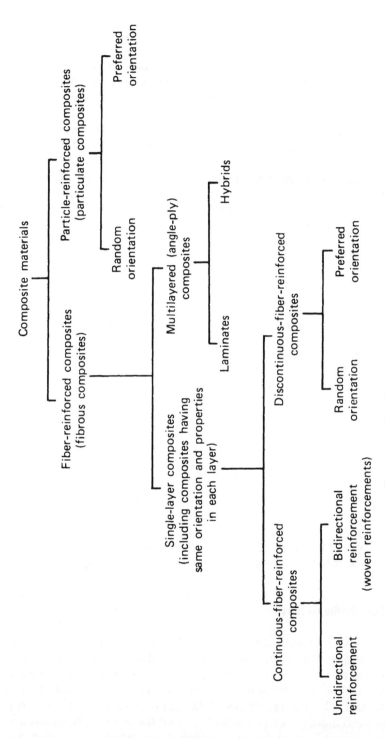

Figure 1-1. Classification of composite materials.

With regard to this classification, the distinguishing characteristic of a particle is that it is nonfibrous in nature. It may be spherical, cubic, tetragonal, a platelet, or of other regular or irregular shape, but it is approximately equiaxed. A fiber is characterized by its length being much greater than its cross-sectional dimensions. Particle-reinforced composites are sometimes referred to as *particulate composites.* Fiber-reinforced composites are, understandably, called *fibrous composites.*

1.4 PARTICULATE COMPOSITES

A composite whose reinforcement may be classified as particles is called a *particulate composite.* A particle, by definition, is nonfibrous and generally has no long dimension, with the exception of platelets. The dimensions of the reinforcement determine its capability of contributing its properties to the composite. Also, a reinforcement having a long dimension discourages the growth of incipient cracks normal to the reinforcement that otherwise might lead to failure, particularly with brittle matrices. Therefore, fibers are very effective in improving the fracture resistance of the matrix. In general, particles are not very effective in improving fracture resistance. However, particles of rubberlike substances in brittle polymer matrices improve fracture resistance by promoting and then arresting crazing in the brittle matrices. Other types of particles, such as ceramic, metal, or inorganic particles, produce reinforcing effects in metallic matrices by different strengthening mechanisms. The particles in a particulate composite place constraints on the plastic deformation of the matrix material between them because of their inherent hardness relative to the matrix. The particles also share the load, but to a much smaller extent than those fibers in fibrous composites that lie parallel to the direction of load. Thus the particles are effective in enhancing the stiffness of the composites but do not offer the potential for much strengthening. For example, hard particles placed in brittle matrices reduce the strength due to stress concentrations in the adjacent matrix material. Particle fillers, however, are used widely to improve the properties of matrix materials, such as to modify the thermal and electrical conductivities, improve performance at elevated temperatures, reduce friction, increase wear and abrasion resistance, improve machinability, increase surface hardness, and reduce shrinkage. In many cases they are used simply to reduce cost.

The particles and matrix material in a particulate composite can be any combination of metallic and nonmetallic materials. The choice of a particular combination depends on the desired end properties. Particles of lead are mixed with copper alloys and steel to improve their machinability. In addition, lead is a natural lubricant in bearings made of copper alloys. Particles of many brittle metals such as tungsten, chromium, and molybdenum are incorporated into ductile metals to improve their elevated temperature performance while maintaining ductile characteristics at room temperature. Composites with particles of tungsten, molybdenum, or their carbides in silver and copper matrices are used widely for electrical-contact applications. These applications require materials with properties such as high thermal and electrical conductivities, high melting point, and low friction and wetting characteristics. These materials are also used for electrodes and related applications in the welding industry.

Cermets are examples of ceramic and metal composites. Oxide-based cermets are used extensively as tool materials for high-speed cutting, thermocouple protection tubes, furnace mufflers, and a variety of high-temperature erosive applications. Carbide-based cermets mostly have particles of tungsten, chromium, and titanium. Tungsten carbide in a cobalt matrix is used in machine parts requiring very high surface hardness such as cutting tools, wire-drawing dies, valve parts, and precision gauges. Chromium carbide in a cobalt matrix is highly resistant to corrosion and abrasion and has a coefficient of thermal expansion close to that of steel. This makes it useful for valve parts, nozzles, and high-load bearings operating at very high temperatures. Titanium carbide in a nickel or cobalt matrix is well suited for high-temperature applications such as turbine parts, torch tips, and hot-mill parts.

Inorganic fillers are used very effectively to improve various properties of plastics, such as to increase surface hardness, reduce shrinkage and eliminate crazing after molding, improve fire retardancy, provide color and improve appearance, modify the thermal and electrical conductivities, and most important, greatly reduce cost without necessarily sacrificing the other desirable properties. Many commercially important elastomers are filled with carbon black or silica to improve their strength and abrasion resistance while maintaining their necessary extensibility. Cold solders consist of metal powders suspended in thermosetting resins. The composite is hard and strong and conducts heat and electricity. Copper in epoxy increases the conductivity immensely. High lead content in plastics acts as a sound deadener and shield against gamma radiation. Fluorocarbon-based plastics are being used as bearing materials. Metallic inclusions are incorporated to increase thermal conductivity, lower the coefficient of expansion, and drastically reduce the wear rate.

Thin flakes offer attractive features for an effective reinforcement. They have a primarily two-dimensional geometry and thus impart equal strength in all directions in their plane compared with fibers that are unidirectional reinforcements. Flakes, when laid parallel, can be packed more closely than fibers or spherical particles. Mica flakes are used in electrical and heat-insulating applications. Mica flakes embedded in a glassy matrix provide composites that can be machined easily and are used in electrical applications. Aluminum flakes are employed commonly in paints and other coatings in which they orient themselves parallel to the surface and give the coating exceptionally good properties. Silver flakes are employed where good conductivity is required. It has not been possible to fully exploit the attractive possibilities of flake composites because of fabrication difficulties.

Nanocomposites, which are emerging new composites, are discussed in Chap. 11. Clay-reinforced nanocomposites are particulate composites. While nanotubes are fibrous in character, their size is very small compared with conventional fibrous reinforcements. Therefore, nanotube-reinforced nanocomposites also may be analyzed as particulate composites, especially since nanotube concentration is very small.

Particulate composites are an important class of composite materials. The discussion in this text, however, deals primarily with fiber composites.

1.5 FIBER-REINFORCED COMPOSITES

It is well known that the measured strengths of most materials are found to be much smaller (by a couple of orders of magnitude) than their theoretical strengths. The discrepancy in strength values is believed to be due to the presence of imperfections or inherent flaws in the material. An attempt to minimize or eliminate flaws enhances the strength of a material. Flaws in the form of cracks that lie perpendicular to the direction of applied loads are particularly detrimental to strength. Therefore, compared with the strength of the bulk material, man-made filaments or fibers of nonpolymeric materials exhibit much higher strengths along their lengths because large flaws that may be present in the bulk material are minimized owing to the small cross-sectional dimensions of the fiber. In the case of polymeric materials, orientation of the molecular structure is responsible for high strength and stiffness. Properties of some common types of fibers as well as some conventional materials are given in Table 1-1, which clearly shows the importance of fibers in achieving higher strengths. The high strength of glass fibers is attributed to a defect-free surface, whereas graphite and aramid fibers attain their strength as a result of improved orientation of their atomic or molecular structure. The most important reinforcement fiber is E-glass because of its relative low cost. However, boron, graphite, and the aramid polymer fibers (Kevlar 49) are most exceptional because of their high stiffness values. Of these, the graphite fibers offer the greatest variety because of the ability to control their structure.

Fibers, because of their small cross-sectional dimensions, are not directly usable in engineering applications. They are, therefore, embedded in matrix materials to form fibrous composites. The matrix serves to bind the fibers together, transfer loads to the fibers, and protect them against environmental attack and damage due to handling. In discontinuous fiber-reinforced composites, the load-transfer function of the matrix is more critical than in continuous-fiber composites. The fibrous composites have become the most important class of composite materials because they are capable of achieving high strengths.

Fibrous composites can be classified broadly as single-layer and multilayer (angle-ply) composites on the basis of studying both the theoretical and ex-

Table 1-1 Properties of fibers and conventional bulk materials

Material	Tensile Modulus (E) (GPa)	Tensile Strength (σ_u) (GPa)	Density (ρ) (g/cm³)	Specific Modulus (E/ρ)	Specific Strength (σ_u/ρ)
Fibers					
E-glass	72.4	3.5^a	2.54	28.5	1.38
S-glass	85.5	4.6^a	2.48	34.5	1.85
Graphite (high modulus)	390.0	2.1	1.90	205.0	1.1
Graphite (high tensile strength)	240.0	2.5	1.90	126.0	1.3
Boron	385.0	2.8	2.63	146.0	1.1
Silica	72.4	5.8	2.19	33.0	2.65
Tungsten	414.0	4.2	19.30	21.0	0.22
Beryllium	240.0	1.3	1.83	131.0	0.71
Kevlar 49 (aramid polymer)	130.0	2.8	1.50	87.0	1.87
Conventional materials					
Steel	210.0	0.34–2.1	7.8	26.9	0.043–0.27
Aluminum alloys	70.0	0.14–0.62	2.7	25.9	0.052–0.23
Glass	70.0	0.7–2.1	2.5	28.0	0.28–0.84
Tungsten	350.0	1.1–4.1	19.30	18.1	0.057–0.21
Beryllium	300.0	0.7	1.83	164.0	0.38

^{a}Virgin strength values. Actual strength values prior to incorporation into composite are approximately 2.1 (GPa).

perimental properties. "Single-layer" composites actually may be made from several distinct layers with each layer having the same orientation and properties, and thus the entire laminate may be considered a "single-layer" composite. In the case of molded composites made with discontinuous fibers, although the planar fiber orientation may not be uniform through the thickness, there are no distinct layers, and they can be classed as single-layer composites. In the case of composites fabricated from nonwoven mats, the random orientation is constant in each layer, and the resulting composite would be considered a single-layer composite even though a resin-rich layer might be found between each reinforcement layer on microscopic examination. Most composites used in structural applications are multilayered; that is, they consist of several layers of fibrous composites. Each layer or lamina is a single-layer composite, and its orientation is varied according to design. Each layer of the composite is usually very thin, typically of a thickness of 0.1 mm, and hence cannot be used directly. Several identical or different layers are bonded together to form a multilayered composite usable for engineering applications. When the constituent materials in each layer are the same, they are called simply *laminates*. *Hybrid laminates* refer to multilayered fibers are used in large concentrations. Thus the principal purpose of a matrix is not to be a load-carrying constituent but essentially to bind the fibers together and protect them. The failure mode of such composites is also generally controlled by the fibers.

The continuous fibers in a "single-layer" composite may be all aligned in one direction to form a unidirectional composite. Such composites are fabricated by laying the fibers parallel and saturating them with resinous material, such as polyester or epoxy resin, that holds the fibers in position and serves as the matrix material. Such forms of preimpregnated fibers are called *prepregs*. Generally, a removable backing is also provided to prevent the layers from sticking together while being stored. The backing provides additional means to hold the fibers in position. The unidirectional composites are very strong in the fiber direction but generally are weak in the direction perpendicular to the fibers. Therefore, unidirectional prepregs are stacked together in various orientations to form laminates usable in engineering applications. However, unidirectionally glass-reinforced adhesive tapes are used widely for heavy-duty sealing applications, and some unidirectional composites are used for fishing poles and other rodlike structures.

The continuous reinforcement in a single layer also may be provided in a second direction to achieve more balanced properties. The bidirectional reinforcement may be provided in a single layer in mutually perpendicular directions as in a woven fabric. The bidirectional reinforcement may be such that the strengths in two perpendicular directions are approximately equal. In some applications, a minimum of reinforcement perpendicular to the primary direction is provided only to prevent damage and fiber separation in handling owing to the poor strength in the transverse direction. In such cases, the transverse strength is much less than the strength in the direction of primary reinforcement.

The orientation of short or discontinuous fibers cannot be controlled easily in a composite material. In most cases, the fibers are assumed to be randomly oriented in the composite. However, in the injection molding of a fiber-reinforced polymer, considerable orientation can occur in the flow direction. Different areas of a single molding can have quite different fiber orientations (see Fig. 4-11 on page 149). Short fibers, sometimes referred to as *chopped fibers,* may be sprayed simultaneously with a liquid resin against a mold to build up a reinforced-plastic structure. Alternatively, chopped fibers may be converted to a lightly bonded preform or mat that can be later impregnated with resin to fabricate single-layer composites. In all these processes, the chopped fibers generally lie parallel to the surface of the mold and are oriented randomly in planes parallel to the surface. Therefore, properties of a discontinuous-fiber-reinforced composite can be isotropic; that is, they do not change with direction within the plane of the sheet.

Chopped fibers also may be blended with resins to make a reinforced molding compound. These fibers tend to become oriented parallel to the direction of material flow during a compression- or injection-molding operation and thus get a preferential orientation. Composites fabricated in this manner are not isotropic. Their properties depend, among other things, on the degree of preferential orientation achieved during the fabrication process.

1.6 APPLICATIONS OF FIBER COMPOSITES

The two outstanding features of oriented-fiber composites are their high strength-weight ratio and controlled anisotropy. The strength and modulus of commonly used bidirectional composites are compared with those of conventional structural materials in Table 1-2. Since polycrystalline metals have equal properties in all directions, for a fair comparison, bidirectional laminate (e.g., cross-ply) properties are used in Table 1-2. Bidirectional laminates have

Table 1-2 Properties of conventional structural materials and bidirectional (cross-ply) fiber composites

Material	Fiber Volume Fraction (V_f) (%)	Tensile Modulus (E) (GPa)	Tensile Strength (σ_u) (GPa)	Density (ρ) (g/cm^3)	Specific Modulus (E/ρ)	Specific Strength (σ_u/ρ)
Mild steel		210	0.45–0.83	7.8	26.9	0.058–0.106
Aluminum						
2024-T4		73	0.41	2.7	27.0	0.152
6061-T6		69	0.26	2.7	25.5	0.096
E-glass–epoxy	57	21.5	0.57	1.97	10.9	0.26
Kevlar 49–epoxy	60	40	0.65	1.40	29.0	0.46
Carbon fiber–epoxy	58	83	0.38	1.54	53.5	0.24
Boron-epoxy	60	106	0.38	2.00	53.0	0.19

strengths and moduli approximately one-half of those of unidirectional laminates. They have equal properties in two principal directions and show smaller property variation with direction. Fiber composites generally are superior to metals with respect to specific strength and modulus (see Table 1-2). However, glass-fiber composites are inferior to both steel and aluminum with respect to specific modulus. In applications where the structure's weight is a factor in the design, comparisons should be made on the basis of *specific* properties of the materials.

Controlled anisotropy means that the ratio of property values in different directions can be varied or controlled. For example, in a unidirectional composite, the longitudinal strength–transverse strength ratio can be changed easily by changing the fiber volume fraction. Similarly, other properties can be altered by altering the material and manufacturing variables. Further, laminates are designed and constructed from unidirectional composites to obtain desired directional properties to match requirements of specific applications.

These two features, high specific strength and controlled anisotropy, make fiber composites very attractive structural materials. Their other advantages include ease of manufacture and structural forms that are otherwise inconvenient or impossible to manufacture. Their use, therefore, in aerospace and transportation industries is increasing continuously.

Fiber-reinforced polymer matrix composites are the most widely used fiber composites. The U.S. polymer composite industry has grown at an average rate of 6.5% since 1960, which is approximately twice the growth rate of the U.S. economy [1]. U.S. growth in composites is compared with that of steel, aluminum, and the U.S. economy (GDP) in Fig. 1-2. Between 1960 and 2004, the U.S. consumption of steel doubled, the U.S. economy or gross domestic product (GDP) tripled, consumption of aluminum quadrupled, and composites consumption grew 16 times (see Fig. 1-2). It is estimated that in 2004, the

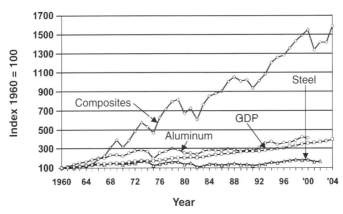

Figure 1-2. Growth of composites in the United States compared with steel, aluminum, and GDP. (Adapted from MacNeil [1].)

tional Science Foundation [4,5]. The bridge can sustain a class H-20 truck passage. This required the design load to be 21 tons. The bridge was constructed from pultruded hollow composite tubes of 76 mm × 76 mm square cross section with a wall thickness to 6.35 mm. Five middle layers of the

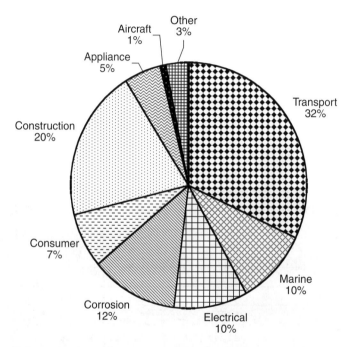

Estimated 2004 composites consumption: 1.8×10^9 kg (4.0 billion lb)

Figure 1-3. Estimated percentage use of composites in different U.S. industries in 2004. (Adapted from MacNeil [1].)

1.6 APPLICATIONS OF FIBER COMPOSITES

Figure 1-4. Cross section of a composite bridge manufactured at the University of Missouri–Rolla.

Figure 1-5. The composite bridge installed at the UMR campus with a truck on it.

bridge, out of a total of seven layers, are made of commercially available glass-vinyl ester tubes. The top and bottom layers are made of carbon-vinyl ester tubes, pultruded in the laboratory. A cross section of the bridge is shown in Fig. 1-4. The bridge, installed in July 2000 on the UMR campus (Fig. 1-5), is being used by pedestrian and bicycle traffic, as well as by campus maintenance vehicles such as snow plows and lawn mowers. The bridge performance is being monitored remotely by means of embedded fiberoptic sensors. Based on the results so far, the bridge is estimated to perform satisfactorily for 75 years. The current estimates also show that such bridges will be cost-effective on a long-term basis, although its initial cost was about 25% higher than the conventional-material bridge.

EXERCISE PROBLEMS

1.1. Think of as many naturally occurring materials as you can that could be classed as composite materials and classify them according to Fig. 1-1.

1.2. Prepare a list of man-made materials—metals, ceramics, and plastics—and also classify them according to Fig. 1-1, provided that they are considered composites.

1.3. Prepare a graph using specific strength and specific modulus as coordinate axes, and using data in Table 1-1, plot the points for various metals, fibers, and bidirectional composites. Also show points for unidirectional composites using data from Table 3-1. Add any other materials you feel are relevant.

1.4. (a) A rectangular cross-sectional beam subjected to a bending moment is made of steel and is 10 cm in width and 6 mm in thickness. If the width of the beam is held constant, calculate the beam thickness if designed from 2024-T4 aluminum and the various composites shown in Table 1-2 to provide the equivalent stiffness (EI) in one case and in another the equivalent strength.

  (b) Calculate the beam weight differences (per unit of beam length) for the preceding cases.

  (c) For all materials considered in part (a), if the beams are to be of identical weight, calculate the stiffnesses and bending strengths relative to those of the steel beam.

1.5. An example of a synergistic property of a composite is the toughness of a glass-fiber-reinforced thermosetting plastic. In other words, the toughness of the composite is much greater than the toughness of both the glass fiber and the thermosetting plastic and cannot be predicted by a volume-fraction law. Why?

REFERENCES

1. R. MacNeil, "U.S. Composites Market Outlook for 2005 and Beyond," *Composites Manufacturing,* January, 16–29 (2005).
2. A. H. Zureick, B. Shih, and E. Munley, "Fiber-Reinforced Polymeric Bridge Decks," *Structural Engineering Review,* **7,** 257–266 (1995).
3. M. Chajes, J. Gillespie, D. Mertz, and H. Shenton, "Advanced Composite Bridges in Delaware," *Proceedings of the Second International Conference on Composites in Infrastructure,* Tuscon, AZ, Vol. 1, 1998, pp. 645–650.
4. P. Kumar, K. Chandrashekhara, and A. Nanni, "Testing and Evaluation of Components for a Composite Bridge Deck," *Journal of Reinforced Composites and Plastics,* **22,** 441–461 (2003).
5. P. Kumar, K. Chandrashekhara, and A. Nanni, "Structural Performance of a FRP Bridge Deck," *Construction and Building Materials,* **8,** 35–47 (2004).

2

FIBERS, MATRICES, AND FABRICATION OF COMPOSITES

A broad overview of materials aspects of fiber composites is presented in this chapter. Important reinforcing fibers and matrix materials are discussed in the first two sections, and composite fabrication processes, in the last section.

2.1 ADVANCED FIBERS

A great majority of materials are stronger and stiffer in the fibrous form than as a bulk material. A high fiber aspect ratio (length–diameter ratio) permits very effective transfer of load via matrix materials to the fibers, thus taking advantage of their excellent properties. Therefore, fibers are very effective and attractive reinforcement materials. Reinforcing fibers used in advanced composites are discussed in this section.

2.1.1 Glass Fibers

Glass fibers are the most common of all the reinforcing fibers for polymer matrix composites. The principal advantages of glass fibers are the low cost and high strength. However, glass fibers have poor abrasion resistance, which reduces their usable strength. They also exhibit poor adhesion to some polymer matrix resins, particularly in the presence of moisture. To improve adhesion, the glass fiber surface often is treated with chemicals called *coupling agents* (mostly silanes). Glass fibers also have a lower modulus compared with the other advanced reinforcing fibers such as Kevlar, carbon, and boron.

2.1.1.1 Production of Glass Fibers Two forms of fiberglass can be produced—continuous fiber and staple (discontinuous) fiber. Both forms are made by the same production method up to the fiber-drawing stage.

Ingredients such as sand, limestone, and alumina are dry-mixed and melted in a refractory furnace. The temperature of the melt varies for each glass composition but generally is about 1260°C. The molten glass flows directly into the fiber-drawing furnace in the direct-melt process or flows into a marble-making machine in the marble process. The marbles are subsequently remelted and drawn into fibers. Most fiberglass is currently produced by the direct-melt process, illustrated schematically in Fig. 2-1.

Continuous fibers are produced by introducing molten glass into a platinum bushing, where the molten glass is gravity-fed through a multiplicity of holes in the base of the bushing. The molten glass exits from each orifice and is gathered together and attenuated mechanically to the proper dimensions, passed through a light water spray (quench), and then traversed over a belt that applies a protective and lubricating binder or size to the individual fibers. These fibers then are gathered together into a bundle of fibers called a *strand* or *end*. The fiberglass strand, typically consisting of 204 filaments, is then wound onto a receiving package (spool) at speeds of up to 50 m/s. This "cake" is then conditioned or dried prior to further processing into other textile forms.

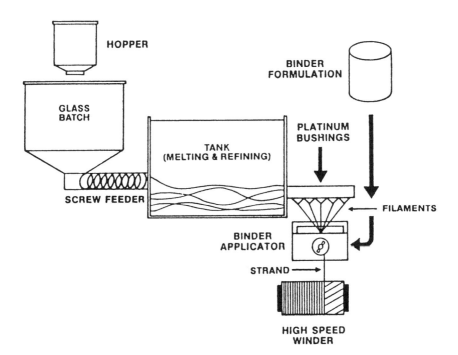

Figure 2-1. Glass-fiber production process.

Staple fibers are produced by passing a jet of air across the orifices in the base of the bushing, thus pulling individual filaments 20–40 cm long from the molten glass exiting from each orifice. These fibers are collected on a rotating vacuum drum, sprayed with a binder, and gathered as a "sliver" that can be drawn and twisted into yarns.

2.1.1.2 Glass Composition and Properties

Glass fibers are amorphous solids. Chemically, glass is composed primarily of a silica (SiO_2) backbone in the form of $(-SiO_4-)_n$ tetrahedra. Modifier ions are added for their contribution to glass properties and manufacturing capability.

For structural composites, the two commonly used types of glass fiber are E-glass and S-glass. Compositions of these are given in Table 2-1 and some important properties in Table 2-2. At present, E-glass constitutes the majority of glass-fiber production.

2.1.1.3 Surface Treatment of Fibers: Sizes and Coupling Agents

The chemical treatments applied during the forming of glass fibers are called *sizes*. These are of two general types: temporary sizes and compatible sizes.

The temporary sizes are applied to minimize the degradation of strength resulting from abrasion of fibers against one another and to bind the fibers together for easy handling in forming woven (twisting and weaving) glass-fiber products. These sizes are often starch-oil sizes. Ingredients of starch-oil sizes interfere with good bonding between the fibers and impregnating resin. Oil, emulsifying agent, and lubricant prevent good fiber wetting by the resin, and the starch, gelatin, and polyvinyl alcohol result in high water absorption and poor fiber–resin adhesion in the presence of moisture. Sizes of this type must be removed and replaced by a finish (coupling agent) before the fibers can be impregnated with resin. The sizes are easily removed by heating the fibers in an air-circulating oven at 340°C or higher temperatures for 15–20 h.

Table 2-1 Typical compositions of E-glass and S-glass fibers

Material	% Weight	
	E-Glass	S-Glass
Silicon oxide	54.3	64.20
Aluminum oxide	15.2	24.80
Ferrous oxide	—	0.21
Calcium oxide	17.2	0.01
Magnesium oxide	4.7	10.27
Sodium oxide	0.6	0.27
Boron oxide	8.0	0.01
Barium oxide	—	0.20
Miscellaneous	—	0.03

Table 2-2 Properties of E-glass and S-glass fibers

Property, units	E-Glass	S-Glass
Density, g/cm^3	2.54	2.49
Tensile strength,[a] MPa	3448	4585
Tensile modulus, GPa	72.4	85.5
Range of diameter, μm	3–20	8–13
Coefficient of thermal expansion, $10^{-6}/°C$	5.0	2.9

[a] Virgin values, immediately on formation. Usable values in finished products may range from 50–75% of virgin values.

The compatible sizes are applied to help improve initial adhesion of resin to glass and to reduce the destructive effects of water and other environmental forces on this bond. The compatible sizes are often called *coupling agents*. The most common coupling agents are organofunctional silanes. Silane coupling agents have the general chemical formula

$$X_3Si(CH_2)_nY$$

where $n = 0$–3
    Y = organofunctional group that is compatible with polymer matrix
    X = hydrolyzable group on silicon

They are generally applied to glass fibers from water solutions and applied from 0.1–0.5% of the weight of glass treated. The hydrolyzable groups are essential for generating intermediate silanols as follows:

$$X_3Si(CH_2)_nY + 3H_2O \rightarrow (HO)_3Si(CH_2)_nY + 3HX$$

A common silane used for epoxy matrix composites is γ-amino propyl triethoxy silane and has the structure

$$H_2NCH_2CH_2CH_2Si(OC_2H_5)_3$$
$$\downarrow \text{Hydrolyzed}$$
$$H_2NCH_2CH_2CH_2Si(OH)_3$$

The silanol functional group establishes hydrogen bonds with the glass surface through hydroxyl (—OH) groups present on the glass surface. The organofunctional group may react with the polymer matrix, forming strong covalent bonds, and/or may form physical bonds or van der Waals bonds. Although the coupling agent may have three reactive silanols per molecule, if the reactive sites on a glass surface are spaced far apart, only one silanol group per molecule may bond to the surface.

20 FIBERS, MATRICES, AND FABRICATION OF COMPOSITES

$$\underset{\underset{\text{OH}}{|}}{\overset{\overset{\text{OH}}{|}}{\text{Y(CH}_2)_n\text{Si}}}-\text{OH} + \text{HO}\{\overset{\overset{\text{Glass}}{|}}{\underset{\text{O}}{\text{Si}}}-\text{O} \longrightarrow \underset{\underset{\text{O}}{|}}{\overset{\overset{\text{O}}{|}}{\text{Y(CH}_2)_n\text{Si}}}-\text{O} \quad \underset{\text{H}}{\overset{\text{H}}{\diamond}} \quad \text{O}-\{\underset{\underset{\text{O}}{|}}{\overset{\overset{|}{}}{\text{Si}}}-\text{O}$$

The remaining silanol groups may condense with adjacent silanols to form a siloxane layer or may remain partly uncondensed at the surface.

Moisture at the glass surface is an important element with regard to the function and success of silanes. Glass surfaces, immediately on forming, absorb water molecules to form hydroxyl groups. The subsequent interaction with silanols was shown earlier. It is thought that coupling agents allow better retention of interfacial strength when composites are subjected to moisture because of their ability to reversibly bond with water molecules at the interface. For example, without coupling agents, water molecules diffusing into a composite could displace the organic polymer functional groups at the interface, thus in effect plasticizing and weakening the interface. The following diagram represents the interaction with penetrating moisture at the interface in the presence of a coupling agent:

Although any individual bond of coupling agent to glass surface is hydrolyzable, the reversible nature of this hydrolysis prevents complete loss of adhesion as long as the silane-modified resin retains its integrity.

The major contribution of silane coupling is to maintain the strength of the interface in the presence of moisture. For polymer matrices that by themselves do not bond well to glass, improvements in dry strengths also may be achieved. The principal composite properties, which will be better preserved, include transverse tensile strength, off-axis tensile strengths, and shear strengths.

2.1.1.4 Forms of Glass Fiber
Glass fibers are commercially available in various forms suitable for different applications. Some of them are described in the following paragraphs and are illustrated in the photographs in Fig. 2-2.

FIBERGLASS ROVING   Fiberglass roving is a collection of parallel continuous ends of filaments. Conventional rovings are produced by winding together the number of single strands necessary to achieve the desired yield (number of meters of roving per kilogram of weight). Generally, rovings are made with fibers of diameter 9 or 13 μm. Roving yields vary from about 3600 to 450 m/kg and typically have 20 strands. Rovings are used directly in pultrusion, filament winding, and prepreg manufacture.

Figure 2-2. Photographs of glass fibers in different forms: (a) roving, (b) chopped strand, (c) chopped-strand mat, (d) woven roving.

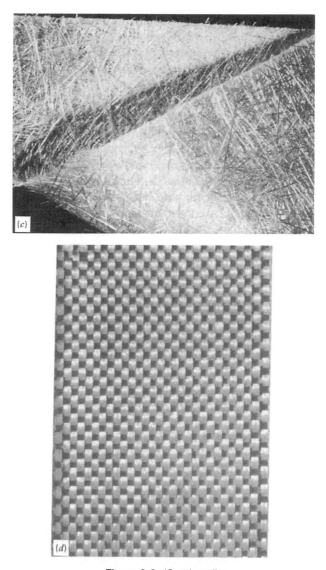

Figure 2-2. (*Continued*)

WOVEN ROVING   Rovings may be woven into a heavy, coarse-weave fabric for applications that require rapid thickness buildup over large areas. This characteristic is especially useful in the manufacture of fiberglass boats, various marine products, and many types of tooling. Woven rovings are available in different widths and weights.

Chopped-Strand Mat and Other Mats   There are three basic forms of fiberglass mat: chopped-strand mat, continuous-strand mat, and surfacing mat or veil.

Chopped-strand mat is a nonwoven material in which the fiberglass strands from roving are chopped into 25–50-mm lengths, evenly distributed at random onto a horizontal plane, and bound together with an appropriate chemical binder. These mats are available in widths of from 5 cm to 2 m and weigh 0.25–0.92 kg/m^2.

Continuous-strand mat consists of unchopped continuous strands of fiberglass deposited and interlocked in a spiral fashion. This mat is open and springy but, as a result of mechanical interlocking, does not require much binder for adequate handling strength.

Surfacing mat or veil is a very thin mat of single continuous filaments often used as a surface reinforcing layer in hand lay-up or molding process to minimize telegraphing the primary reinforcement through to the finished surface of a component, thus providing a smoother surface.

TEXTILE FIBERGLASS YARN   A *yarn* is a combination of strands that can be woven suitably into textile materials. The continuous, individual strand as it comes from the bushing represents the simplest form of textile fiberglass yarn and is referred to as a *single yarn*. In order for this yarn to be used properly and efficiently in a weaving operation, additional strand integrity is introduced by twisting it slightly, usually less than 40 turns per meter.

However, many woven fabrics require yarns that are heavier than can be conveniently drawn from a bushing. These can be produced by combining single strands via twisting and plying operations. Typically, this involves twisting two or more strands together and subsequently plying (i.e., twisting two or more of the twisted strands together).

FIBERGLASS FABRIC   Fiberglass yarn is woven into fabric by standard textile operations. The properties and contribution to product performance of fiberglass fabric depend on the fabric construction, that is, the number of yarns per inch in each direction, weave pattern, and yarn type.

CHOPPED-STRAND MILLED FIBERS   Continuous fiberglass strands can be chopped to specific lengths or hammer-milled into very short fiber lengths (generally 0.4–6.5 mm). The actual lengths are determined by the diameter of the screen openings through which the fibers pass during the milling. Milled fibers are used as reinforcements and fillers for thermoplastic and thermosetting resins.

2.1.2 Carbon and Graphite Fibers

Carbon/graphite fibers are the predominant high-strength, high-modulus reinforcement used in the fabrication of high-performance polymer-matrix composites. Their use is growing rapidly owing to a significant reduction in their price in the 1990s and an increase in their availability. Besides aerospace applications, they are now being used in sporting goods, automotive, civil infrastructure, offshore oil, and many other consumer applications.

In the graphite structure, the carbon atoms are arranged in the form of hexagonal layers with a very dense packing in the layer planes. The high-strength bond between carbon atoms in the layer plane results in an extremely high modulus, whereas the weak van der Waals–type bond between the neighboring layers results in a lower modulus in that direction. Strictly speaking, the term "graphite fibers" is a misnomer because there is no true graphite crystal structure in the fibers. The term "graphite fiber" is used to describe fibers that have a carbon content in excess of 99%, whereas the term "carbon fiber" describes fibers that have a carbon content of 80–95%. The carbon content is a function of the heat-treatment temperature.

The current technology for producing carbon fibers generally centers on the thermal decomposition of various organic precursors. However, currently available carbon fibers are made using one of the three precursor materials: polyacrylonitrile (PAN), pitch, and rayon.

Candidate organic materials for pyrolysis into carbon fibers having good properties should satisfy four criteria. First, the precursor should possess the appropriate strength and handling characteristics needed "to hold the fibers together" during all stages of the conversion process to carbon. Second, the precursor should not melt during any stage of the conversion process. This can be accomplished by either selecting infusible precursor material or by stabilizing a thermoplastic precursor prior to the conversion process. Third, the precursor material must not volatilize completely during the pyrolysis process; that is, the carbon yield of the precursor fiber after pyrolysis should be appreciable enough to justify its use on an economic basis. Furthermore, in order to obtain optimal properties, the carbon atoms should tend to array themselves in an aligned graphite structure during pyrolysis. In general, the more highly graphitic and oriented the fibers, the better are the mechanical properties. Finally, the precursor material should be as inexpensive as possible. One of the production methods for graphite fiber production is described below.

Graphite Fibers from PAN The process by which PAN* is converted to carbon fibers involves five steps (Fig. 2-3):

1. Spinning the PAN into a precursor fiber.
2. Stretching the precursor.
3. Stabilization by holding the prestretched polymer under tension at a temperature of 205–240°C for up to 24 h in an oxidizing atmosphere (air).

---

*Chemical structure of PAN is

$$\left(-\underset{\underset{H}{|}}{\overset{\overset{H}{|}}{C}}-\underset{\underset{C\equiv N}{|}}{\overset{\overset{H}{|}}{C}}-\right)_n$$

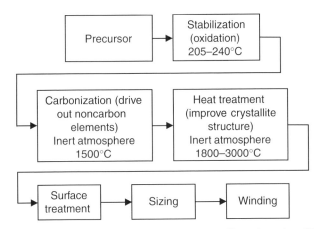

Figure 2-3. Process of converting PAN precursor fibers to carbon fibers.

4. Carbonization at approximately 1500°C in an inert atmosphere. Carbonization is the process of pyrolizing stabilized PAN fibers to drive out most (if not all) noncarbon elements of the precursor fibers until they are essentially transformed into carbon fibers. It is during this stage that the high mechanical properties found in most commercially available carbon fibers are developed.

5. Graphitization at approximately 3000°C in inert atmosphere. Graphitization heat treatments are carried out at temperatures in excess of 1800°C in order to improve the tensile modulus of elasticity of the fiber by improving the crystallite structure and preferred orientation of the graphitelike crystallite within each individual fiber.

Carbon fibers produced from each precursor have distinct advantages and drawbacks in terms of both cost and properties. The PAN-based carbon fibers are lower in cost and have good properties. They are the dominant class of structural carbon fibers and are used widely in military aircraft, missiles, and spacecraft structures. Pitch-based carbon fibers generally have higher stiffness and thermal conductivities, which make them useful in satellite structures and thermal-management applications, such as space radiators and electronic enclosures. Rayon-based carbon fibers are not used for structural applications, but their low thermal conductivity makes them useful for insulating and ablative applications such as rocket nozzles, missile reentry vehicle nosecones, and heat shields.

Typical property ranges of each class of carbon fibers are given in Table 2-3. Carbon fibers are available in a great variety of property combinations because their properties can be altered easily by controlling their structure through manufacturing process (e.g., by heat-treatment temperature). Fiber properties change from batch to batch and also as manufacturers improve

Table 2-3 Properties of carbon fibers

Property, units	Precursor		
	PAN	Pitch	Rayon
Tensile strength, MPa	1925–6200	2275–4060	2070–2760
Tensile modulus, GPa	230–595	170–980	415–550
Density, g/cm^3	1.77–1.96	2.0–2.2	1.7
Elongation, %	0.4–1.2	0.25–0.7	—
Coefficient of thermal expansion			
Axial, 10^{-6}/°C	−0.75 to −0.4	−1.6 to −0.9	—
Transverse, 10^{-6}/°C	7–10	7.8	—
Thermal conductivity, W/mK	20–80	400–1100	—
Fiber diameter, μm	5–8	10–11	6.5

fiber technology. The latest data from the manufacturer should be consulted for specific fibers and their actual properties.

Carbon fibers are available in various forms: continuous, chopped, woven fabric, or mat. Tows, yarn, rovings, and tape are the most common forms of continuous-graphite fibers sold today. A tow consists of numerous filaments in a straight-laid bundle and is specified by their number. Depending on the organic precursor and manufacturer, typical filament counts vary from 400–10,000 or as high as 160,000. A yarn is a twisted tow, whereas a roving is a number of ends or strands collected in a parallel bundle with little or no twist and is specified by the number of ends. Finally, a tape consists of numerous tows or yarns (e.g., 300) side by side on a backing or stitched together.

2.1.3 Aramid Fibers*

Various types of polymer fibers (e.g., nylon, polyester, rayon) have been in use for many years as reinforcements in automobile tires, large balloons and dirigibles, body armor, and rubber-coated fabrics.

Polymer aramid fibers (Kevlar) were first introduced in 1971. The aramid fiber–forming polymers, that is, the aromatic polyamides, are believed to be made by solution–polycondensation of diamines and diacid halides at low temperatures. The polymers are spun from strong acid solutions (e.g., concentrated H_2SO_4) by a dry-jet wet spinning process. The polymers are made

---

*Typical chemistry of fiber:

by rapidly adding a diacid chloride to a cool (5–10°C) amine solution while stirring. The polymer thus formed is recovered from the crumbs or gel by pulverizing, washing, and drying. To form filaments, the clean polymer, mixed with a strong acid, is extruded from spinnerets at an elevated temperature (51–100°C) through a 0.5–1.9-cm layer of air into cold water (0–4°C). The fibers are then washed thoroughly in water and dried on bobbins.

Fiber properties can be altered by using solvent additives, varying the spinning conditions, and using postspinning heat treatments.

Kevlar fibers possess unique properties. Tensile strength and modulus are substantially higher and fiber elongation is significantly lower for Kevlar fibers than for other organic fibers. Kevlar fibers have poor characteristics in compression, with compressive strength being only one-eighth the tensile strength. This results from their anisotropic structure, which permits rather easy local yielding, buckling, and kinking of the fiber in compression. They are not as brittle as glass or graphite fibers and can be readily woven on conventional fabric looms. Representative properties of Kevlar fibers are given in Table 2-4.

2.1.4 Boron Fibers

Boron filaments are produced by chemical vapor deposition (CVD) from the reduction of boron trichloride (BCl_3) with hydrogen on a tungsten or carbon monofilament substrate. The substrate is resistively heated to a temperature of 1260°C and pulled continuously through a reactor to obtain the desired boron coating thickness. Currently, boron filaments are produced with diameters of 100, 140, and 200 μm (4, 5.6, and 8 mils), in descending order of production quantity; however, both smaller- and larger-diameter fibers have been produced in experimental quantities.

The tensile strength of boron–tungsten filaments has improved steadily over the past decade from an average of under 2750 MPa to over 3445 MPa.

Table 2-4 Typical properties of Kevlar fibers

Property, units	Kevlar 29	Kevlar 49	Kevlar 129	Kevlar 149
Diameter, μm	12	12	—	—
Density, g/cm^3	1.44	1.44	1.44	1.44
Tensile strength, MPa	2760	3620	3380	3440
Tensile modulus, GPa	62	124	96	186
Elongation, %	3.4	2.8	3.3	2.5
Coefficient of thermal expansion (0–100°C), m/m/°C				
In axial direction	-2×10^{-6}	-2×10^{-6}	-2×10^{-6}	-2×10^{-6}
In radial direction	60×10^{-6}	60×10^{-6}	—	—

The tensile strength of boron–tungsten filaments can be increased by etching away part of the outer portion of the filament. This improvement in tensile strength is attributed to the decrease in residual tensile stresses at the inner surface in the core owing to removal of the outer region of the filament, which contains a compressive residual stress. Typical properties of boron–tungsten filaments are given in Table 2-5.

2.1.5 Other Fibers

The need for high-temperature reinforcing fibers has led to the development of ceramic fibers. The ceramic fibers combine high strength and elastic modulus with high-temperature capability and a general freedom from environmental attack. Alumina fibers and silicon carbide fibers are among the important ceramic fibers. Alumina fibers marketed by Du Pont (E.I. Du Pont de Nemours & Co.) with the trade name "Fiber FP" are a continuous α-alumina yarn with a 98% theoretical density. These fibers are made by spinning of an aqueous slurry and a two-step firing. As-produced Fiber FP surface is very rough. A thin silica coating enhances tensile strength by about 50%. An excellent feature of Fiber FP is its strength retention at high temperatures. They retain strength up to about 1370°C. Properties of Fiber FP are given in Table 2-6.

Silicon carbide (SiC) fibers are produced by a chemical vapor deposition (CVD) process (AVCO Specialty Materials Co.), as well as by controlled pyrolysis of a polymeric precursor (Nippon Carbon Co.—Nicalon fibers). Properties of SiC fibers are also given in Table 2-6. Silicon carbide fibers retain tensile strength well above 650°C. Alumina and SiC fibers are suitable for reinforcing metal matrices, in which carbon and boron fibers exhibit adverse reactivities. In addition, alumina has an inherent resistance to oxidation that is desirable in applications such as gas-turbine blades.

High-Performance Polyethylene (HPPE) Fibers: Ultrastrong and high-modulus fibers can be produced from the polyethylene molecule. For this purpose, an ultra-high-molecular-weight polyethylene (UHMW-PE) is dis-

Table 2-5 Properties of boron fiber (with tungsten, core)

Property, units	Fiber Diameter		
	100 μm	140 μm	200 μm
Tensile strength, MPa	3450	3450	3450
Tensile modulus, GPa	400	400	400
Coefficient of thermal expansion, m/m/°C	4.9×10^{-6}	4.9×10^{-6}	4.9×10^{-6}
Density, g/cm^3	2.61	2.47	2.39

Table 2-6 Properties of ceramic fibers

	Fiber		
Property, units	Alumina (Fiber FP)	SiC (CVD)	SiC (Pyrolysis)
Diameter, μm	20 ± 5	140	10–20
Density, g/cm³	3.95	3.3	2.6
Tensile strength, MPa	1380	3500	2000
Tensile modulus, GPa	379	430	180

solved in a solvent, spun through a spinneret, and cooled to obtain filaments. These PE filaments have an ultrahigh molecular weight and a low degree of molecular entanglement and are capable of being superdrawn. The PE fibers thus obtained have very long molecular chains, oriented and crystallized in the fiber direction, that together impart exceptional properties to the fibers. These fibers are called *high-performance polyethylene* (HPPE), *high-modulus polyethylene* (HMPE), or *extended-chain polyethylene* (ECPE) fibers. These fibers are produced commercially under the trade name "Dyneema" by DSM in the Netherlands, and "Spectra" by Honeywell (formerly Allied Signal) in the United States. Several grades of fibers are marketed with different properties for different applications. The range of properties of the HPPE fibers, as reported by manufacturers, is given in Table 2-7. The reported properties often are influenced by the testing method, and actual usable properties of fibers may be lower than the reported values.

HPPE fibers have a density of only 0.97 g/cm³. Their modulus and strength are slightly lower than those of Kevlar fibers but on a per-unit-weight basis, HPPE fibers have 30–40% higher strength and modulus than Kevlar fibers. High-energy absorption of HPPE fibers makes them suitable for use in ballistic protection applications. HPPE fiber–based composites perform exceptionally well against high-velocity impacts such as the ones produced by rifle rounds and shock waves from an explosion. This is so because the high-velocity impact produces a high strain-rate loading on the composite, and the strength and stiffness of the PE fibers increase at high strain rate. High-performance body armors (concealed ballistic vests) made from these fibers

Table 2-7 Properties of high-performance polyethylene (HPPE) fibers

Diameter, μm	38
Density, g/cm³	0.97
Tensile strength, MPa	2180–3600
Tensile modulus, GPa	62–120
Elongation, %	2.8–4.4

30 FIBERS, MATRICES, AND FABRICATION OF COMPOSITES

provide comfort and maneuverability. They also find use in vehicles for security and law enforcement, blast-containment applications, and reinforced cockpit doors on commercial airliners.

Major limitations in the application of HPPE fibers come from their low melting point of 150°C, an inert surface that makes it difficult to bond to other polymers, and poor creep characteristics.

2.2 MATRIX MATERIALS

Fibers, because of their small cross-sectional dimensions, cannot be loaded directly. Further, fibers, acting alone, cannot transmit loads from one to another to be able to share a load. This severely limits their direct use in load-bearing engineering applications. This limitation is overcome by embedding them in a matrix material to form a composite. The matrix binds the fibers together, transfers loads between them, and protects them against environmental attack and damage due to handling. The matrix has a strong influence on several mechanical properties of the composite, such as transverse modulus and strength, shear properties, and properties in compression. The matrix material frequently limits a composite's service temperature. Temperature ranges at which composites with different types of matrices can be used are shown in Fig. 2-4. Physical and chemical characteristics of the matrix, such as melting or curing temperatures, viscosity, and reactivity with fibers, influence the choice of fabrication process. The matrix material for a composite system is selected keeping in view all these factors. Commonly used matrix materials are described in this section.

2.2.1 Polymers

Polymers (commonly called *plastics*) are the most widely used matrix material for fiber composites. Their chief advantages are low cost, easy processability,

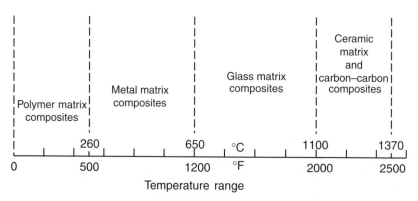

Figure 2-4. Usable temperature ranges for composites with different types of matrix materials.

good chemical resistance, and low specific gravity. On the other hand, low strength, low modulus, and low operating temperatures limit their use. They also degrade by prolonged exposure to ultraviolet light and some solvents.

2.2.1.1 Thermosetting and Thermoplastic Polymers

According to their structure and behavior, polymers can be classified as thermoplastics or thermosets. The polymers that soften or melt on heating, called *thermoplastic polymers,* consist of linear or branched-chain molecules having strong intramolecular bonds but weak intermolecular bonds. Melting and solidification of these polymers are reversible, and they can be reshaped by application of heat and pressure. They are either semicrystalline or amorphous in structure. Examples include polyethylene, polystyrene, nylons, polycarbonate, polyacetals, polyamide–imide, polyether–ether ketone (PEEK), polysulfone, polyphenylene sulfide (PPS), and polyether imide. Thermosetting plastics have cross-linked or network structures with covalent bonds between all molecules. They do not melt but decompose on heating. Once solidified by a cross-linking (curing) process, they cannot be reshaped. Common examples of thermosetting polymers include epoxides, polyesters, phenolics, ureas, melamine, silicone, and polyimides.

2.2.1.2 Polymer Properties of Importance to the Composite

Certain physical and chemical properties of a polymer have particular significance to the properties of a composite. Further, polymers have unique characteristics that set them quite apart from metals and ceramics. Of particular importance are the properties shown in Table 2-8. This table shows the influence of external variables on the properties. Unlike metals and ceramics, polymers may be considerably influenced by external variables. In contrast, the mechanical properties of metals typically are influenced only near the melt temperature.

Table 2-8 Effect of external variables on polymer properties

	Temperature	Environment	Strain Rate
Strength	X	X	X
Stiffness	X	X	X
Thermal expansion	X		
Thermal conductivity	X		
Permeability	X		
Solubility	X		
Environmental aging (ultraviolet)	X	X	
Melt temperature (semicrystalline polymer)			
Glass transition temperature (amorphous polymer)	X	X	

An X indicates a strong interaction between the external variable and the property.

The temperature limitations of a thermoplasic depend on whether it is semicrystalline or amorphous. Thermosetting plastics typically have amorphous structures, but thermoplastics may be either semicrystalline (they are never 100% crystalline) or amorphous. The amorphous state is characterized by a glass transition temperature (T_g) only, whereas the semicrystalline polymer has a crystalline melting point (T_m) as well as a glass transition temperature. These transition temperatures are illustrated in Fig. 2-5 as measured by specific volume changes with temperature. Transition temperatures (T_g and T_m) of some polymers are given in Table 2-9.

The temperature for processing of thermoplastics is governed by either the melt temperature or glass transition temperature. For example, an amorphous thermoplastic must be molded well above its T_g in order to reduce its melt viscosity sufficiently.

An understanding of the effect of these temperatures on the mechanical behavior of polymers is best seen by the behavior of modulus of elasticity (E) with temperature (Fig. 2-6). An amorphous thermoplastic (e.g., polystyrene, polycarbonate, or polymethylmethacrylate) has a significant change of mechanical properties at the glass transition temperature. Hence maximum use temperatures must be less than the glass transition temperatures. A thermoset (e.g., epoxy, polyester, or phenolic) has a much reduced change in properties at the glass transition temperature because of its high degree of cross-linking. However, their maximum use temperatures should not exceed T_g. Semicrystalline thermoplastics also have a modest change in properties at the glass transition temperature owing to the presence of the crystalline regions. Their maximum use temperatures are more dictated by the melting points, as in the case of metals and ceramics.

Examples of stress–strain curves for thermoplastics are shown in Fig. 2-7. These general shapes are applicable for crystalline or amorphous materials. It is important to recognize that these large variations in behavior can occur

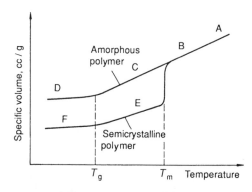

Figure 2-5. Specific volume of amorphous and semicrystalline polymers: variation with temperature.

Table 2-9 Transition temperatures for polymers

Polymer	T_g (°C)	T_m (°C)
Thermosets		
Epoxy	100–250	—
Polyester	75–150	—
Thermoplastics		
Polystyrene	100	—
Polyethylene (HD)	−80	137
Polycarbonate	150	—
Nylon (6,6)	50	255–265
Polymethylmethacrylate	105	—
Polyetherether ketone (PEEK)	143	334
Polyphenylene sulfide (PPS)	85	285
Polyether sulfone	225	—

over a temperature range of only 100–200°C for some materials. Further, the rate of strain has a comparable effect, although much greater rate changes must occur to alter the behavior substantially. Thermosets also are not as affected by temperature and strain rate, the range of behavior being limited to approximately the higher three curves shown in Fig. 2-7.

Environmental influences (e.g., moisture absorption) also have a significant effect on the behavior of polymers, particularly relative to metals and ceramics. For example, epoxy or polyester resins can absorb up to 4–5% by weight of water if exposed to 100% relative humidity or immersed in water. The mechanical properties of nylon thermoplastics are influenced significantly by their moisture contents. Polymers are also susceptible to deterioration as a result of exposure to ultraviolet radiation.

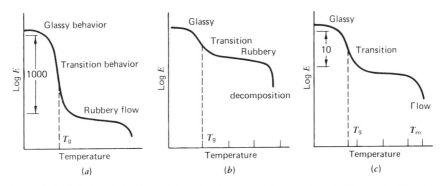

Figure 2-6. Variation of elastic modulus of polymers with temperature: (a) thermoplastic, amorphous (high molecular weight or lightly cross-linked); (b) thermoset, highly cross-linked; (c) semicrystalline.

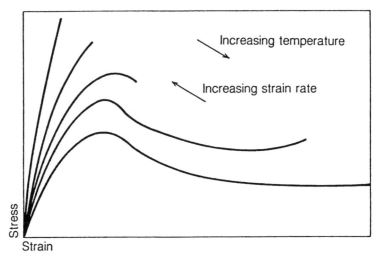

Figure 2-7. Tensile stress–strain curves of a thermoplastic at different strain rates and temperatures.

These effects must be properly taken into consideration when the matrix is selected for a composite.

2.2.1.3 Common Polymeric Matrix Materials Polyester and epoxy resins are the most common polymeric matrix materials used with high-performance reinforcing fibers. Both are thermosetting polymers. Easy processibility and good chemical resistance are their chief advantages.

Polyester Resin A polyester resin is an unsaturated (reactive) polyester solid dissolved in a polymerizable monomer. Unsaturated polyesters are long-chain linear polymers containing a number of carbon double bonds. They are made by a condensation reaction between a glycol (ethylene, propylene, or diethylene glycol) and an unsaturated dibasic acid (maleic or fumaric). A typical polyester resin made from reaction of maleic acid and diethylene glycol is shown below:

$$\underset{\text{Maleic acid}}{\text{HO}-\overset{\overset{\displaystyle O}{\|}}{C}-CH=CH-\overset{\overset{\displaystyle O}{\|}}{C}-OH} \;+\; \underset{\text{Diethylene glycol}}{HOCH_2CH_2OCH_2CH_2OH} \longrightarrow$$

$$HO\left[CH_2CH_2\,O\,CH_2CH_2\,O\,\overset{\overset{\displaystyle O}{\|}}{C}-CH=CH\overset{\overset{\displaystyle O}{\|}}{C}-O\right]_n -H + H_2O$$

2.2 MATRIX MATERIALS

The length of the molecule or degree of polymerization n may vary. The resin generally will be a solid. It is dissolved in a polymerizable (reactive) monomer such as styrene, which also contains carbon double bonds and acts as a cross-linking agent by bridging adjacent polyester molecules at their unsaturation points. The monomer also acts as a diluent, reducing the viscosity of the polyester, and makes it easier to process. The curing or cross-linking process is initiated by adding a small quantity of a free-radical initiator/curing agent such as an organic peroxide (e.g., benzoyl peroxide) or an aliphatic azo compound.

The styrene monomer cross-links or reacts with the double bond in the polyester backbone above to form a network polymer, as shown below:

Styrene monomer: $\quad CH_2\!=\!CH\!-\!C_6H_5$

Styrene link (St): $\quad -CH_2-CH(C_6H_5)-$

Cross-linked polyester:

```
          |           |           |           |
         (St)        (St)        (St)        (St)
          |           |           |           |
—R—OOCC—CCOO—OOCC—CCOO—OOCC—CCOO—OOCC—CCOO—
          |           |           |           |
         (St)        (St)        (St)        (St)
          |           |           |           |
—R—OOCC—CCOO—OOCC—CCOO—OOCC—CCOO—OOCC—CCOO—
          |           |           |           |
         (St)        (St)        (St)        (St)
          |           |           |           |
```

This reaction does not produce a by-product and is an exothermic reaction. Thus the curing process is accompanied by shrinkage as well as a temperature increase. The curing is done at room temperature with or without the application of pressure.

Capabilities of modifying or tailoring the chemical structure of polyesters by processing techniques and raw-materials selection make them versatile. For example, the starting acids and glycols, as well as solvent monomers, all can be varied. Typical properties of cast thermosetting polyesters are given in Table 2-10.

Table 2-10 Typical properties of cast thermosetting polyesters

Density, g/cm^3	1.1–1.4
Tensile strength, MPa	34.5–103.5
Tensile modulus, GPa	2–4.4
Thermal expansion, 10^{-6}/°C	55–100
Water absorption, % in 24 h	0.15–0.6

Epoxy Resins Epoxy resins are low-molecular-weight organic liquids containing a number of epoxide groups, which are three-membered rings with one oxygen and two carbon atoms:

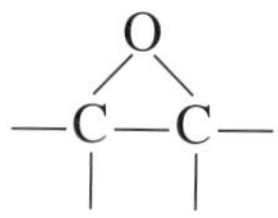

$$-\overset{|}{\underset{|}{C}}-\overset{\overset{O}{\triangle}}{\underset{|}{C}}-$$

The most common process for producing epoxies is the reaction of epichlorohydrin with bisphenol-A and obtaining cross-linking by introducing chemicals that react with the epoxy groups between the adjacent chains.

The chemical reaction to form the epoxy resin prepolymer is

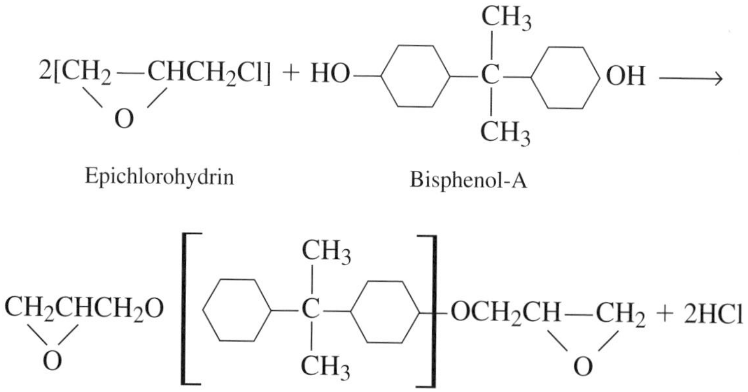

2[CH$_2$—CHCH$_2$Cl] + HO—⟨◯⟩—C(CH$_3$)(CH$_3$)—⟨◯⟩—OH ⟶
  \O/
  Epichlorohydrin   Bisphenol-A

CH$_2$CHCH$_2$O—[⟨◯⟩—C(CH$_3$)(CH$_3$)—⟨◯⟩]—OCH$_2$CH—CH$_2$ + 2HCl
  \O/   \O/

The epoxy resin is a viscous liquid, and the viscosity is a function of the degree of polymerization n. Each epoxy molecule is end-capped with the epoxy group. A curing agent is mixed into the liquid epoxy to polymerize the polymer and form a solid network cross-linked polymer. For example, diethylene triamine, as shown below, achieves rapid cure at room temperature:

Diethylene triamine: NH$_2$(CH$_2$)$_2$NH(CH$_2$)$_2$NH$_2$

Five molecules of epoxy can react with each amine molecule through the active hydrogen on the nitrogen atom. A polymerization reaction involving the breakage of the epoxy ring occurs as follows:

$$\sim\!\!\sim\!\text{CH}-\text{CH}_2 + \text{H}-\text{N}\!\sim\!\!\sim \longrightarrow$$
$$\diagdown\!\text{O}\!\diagup$$

$$\sim\!\!\sim\!\text{CH}-\text{CH}_2-\text{N}\!\sim\!\!\sim$$
$$\quad\quad |$$
$$\quad\text{OH}$$

Thus a segment of the fully cured structure is as follows:

```
       OH                                              OH
       |                                               |
     —C—CH₂                              CH₂—C—
              \   NCH₂CH₂NCH₂CH₂N   /
     —C—CH₂  /           |           \  CH₂—C—
       |                CH₂                          |
       OH                |                           OH
                       C—OH
                        |
```

This reaction does not produce a by-product but does produce heat accompanied with chemical shrinkage.

Epoxy systems, like polyesters, can be cured at room temperature. The choice of curing agent dictates whether a room-temperature or elevated-temperature cure is required. Heat is added quite often to accelerate curing and to achieve a higher degree of cure.

The properties of a cured epoxy resin depend on the chemical composition of the epoxy prepolymer, which can be modified greatly, as well as on the curing-agent molecule. Typical properties of cast epoxy resin are given in Table 2-11.

Table 2-11 Typical properties of cast epoxy resins (at 23°C)

Density, g/cm^3	1.2–1.3
Tensile strength, MPa	55–130
Tensile modulus, GPa	2.75–4.10
Thermal expansion, $10^{-6}/°C$	45–65
Water absorption in 24 h, %	0.08–0.15

Epoxy systems are superior to polyesters, particularly with regard to adhesion with a wide variety of fibers, moisture resistance, and chemical resistance.

Other Thermosetting Polymers Vinyl esters, polyimides, and phenolics are among the other thermosetting polymers that are used as matrix materials for composites. Vinyl esters are closely related to the unsaturated polyesters. Like unsaturated polyesters, they possess low viscosity and cure fast but are slightly more expensive. They possess exceptional mechanical and chemical performance characteristics and thus provide a transition in mechanical properties and cost between polyesters and the high-performance epoxy resins. Vinyl esters often are used because of their ease and speed of processing and their good resistance to wet environments. Their applications include high-performance gel coats, pipes, and reaction vessels.

Polyimides have a relatively high service temperature range (250–300°C). They also possess excellent chemical and solvent resistance. Bismaleimides (BMIs) are the most widely used thermosetting polyimides for composites employed at high temperatures. However, these materials are inherently very brittle, and thus they are often combined with polysulfone, polyetherimide, or other thermoplastics. The handling and processing techniques for BMI resins are similar to those for epoxy resins. Their applications include aerospace wing-skin ribs, helicopter firewalls, and printed wiring boards.

Phenolic resins have low flammability, low smoke production, good dimensional stability under temperature fluctuations, and good adhesive properties. Phenolics are attractive for aircraft and mass-transit vehicles and as interior construction materials where outgassing due to fire must be extremely low. Phenolics also are used for rocket nozzle ablative and insulation liners. Chopped-fiber molding compounds of phenolic resin are used mostly in the automotive, appliance, and electrical component markets.

Typical properties of vinyl esters, polyimides and phenolic resins are given in Table 2-12.

Thermoplastic polymers are used extensively for short-fiber composites in large-volume applications. The manufacturing cost usually is lower because the composite can be manufactured by mass-production methods much more quickly. Most of the manufacturing processes (e.g., injection molding) that

Table 2-12 Typical properties of vinyl esters, polyimides, and phenolics

Property, units	Vinyl Esters	Polyimides	Phenolics
Density, g/cm^3	1.12–1.32	1.46	1.30
Tensile strength, MPa	73–81	120	50–55
Tensile modulus, GPa	3.0–3.5	3.5–4.5	2.7–4.1
Coefficient of thermal expansion, 10^{-6}/°C	53	90	45–110
Water absorption in 24 h, %	—	0.3	0.1–0.2

are used for unreinforced thermoplastics also can be adopted to manufacture short-fiber-reinforced thermoplastics. Such composites generally are viewed as higher-strength and higher-stiffness replacements for plastics rather than the high-performance load-bearing composites competing with such structural materials as metals. However, some thermoplastic resins have much higher glass transition temperatures and maximum use temperatures than epoxies and BMIs. Such resins are used in high-temperature applications of high-performance composites. Properties of some of the high-temperature thermoplastic resins are given in Table 2-13. Standard references can be consulted for properties and performance of more common thermoplastics.

2.2.1.4 Fillers Fillers are used widely in polymeric composites primarily to reduce cost with some sacrifice in physical properties. They also are used to reduce shrinkage, control viscosity, and improve part stiffness. Commonly used fillers include calcium carbonate, kaolin (china clay), silica (sand), feldspar, talc, and glass microspheres. Fillers are not as common in high-performance composites because they may adversely affect the fiber–resin load transfer and decrease the toughness of the resin at high filler content.

Calcium carbonate ($CaCO_3$) is a widely used filler for both economic and performance considerations. Glass-fiber-reinforced polyesters from sheet-molding compounds or bulk-molding compounds may contain substantial amounts of $CaCO_3$ for cost reduction and shrinkage control. High surface smoothness may be achieved and sink marks eliminated. Kaolin is used to increase resin viscosity to prevent fibers from extruding from molded surfaces. It also improves the fire resistance of the compound. Natural silicas are used in thermoset resins for dimensional stability, good electrical insulation, and improved thermal conductivity. Talc (hydrated magnesium silicate), in the form of finely ground thin platelets, is added to resins to improve stiffness and creep resistance. Some natural organic materials such as wood flour, shell fibers (e.g., shell of almond, coconut, peanut, walnut, etc.), and cotton and vegetable fibers (e.g., hemp, jute, ramie, and sisal) are also used as fillers. Hollow microspheres made from glass or polymers can be used to reduce the density of the resin significantly.

Additives used for fire resistance (e.g., antimony oxide), chemical thickening (e.g. magnesium oxide and calcium hydroxide), and for lowering shrinkage (e.g., fine-powdered polyethylene) are also common.

For more detailed information on fillers and additives, the reader should refer to the books suggested at the end of this chapter.

2.2.2 Metals

Metals are by far the most versatile engineering materials. The properties that are particularly important for their use as matrix materials in composites include high strength, high modulus, high toughness and impact resistance, and relative insensitivity to temperature changes. Their greatest advantage over

Table 2-13 Typical properties of some thermoplastic resins

Property, units	PEEK	Polyamide–imide	Polyetherimide	Polysulfone	Polyphenylene Sulfide
Density, g/cm^3	1.30	1.38	1.24	1.25	1.32
Tensile strength, MPa	92	95	105	75	70
Tensile modulus, GPa	3.24	2.76	3.0	2.48	3.3
Continuous service temperature, °C	310	190	170	175–190	260
Coefficient of thermal expansion, 10^{-6}/°C		63	56	94–100	99
Water absorption in 24 h, %	0.1	0.3	0.25	0.2	0.2

the polymer matrices is in applications that require exposures to high temperatures and other severe environmental conditions. The factors that limit their use include their high density, high processing temperatures (due to a high melting point), reactivity with fibers, and attack by corrosion.

The most commonly used metal matrices are based on aluminum and titanium. Both these metals have low densities (aluminum, 2.7 g/cm^3; titanium, 4.5 g/cm^3) and are available as alloys. Magnesium, although lighter (density 1.74 g/cm^3), is unsuitable because of its great affinity to oxygen, which promotes corrosion. Nickel- and cobalt-based superalloys have been used as a matrix; however, some alloying elements in them tend to accentuate the oxidation of fibers at elevated temperatures.

Aluminum and its alloys are the most widely used metal matrices. Mechanical properties of some aluminum-alloy matrix materials are given in Table 2-14. Commercially pure aluminum has good corrosion resistance. Aluminum alloys such as 6061 and 2024 have been used for their higher strength–weight ratios. Carbon is the most common fiber used with aluminum alloys. However, carbon reacts with aluminum at typical fabrication temperatures of 500°C or higher. This severely degrades the mechanical properties of the composite. Protective coatings are used often on carbon fibers to reduce this degradation of fibers, as well as to improve fiber wetting by the aluminum-alloy matrix.

Titanium alloys used as matrices include α and β alloys (e.g., Ti–6A1–9V) and metastable β alloys (e.g., Ti–10V–2Fe–3Al). These alloys have higher strength–weight ratios and are superior to aluminum alloys in strength retention at 400–500°C. One problem with titanium alloys is their high reactivity with boron and alumina fibers at normal fabrication temperatures. Silicon carbide (SiC) and borsic (boron fibers coated with silicon carbide) fibers show less reactivity with titanium.

2.3 FABRICATION OF COMPOSITES

Finished products are formed from materials such as plastics and metals by molding or shaping methods. The material is first created and then processed

Table 2-14 Properties of some aluminum-alloy matrix materials

Alloy	Tensile Modulus (GPa)	Yield Stress (0.2% Offset) (MPa)	Ultimate Tensile Strength (MPa)	Strain to Failure (%)
1100	63	43	86	20
2024	71	128	240	13
5052	68	135	265	13
6061	70	77	136	16
Al–7Si	72	65	120	23

at a later stage by companies specializing in forging, sheet forming, injection molding, etc. However, products consisting of composites can be created simultaneous with the creation of the material. Such is the case when filament winding a pipe from a polymer and glass fiber strands.

With regard to polymeric matrix composites, the processing methods for thermosetting materials typically involve material formation during final molding (e.g., hand lay-up, spray-up, and vacuum-bag molding). In some cases material formation is accomplished separately from forming or shaping, but because of the curing nature of thermosetting resins, final curing occurs during final formation. In thermoplastic matrix composites, it is more common to fabricate the composite first and form or mold a shape in a second operation. However, in this latter step, the composite properties still can be influenced (e.g., fiber length reduction or fiber orientation during molding).

The choice of a fabrication process is strongly influenced by the chemical nature of the matrix (e.g., thermoset or thermoplastic in case of a polymer) and the temperature required to form, melt, or cure the matrix. The various composite forming and fabrication methods are described in the following subsections.

2.3.1 Fabrication of Thermosetting Resin Matrix Composites

Monomers or *prepolymers* of thermosetting resin systems are usually in a fluid state. They become solid as a result of a chemical reaction. During this chemical reaction, molecules of monomers or prepolymers are linked together to form polymer networks. This process of linking the molecules is called *polymerization* and *cross-linking* in polymers. The cross-linking is accomplished by catalysts or curing agents usually selected to give a desired combination of time and temperature to complete the reaction suitable for a particular product. The curing and accompanying hardening are irreversible. Further heating does not melt or soften them for molding or reshaping. However, the curing can be staged so that formation of the composite can be accomplished separate from the final stage of hardening.

Fabrication processes for thermosetting resin matrix composites can be broadly classified as wet-forming processes and processes using premixes or prepregs. In the wet-forming processes, the final product is formed while the resin is quite fluid, and then the curing process is usually completed by heating. The wet processes include hand lay-up, bag molding, resin-transfer molding, filament winding, and pultrusion. In the processes using premixes, as the name suggests, material preparation is separated from lay-up or molding. Premixes such as bulk molding compounds (BMCs) and sheet molding compounds (SMCs) are compounded from resin, fillers, and fibers and partially cured. Prepregs are usually partially cured sheets of oriented fibers or fabrics. The matrix material in some of the premixes is thickened so that it is tack-free or slightly tacky, does not flow, and can be handled easily. Thickening is achieved by the use of a thickening agent and by advancing the cure of the resin. In the latter case, the resins must be stored and transported at low temperatures (10–15°C). Premixes are used subsequently for product lay-up, and final curing completed under heat and pressure. The use of premixes makes manufacturing more simple and increases the possibility of automation. High-fiber-volume fractions can be achieved with uniform fiber distribution.

Different wet-forming processes and compounding of premixes and their subsequent use for final product fabrication are described in the following subsections.

2.3.1.1 Hand Lay-up Technique The hand lay-up technique is the oldest, simplest, and most commonly used method for the manufacture of both small and large reinforced products. A flat surface, a cavity (female) or a positive (male) mold, made from wood, metal, plastics, reinforced plastics, or a combination of these materials may be used. Fiber reinforcements and resin are placed manually against the mold surface. Thickness is controlled by the layers of materials placed against the mold.

This technique, also called *contact lay-up,* is an open-mold method of molding thermosetting resins (polyesters and epoxies) in association with fibers (usually glass-fiber mat, fabric, or woven roving). A chemical reaction initiated in the resin by a catalytic agent causes hardening to a finished part. Hand lay-up techniques are best used in applications where production volume is low and other forms of production would be prohibitive because of costs or size requirements. Typical applications include boat and boat hulls, radomes, ducts, pools, tanks, furniture, and corrugated and flat sheets.

The following operations are involved in a typical hand lay-up process:

Mold preparation This is one of the most important functions in the molding cycle. If it is done well, the molding will look good and separate from the mold easily. Production mold preparation requires a thorough machine buffing and polishing of the mold. After the desired finish has been attained, several coats (usually three or four) of paste wax are applied for the purpose of mold release. Many different release systems are available, such as wax, polyvinyl alcohol (PVA), fluorocarbons, silicones, release papers and release films, and liquid internal releases. The choice of release agent depends on the type of surface to be molded, the degree of luster desired on the finished product, and whether painting or other secondary finishing will be required.

Gel coating When good surface appearance is desired, the first step in the open-mold processes is the application of a specially formulated resin layer called the *gel coat.* It is normally a polyester, mineral-filled, pigmented, nonreinforced layer or coating. It is applied first to the mold and thus becomes the outer surface of the laminate when complete. This produces a decorative, protective, glossy, colored surface that requires little or no subsequent finishing. The gel coating may be painted on, air-atomized with gravity or pressure feeding, or sprayed by an airless sprayer.

Hand lay-up After properly preparing the mold and gel-coating it, the next step in the molding process is material preparation. In hand lay-up, the fiberglass is applied in the form of chopped strand mat, cloth, or woven roving. Premeasured resin and catalyst (hardener) are then thoroughly mixed together. The resin mixture can be applied to the glass either outside of or on the mold. To ensure complete air removal and wet-out, serrated rollers are used to compact the material against the mold to remove any entrapped air. The resin–catalyst mixture can be deposited on the glass via a spray gun, which automatically meters and combines the ingredients. The first layer of reinforcement is usually a thin, randomly oriented fiber mat designed to reinforce the resin-rich surface of the moldings and improve surface finish. Such a reinforcement, called *surfacing mat* or *veil* and made with a weight of about 30 g/m^2, also may be made from a chemically resistant type of glass if corrosion resistance is required. Extra care must be given to this surfacing mat to ensure that no air bubbles are left between the glass and the gel coat.

Spray-up This is a partially automated form of hand lay-up. Chopped glass fibers and resin are deposited simultaneously on an open mold. Fiberglass roving is fed through a chopper on the spray gun and blown into a resin stream that is directed at the mold by either of two spray systems:

(i) The external mixing system has two nozzles. One of the nozzles ejects resin premixed with catalyst or catalyst alone, whereas the other nozzle ejects resin premixed with accelerator.

(ii) The internal mixing system has only one nozzle. Resin and catalyst are fed into a single gun mixing chamber ahead of the spray nozzle.

By either method, the resin mix precoats the strands of glass, and the merged spray is directed at the mold in an even pattern by the operator. A typical spray-up process is shown schematically in Fig. 2-8. After the resin and glass mix is deposited, it is rolled by hand to remove air, to compact the fibers, and to smooth the surface. The spray-up process relies heavily on operator skill for product quality. It is faster and hence less costly than hand lay-up. In practice, combinations of hand lay-up and spray-up are often used in which layers of sprayed and/or chopped fibers are alternated with cloth or woven roving. It should be noted that the spray-up method produces short-fiber composites, whereas the hand lay-up method can be used with continuous woven fibers as well. Advantages and disadvantages of these processes are as follows:

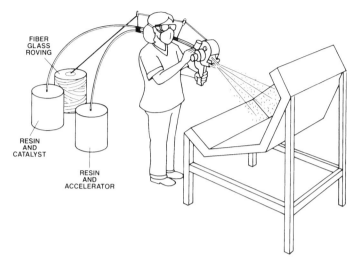

Figure 2-8. Spray-up process. (Courtesy of CertainTeed Corp.)

Advantages of Hand Lay-up and Spray-up

1. Large and complex items can be produced.
2. Minimum equipment investment is necessary.
3. The startup lead time and cost are minimal.
4. Tooling cost is low.
5. Semiskilled workers are easily trained.
6. Design flexibility.
7. Molded-in inserts and structural changes are possible.
8. Sandwich constructions are possible.

Disadvantages of Hand Lay-up and Spray-up

1. The process is labor-intensive.
2. It is a low-volume process.
3. Longer cure times may be required because room-temperature curing agents generally are used.
4. Quality is related to the skill of the operator.
5. Product uniformity is difficult to maintain within a single part or from one part to another.
6. Only one good (molded) surface is obtained.
7. The waste factor is high.

2.3.1.2 Bag Molding Processes

Bag molding is one of the oldest and most versatile of the processes used in manufacturing composite parts. The laminae (preimpregnated or freshly impregnated with wet resin) are laid up in a mold, covered with a flexible diaphragm or bag, and cured with heat and pressure. After the required cure cycle, the materials become an integrated molded part shaped to the desired configuration.

A cross section of a typical lay-up of a composite structure is shown in Fig. 2-9. In addition to the actual composite laminate, the lay-up includes release coatings, peel plies, release films, bleeder plies, breather plies, vacuum bags, sealant tape, and damming material. Each of these materials serves a specific function. Release agents are used to prevent the composite material from bonding to the mold. Peel plies protect the surface of the molded part from contamination. Release films are used to separate the bleeder or breather materials from the composite laminate. In some cases the release film is porous so that resin can flow through the film. Bleeder and breather plies are porous, high-temperature fabrics that are used to absorb excess resin during processing. Breather plies provide a pathway into the composite laminate and act as a conduit for the removal of air and volatiles during curing. Bagging films form a barrier between the composite laminate and the oven or autoclave environment.

Laying up and bagging are critical steps influencing the quality of part production. Thus a worker's skill and know-how play an important role in the quality of a part. The size of a part that can be produced by bag molding is limited only by the curing equipment, specifically, the size of the curing oven or autoclave.

The general process of bag molding can be divided, on the basis of pressure and heat applied to the laminate during curing, into pressure bag, vacuum bag, and autoclave. In pressure-bag molding, pressures above atmospheric are applied on the laminate inside the closed mold, as shown in Fig. 2-10a. The curing is accomplished by heating the mold in an oven. In vacuum-bag molding, air and other volatiles between the bag and laminate are removed by a vacuum pump (see Fig. 2-10b). Thus the laminate is subjected to atmospheric

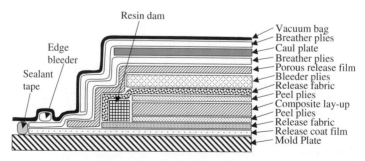

Figure 2-9. Typical bagging lay-up.

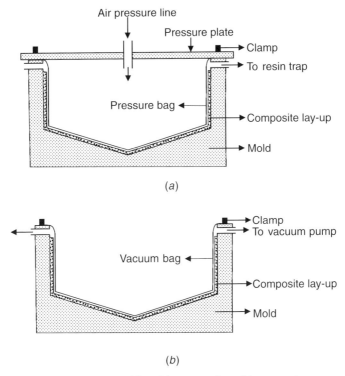

Figure 2-10. Bag molding: (a) pressure bag, (b) vacuum bag.

pressure when it is being cured in an oven. Autoclave processing (Fig. 2-11) of composites is an extension of the vacuum-bag technique. Autoclaves can be pressurized during processing of the composites. A picture of an autoclave is shown in Fig. 2-12. In this method, the composite part is laid up and enclosed in a vacuum bag. Full or partial vacuum is drawn under the bag, and gas pressure greater than atmospheric pressure is applied on the exterior of the bag. Curing of the polymer is initiated by raising the part temperature in the autoclave chamber. Augmented pressure exerts higher mechanical forces on the lay-up, increases the efficiency of transport of volatiles to the vacuum ports, and results in increased wetting and flow of the resin. Reduction in volume of trapped air and released volatiles results in lower void content.

Vacuum-bag and autoclave methods are used to produce most bag-molded parts. Their main advantages are the relatively inexpensive tooling and use of the basic curing equipment (oven and autoclave) for an unlimited variety of shaped parts. The disadvantage of the pressure-bag system is the relatively expensive tooling because it is combined with the curing pressure system. Further, the tooling can be used only for a specific part for which it is designed.

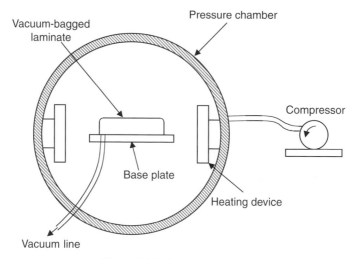

Figure 2-11. Autoclave molding.

Figure 2-12. A 3-ft-diameter and 6-ft-long autoclave at the University of Missouri–Rolla.

2.3.1.3 Resin Transfer Molding Resin transfer molding (RTM) is a wet impregnation process in which fibers and resin are introduced separately into the mold. It requires placing dry-fiber reinforcement in the mold and introducing liquid resin in the closed mold to impregnate the fibers and fill the mold cavity. RTM uses a mold with inlets for resins and outlets for air to escape (Fig. 2-13). When the mold is full, the resin supply is removed, the mold inlets and outlets are sealed, and heat is applied to cure the resin. Continuous-strand mat, woven roving, or cloth can be used as reinforcement in RTM. Typically, a thermosetting polymer of relatively low viscosity is used in the RTM process. Large continuous-fiber-reinforced composites can be produced by RTM with relatively short cycle times. The RTM allows for better control over the orientation of the fibers, thus improving material properties.

A variation of the RTM process is vacuum-assisted resin transfer molding (VARTM). In VARTM, vacuum is applied to the outlet of the mold, and the resin is drawn into the mold by vacuum only. Since vacuum is applied instead of pressure, half the mold may be replaced by a vacuum bag. Also, since the pressure differential is much lower than the pressure used in conventional RTM, the cost of the mold can be reduced substantially.

2.3.1.4 Filament Winding Filament winding is a technique used for the manufacture of surfaces of revolution such as pipes, tubes, cylinders, and spheres and is used frequently for the construction of large tanks and pipework for the chemical industry. High-speed precise lay-down of continuous reinforcement in predescribed patterns is the basis of the filament-winding method. Continuous reinforcements in the form of rovings are fed from a multiplicity of creels. A *creel* is a metallic shelf holding roving packages at desired intervals and designed for pulling roving from the inside of the package. The reinforcement goes from the creels to a resin bath and may be gathered into a band of given width and wound over a rotating male mandrel. The winding angles and the placement of the reinforcements are controlled through specially designed machines traversing at speeds synchronized with

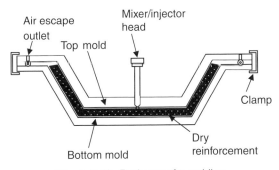

Figure 2-13. Resin-transfer molding.

the mandrel rotation. The reinforcements may be wrapped in adjacent bands that are stepped the width of the band and which eventually cover the entire mandrel surface. The technique has the capacity to vary the winding tension, wind angle, and resin content in each layer of reinforcement until the desired thickness and resin content of the composite are obtained. A diagram of a filament winding operation is given in Fig. 2-14.

The winding angle used for construction of pipes or tanks depends on the strength–performance requirements and may vary from longitudinal through helical to circumferential, as shown in Fig. 2-15. Fibers aligned perfectly in the longitudinal direction are difficult to place and depend on the mandrel ends and machine design. Often a combination of different winding patterns is used to give optimal performance.

In addition to the wet winding described earlier, filament-wound vessels can be produced from prepreg tapes and rovings (fabrication of prepreg tapes and rovings is discussed in Section 2.3.1.6). This technique reduces fiber damage during the winding operation and permits the use of resin systems that cannot be handled by wet lay-up techniques. Resin content of the laminate can be controlled more accurately with prepregs. The use of prepregs also makes for a cleaner operation.

Mandrel design is comparatively simple for open-end structures such as cylinders or conical shapes. Either cored or solid steel or aluminum serves satisfactorily. When end closures are integrally wound, as in pressure vessels, a careful consideration must be given to mandrel design and selection of a suitable material. Concepts frequently used for the construction of mandrels include segmented collapsible metal, low-melting alloys, eutectic salts, soluble plasters, frangible or break-out plasters, and inflatables. References are given at the end of this chapter for a more detailed discussion on the subject.

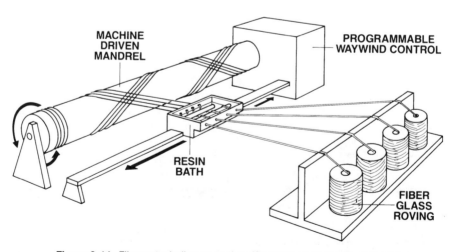

Figure 2-14. Filament winding operation. (Courtesy of CertainTeed Corp.)

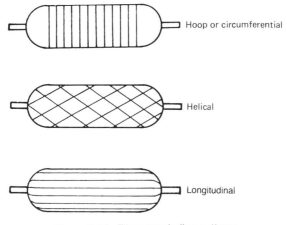

Figure 2-15. Filament winding patterns.

The filament winding process has the following advantages:

1. The process may be automated and provides high production rates.
2. Highest-strength products are obtained because of fiber placement control.
3. There is versatility of sizes.
4. Control of strength in different directions possible.

The following are limitations of filament winding:

1. Winding reverse curvatures is difficult.
2. Winding at low angles (parallel to rotational axis) is difficult.
3. Complex (double-curvature) shapes are difficult to obtain.
4. There is poor external surface.

2.3.1.5 Pultrusion Pultrusion is an automated process for manufacturing composite materials into continuous, constant-cross-section profiles. This technique has some similarities to aluminum extrusion or thermoplastic extrusion. In pultrusion, however, the product is pulled from the die rather than forced out by pressure. A large number of profiles such as rods, tubes, and various structural shapes can be produced using appropriate dies. A photograph showing several pultruded shapes is given in Fig. 2-16. Profiles may have high strength and stiffness in the length direction, with fiber content as high as 60–65% by volume.

The pultrusion process generally consists of pulling continuous rovings and/or continuous glass mats through a resin bath or impregnator and then into preforming fixtures, where the section is partially shaped, and excess

Figure 2-16. Pultruded samples made at the University of Missouri–Rolla.

resin and/or air are removed. Then it goes into a heated die, where the section is cured continuously. The basic pultrusion machine consists of the following elements: (1) creels, (2) resin bath or impregnator, (3) heated dies, (4) puller or driving mechanism, and (5) cutoff saw. A diagram of a pultrusion scheme is given in Fig. 2-17. A picture of a pultrusion machine is shown in Fig. 2-18.

The pultrusion process is most suitable for thermosetting resins that cure without producing a condensation by-product (polyester and epoxy). The reinforcements used consist of continuous fibers such as rovings or chopped-

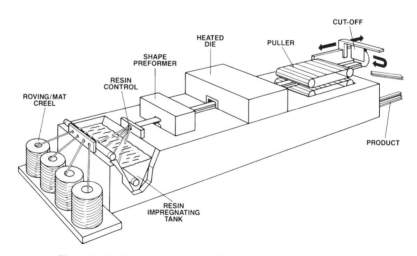

Figure 2-17. Pultrusion scheme. (Courtesy of CertainTeed Corp.)

Figure 2-18. A Labstar pultrusion machine at the University of Missouri–Rolla.

strand mat or a combination of the two, depending on the strength and rigidity required in the molded profile. Thermoplastic resins may be used, but special impregnation equipment is required to melt or soften the resin.

References are given at the end of this chapter for a more detailed discussion on pultrusion machines, pultrusion part design, and other related topics.

2.3.1.6 Preformed Molding Compounds A large number of reinforced thermosetting resin products are made by matched-die molding processes such as hot-press compression molding, injection molding, and transfer molding. Matched-die molding can be a wet process, but it is most convenient to use a preformed molding compound or premix to which all necessary ingredients have been added. This enables faster production rates to be achieved. Molding compounds can be divided into three broad categories: bulk or dough molding compounds (BMC, DMC), sheet molding compound (SMC), and prepregs. The following paragraphs briefly describe these premixes and their manufacturing methods.

BULK OR DOUGH MOLDING COMPOUNDS (BMC, DMC) These compounds consist of a doughlike or puttylike mixture of a resin, fiber reinforcement, and filler to which pigments and other materials may be added. The reinforcement used may be glass, cellulose, cotton, or other fibrous material. Manufacture usually is carried out in a high-shear Z-blade mixer, after which the compound is often extruded into rope form. Usually polyester DMC contains glass fiber strands, with fiber lengths ranging from 3–12 mm ($\frac{1}{8}$–$\frac{1}{2}$ in.) and a fiber content of between 15 and 20% by weight. These limits are dictated by the high shear loads created during manufacture and extrusion into usable forms.

SHEET MOLDING COMPOUNDS (SMC) These compounds are produced as flat sheets and invariably are based on unsaturated polyester resin systems reinforced with chopped-glass fibers, although carbon and/or aramid fibers also can be used separately or as hybrids. However, SMC differs from DMC in that it has a higher fiber content, usually 20–35%, and longer fibers, 21–55 mm ($\frac{7}{8}$–2 in.). Moldings made from SMC therefore have somewhat higher mechanical properties.

To manufacture SMC, a continuous polyethylene or cellophane film is coated with a suitably formulated polyester resin system into which is deposited a layer of either a chopped-strand mat (450 g/m^2) or chopped rovings. A second layer of polyethylene film, similarly coated with the resin system, is placed over the reinforcement, and the sandwich thus formed is passed through a series of rollers to press the glass fibers into the resin and ensure thorough wetting. The sandwich is then wound into a roll and allowed to stand while the resin thickens (Fig. 2-19).

Resins for SMC need to be of low initial viscosity to ensure thorough wetting of the glass reinforcement. However, once the glass has been wetted, the resin must thicken so that a relatively tack-free, easy-to-handle sheet of molding material is produced. This thickening is usually brought about by the addition of a thickening agent such as calcium oxide or magnesium oxide to the resin. Prior to molding, the SMC is cut to the required size, and both layers of polyethylene films are removed. The SMC then can be placed in the mold, pressed, and cured.

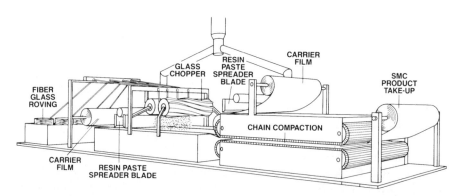

Figure 2-19. Manufacture of sheet molding compound. (Courtesy of CertainTeed Corp.)

PREPREGS   This is the short form for preimpregnated fiber-reinforced plastics. Prepregs consist of roving, woven fabric, continuous unidirectional fiber reinforcement sheets, or random chopped-fiber sheets impregnated with a partially cured resin system. These differ from sheet molding compounds in that thickening agents, fillers, pigments, and additives are rarely, if ever, present. Most prepregs are based on epoxy resin systems, and reinforcements usually include glass, carbon, and aramid fibers.

Prepregs are manufactured from either a resin solution or a solvent-free resin system. In the first case, the fiber reinforcement is drawn through a bath of resin solution and then through a "doctor" blade or metering roller assembly to control resin pickup. The impregnated reinforcement then passes through a vertical heating zone to evaporate the solvent and partially advance the cure of the resin system. The prepreg is then cooled and sandwiched between two layers of release film, such as silicone-impregnated paper or polyethylene film, prior to winding into a roll or cutting into sheets.

Prepregs made from solvent-free resin systems are produced in a similar way to sheet molding compounds in that the reinforcement is sandwiched between two layers of a resin system applied to suitable release film. With epoxy resin systems, heated rollers may be used to melt or lower the viscosity of the resin and ensure adequate wetting of the fibers. The solvent-free impregnation is claimed to produce superior composites with no possibility of any solvent entrapment. Thickening of the epoxy resin in prepregs is accomplished by advancing the cure of the resin system with heat until an only slightly tacky or tack-free prepreg is obtained. Cooling the prepreg to room temperature then reduces the rate of further cure. Such prepregs have shelf lives of from several weeks to several months at 20°C depending on the particular curing agent used.

The preformed molding compounds discussed earlier are very convenient to use in various molding processes for producing composite-materials parts and structures. The BMC, DMC, and SMC compounds are used with matched-die molding processes (Fig. 2-20), whereas prepregs are used widely in hand lay-up, bag-molding, and winding processes, as already discussed.

2.3.2 Fabrication of Thermoplastic–Resin Matrix Composites (Short-Fiber Composites)

The principal method used for the production of parts with short-fiber-reinforced thermoplastics is injection molding. Conventional mold-and-plunger or reciprocating screw-type machines are used for this purpose. The normal molding cycle used for unfilled thermoplastics is also used for the reinforced material, but the details of processing conditions employed are quite different. Further details of injection-molding processes can be seen in any standard text on processing of thermoplastics. It may be mentioned, however, that when fibers are introduced into a thermoplastic, the rheologic properties of the melt are modified significantly. Furthermore, the thermal

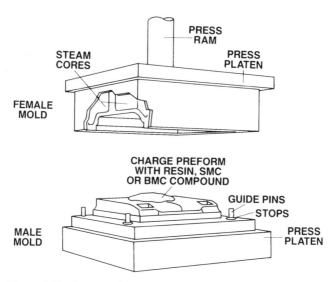

Figure 2-20. Compression molding. (Courtesy of CertainTeed Corp.)

conductivity of the melt usually is increased by the presence of the fibers. Hence both the flow fields and thermal conditions will be quite different compared with an unfilled thermoplastic. Another important point that must be considered is the fact that the properties of a short-fiber-reinforced thermoplastic are very dependent on fiber length and orientation, which can be greatly affected by the molding conditions. It is therefore important that both these parameters be controlled in the final molding by an appropriate choice of molding conditions.

The raw material used for injection molding of reinforced thermoplastics is a molding compound of the resin and fibers in a pelletized form. The compounding is carried out as a separate process prior to the injection molding. A compounding method aims at achieving the following:

1. Total enclosure of each fiber by the matrix
2. Uniform dispersion of fibers throughout the matrix
3. Low fiber breakage so that a high aspect ratio (ratio of fiber length to diameter) is maintained for effective stress transfer

In principle, the compounding and fabrication stages can be combined in one operation; that is, a dry blend of polymer and fibers is fed to the molding machine. However, injection molding of dry blends often results in poor surface finish and variable strength owing to the presence of undispersed fiber bundles in the finished product. The use of pelletized compound is generally more convenient. It eliminates the need for special material-handling equipment and minimizes the health hazards associated with airborne fibers.

The choice of a compounding technique depends on the requirements of fiber length, volume fraction, and degree of dispersion of the fibers throughout the matrix. The two most common compounding methods are extruder compounding and strand coating. In the extruder-compounding process, the fibers and resin are fed directly into an extruder for mixing. The fiber feed normally is introduced in the form of chopped fibers. Certain twin-screw extruders are designed so that continuous rovings enter at a downstream point where the polymer is already fully molten. The addition of fibers to the premelted polymer has the advantage of less fiber breakage. It also improves fiber dispersion and reduces wear in the working parts of the extruder.

The strand-coating method consists of passing the rovings or tows of fibers through a specially designed extruder die head so that they are coated and partially impregnated by the molten polymer (Fig. 2-21). The impregnated fiber tow is cooled in a water bath and then chopped into desired lengths. The molding pellets produced by this method contain a high volume fraction of long fibers. The fibers are contained in the pellets as a concentrated core surrounded by the resin. This may result in the uneven penetration of the fiber roving by the polymer and therefore an incomplete wetting of the fibers in the core. However, further dispersion of the fibers will occur during subsequent molding.

In addition to the injection molding of a pelletized molding compound, reinforced thermoplastic sheets are produced for stamping or thermoforming. These materials are made by lamination of a chopped-strand mat into a thermoplastic matrix. The process of producing these materials is similar to that of producing SMC. The important difference is that in this case the matrix material is a thermoplastic that is melted in an extruder before being incor-

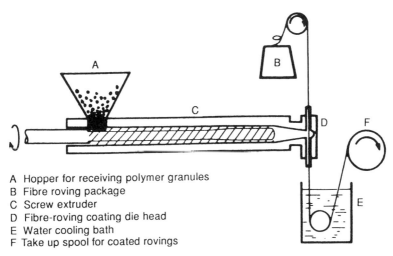

A Hopper for receiving polymer granules
B Fibre roving package
C Screw extruder
D Fibre-roving coating die head
E Water cooling bath
F Take up spool for coated rovings

Figure 2-21. Production of polymer-coated continuous-fiber roving.

porated with the fibers on a conveyor belt. The fibers contained in these materials are longer than those contained in pelletized compounds. In some cases the reinforcement is in the form of continuous roving. The AZDEL material of AZDEL, Inc., consists of 40% glass mat by weight in a polypropylene matrix, and AZMET has 35% glass mat by weight in a polybutylene terephthalate matrix. The reinforced-plastic sheets are thermoformed to the desired shape. A schematic of the process is shown in Fig. 2-22.

2.3.3 Fabrication of Metal Matrix Composites

Metals require significantly higher process temperatures than do polymers. At these higher temperatures, fibers may react with the metal matrix material, which invariably has a detrimental effect on the properties of the composite. Sufficient care must be exercised to limit the interaction between the fibers and the metal matrix. Metal matrix composites are fabricated most often by liquid infiltration or hot-pressing of solid matrix on fibers.

The simplest method of liquid infiltration is to pour the molten matrix into a vessel containing the fibers. The method is quite suitable if the fiber–matrix combination is nonreactive. In the case of a mutually reactive matrix–fiber combination (e.g., aluminum and silica), each individual fiber may be coated by drawing it singly through a bead of the molten metal. Continuous single fibers can be moved very quickly through small metal beads to give adequate coating thicknesses with very little time for the chemical reaction because the coating on a single fiber will cool extremely quickly. With this method, the coated fibers must be hot-pressed to make the composite.

With less reactive systems (e.g., carbon aluminum), a fiber tow can be drawn through a crucible containing the molten metal. The fibers can be protected against attack to some degree by a suitable metallic coating (e.g., nickel for carbon–aluminum system) or by reducing the rate of reaction

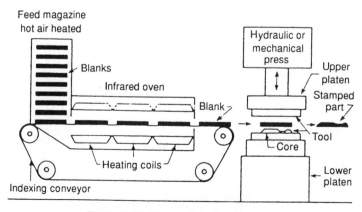

Figure 2-22. Thermostamping process.

through lowering of processing temperature obtained as a result of alloying the metal with a substance that reduces its melting point (e.g., 12% silicon in the molten aluminum reduces the processing temperature from just over 660 to 580°C). If the reinforced-rod preforms produced by this process are not thick enough for immediate use in a structure, a large number of them can be combined by hot-pressing to produce the desired cross section.

Plasma spraying is another method of liquid infiltration. The matrix is sprayed onto the fibers, which are supported on a foil, also made of the matrix material. The resulting tape is very porous, easily deformable, and suitable for cutting to the shape and size required for hot-pressing to the finished product. In the plasma-spraying technique, the metal cools rapidly and freezes in contact with the fibers, thus minimizing the undesirable interaction between the fibers and the matrix.

In the fabrication of metal matrix composites by hot-pressing of solid matrix on fibers, the matrix can be in the form of a sheet or powder. In the first case, fibers are laid between thin sheets of matrix foil in a mold, and the material is then consolidated by hot-pressing. The method is suitable only for relatively large-diameter fibers (e.g., boron) or wires. The temperature and pressure must be controlled very carefully to ensure adequate consolidation without too much chemical interaction or mechanical damage. Tapes can be made up with sheet matrix on either side of a layer of fibers, and the fibers can be held together by a resin binder (e.g., polystyrene) that evaporates during the first stage of the consolidation process. The tape can be used in the same way as polymer prepreg tape in processes such as filament winding.

When the matrix is in a powder form, the fibers and matrix powder are combined and held together by a volatile solid binder. When the mixture is hot-pressed, the binder escapes by evaporation. Often the pressing is done in two stages. Stage 1 results in removal of this binder and sufficient consolidation to hold the matrix and fibers together. Stage 2 results in consolidation of the material to the practical limit consistent with acceptable chemical and mechanical damage to the fibers.

Metal matrix composites also can be formed by a chemical vapor deposition process, coextrusion of matrix and pellet-shaped particles producing fibers, or forming fibers *in situ* during the controlled solidification of some off-eutectic or eutectic alloys.

2.3.4 Fabrication of Ceramic Matrix Composites

Ceramic matrix composites, glass, glass–ceramic, and oxide–ceramic matrices generally are fabricated by a two-stage process. In the first stage, fibers are incorporated into an unconsolidated matrix. The most common technique for this purpose is the slurry infiltration process, in which a fiber tow is passed through a slurry tank (containing the matrix powder, a carrier liquid, and an organic binder) and wound on a drum and dried. The second stage consists of cutting and stacking of tows and consolidation. Hot-pressing or firing at temperatures in excess of 1200°C is the most common technique for consolidating the ceramic matrix composites. The high temperatures are needed to promote rapid diffusion and recrystallization so that the densification proceeds to the required extent in a reasonable time. Porosity in a ceramic material is a common and serious flaw. To minimize the porosity, it is essential to remove the fugitive binder completely and have the matrix powder particle smaller than the fiber diameter.

SUGGESTED READING

1. D. V. Rosato and C. S. Grove, Jr., *Filament Winding,* Wiley, New York, 1964.
2. M. W. Gaylord, *Reinforced Plastics,* Chaners Publishing, Boston, MA, 1974.
3. M. R. Piggot, *Load Bearing Fibre Composites,* Pergamon Press, Oxford, U.K., 1980.
4. R. G. Weatherhead, *FRP Technology,* Applied Science Publishers, London, 1980.
5. J. Delmonte, *Technology of Carbon and Graphite Fiber Composites,* Van Nostrand, New York, 1981.
6. D. Hull, *An Introduction to Composite Materials,* Cambridge University Press, Cambridge, U.K., 1981.
7. M. J. Folkes, *Short Fibre Reinforced Thermoplastics,* Wiley, Chichester, U.K., 1982.
8. G. Lubin, ed., *Handbook of Composites,* Van Nostrand, New York, 1982.
9. R. W. Meyer, *Pultrusion Technology,* Chapman and Hall, New York, 1985.
10. S. W. Tsai, *Composites Design,* Think Composites Publishing, Dayton, OH, 1988.
11. A. B. Strong, *Fundamentals of Composites Manufacturing,* Society of Manufacturing Engineers, Dearborn, MI, 1989.
12. J. B. Donnet and R. C. Bansal, *Carbon Fibers,* 2d ed., Marcel Dekker, New York, 1990.
13. K. Ashbee, *Fundamental Principles of Fiber Reinforced Composites,* Technomic Publishing, Lancaster, PA, 1993.
14. P. K. Mallick, *Fiber-Reinforced Composites,* 2d ed., Marcel Dekker, New York, 1993.
15. R. F. Gibson, *Principles of Composite Material Mechanics,* McGraw-Hill, New York, 1994.
16. I. M. Daniel and O. Ishai, *Engineering Mechanics of Composite Materials,* 2d ed., Oxford University Press, New York, 2005.
17. M. M. Gauthier, *Engineered Materials Handbook,* ASM International Handbook Committee, CRC Press, Boca Raton, FL, 1995.
18. P. K. Mallick, ed., *Composites Engineering Handbook,* Marcel Dekker, New York, 1997.
19. A. K. Kaw, *Mechanics of Composite Materials,* 2d ed., CRC Press, Boca Raton, FL, 2005.
20. M. M. Schwartz, *Composite Materials,* Prentice-Hall, Upper Saddle River, NJ, 1997.

21. S. T. Peters, ed., *Handbook of Composites*, 2d ed., Kluwer Academic Publishers, Norwell, MA, 1998.
22. K. K. Chawla, *Composite Materials*, 2d ed., Springer-Verlag, New York, 1998.
23. C. T. Herakovich, *Mechanics of Fibrous Composites*, Wiley, New York, 1998.
24. M. W. Hyer, *Stress Analysis of Fiber-Reinforced Composite Materials*, WCB/McGraw-Hill, New York, 1998.
25. E. J. Barbero, *Introduction to Composite Materials Design*, Taylor and Francis, Philadelphia, 1999.
26. G. H. Staab, *Laminar Composites*, Butterworth-Heinemann, Woburn, MA, 1999.
27. R. M. Jones, *Mechanics of Composite Materials*, 2d ed., Taylor and Francis, Philadelphia, 1999.
28. S. K. Mazumdar, *Composites Manufacturing*, CRC Press, Boca Raton, FL, 2002.
29. B. T. Astrom, *Manufacturing of Polymer Composites*, Nelson Thornes, Cheltenham, U.K., 2002.
30. D. Gay, S. V. Hoa, and S. W. Tsai, *Composite Materials: Design and Applications*, CRC Press, Boca Raton, FL, 2003.
31. F. C. Campbell, *Manufacturing Processes for Advanced Composites*, Elsevier, New York, 2004.
32. M. Biron, *Thermosets and Composites*, Elsevier, New York, 2004.

3

BEHAVIOR OF UNIDIRECTIONAL COMPOSITES

3.1 INTRODUCTION

Fiber-reinforced composites are certainly one of the oldest and most widely used composite materials. Their study and development have been carried out largely because of their vast structural potential. Most of the structural elements or laminates made of fibrous composites consist of several distinct layers. Each layer or lamina is usually made of the same constituent materials (e.g., resin and glass). But an individual layer may differ from another layer in (1) relative volumes of the constituent materials, (2) form of the reinforcement used such as continuous or discontinuous fibers, woven or nonwoven reinforcement, and (3) orientation of fibers with respect to common reference axes. Furthermore, hybrid laminates can be made, consisting of layers having different fibers and/or matrix material. Thus the directional properties of the individual layers may be quite different from each other. Analysis and design of any structural element would require a complete knowledge of the properties of individual layers. A unidirectional composite, which consists of parallel fibers embedded in a matrix, represents a basic building block for the construction of laminates or multilayered composites. The properties and behavior of unidirectional composites are described in this chapter. An understanding of the behavior of unidirectional composites is essential to better understand the behavior of discontinuous or short-fiber composites discussed in the next chapter.

3.1.1 Nomenclature

A unidirectional composite is shown schematically in Fig. 3-1. Several unidirectional layers can be stacked in a specified sequence of orientation to

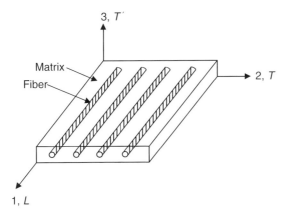

Axis 1, L – longitudinal direction
Axis 2, T – transverse direction (in lamina plane)
Axis 3, T' – transverse direction (perpendicular to lamina plane)

Figure 3-1. Schematic representation of a unidirectional composite.

fabricate a laminate that will meet design strength and stiffness requirements. Each layer of a unidirectional composite may be referred to as simply a *layer, ply,* or *lamina*. The direction parallel to fibers generally is called the *longitudinal direction* (axis 1). The direction perpendicular to the fibers is called the *transverse direction* (any direction in the 2–3 plane). These axes are also referred to as the *material axes* of the ply.

The ply depicted schematically in Fig. 3-1 shows only one fiber through the ply thickness. In practice, this may be true only for large-diameter fibers such as boron. Plies from other fibers have several fibers through the actual ply thickness. Typical cross sections of composites taken from a single ply are shown in Fig. 3-2. The fibers are distributed randomly throughout the cross section and may be in contact with each other in some locations. This type of fiber distribution in the ply is typical of several fiber–resin systems. These plies generally are constructed from single-end or multiple-end rovings impregnated with the matrix (see Chap. 2). The thickness of a single ply is often greater than 0.1 mm (0.005 in.), whereas the fiber diameter is typically 10 μm. Thus the ply thickness–fiber diameter ratio is typically 10.

Because of the structure of the composite, a unidirectional composite shows different properties in the longitudinal and transverse directions. Thus the unidirectional composites are orthotropic with the axes 1, 2, and 3 as the axes of symmetry (see Fig. 3-1). A unidirectional composite has the strongest properties in the longitudinal direction. Because of the random fiber distribution in the cross section, material behavior in the other two directions (2, 3) is nearly identical. Therefore, a unidirectional composite or ply can be considered to be transversely isotropic; that is, it is isotropic in the 2–3 plane.

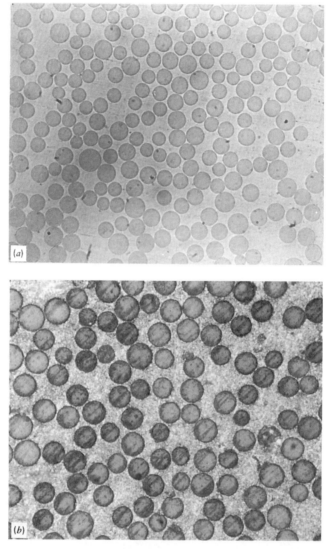

Figure 3-2. Cross section of a unidirectional composite: (a) glass–epoxy (from a single ply); (b) Fiber FP (alumina fiber) in aluminum matrix.

3.1.2 Volume and Weight Fractions

One of the most important factors determining the properties of composites is the relative proportions of the matrix and reinforcing materials. The relative proportions can be given as the weight fractions or the volume fractions. The weight fractions are easier to obtain during fabrication or by one of the experimental methods after fabrication. However, the volume fractions are used

3.1 INTRODUCTION

exclusively in the theoretical analysis of composite materials. It is thus desirable to determine the expressions for conversion between the weight fractions and volume fractions. These expressions are derived for a two-phase material and then generalized to a multiphase material.

Consider a volume v_c of a composite material that consists of volume v_f of the fibers and volume v_m of the matrix material. Let w_c, w_f, and w_m represent the corresponding weights of the composite material, fibers, and the matrix material, respectively. (Throughout this book, the subscripts c, f, and m are used consistently to represent the composite material, fibers, and the matrix material, respectively.) Let the volume fraction and weight fraction be denoted by the capital letters V and W, respectively. The volume fractions and weight fractions are defined as follows:

$$v_c = v_f + v_m \tag{3.1a}$$

$$V_f = \frac{v_f}{v_c} \qquad V_m = \frac{v_m}{v_c} \tag{3.1b}$$

and

$$w_c = w_f + w_m \tag{3.1c}$$

$$W_f = \frac{w_f}{w_c} \qquad W_m = \frac{w_m}{w_c} \tag{3.1d}$$

To establish conversion relations between the weight fractions and the volume fractions, the density ρ_c of the composite material must be obtained. The density of the composite material can be obtained easily in the terms of the densities of the constituents and their volume fractions or weight fractions. To that end, the weights in Eq. (3.1c) can be replaced by the products of corresponding density and volume and the equation rewritten as

$$\rho_c v_c = \rho_f v_f + \rho_m v_m \tag{3.2}$$

Dividing both sides of Eq. (3.2) by v_c and substituting the definition for the volume fractions from Eq. (3.1b) yields

$$\rho_c = \rho_f \frac{v_f}{v_c} + \rho_m \frac{v_m}{v_c}$$
$$\rho_c = \rho_f V_f + \rho_m V_m \tag{3.3}$$

By similar manipulations in Eq. (3.1a), the density of composite materials in terms of weight fractions can easily be obtained as

$$\rho_c = \frac{1}{(W_f/\rho_f) + (W_m/\rho_m)} \quad (3.4)$$

Now the conversion between the weight fraction and volume fraction can be obtained by considering the definition of weight fraction and replacing in it the weights by the products of density and volume as follows:

$$W_f = \frac{w_f}{w_c} = \frac{\rho_f v_f}{\rho_c v_c} = \frac{\rho_f}{\rho_c} V_f$$

$$W_f = \frac{\rho_f}{\rho_c} V_f \quad (3.5)$$

$$W_m = \frac{\rho_m}{\rho_c} V_m$$

The inverse relations can be obtained similarly from Eq. (3.1) or Eq. (3.5) by multiplying both sides of the equation by an appropriate ratio of densities. The inverse relations are

$$V_f = \frac{\rho_c}{\rho_f} W_f$$

$$V_m = \frac{\rho_c}{\rho_m} W_m \quad (3.6)$$

Equations (3.3)–(3.6) have been derived for a composite material with only two constituents but can be generalized for an arbitrary number of constituents. The generalized equations are

$$\rho_c = \sum_{i=1}^{n} \rho_i V_i$$

$$\rho_c = \frac{1}{\sum_{i=1}^{n} (W_i/\rho_i)}$$

$$W_i = \frac{\rho_i}{\rho_c} V_i \quad (3.7)$$

$$V_i = \frac{\rho_c}{\rho_i} W_i$$

The validity of generalized equations may be verified easily by the reader.

It may be pointed out here that the composite density calculated theoretically from weight fractions using Eq. (3.5) may not always be in agreement with the experimentally determined density. This will happen when voids are present in the composite. The difference in densities indicates the void content. It can be shown easily that if the theoretical composite density is denoted by ρ_{ct} and the experimentally determined density by ρ_{ce}, the volume fraction of voids V_v is given by

$$V_v = \frac{\rho_{ct} - \rho_{ce}}{\rho_{ct}} \tag{3.8}$$

In an actual composite, the void content may be determined by following ASTM (American Society for Testing and Materials) Standard D2734-94 (reapproved 2003). The density of the resin in this method is assumed to be the same in the composite as it is in an unreinforced bulk state. Although it is necessary to use this assumption, it may not be correct. Differences in curing, heat and pressure, and interaction with the reinforcement surface may affect the in situ resin density. It is thought that the bulk density is lower, making the void content seem lower than it really is.

The void content of a composite may affect some of its mechanical properties significantly. Higher void contents usually mean lower fatigue resistance, greater susceptibility to water penetration and weathering, and increased variation or scatter in strength properties. The knowledge of void content is desirable for estimation of the quality of composites. A good composite should have less than 1% voids, whereas a poorly made composite can have up to 5% void content.

3.2 LONGITUDINAL BEHAVIOR OF UNIDIRECTIONAL COMPOSITES

The properties of a composite material depend on the properties of its constituents and their distribution and physical and chemical interactions. Properties of composites can be determined through experimental measurements. Experimental methods may be simple and direct. However, one set of experimental measurements determines the properties of a fixed fiber matrix system produced by a single fabrication process. Additional measurements are required when any change in the system variables occurs, such as relative volumes of the constituents, constituent properties, and fabrication process. Thus experiments may become time-consuming and cost-prohibitive. Theoretical and semiempirical methods of determining composite properties can be used to predict the effects of a large number of system variables. These methods may not be reliable for component design purposes and present difficulty in selecting a representative but tractable mathematical model for some cases, such as the transverse properties of unidirectional composites. However, the mathematical model for studying some of the longitudinal properties (tensile strength and modulus of elasticity) of a unidirectional composite is quite accurate. This model and its predictions are discussed in the following subsections.

3.2.1 Initial Stiffness

A unidirectional composite may be modeled by assuming fibers to be uniform in properties and diameter, continuous, and parallel throughout the composite (Fig. 3-3). It may be further assumed that a perfect bonding exists between the fibers and the matrix so that no slippage can occur at the interface, and the strains experienced by the fiber, matrix, and composite are equal:

$$\epsilon_f = \epsilon_m = \epsilon_c \tag{3.9}$$

For this model, the load P_c carried by the composite is shared between the fibers P_f and the matrix P_m so that

$$P_c = P_f + P_m \tag{3.10}$$

The loads P_c, P_f, and P_m carried by the composite, the fibers, and the matrix, respectively, may be written as follows in terms of stresses σ_c, σ_f, and σ_m experienced by them and their corresponding cross-sectional areas A_c, A_f, and A_m. Thus

$$P_c = \sigma_c A_c = \sigma_f A_f + \sigma_m A_m$$

or

$$\sigma_c = \sigma_f \frac{A_f}{A_c} + \sigma_m \frac{A_m}{A_c} \tag{3.11}$$

But for composites with parallel fibers, the volume fractions are equal to the area fractions such that

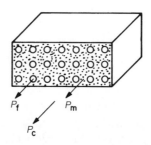

Figure 3-3. Model for predicting longitudinal behavior of unidirectional composites.

3.2 LONGITUDINAL BEHAVIOR OF UNIDIRECTIONAL COMPOSITES

$$V_f = \frac{A_f}{A_c} \qquad V_m = \frac{A_m}{A_c} \tag{3.12}$$

Thus

$$\sigma_c = \sigma_f V_f + \sigma_m V_m \tag{3.13}$$

Now Eq. (3.13) can be differentiated with respect to strain, which is the same for the composite, the fibers, and the matrix. The differentiation yields

$$\frac{d\sigma_c}{d\epsilon} = \frac{d\sigma_f}{d\epsilon} V_f + \frac{d\sigma_m}{d\epsilon} V_m \tag{3.14}$$

where $(d\sigma/d\epsilon)$ represents the slope of the corresponding stress–strain diagrams at the given strain. If the stress–strain curves of the materials are linear, the slopes $(d\sigma/d\epsilon)$ are constants and can be replaced by the corresponding elastic modulus in Eq. (3.14). Thus

$$E_c = E_f V_f + E_m V_m \tag{3.15}$$

Equations (3.13)–(3.15) indicate that the contributions of the fibers and the matrix to the average composite properties are proportional to their volume fractions. Such a relationship is called the *rule of mixtures*. Equations (3.13) and (3.15) can be generalized as

$$\sigma_c = \sum_{i=1}^{n} \sigma_i V_i \tag{3.16}$$

$$E_c = \sum_{i=1}^{n} E_i V_i \tag{3.17}$$

The following numerical example illustrates the influence of elastic modulus and volume fraction of the fibers on the longitudinal modulus of the composite.

Example 3-1: Calculate the ratios of longitudinal modulus of the composite to the matrix modulus for glass–epoxy and carbon–epoxy composites with 10% and 50% fibers by volume. Elastic moduli of glass fibers, carbon fibers, and epoxy resin are 70, 350, and 3.5 GPa, respectively.

Equation (3.15) can be written as

$$\frac{E_c}{E_m} = \left(\frac{E_f}{E_m} - 1\right) V_f + 1$$

Calculations will give the following results:

System (E_f/E_m)	E_c/E_m	
	$V_f = 10\%$	$V_f = 50\%$
Glass–epoxy (20)	2.9	10.5
Carbon–epoxy (100)	10.9	50.5

It may be observed that as the fiber volume fraction increases by a factor of 5, the ratio of E_c/E_m also increases by a similar factor (3.62 for glass–epoxy and 4.63 for carbon–epoxy). Further, as the fiber modulus increases by a factor 5, the ratio of E_c/E_m again increases by a similar factor (3.76 at $V_f = 10\%$ and 4.81 at $V_f = 50\%$). These calculations show that fibers are very effective in increasing the composite modulus in the longitudinal direction. Further, the elastic modulus of fibers has a significant influence on the composite modulus. This behavior will be compared with the influence of these factors on the composite transverse modulus in a later section.

The predictions of Eq. (3.13) can be explained by considering the stress–strain diagrams for the fibers and the matrix. Let us consider two composites. The fibers in both composites have linear stress–strain curves up to their fracture. The matrix material of one of the composites also has a linear stress–strain curve, but that of the other has a nonlinear stress–strain curve (Fig. 3-4a,b). The stress in the composite at a given strain can be calculated according to Eq. (3.13) by first finding the matrix stress and the fiber stress at the given strain from the corresponding stress–strain diagrams and then adding them proportional to their volume fractions. This process can be repeated for a number of strain values up to the fiber fracture strain. Thus a complete stress–strain diagram for the composite may be obtained. It may be noted that this procedure is applicable to both the composites being considered here because

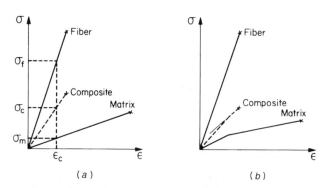

Figure 3-4. Longitudinal stress–strain diagrams for a composite with (a) linear and (b) nonlinear matrix material.

Eq. (3.13) has been derived without making any assumption about the properties of the constituent materials. Thus the stress–strain curve of composite (*a*) will be linear, whereas that of the composite (*b*) will be nonlinear. The composite strain at which the stress–strain curve for composite (*b*) becomes nonlinear will be the strain at which the matrix stress–strain curve becomes nonlinear. However, because of the predominance of fiber properties, the nonlinearity of the composite stress–strain curve may not be distinct, especially at higher fiber volume fractions. In any case, the composite stress–strain curves would lie between the stress–strain curves of the fibers and the matrix. The actual location of the composite stress–strain curve will depend on the relative volume fractions of the constituents. If the fiber volume fraction is high, the composite stress–strain curve will be closer to the fiber stress–strain curve. On the other hand, the composite stress–strain curve may be closer to the matrix stress–strain curve for a higher matrix volume fraction. Thus it can be seen that the assumption of a linearly elastic stress–strain curve for polymeric matrices will not cause large errors in predicted values of composite stress.

The predictions of Eqs. (3.14) and (3.15) are quite accurate when the applied load is tensile and agree very well with experimental results. However, when the applied load is compressive, the experimental observations may deviate from the theoretical predictions. This may be attributed to the fact that the behavior of the fibers in the composite subjected to compressive loads is analogous to the behavior of columns on an elastic foundation. Thus the response of the composite to compressive load is strongly dependent on matrix properties such as its shear stiffness. This observation is different from the response of the composite to longitudinal tensile loads, which is governed primarily by the fibers.

3.2.2 Load Sharing

It is of considerable interest to know how the load is shared between the constituents of a composite and the stresses to which they may be subjected. To that end, Eq. (3.9), for linearly elastic fibers and matrix material, can be written in terms of stresses and elastic moduli as

$$\frac{\sigma_f}{E_f} = \frac{\sigma_m}{E_m} = \frac{\sigma_c}{E_c} \tag{3.18}$$

Thus

$$\frac{\sigma_f}{\sigma_m} = \frac{E_f}{E_m} \qquad \frac{\sigma_f}{\sigma_c} = \frac{E_f}{E_c} \tag{3.19}$$

Equation (3.19) indicates that the ratio of stresses is the same as the ratio of corresponding elastic moduli. Thus, to attain high stresses in the fibers and

72 BEHAVIOR OF UNIDIRECTIONAL COMPOSITES

thereby use high-strength fibers most efficiently, it is necessary for the fiber modulus to be much greater than the matrix modulus.

The ratio of loads now can be obtained in terms of elastic moduli and volume fractions as follows:

$$\frac{P_f}{P_m} = \frac{\sigma_f A_f}{\sigma_m A_m} = \frac{E_f}{E_m} \frac{V_f}{V_m} \tag{3.20}$$

$$\frac{P_f}{P_c} = \frac{\sigma_f A_f}{\sigma_f A_f + \sigma_m A_m} = \frac{E_f/E_m}{(E_f/E_m) + (V_m/V_f)} \tag{3.21}$$

The load carried by the fibers is plotted as a fraction of the composite load in Fig. 3-5. The percentage of load carried by the fibers becomes higher for a higher ratio of elastic moduli of fibers and matrix and a higher volume content of fibers. Thus, for a given fiber matrix system, the volume fraction of fibers in the composite must be maximized if the fibers are to carry a higher proportion of the composite load. Although the maximum volume percent of cylindrical fibers that can be packed into a composite is almost 91%,

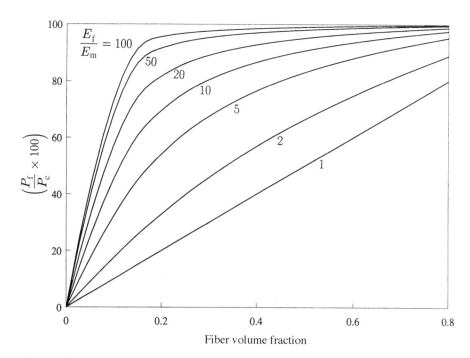

Figure 3-5. Percentage load carried by the fibers in a unidirectional composite loaded in the longitudinal direction.

above 80% the composite properties usually begin to decrease because of the inability of the matrix to wet and infiltrate the bundles of fibers. This results in poorly bonded fibers and voids in the composite.

The excellent strengths and strength–weight ratios achieved by glass-fiber-reinforced plastics are a result of the high strength of the glass fibers and the ability of the composite to use this strength because the ratio E_f/E_m is approximately 20. Even at 10% by volume of glass, the fiber will account for 70% of the total load.

Example 3-2: Calculate the fraction of load carried by the fibers in the composites indicated in Example 3-1.

The desired fractions can be obtained easily using Eq. (3.21). The results are

	P_f/P_c	
System (E_f/E_m)	$V_f = 10\%$	$V_f = 50\%$
Glass–epoxy (20)	0.69	0.952
Carbon–epoxy (100)	0.917	0.99

3.2.3 Behavior beyond Initial Deformation

The rule of mixtures accurately predicts the stress–strain behavior of a unidirectional composite subjected to longitudinal loads, provided that Eq. (3.13) is used for the stress and Eq. (3.14) for the slope of the stress–strain curve. However, the simplification of Eq. (3.14) to Eq. (3.15) through the replacement of slopes by the elastic moduli is possible only when both the constituents deform elastically. This may constitute only a small portion of the composite stress–strain behavior and is applicable primarily for glass- or ceramic-fiber-reinforced thermosetting plastics. In general, the deformation of a composite may proceed in four stages [1], summarized as follows:

1. Both the fibers and the matrix deform in a linear elastic fashion.
2. The fibers continue to deform elastically, but the matrix now deforms nonlinearly or plastically.
3. The fibers and the matrix both deform nonlinearly or plastically.
4. The fibers fracture followed by the composite fracture.

Stage 2 may occupy the largest portion of the composite stress–strain curve, particularly for a metal matrix composite, and in this stage the matrix stress–strain curve is no longer linear, so the composite modulus must be predicted at each strain level by

$$E_c = E_f V_f + \left(\frac{d\sigma_m}{d\epsilon_m}\right)_{\epsilon_c} V_m \tag{3.22}$$

where $(d\sigma_m/d\epsilon_m)_{\epsilon_c}$ is the slope of stress–strain curve of the matrix at the strain ϵ_c of the composite.

Although stage 3 is not observed with brittle fibers, the elastic modulus for composites with ductile fibers must be predicted by Eq. (3.14). Furthermore, for ductile fibers that fail by necking, additional factors such as the hydrostatic lateral restraint exerted by the matrix to prevent necking of the fibers will cause deviations from the simple rule of mixtures.

The stress–strain curves for hypothetical composite materials with ductile and brittle fibers and a typical metal or ductile matrix are shown in Fig. 3-6. The stress–strain curves of the composite fall between those of the fiber and the matrix. It is generally observed that the composites with brittle fibers fracture at the fracture strain of the fibers. However, if the fibers are capable of deforming plastically within the matrix, the fracture strain of fibers in the composite may be larger than the fracture strain of fibers when tested separately (without the matrix). Thus the fracture strain of the composite may exceed that of the fibers. This situation has been shown in Fig. 3-6. The difference between the two fracture strains increases as V_f decreases and as the matrix strength–fiber strength ratio increases.

3.2.4 Failure Mechanism and Strength

In a unidirectional composite subjected to a longitudinal load, failure initiates when the fibers are strained to their fracture strain. This assumes that the

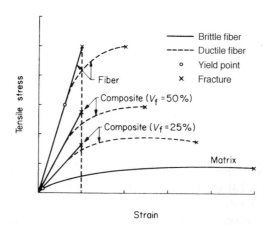

Figure 3-6. Longitudinal stress–strain curves for unidirectional composites with ductile and brittle fibers and a typical ductile matrix.

3.2 LONGITUDINAL BEHAVIOR OF UNIDIRECTIONAL COMPOSITES

failure strain of the fibers is less than that of the matrix. It is generally assumed, for theoretical predictions, that all the fibers fail at the same strain. If the fiber volume fraction is large enough (above a certain minimum, V_{min}, defined later), the matrix will not be able to support the entire load when all the fibers break, and composite failure then will take place instantly. Under these conditions, the ultimate longitudinal tensile strength of the composite can be assumed equal to the composite stress at the fiber fracture strain ε_f^*. The rule of mixtures [Eq. (3.13)] therefore can be used to obtain

$$\sigma_{cu} = \sigma_{fu} V_f + (\sigma_m)_{\varepsilon_f^*}(1 - V_f) \tag{3.23}$$

where σ_{cu} is the longitudinal strength of the composites, σ_{fu} is the ultimate strength of the fibers, and $(\sigma_m)_{\varepsilon_f^*}$ is the matrix stress at the fiber fracture strain ε_f^*.

If the fiber volume fraction is small, that is, below V_{min}, the matrix will be able to support the entire composite load when all the fibers break. Further, the matrix will be able to take additional load with increasing strain. It is generally assumed that the fibers do not support any load (i.e., $\sigma_f = 0$) at composite strains higher than the fiber fracture strain. The composite eventually fails when the matrix stress equals its ultimate strength (i.e., $\sigma_m = \sigma_{mu}$). Thus the ultimate strength of a composite with the fiber volume fraction less than V_{min} is given by

$$\sigma_{cu} = \sigma_{mu}(1 - V_f) \tag{3.24}$$

Now V_{min} can be defined as the minimum fiber volume fraction that ensures fiber-controlled composite failure. It can be seen easily that V_{min} is obtained by equating the right-hand sides of Eqs. (3.23) and (3.24). Thus

$$V_{min} = \frac{\sigma_{mu} - (\sigma_m)_{\varepsilon_f^*}}{\sigma_{fu} + \sigma_{mu} - (\sigma_m)_{\varepsilon_f^*}} \tag{3.25}$$

The longitudinal composite strengths, as predicted by Eqs. (3.23) and (3.24), have been plotted against fiber volume fraction in Fig. 3-7. The matrix may be a strain hardening metal or an inelastic polymer, and a typical stress–strain curve is shown in Fig. 3-7. The solid portions of the lines represent the range of their applicability, and their intersection defines V_{min}. It may be noticed that Eq. (3.24) predicts composite strength that is always less than the strength of unreinforced matrix. On the other hand, Eq. (3.23) predicts composite strength that can be lower or higher than the matrix strength depending on the fiber volume fraction. A critical fiber volume fraction V_{crit} that must be exceeded for strengthening therefore can be defined as follows:

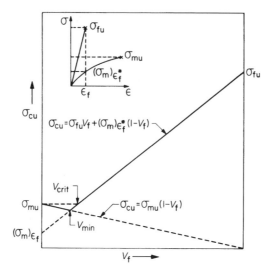

Figure 3-7. Longitudinal strength of a unidirectional composite as a function of fiber volume fraction.

$$\sigma_{cu} = \sigma_{fu} V_f + (\sigma_m)_{\varepsilon_f^*}(1 - V_f) \geq \sigma_{mu} \tag{3.26}$$

or

$$V_{crit} = \frac{\sigma_{mu} - (\sigma_m)_{\varepsilon_f^*}}{\sigma_{fu} - (\sigma_m)_{\varepsilon_f^*}} \tag{3.27}$$

Thus V_{crit} is obviously a more important system property than V_{min}. An examination of Eqs. (3.25) and (3.27) shows that V_{crit} and V_{min} arise as a result of the strain hardening and plastic flow of the matrix and the matrix strength being higher than the matrix stress at the fiber fracture strain. In metal matrix composites, V_{crit} and V_{min} increase as the degree of strain hardening of the matrix increases and also as the matrix strength approaches the fiber strength. Thus, when a strong matrix is to be reinforced by fibers of marginal strength, large volume fractions will be required before strengthening can be observed. In composites with polymeric matrices, V_{crit} and V_{min} are very small because most polymers exhibit only a limited amount of plastic flow and strain hardening. For example, if it is assumed that the strength of a glass-fiber reinforcement is 2.8 GPa ($E_f = 70$ GPa), the maximum composite strain at failure would be 4%, so $\sigma_{mu} - (\sigma_m)_{\varepsilon_f^*}$ for typical epoxy resins may range 7–28 MPa. Therefore, V_{min} would range between 0.25% and 1.00%.

3.2.5 Factors Influencing Longitudinal Strength and Stiffness

While deriving stiffness and strength of unidirectional composites in the preceding subsection, many simplifying assumptions regarding the physical variables of the system were made. Only some of these assumptions were stated explicitly, whereas others were implied. It is very rare that all the assumptions are met completely in an actual situation. This results in deviations of composite properties from the derived equations. Many of these deviations may be small and may not require any correction. Corrections may be necessary in many other situations; therefore, it is of interest to study the factors influencing strength and stiffness. Some of these factors are discussed in the following paragraphs. A description of how these factors may influence the properties and how corrections are to be incorporated is provided. Actual correction terms are not given, but appropriate references have been supplied for further study. The factors influencing the strength and stiffness of composites are (1) misorientation of fibers, (2) fibers of nonuniform strength, (3) discontinuous fibers, (4) interfacial conditions, and (5) residual stresses.

Orientation of fibers with respect to the loading axis is an important parameter. Fiber orientation directly affects the distribution of load between the fibers and the matrix. The contribution of the fibers to the composite properties is maximum only when they are parallel to the loading direction. Composite strength and stiffness will be reduced when the fibers are not parallel to the loading direction. The extent to which the strength and stiffness may be reduced depends on the angle of the fibers to the loading axis or the number of fibers that are not parallel to the loading direction. In practice, all the fibers cannot be aligned perfectly while making composites. No correction is necessary when the misorientation is limited to only a few degrees. In the case of laminated composites, the loads in some of the plies may not be in the fiber direction. Appropriate theories of behavior are necessary for such situations. Methods of analysis for laminated composites are discussed in later chapters.

The strength of fibers affects the strength of composites in a very direct manner [see Eq. (3.23)]. Any reduction in fiber strength will result in lowering of the composite strength. A high-strength composite will be obtained when all the fibers are uniform in their strength values. Individual metal fibers generally exhibit reproducible strengths [9], particularly ductile wires and to some extent uniform brittle wires. Drawn glass or silica fibers show a considerable spread in strengths about a mean value. There are two possible causes for this variation: (1) The variation can occur as a result of variations in fiber diameter with length resulting from the manufacturing process, and (2) a variation can occur as a result of the handling of fibers and from their surface treatment because of differences in the nature and intensity of chemical action at the coating–fiber interface. Whenever such nonuniformity in fiber strength exists, this should be taken into account through appropriate statistical models if a detailed understanding of strength is desired.

The statistical models of composite strength have been developed on the basis of experimental observations of the failure of the composites. It is observed that the fracture of individual fibers in a composite starts at loads much smaller than the composite failure load. More fibers break with increasing load, and some fibers break at many different cross sections. Thus the fiber breaks accumulate as the load increases. This results in a statistical weakening of the cross sections of the composite. The model, based on this weakening of the cross sections resulting from a statistical accumulation of fiber fractures, is usually referred to as the *cumulative-weakening failure model*. This model takes into account, through appropriate statistical parameters, the effects of stress redistributions in the vicinity of an individual fiber break.

The second consideration in the statistical models is the fact that fiber strength depends on fiber length. This length dependence can be explained by considering a fiber as a long chain of links. This chain breaks at the weakest link. Moreover, a longer chain has a higher probability of having a very weak link; thus longer fibers have smaller strengths. Another important aspect is the fact that composites actually contain bundles of fibers of nonuniform strength. The fiber strength usually is assumed to follow a Weibull distribution. In general, the average strength of a bundle is somewhat less than the average strength of the fibers when tested individually. The statistical tensile failure model takes into account the length–strength relationship, the statistical variation of fiber strength, and the difference between the strength of the bundle and an average strength of the fibers. Thus this model predicts the strength of the composite by taking into consideration the in situ fiber strength. Detailed discussions and the quantitative approach of the statistical models have been presented in other works [1–3].

In composites, load is not directly applied to the fibers but to the matrix material and is transferred to the fibers through the fiber ends and small fiber lengths near the end. When the length of a fiber is much greater than the length over which the transfer of stress takes place, the end effects can be neglected, and the fiber may be considered to be infinite in length or continuous. The stress on a continuous fiber therefore can be assumed constant over its entire length. In the case of short-fiber composites, the end effects cannot be neglected. Their behavior cannot, in general, be described by relations such as Eqs. (3.13) and (3.15). Some corrections in the values of σ_f or V_f in Eqs. (3.13) and (3.23) will be needed to account for the fact that a portion of the end of each finite-length fiber is stressed at less than the maximum fiber stress. The extent of correction depends on the length of fibers over which the load gets transferred from the matrix. This adjustment or correction becomes negligible when the fiber length is much greater than a critical length. However, the properties of discontinuous-fiber-reinforced composites are lowered to a greater degree because of the difficulty in controlling fiber alignment. The subject of discontinuous-fiber reinforcement is treated in a later chapter and has been discussed elsewhere in the literature [1,4–7].

Another important consideration in the behavior of discontinuous-fiber-reinforced composites is that the fiber ends cause stress concentrations. This is particularly important in the case of the failure of composites with brittle matrices. As a result of the stress concentrations, the fiber ends become separated from the matrix at a very small load, thus producing a microcrack in the matrix. A similar situation arises in a continuous-fiber-reinforced composite when a fiber breaks at its weakest cross section. The first microcrack at the fiber end may result in several alternate effects. The interface shear stresses may separate the fiber from the remaining composite by propagating the crack along the length of the fiber. When this happens, the fiber becomes totally ineffective, and the composite is acting as if it were a bundle of fibers. For this case, the strength of the composite is not enhanced by the presence of a matrix material. Alternatively, the crack may propagate out in a direction normal to the fibers across the other fibers, as a result of the local stress concentrations, leading to immediate composite fracture. If both these crack-propagation effects are suppressed, an increased load results in further separation of fiber ends from the matrix in a discontinuous-fiber composite and the fracture of fibers at additional sites in a continuous-fiber composite. Each crack thus formed causes a stress redistribution in its vicinity and changes the relative probability of modes of crack propagation. Such probabilistic behavior is considered in the statistical models, as indicated earlier. However, important considerations in the statistical models are the stress distribution in the vicinity of the crack at the fiber ends and the mechanism of load transfer from the matrix to the fibers. A discussion on these subjects may be found in other works [2,8–15].

The interfacial bond between the matrix and the fibers is an important factor influencing the mechanical properties and performance of composites. The interface is responsible for transmitting the load from the matrix to the fibers, which contribute the greater portion of the composite strength. Thus the composite strength is affected by the interfacial condition. The mechanism of load transfer through the interface becomes more important in the discontinuous-fiber-reinforced composites and in the continuous-fiber-reinforced composites when the individual fibers fracture prior to ultimate failure of the composite. The interfacial condition controls the mode of propagation of microcracks at the fiber ends. When a strong bond exists between the fibers and the matrix, the cracks do not propagate along the length of the fibers. Thus the fiber reinforcement remains effective even after the fiber breaks at several points along its length. A strong bond is also essential for higher transverse strengths and for good environmental performance of composites. Improved adhesion usually enhances water resistance of polymer matrix composites. The detrimental effect of an adverse environment becomes severe if the adhesion is inadequate, especially when the composite is under load. However, there is at least one composite property, namely, the fracture toughness, that may be improved by decreasing adhesion. Thus maximum use of fiber properties requires optimal bonding across the interface, although it is difficult to relate other composite properties (e.g., fatigue and creep behavior) to the interface properties. Some aspects of the interfacial conditions are discussed in Chapter 9, on the performance of composites. As such, the subject of interfaces is quite vast, and there are books [16,17] devoted exclusively to the subject.

The fabrication process used to make fibrous composites inherently produces residual stresses in the constituents and at the interface. The residual stresses are caused by two primary reasons: (1) the difference in the coefficients of thermal expansion for the constituents and (2) the difference in fabrication temperature and the temperatures at which they are used. Moreover, in laminates, residual stresses are present because of the difference in thermal expansion of the individual plies. These residual stresses may be caused even if the plies are identical and differ only in their relative orientation. Like other properties, thermal expansion of unidirectional composites is also orientation-dependent. The residual stresses affect the in situ properties of the matrix and the actual state of stress resulting from the service loads. Thus the residual stresses affect the strength of the composite as well as its response to mechanical loads. The residual stresses should not be neglected for an accurate analysis of the laminated composites using advanced methods. Detailed discussion on the residual stresses and the methods for the analysis of laminated anisotropic materials subjected to combined states of stress resulting from applied loads and the residual stresses have been presented by Chamis [18] and Tsai and Azzi [19].

3.3 TRANSVERSE STIFFNESS AND STRENGTH

3.3.1 Constant-Stress Model

A simple mathematical model may be constructed for studying the transverse properties of composites in the same manner as the one constructed earlier for studying the longitudinal properties in Sec. 3.2.1. The fibers may be assumed to be uniform in properties and diameter, continuous and parallel throughout the composite. The composite is stressed in the transverse direction, that is, the direction perpendicular to the parallel fibers. This model may be thought of as made up of layers representing fibers and matrix material shown in Fig. 3-8. Each layer is perpendicular to the direction of loading and has the same area on which the load acts. It is clear that each layer will carry the same load and experience equal stress, that is,

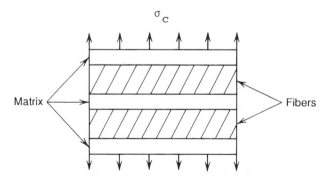

Figure 3-8. Model for predicting transverse properties of unidirectional composites.

3.3 TRANSVERSE STIFFNESS AND STRENGTH

$$\sigma_f = \sigma_m = \sigma_c \qquad (3.28)$$

Each layer is also assumed uniform in thickness so that the cumulative thickness of fiber layers and the matrix layers will be proportional to their respective volume fractions. In this case, the composite elongation δ_c in the direction of the load is the sum of the fiber elongation δ_f and the matrix elongation δ_m:

$$\delta_c = \delta_f + \delta_m \qquad (3.29)$$

The elongation in the material can be written as the product of the strain and its cumulative thickness, so

$$\delta_c = \epsilon_c t_c$$
$$\delta_f = \epsilon_f t_f \qquad (3.30)$$
$$\delta_m = \epsilon_m t_m$$

Substituting Eq. (3.30) in Eq. (3.29) gives

$$\epsilon_c t_c = \epsilon_f t_f + \epsilon_m t_m \qquad (3.31)$$

Dividing both sides of Eq. (3.31) by t_c and recognizing that the thickness is proportional to the volume fraction yields

$$\epsilon_c = \epsilon_f \frac{t_f}{t_c} + \epsilon_m \frac{t_m}{t_c}$$
$$= \epsilon_f V_f + \epsilon_m V_m \qquad (3.32)$$

If the fibers and the matrix are now assumed to deform elastically, the strain can be written in terms of the corresponding stress and the elastic modulus as follows:

$$\frac{\sigma_c}{E_c} = \frac{\sigma_f}{E_f} V_f + \frac{\sigma_m}{E_m} V_m \qquad (3.33)$$

In view of Eq. (3.28), Eq. (3.33) can be simplified as

$$\frac{1}{E_c} = \frac{V_f}{E_f} + \frac{V_m}{E_m} \qquad (3.34)$$

The transverse modulus of a composite with n number of materials may be obtained by generalizing Eq. (3.34):

$$E_c = \frac{1}{\sum_{i=1}^{n}(V_i/E_i)} \qquad (3.35)$$

The transverse modulus of a unidirectional composite as predicted by Eq. (3.34) has been plotted in Fig. 3-9 as a function of the fiber volume fraction. The longitudinal modulus as predicted by the rule of mixtures [Eq. (3.15)] also has been shown in Fig. 3-9. It may be noted that the fibers are much less effective in raising the composite modulus in the transverse direction than in the longitudinal direction.

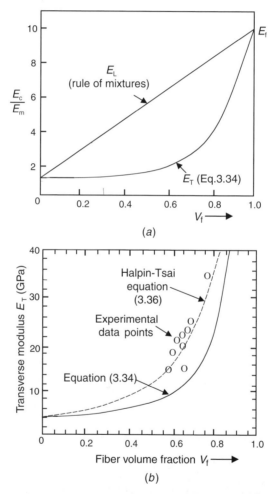

Figure 3-9. Transverse modulus of a unidirectional composite as a function of fiber volume fraction: (a) predictions, (b) comparison of predictions with the experimental measurements on a boron–epoxy lamina (E_f = 414 GPa, v = 0.2, E_m = 4.14 GPa, v_m = 0.35). (Experimental data from ref. 29.)

As an example, observe that if $(E_f/E_m) = 10$, more than 55% by volume of fibers is required to raise the transverse composite modulus to twice the matrix modulus, whereas only 11% by volume of fibers is required to raise the longitudinal modulus to the same value. Theoretically, the transverse modulus can be raised to five times the matrix modulus by providing 90% fibers, which, as stated previously, is not practical. Thus the fibers do not contribute much to the transverse modulus unless the percentage of fibers is very high. Further, the use of higher-modulus fibers also does not enhance the transverse modulus significantly.

Experimentally measured transverse modulus E_T of a boron–epoxy unidirectional composite is compared with the predictions of Eq. (3.34) (and the Halpin-Tsai equations to be discussed later) in Fig. 3-9b. It is noted that the measured values are significantly higher than the predictions of Eq. (3.34). This disagreement in moduli values highlights the fact that the model used to derive Eq. (3.34) does not properly simulate the deformation behavior of a unidirectional composite subjected to a transverse load.

The simple model described in the preceding paragraphs is not mathematically rigorous. In a real composite, the parallel fibers are dispersed in the matrix material in a random fashion, as shown in Fig. 3-2. Generally, both the fibers and the matrix will be present at any section perpendicular to the load, especially at the higher fiber volume fractions. Thus the load is shared between the fibers and the matrix, and the assumption that the stresses in the fibers and the matrix are equal is inaccurate. The assumption of equal stresses also results in a mismatch of strains in the loading direction at the fiber–matrix interface. Another inaccuracy in the solution arises owing to the mismatch of Poisson ratios of the fibers and the matrix, which induces stresses in the fibers and matrix perpendicular to the load with no net resulting force on the composite in that direction. A mathematically rigorous solution with a complete match of displacements across the boundary between the fiber and the matrix is accomplished through the use of the theory of elasticity.

3.3.2 Elasticity Methods of Stiffness Prediction

The methods of predicting composite stiffness using elasticity principles can be divided into three categories: (1) the bounding techniques, (2) the exact solutions, and (3) the self-consistent model. In the first of these methods, the energy theorems of classical elasticity are used to obtain bounds on the elastic properties. The minimum-complementary energy theorem yields the lower bound, whereas the minimum-potential-energy theorem yields the upper bound. This method has been used, among others, by Paul [20] and Hashin and Rosen [21]. Paul's bounds are too far apart to be of much practical utility, particularly at intermediate fiber volume fractions. The bounds obtained by Hashin and Rosen are much improved. A comparison of their predictions with experimental data shows that the experimental results on the transverse modulus lie close to the upper bound.

An exact method of analyzing fibrous composites consists of assuming the fibers to be arranged in a regular periodic array. The resulting elasticity problem has to be solved using an appropriate technique. There are few closed-form solutions using classical methods because of the difficulties arising as a result of the complex geometries of the reinforcement. The method of a series development [22–24] and the complex-variable technique [25] have been used as alternative approaches for the solution. However, the numerical solution techniques are the best to analyze the complex geometries with ease. Adams and Doner [26] used the finite-difference method to predict the transverse properties of a fibrous composite, whereas Chen and Lin [27] used the finite-element method for the purpose. Numerical results of Adams and Doner for transverse modulus are shown in Fig. 3-10. It is assumed here that fibers are packed in a square array. It has been found that good agreement exists between experiment and their numerical results.

In the self-consistent model, a single fiber is assumed to be embedded in a concentric cylinder of matrix material. This outer cylinder, in turn, is embedded in a homogeneous material that is macroscopically the same as the composite being studied. The ratio of the volume of the fiber to that of the cylinder containing the matrix and fiber is assumed to be equal to the volume fraction of the fibers in the composite. This model has the advantage that its results are applicable to any regular or irregular packing of fibers. This method has been used by Hill [28], Whitney and Riley [29], and Hermans [30]. A more detailed discussion of these methods and various models has been presented in a review article by Chamis and Sendeckyj [31].

Although a large amount of useful data has been generated through the procedures, some of the results are in the form of curves and others in the

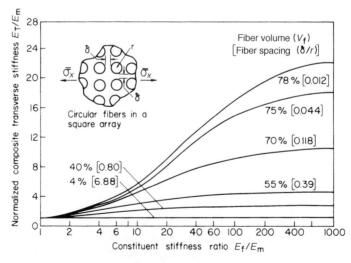

Figure 3-10. Transverse modulus predicted through numerical calculations. (From Adams and Doner [26].)

form of complicated equations, thus limiting their adoptability to design procedures. For design purposes, it is often desirable to have simple and rapid computational procedures for estimating composite properties even though the estimations may be only approximate.

3.3.3 Halpin–Tsai Equations for Transverse Modulus

Halpin and Tsai [32] have developed simple and generalized equations to approximate the results of more exact micromechanics analyses. These equations are simple and can be used readily in the design process. Moreover, the predictions of these equations are quite accurate if the fiber volume fraction does not approach 1. The Halpin–Tsai equation for transverse composite modulus can be written as

$$\frac{E_T}{E_m} = \frac{1 + \xi \eta V_f}{1 - \eta V_f} \quad (3.36)$$

where

$$\eta = \frac{(E_f/E_m) - 1}{(E_f/E_m) + \xi} \quad (3.37)$$

in which ξ is a measure of reinforcement and depends on the fiber geometry, packing geometry, and loading conditions. The values of ξ are obtained by comparing Eqs. (3.36) and (3.37) with exact elasticity solutions through curve-fitting techniques. Halpin and Tsai have suggested that a value of $\xi = 2$ may be used for fibers with circular or square cross sections. For rectangular cross-section fibers, ξ may be calculated as

$$\xi = 2\frac{a}{b} \quad (3.38)$$

where a/b is the rectangular cross-section aspect ratio with the dimension a taken in the direction of the loading. Predictions of the Halpin–Tsai equation for transverse composite modulus have been shown as a function of fiber volume fraction in Fig. 3-11 for different constituent modulus ratios. Halpin and Tsai [32] have demonstrated the applicability of these equations by showing that the predictions of Eq. (3.36) agree very well with some of the more exact solutions. A more thorough discussion on the comparison of the Halpin–Tsai equations and the exact elasticity solution has been presented in the original reference [32]. Also given in this reference are similar simple relations for other composite properties, some of which will be discussed later.

Predictions of the Halpin–Tsai approach [Eq. (3.36)] are compared with the experimental measurements in Fig. 3-9b, along with the predictions of Eq. (3.34). Experimental results clearly are much closer to the Halpin–Tsai

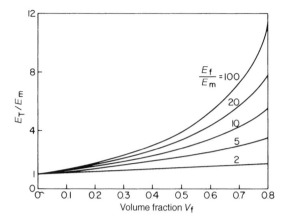

Figure 3-11. Transverse modulus predicted by Halpin–Tsai equation [Eq. (3.36)].

equation than they are to Eq. (3.34). This comparison further demonstrates applicability of the Halpin–Tsai equations for transverse modulus prediction.

It is suggested that the Halpin–Tsai equations are quite adequate to satisfy the practical requirements for the predictions of transverse composite modulus, particularly because the variations in composite materials manufacturing processes always cause a variation in the composite moduli. Therefore, one cannot hope to precisely predict composite moduli.

Example 3-3: Calculate, using the Halpin–Tsai equation, the ratios of transverse modulus of the composite to the matrix modulus for the composites given in Example 3-1. Compare these ratios with those obtained in Example 3-1.

$$\xi = 2 \quad \text{for all cases}$$

Glass–Epoxy System

$$\frac{E_f}{E_m} = 20 \qquad \eta = \frac{20 - 1}{20 + 2} = \frac{19}{22}$$

$$V_f = 10\% \qquad \frac{E_T}{E_m} = \frac{1 + 2 \times (19/22) \times 0.1}{1 - (19/22) \times 0.1} = 1.28$$

$$V_f = 50\% \qquad \frac{E_T}{E_m} = \frac{1 + 2 \times (19/22) \times 0.5}{1 - (19/22) \times 0.5} = 3.28$$

Carbon–Epoxy System

$$\frac{E_f}{E_m} = 100 \qquad \eta = \frac{100 - 1}{100 + 2} = \frac{99}{102}$$

$$V_f = 10\% \qquad \frac{E_T}{E_m} = \frac{1 + 2 \times (99/102) \times 0.1}{1 - (99/102) \times 0.1} = 1.32$$

$$V_f = 50\% \qquad \frac{E_T}{E_m} = \frac{1 + 2 \times (99/102) \times 0.5}{1 - (99/102) \times 0.5} = 3.83$$

For a better comparison, these results, along with the results of Example 3-1, can be tabulated as follows:

System (E_f/E_m)	$V_f = 10\%$		$V_f = 50\%$	
	E_L/E_m	E_T/E_m	E_L/E_m	E_T/E_m
Glass–epoxy (20)	2.9	1.28	10.5	3.28
Carbon–epoxy (100)	10.9	1.32	50.5	3.83

It is easily observed that under these conditions, the transverse modulus of a unidirectional composite is much smaller than its longitudinal modulus. Further, while an increase in fiber volume fraction results in an increase of transverse modulus similar to the longitudinal modulus, an increase in fiber modulus only marginally increases the transverse modulus, unlike the longitudinal modulus.

3.3.4 Transverse Strength

So far in this discussion it is seen that the composite longitudinal strength and stiffness and transverse stiffness are improvements over the corresponding matrix properties owing to the presence of fibers. The longitudinal strength and stiffness are improved as a result of the predominant role played by the fibers. The response of composites to longitudinal loading is determined by the fact that the load is shared between the fibers and the matrix. However, because of their higher strength and stiffness, fibers carry a major portion of the load and thus cause composite properties that are significantly improved over the matrix properties.

When a unidirectional composite is subjected to transverse loads, the fibers, as a result of the geometry, are unable to take as large a proportion of the load as they do in the case of longitudinal loading. The high-modulus fibers serve as effective constraints on the deformation of the matrix, which results in the transverse composite modulus being higher than the matrix modulus, although only marginally unless the fiber volume fraction is very high. In terms of the transverse strength, the constraints placed on the matrix by the fibers cause strain and stress concentrations in the matrix adjacent to the fibers and thus result in composite failure at a much lower strain than the strain at which the unrestrained matrix material fails. Therefore, unlike the longitudinal strength and stiffness and transverse modulus, the transverse strength is reduced because of the presence of fibers.

3.3.4.1 Micromechanics of Transverse Failure

Failure is a process that is initiated by localized conditions. State of stress is the most important condition influencing initiation of failure. The failure of structures and components generally is initiated at the locations of highest stress produced by geometric or material discontinuities. Geometric discontinuities result from the shape of the structure or from holes and cutouts made for assembly purposes. These geometric discontinuities reduce the strength of structures on a macroscopic level as a result of the stress concentrations produced by the discontinuity.

In the case of composite materials, material discontinuities are always present and influence local stress states on a microscopic level, which then control the microscopic failure events (i.e., matrix cracking, interface failure, fiber break, etc.) and eventually the macroscopic failure. It is therefore important to understand and appreciate the internal stresses and their influence on initiation of failure.

Internal stresses in a unidirectional composite subjected to a transverse load can be explained through the results of classical analysis by Goodier. He analyzed the stresses in an elastic matrix surrounding a single cylindrical inclusion (e.g., a single fiber). Variations in radial and tangential stresses are shown in Fig. 3-12a. It is observed that near the inclusion, both these stress components are significantly greater than the respective applied stresses. Thus such inclusions cause stress concentrations in the matrix. The *stress-concentration factor* (SCF) is defined as the ratio of maximum internal stress to the applied stress. When these stresses are sufficiently high, they cause failure initiation in the matrix material. The applied stress required to cause failure can be predicted from knowledge of matrix strength and the stress concentration factor. Since the state of stress near the inclusion is triaxial, a suitable failure criterion is used to predict the failure stress. For example, in the case of brittle materials, the maximum principal stress may be used to predict failure. Therefore, a brittle material with an inclusion will fail at an applied stress lower than the failure stress of the material by a factor equal to the SCF.

It may be further observed in Fig. 3-12a that the tangential stress σ_θ diminishes rapidly and is small for $(r/a) = 2$. However, the influence of the radial stress σ_r extends to $(r/a) = 3$ or 4. Clearly, two inclusions centered on a line along the direction of applied load that are separated by a center-to-center distance of less than $3a$ produce significant interaction, and some in-

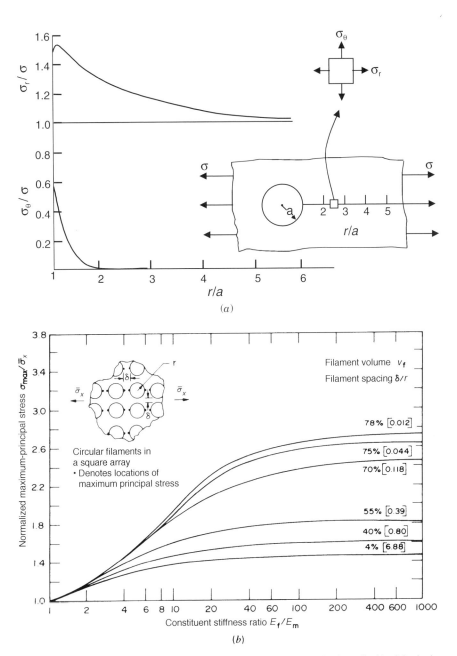

Figure 3-12. (a) Stress distributions in matrix surrounding a single cylindrical inclusion: $E_f/E_m = 10$, $v_m = 0.35$, $v_f = 0.30$; (b) principal stress in matrix surrounding multiple fibers: $v_m = 0.35$, $v_f = 0.20$.

teraction persists for separation distances of up to $6a$. Therefore, stress concentrations and stress distributions in composites depend on volume fraction of fibers, in addition to the elastic properties of fibers and matrix materials. The complete state of stress in the matrix material may be obtained by using numerical-solution techniques such as the finite-difference method [26] or the finite-element method [27]. The results of a finite-element analysis are shown in Fig. 3-12b for stress concentrations in unidirectional composites loaded in a transverse direction. Such results, in combination with the knowledge of matrix strength and suitable failure criterion, are used to predict the composite transverse strength, as discussed in the next section.

3.3.4.2 Prediction of Transverse Strength

The transverse tensile strength of the composite can be predicted using one of two methods: (1) the strength-of-materials method or (2) the advanced elasticity method using numerical-solution techniques. In both methods it is assumed that the composite transverse strength σ_{TU} is controlled by the matrix ultimate strength σ_{mu}. It is further assumed that the composite strength is lower than the matrix strength by a factor S known as the *strength-reduction factor*, which depends on the relative properties of the fibers and the matrix and their volume fractions. Thus the composite transverse strength can be written as

$$\sigma_{TU} = \frac{\sigma_{mu}}{S} \tag{3.39}$$

where S is determined using one of the preceding methods.

In the strength-of-materials method, the factor S is assumed to be the stress-concentration factor (SCF) [33] or the strain-magnification factor (SMF) [34,35].* These factors are calculated using simple mathematical models. When the Poisson effects are neglected, the equations for the SCF and SMF take the following simplified forms:

$$\text{SCF} = \frac{1 - V_f[1 - (E_m/E_f)]}{1 - (4V_f/\pi)^{1/2}[1 - (E_m/E_f)]} \tag{3.40}$$

$$\text{SMF} = \frac{1}{1 - (4V_f/\pi)^{1/2}[1 - (E_m/E_f)]} \tag{3.41}$$

Thus, once the SCF or SMF is known from Eq. (3.40) or Eq. (3.41), the transverse strength in terms of stresses or strains can be calculated easily.

In the advanced methods the factor S is calculated from the complete knowledge of state of stress or strain in the composite. The failure of the matrix can be predicted by a suitable failure criterion. The maximum-

---

*The strain-magnification factor can be used in the expression $\epsilon_{TU} = (\epsilon_{mu}/\text{SMF})$.

distortion-energy criterion is the most commonly used failure criterion. According to this criterion, the material failure occurs when the distortion energy at any point reaches a critical value. It can be shown easily that according to this criterion, the strength-reduction factor S can be written as

$$S = \frac{(U_{max})^{1/2}}{\sigma_c} \qquad (3.42)$$

where U_{max} is the maximum normalized distortional energy at any point in the matrix, and σ_c is the applied stress on the composite. For a given composite stress σ_c, the distortional energy U_{max} is a function of fiber volume fraction, fiber packing, condition at the fiber–matrix interface, and constituent properties. This method is more accurate and rigorous and hence is expected to yield more reliable results.

A further empirical approach for the prediction of transverse tensile strength of fibrous composites, which can be modeled in a fashion similar to particulate composites, has been described by Nielsen [36]. The composite strain to failure may be approximated as follows:

$$\epsilon_{cB} = \epsilon_{mB}(1 - V_f^{1/3}) \qquad (3.43)$$

where ϵ_{cB} is the breaking strain of the composite transverse to the fibers, ϵ_{mB} is the matrix breaking strain, and V_f is the volume fraction of fibers. If the matrix and composite have linear elastic stress–strain curves, Eq. (3.43) can be stated in terms of stress:

$$\sigma_{cB} = \sigma_{mB} \frac{E_T}{E_m} (1 - V_f^{1/3}) \qquad (3.44)$$

where E_T is the transverse modulus of composite, and E_m is the matrix modulus. The preceding equations assume perfect adhesion between phases, and thus failure occurs by matrix fracture at or near the interface.

3.4 PREDICTION OF SHEAR MODULUS

The in-plane shear modulus of a unidirectional composite may be predicted by the same model used for transverse modulus in Sec. 3.3.1. This model with the shear loading is shown in Fig. 3-13. Shearing stress on the fibers and the matrix are equal. Thus

$$\tau_f = \tau_m = \tau_c \qquad (3.45)$$

The total shear deformation of the composite Δ_c is the sum of the shear deformations of the fibers Δ_f and the matrix Δ_m:

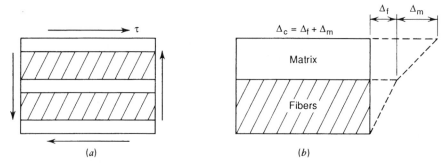

Figure 3-13. (a) Model for predicting shear modulus of a unidirectional composite and (b) shear deformations in the model.

$$\Delta_c = \Delta_f + \Delta_m \tag{3.46}$$

The shear deformation in each material can be written as the product of corresponding shear strain γ and the cumulative thickness of the material:

$$\Delta_c = \gamma_c t_c$$
$$\Delta_f = \gamma_f t_f \tag{3.47}$$
$$\Delta_m = \gamma_m t_m$$

Substitution of Eq. (3.47) in Eq. (3.46) gives

$$\gamma_c t_c = \gamma_f t_f + \gamma_m t_m \tag{3.48}$$

Dividing both sides of Eq. (3.48) by t_c and recognizing that the thickness is proportional to the volume fraction yields

$$\gamma_c = \gamma_f \frac{t_f}{t_c} + \gamma_m \frac{t_m}{t_c}$$
$$= \gamma_f V_f + \gamma_m V_m \tag{3.49}$$

If the shear stress–shear strain behavior of fibers and matrix is assumed linear, the shear strains in Eq. (3.49) can be replaced by the ratios of shear stress and appropriate shear modulus as follows:

$$\frac{\tau_c}{G_{LT}} = \frac{\tau_f}{G_f} V_f + \frac{\tau_m}{G_m} V_m \tag{3.50}$$

where G_{LT} is the in-plane shear modulus of the composite, and G_f and G_m are the shear modulus of fibers and matrix, respectively. In view of Eq. (3.45), Eq. (3.50) can be simplified as

$$\frac{1}{G_{LT}} = \frac{V_f}{G_f} + \frac{V_m}{G_m} \qquad (3.51)$$

or

$$G_{LT} = \frac{G_f G_m}{G_m V_f + G_f V_m} \qquad (3.52)$$

This model for predicting shear modulus of a unidirectional composite also suffers from the same limitations as were pointed out during the discussion on transverse modulus in Sec. 3.3.1. It is therefore desirable to use either a more rigorous mathematical model or a proven empirical method to predict shear modulus. Numerical results of Adams and Doner for shear modulus are shown in Fig. 3-14. As was pointed out in Sec. 3.3.3, Halpin and Tsai [32] have developed simple equations to approximate the results of more exact micromechanics analyses. The Halpin–Tsai equations for in-plane shear modulus of a unidirectional composite can be written as

$$\frac{G_{LT}}{G_m} = \frac{1 + \xi \eta V_f}{1 - \eta V_f} \qquad (3.53)$$

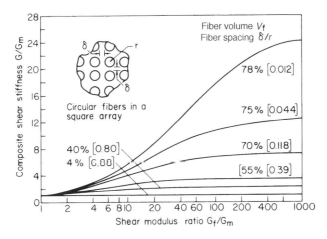

Figure 3-14. Shear modulus predicted through numerical calculations. (From Adams and Doner [26].)

where

$$\eta = \frac{(G_f/G_m) - 1}{(G_f/G_m) + \xi} \quad (3.54)$$

in which Halpin and Tsai have suggested that $\xi = 1$. Shear modulus as predicted by Eq. (3.53) has been shown as a function of fiber volume fraction in Fig. 3-15. It may be pointed out that G_m (or E_m) has as significant an influence on G_{LT} as E_m has on E_T.

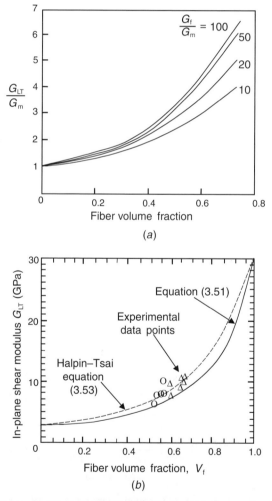

Figure 3-15. (a) Shear modulus of a unidirectional composite as predicted by Halpin–Tsai equation [Eq. (3.53)]; (b) comparison of predictions with the experimental measurements on a glass–epoxy unidirectional lamina ($G_f = 30.19$ GPa, $G_m = 1.83$ GPa). (Experimental data from ref. 29.)

Experimentally measured values of in-plane shear modulus of a glass–epoxy lamina are compared with the predictions of the Halpin–Tsai equations [Eq. (3.53)] and Eq. (3.51) in Fig. 3-15b. Experimental results are closer to the predictions of the Halpin–Tsai equations than they are to Eq. (3.51). As with transverse modulus, this comparison demonstrates the applicability of the Halpin–Tsai equations for predicting shear modulus.

3.5 PREDICTION OF POISSON'S RATIO

For in-plane loading of a unidirectional composite, two Poisson ratios are defined. The first of these relates the longitudinal stress to the transverse strain and is denoted by ν_{LT}. It is normally referred to as the *major Poisson ratio*. The second one, called the *minor Poisson ratio* (ν_{TL}), relates the transverse stress to the longitudinal strain.

The major Poisson ratio can be predicted using the same model as that used for predicting E_T. However, the load is applied parallel to the fibers, that is, parallel to the layers in the model. The deformation pattern is shown in Fig. 3-16 for cumulative thicknesses of layers.

Transverse strains in the fibers, matrix, and composite can be written in terms of longitudinal strains and the Poisson ratios as follows:

$$(\varepsilon_T)_f = -\nu_f (\varepsilon_L)_f$$
$$(\varepsilon_T)_m = -\nu_m (\varepsilon_L)_m \quad (3.55)$$
$$(\varepsilon_T)_c = -\nu_{LT} (\varepsilon_L)_c$$

where ν_f and ν_m are the Poisson's ratios of the fibers and matrix, respectively. Transverse deformations can be written as the product of strain and cumulative thickness:

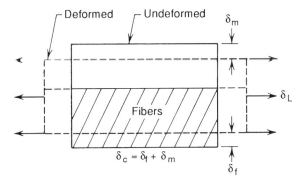

Figure 3-16. Model for predicting Poisson's ratio of a unidirectional composite.

$$\delta_f = t_f(\varepsilon_T)_f = -t_f \nu_f (\varepsilon_L)_f$$
$$\delta_m = t_m(\varepsilon_T)_m = -t_m \nu_m (\varepsilon_L)_m \qquad (3.56)$$
$$\delta_c = t_c(\varepsilon_T)_c = -t_c \nu_{LT} (\varepsilon_L)_c$$

The deformation of the composite is the sum of deformations of the fibers and the matrix. Therefore,

$$-t_c \nu_{LT}(\varepsilon_L)_c = -t_f \nu_f(\varepsilon_L)_f - t_m \nu_m(\varepsilon_L)_m \qquad (3.57)$$

Since the longitudinal strains in the fibers, matrix, and composite owing to the longitudinal stress are equal, Eq. (3.57) becomes

$$t_c \nu_{LT} = t_f \nu_f + t_m \nu_m \qquad (3.58)$$

Dividing both sides of Eq. (3.58) by t_c and recognizing that the thickness is proportional to the volume fraction yields

$$\nu_{LT} = \nu_f V_f + \nu_m V_m \qquad (3.59)$$

This is the rule of mixtures for the major Poisson ratio of a unidirectional composite.

The minor Poisson ratio can be obtained from the knowledge of values for E_L, E_T, and ν_{LT}. Derivation of the following relation between these four elastic constants will be given in Chap. 5:

$$\frac{\nu_{LT}}{E_L} = \frac{\nu_{TL}}{E_T} \qquad (3.60)$$

3.6 FAILURE MODES

In a very broad sense, failure of a structural element can be stated to have taken place when it ceases to perform satisfactorily. Therefore, the definition of failure will change from one application to another. In some applications a very small deformation may be considered failure, whereas in others only total fracture or separation constitutes failure. In the case of composite materials, internal material failure generally initiates much before any change in its macroscopic appearance or behavior is observed. The internal material failure may be observed in many forms, separately or jointly, such as (1) breaking of the fibers, (2) microcracking of the matrix, (3) separation of fibers from the matrix (called *debonding*), and (4) separation of laminae from each other in a laminated composite (called *delamination*). The photomicrographs in Fig. 3-17 illustrate some of the types of internal material failures. The

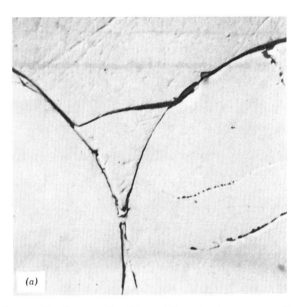

Figure 3-17. Microscopic failure events in fiber-reinforced polymer composites: (a) separation of fibers from matrix in a glass–epoxy composite (5000×); (b) microcracking of a glass–epoxy laminate during fatigue (800×); (c) separation of laminae in a glass–epoxy laminate (200×).

effect of internal damage on macroscopic material response is observed only when the frequency of internal damage is sufficiently high.

In many cases the macroscopic material response changes well before the macroscopic failure. Thus, depending on the application or design procedure, the failure load of a unidirectional composite could be considered as the load at which material behavior deviates from linear stress–strain response or the load at fracture. In the first case, a ply or lamina is considered to have failed when the stress exceeds the proportional limit so that its subsequent stress–strain behavior cannot be predicted using initial properties. The second definition of failure load permits maximal material utilization assuming that an appropriate safety factor is considered. However, most unidirectional materials exhibit a linear stress–strain behavior up to fracture. In such cases, the two definitions yield the same failure load. Figure 3-18 shows longitudinal stress–strain diagrams for unidirectional boron–epoxy and graphite–epoxy systems. The only unidirectional composites in which a nonlinear curve would be expected (assuming that fiber behavior is elastic) would be very low volume fraction fiber composites having a matrix capable of plastic deformation. The nonlinearity would be more pronounced if the matrix also had a high value of Young's modulus.

In the following subsections, the fracture modes of unidirectional composites subjected to different loading conditions are described.

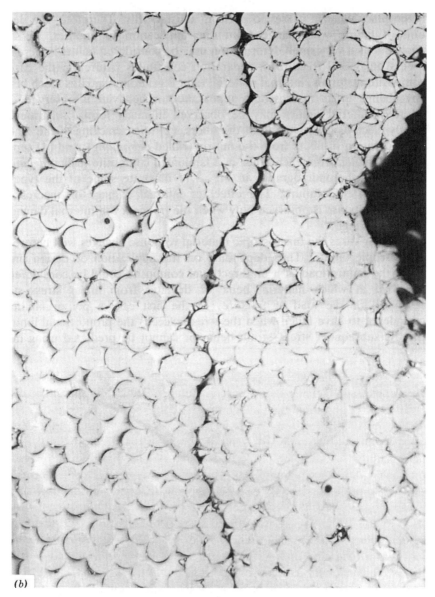

Figure 3-17. (*Continued*)

3.6 FAILURE MODES 99

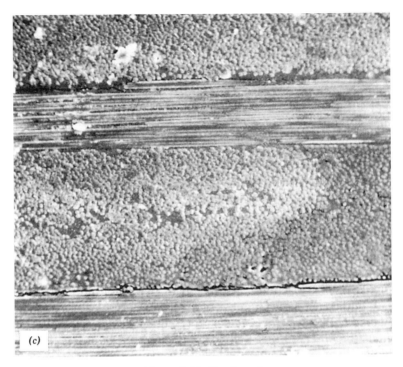

Figure 3-17. (Continued)

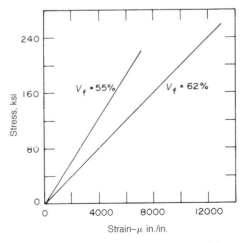

Figure 3-18. Longitudinal tensile stress–strain curves for graphite–epoxy ($V_f = 62\%$) and boron–epoxy ($V_f = 55\%$) composites.

3.6.1 Failure under Longitudinal Tensile Loads

In a unidirectional composite (consisting of brittle fibers) subjected to increasing longitudinal tensile load, failure initiates by fiber breakage at their weakest cross sections. As the load increases, more fibers break. Variation in the cumulative number of fiber breaks is shown as a function of applied load in Fig. 3-19 for a model representing a unidirectional composite [37]. It can be observed that the individual fibers break at less than 50% of the ultimate load. Breaking of the fibers is a completely random process. As the number of broken fibers increases, some cross section of the composite may become too weak to support an increased load, thus causing a complete rupture of the composite. The interfaces of broken fibers may become debonded because of stress concentrations created at the fiber ends and thus may contribute to the separation of the composite at a given cross section. In other cases, cracks at different cross sections of the composite may join up by debonding of the fibers along their length or by shear failure of the matrix. Therefore, a unidirectional composite can fail in at least three modes under longitudinal tensile load. These modes are (1) brittle, (2) brittle with fiber pullout, and (3) brittle failure with fiber pullout and (*a*) interface–matrix shear failure and (*b*) constituent debonding (i.e., matrix breaking away from the fibers). These three modes are illustrated schematically in Fig. 3-20. Photographs of specimens failed under longitudinal tensile load are shown in Fig. 3-21 [38]. The photographs show how the cracks at different cross sections join up to cause

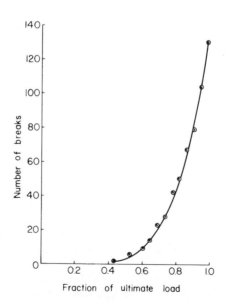

Figure 3-19. Cumulative number of fiber breaks with increasing longitudinal load. (From Rosen [37].)

3.6 FAILURE MODES   101

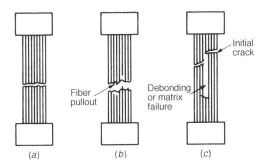

Figure 3-20. Failure modes of a unidirectional composite subjected to longitudinal tensile load: (a) brittle failure, (b) brittle failure with fiber pullout, and (c) brittle failure with debonding and/or matrix failure.

ultimate failure of the specimen. The pullout of fibers from the matrix depends on the bond strength and the load-transfer mechanism from matrix to fiber.

Interfiber matrix shear failure and constituent debonding could occur either independently or combined; that is, portions of the failure path in Fig. 3-20c occur by debonding and other portions by matrix shear failure. Glass–fiber composites having low fiber volume fractions ($V_f < 0.40$) exhibit predominantly the brittle-type failure mode. Composites with intermediate fiber vol-

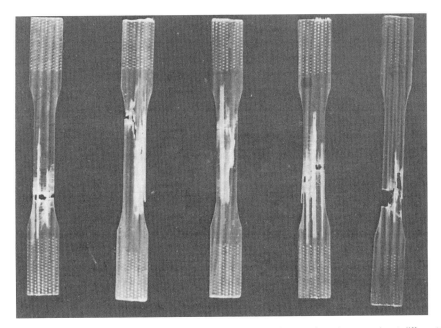

Figure 3-21. Photographs of unidirectional composite specimens show how cracks at different cross sections join up to cause failure under longitudinal tensile loads.

ume fraction ($0.40 < V_f < 0.65$) exhibit brittle failure with fiber pullout. Composites with high fiber volume fraction ($V_f > 0.65$) exhibit brittle failure with fiber pullout and debonding or matrix shear failure. These ranges are applicable if the void content in the composite is negligible. Graphite–fiber composites generally fail as shown by Fig. 3-20a,b.

3.6.2 Failure under Longitudinal Compressive Loads

When composites are subjected to compressive loads, continuous fibers act as long columns, and microbuckling of the fibers can occur. In composites with very low fiber volume content, fiber microbuckling may occur even when the matrix stresses are in the elastic range. However, at practical fiber volume fractions ($V_f > 0.40$), fiber microbuckling generally is preceded by matrix yield and/or constituent debonding and matrix microcracking. Compressive failure of a unidirectional composite loaded in the fiber direction may be initiated by transverse splitting or failure of the composite [39,40]. In other words, the transverse tensile strain resulting from the Poisson ratio effect can exceed the ultimate transverse strain capability of the composite, resulting in cracks at the interface. Shear failure is another mode of gross failure of composites subjected to longitudinal compressive loads. Failure modes for composites subjected to longitudinal compressive load may be listed as follows: (1) transverse tensile failure, (2) fiber microbuckling (*a*) with matrix still elastic, (*b*) preceded by matrix yielding, and (*c*) preceded by constituents debonding, and (3) shear failure.

Transverse failure and fiber microbuckling are illustrated in Fig. 3-22. The adjacent fibers in a composite may buckle independently of each other or may buckle in a cooperative manner. In the first case, the transverse deformation of the fibers is out of phase relative to each other (Fig. 3-22b). The resulting strains in the matrix are predominantly *extensional*. This mode of buckling therefore is referred to as the *extension mode*. This buckling mode is possible

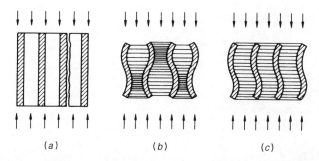

Figure 3-22. Failure modes of unidirectional composite subjected to longitudinal compressive load: (a) transverse tensile failure, (b) fiber microbuckling in extensional mode, and (c) fiber microbuckling in shear mode.

only when the interfiber distance is quite large, that is, when the fiber volume fraction is very small. The second buckling mode is more common and may occur in most of the practical fiber volume fractions. In this case, the transverse deformations of the adjacent fibers are in phase (Fig. 3-22c) with each other. The resulting strains in the matrix are predominantly shearing strains. This mode therefore may be referred to as the *shear mode*. The theoretical analyses for prediction of compressive strength have been carried out on the basis of these modes of buckling [41]. Photographs in Fig. 3-23 show that the two failure modes really do occur [42].

The shear failure mode in a compression test is illustrated in Fig. 3-24. Experimental results of Hancox [43] indicate that carbon fiber composites fail in a shear mode at approximately 45° to the loading axis. There is evidence (Fig. 3-25) of localized fiber rotation that may occur before or during failure.

The preceding observations of failure modes in compression have been used to formulate theoretical expressions for predicting longitudinal com-

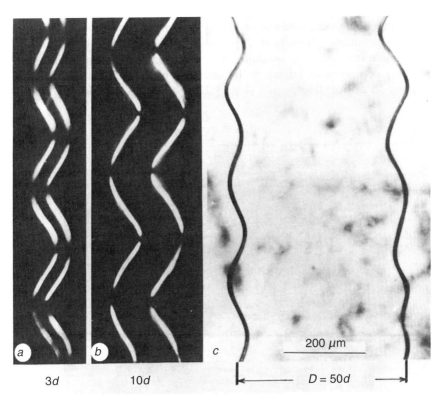

Figure 3-23. Microbuckling of fibers in unidirectional composites owing to shrinkage during curing. Parallel fibers buckle in phase (i.e., in shear mode) at separations of up to 10 fiber diameters (a,b). Only at very large separations of 50 fiber diameters do they buckle independently in extension mode (c). (From Dale and Baer [42].)

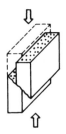

Figure 3-24. Shear failure of a unidirectional composite under longitudinal compressive load.

pressive strength. For example, Rosen [44] has derived an equation that considers compressive failure to be initiated by fiber microbuckling. If a shear mode of buckling occurs (Fig. 3-22c), a simple equation results:

$$\sigma'_{LU} = \frac{G_m}{1 - V_f} \tag{3.61}$$

where σ'_{LU} is the longitudinal compressive strength, G_m is the matrix shear modulus, and V_f is the fiber volume fraction. This equation, at reasonable fiber fractions, predicts a lower strength value than does the equation resulting from the assumption that fiber buckling is in an extensional mode. This is shown in Fig. 3-26. However, Eq. (3.61) predicts values much greater than measured values [45], and even when it is assumed that the matrix can behave inelastically (see Fig. 3-26), the predictions are still too great.

Figure 3-25. Photograph shows localized rotation of carbon fibers that may occur before or during shear failure under longitudinal compressive load (50×). (From Hancox [43].)

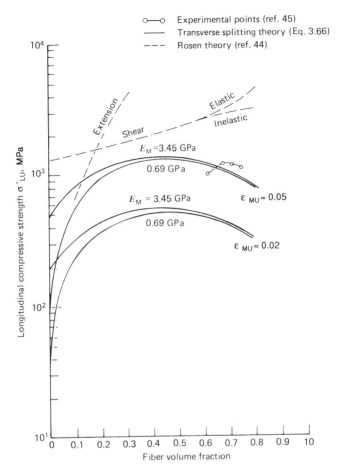

Figure 3-26. Comparison of predicted longitudinal compressive strength with experimental data.

The observation that transverse splitting or debonding might be the failure-initiating event [39,40] also can be formulated as a simple theoretical expression for the longitudinal compressive strength. In this case, failure is assumed to occur when the transverse tensile strain, produced as a result of longitudinal compression, exceeds the ultimate transverse strain capability of the composite. The expression may be derived as follows.

A longitudinal compressive stress σ'_L produces the longitudinal strain

$$\varepsilon_L = -\frac{\sigma'_L}{E_{L,c}} \tag{3.62}$$

where $E_{L,c}$ is the longitudinal modulus in compression. The transverse tensile strain owing to Poisson's effect is given as

$$\varepsilon_T = -\nu_{LT}\varepsilon_L = \nu_{LT}\frac{\sigma'_L}{E_{L,c}} \qquad (3.63)$$

At failure, σ'_L is the ultimate longitudinal compressive strength (σ'_{LU}) such that ε_T equals the ultimate transverse tensile strain (ε_{TU}). Thus

$$\nu_{LT}\frac{\sigma'_{LU}}{E_{L,c}} = \varepsilon_{TU} \qquad (3.64)$$

or

$$\sigma'_{LU} = \frac{E_{L,c}\varepsilon_{TU}}{\nu_{LT}} \qquad (3.65)$$

Use of Eq. (3.44) for ε_{TU} and the rule of mixtures for $E_{L,c}$ and ν_{LT} give the desired expression for the longitudinal compressive strength of a unidirectional composite:

$$\sigma'_{LU} = \frac{(E_f V_f + E_m V_m)(1 - V_f^{1/3})\varepsilon_{mu}}{\nu_f V_f + \nu_m V_m} \qquad (3.66)$$

where ε_{mu} is the matrix ultimate strain. Predictions of Eq. (3.66) for the longitudinal compressive strength are also shown in Fig. 3-26 for two values of ε_{mu}, namely, $\varepsilon_{mu} = 0.02$ and 0.05. Experimental points showing strength are in much better agreement with the predictions of Eq. (3.66) than with the predictions based on microbuckling of fibers. The agreement is particularly good when $\varepsilon_{mu} = 0.05$, which is a reasonable value for the epoxy resin used in the composite tested. As shown, the predicted strength is greatly influenced by the matrix ultimate strain (or transverse composite ultimate strain if interface failure or fiber failure occurs prior to matrix failure). Also, there is an optimal value of fiber volume fraction at which strength will be maximum.

3.6.3 Failure under Transverse Tensile Loads

Fibers perpendicular to the loading direction act essentially to produce stress concentrations at the interface and in the matrix. Therefore, unidirectional composites subjected to transverse tensile loads fail because of matrix or interface tensile failure, although in some cases they may fail by fiber transverse tensile failure if the fibers are highly oriented and weak in the transverse direction. Thus the composite failure modes under transverse tensile load may be described as (1) matrix tensile failure and (2) constituent debonding and/or fiber splitting. Matrix tensile failure is illustrated in Fig. 3-27. Matrix ten-

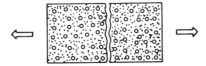

Figure 3-27. Failure of unidirectional composite under transverse tensile load.

sile failure with constituent debonding means that some portions of the fracture surface are formed because of failure of interfacial bonds between the fibers and the matrix. Figure 3-28a,b shows the fracture surface of high-modulus graphite and high-tensile-strength carbon composites [46]. These photographs indicate that constituent debonding and fiber splitting both may take place.

3.6.4 Failure under Transverse Compressive Loads

A unidirectional composite subjected to transverse compressive load generally fails by shear failure of the matrix, which may be accompanied by constituent debonding or fiber crushing. Therefore, the failure modes of a unidirectional composite under transverse compressive loads may be listed as (1) matrix shear failure or (2) matrix shear failure with constituent debonding and/or fiber crushing. The failure modes are illustrated in Fig. 3-29, where some portions of the failure surface are created by constituent debonding. Experimental investigations of Collings [47] with carbon-fiber-reinforced plastics indicate that when the composites are subjected to transverse compressive loads, failure occurs by shear in a direction normal to the fibers on planes parallel to them at the expected angles. The photographs of two of the failed specimens in Fig. 3-30 clearly illustrate the failure mode. It has been suggested that the failure is precipitated by failure of the fiber–resin bond. Thus the transverse compressive strength is lower than the longitudinal compressive strength. However, if constraints are placed on the specimen to prevent its deformation in the direction perpendicular to the plane of load–fiber axes, it is possible to achieve a transverse compressive strength comparable with the longitudinal compressive strength. The increase in strength is observed because the failure now occurs by the shear failure of the fibers, whose strength is higher than the matrix strength or the bond strength. In this mode of failure, that is, the shear failure of the fibers, it has been observed, as expected, that the transverse compressive strength of the composite increases with an increase in the fiber volume fraction.

3.6.5 Failure under In-Plane Shear Loads

In this case the failure could take place by matrix shear failure, constituent debonding, or a combination of the two. Thus the failure modes are (1) matrix shear failure, (2) matrix shear failure with constituent debonding, and (3)

Figure 3-28. Photomicrographs of fracture surfaces of unidirectional graphite–fiber composites failed under transverse tensile loads with fibers being (a) high modulus and (b) high tensile strength. (From Chamis et al. [46].)

constituent debonding. These failure modes are illustrated in Fig. 3-31, where the failure surface could include debonded portions as well.

3.7 EXPANSION COEFFICIENTS AND TRANSPORT PROPERTIES

3.7.1 Thermal Expansion Coefficients

A change in temperature of a body causes a change in its dimensions proportional to the change in temperature and its initial dimensions. For equal

3.7 EXPANSION COEFFICIENTS AND TRANSPORT PROPERTIES 109

Figure 3-28. (*Continued*)

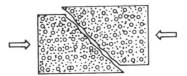

Figure 3-29. Shear failure of unidirectional composite subjected to transverse compressive load.

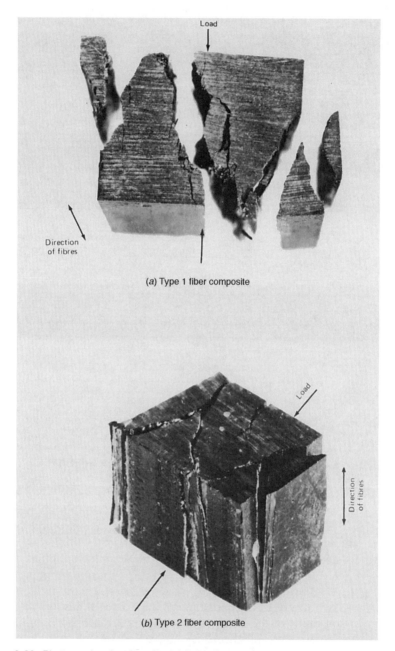

Figure 3-30. Photographs showing shear failure of unconstrained unidirectional carbon-fiber-reinforced plastics subjected to transverse compressive loads. (From Collings [47].)

3.7 EXPANSION COEFFICIENTS AND TRANSPORT PROPERTIES

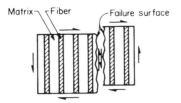

Figure 3-31. Failure of a unidirectional composite subjected to in-plane shear load.

change in temperature and initial dimensions, bodies made of different materials change in dimensions by different amounts. The actual change depends on the material property called *coefficient of thermal expansion,* which, is defined as the change in linear dimension per unit initial length per unit change in temperature. For isotropic bodies, the coefficient of thermal expansion is the same in all directions. For composite materials, the coefficient of thermal expansion, like other properties, changes with direction.

Unidirectional composites have two principal coefficients of thermal expansion, the longitudinal coefficient of expansion α_L and the transverse coefficient of expansion α_T. The longitudinal coefficient α_L generally is small because the fibers, which usually have a smaller coefficient than that for the polymer matrices, impose a mechanical restraint on the matrix material. The transverse coefficient α_T is larger and at low volume fractions of fibers even can be greater than that of the unreinforced polymer because the matrix, which is prevented from much expansion in the longitudinal direction, is forced to expand more than usual in the transverse direction.

Schapery [48] has derived the following simple expressions for the longitudinal and transverse coefficients of thermal expansion:

$$\alpha_L = \frac{1}{E_L} (\alpha_f E_f V_f + \alpha_m E_m V_m) \tag{3.67}$$

$$\alpha_T = (1 + \nu_f)\alpha_f V_f + (1 + \nu_m)\alpha_m V_m - \alpha_L \nu_{LT} \tag{3.68}$$

where α_f and α_m are coefficients of thermal expansion for fibers and matrix, E_L is the elastic modulus of the composite in the longitudinal direction and can be evaluated by the rule of mixtures [Eq. (3.15)], and ν_{LT} is the major Poisson ratio of the composite and also can be approximated by the rule of mixtures as [Eq. (3.59)]. In the derivation of Eq. (3.68) it has been assumed that the Poisson ratios of the fibers and the matrix are not too far apart. For fiber volume fractions greater than 0.25, Eq. (3.68) can be closely approximated by

$$\alpha_T = \alpha_f V_f + (1 + \nu_m)\alpha_m V_m \tag{3.69}$$

The coefficients of thermal expansion for a typical glass–epoxy system are shown in Fig. 3-32 as functions of fiber volume fraction. The following constituent properties have been assumed:

$$\alpha_f = 0.5 \times 10^{-5}/°C \qquad \alpha_m = 6.0 \times 10^{-5}/°C$$

$$E_f = 70 \text{ GPa} \qquad E_m = 3.5 \text{ GPa}$$

$$\nu_f = 0.20 \qquad \nu_m = 0.35$$

It may be observed that for unidirectional composites having fiber volume fractions of practical importance, the coefficients of thermal expansion in the longitudinal and transverse directions are quite different. This thermal-expansion anisotropy of unidirectional laminae causes residual thermal stresses in laminates. Transformation of coefficients of thermal expansion in an arbitrary direction and a procedure for calculating residual stresses in the laminates are discussed in Chap. 6. As shown in Table 3-1, the longitudinal thermal-expansion coefficient of a graphite–fiber composite is quite small and, in some cases, negative, depending on the specific value for the fiber. The graphite crystallographic structure can produce negative thermal-expansion coefficients in certain crystallographic directions. Moreover, the thermal-expansion coefficients usually are a function of temperature, as shown in Fig. 3-33 [49]. Similar results occur for the highly oriented polymeric Kevlar fibers. In both cases the fibers are highly anisotropic, and the transverse expansion coefficients of the fiber and their composites are much larger. The near-zero thermal expansions in the fiber axis direction are very significant because laminates thus can be designed having near-zero expansions in certain directions. This is a very powerful consideration for structures requiring great

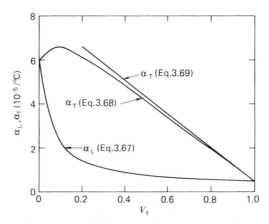

Figure 3-32. Coefficients of thermal expansion for a unidirectional composite.

3.7 EXPANSION COEFFICIENTS AND TRANSPORT PROPERTIES

Table 3-1 Typical properties of unidirectional-fiber-reinforced epoxy resins

Property	E-Glass	Kevlar 49	Graphite (Thornel 300)
Fiber volume fraction	46	60–65	63
Specific gravity	1.80	1.38	1.61
Tensile strength, 0° (MPa)	1104	1310	1725
Tensile modulus, 0° (GPa)	39	83	159
Tensile strength, 90° (MPa)	36	39	42
Tensile modulus, 90° (GPa)	10	5.6	10.9
Compression strength, 0° (MPa)	600	286	1366
Compression modulus, 0° (GPa)	32	73	138
Compression strength, 90° (MPa)	138	138	230
Compression modulus, 90° (GPa)	8	5.6	11
In-plane shear strength (MPa)	—	60	95
In-plane shear modulus (GPa)	—	2.1	6.4
Longitudinal Poisson ratio (ν_{LT})	0.25	0.34	0.38
Interlaminar shear strength (MPa)	31	69	113
Longitudinal coefficient of thermal expansion (10^{-6}/°C)	5.4	-2.3 to -4.0[a]	0.045
Transverse coefficient of thermal expansion (10^{-6}/°C)	36	35[b]	20.2

[a] -79 to $+100$°C.
[b] -195 to $+120$°C.

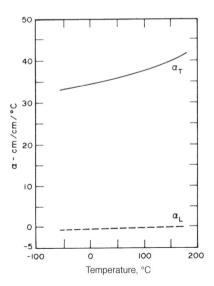

Figure 3-33. Coefficients of thermal expansion of a unidirectional high-modulus graphite–epoxy composite as a function of temperature. (From Northrop Corporation [49].)

3.7.2 Moisture Expansion Coefficients

Moisture absorption by a body (e.g., resin matrix in composite materials) causes a volumetric change (or swelling) in the body. A coefficient of moisture expansion β can be defined as the change in linear dimension of the body per unit initial length per unit change in moisture concentration (moisture concentration may be defined as the weight of moisture present per unit weight of the body). The following expression for the moisture expansion coefficient of a body can be derived easily by converting the weight of moisture content to its volume and considering that linear strain is only one-third of the volumetric strain:

$$\beta = \frac{1}{3}\frac{\rho}{\rho_w} \tag{3.70}$$

where ρ and ρ_w are the densities of the body and water, respectively. It may be pointed out here, however, that Eq. (3.70) is applicable when there are no voids in the body. When voids are present, the actual expansion of the body owing to moisture will be less than that indicated by Eq. (3.70) because a part of the moisture absorbed will fill some voids and thus will not contribute to swelling.

It is well known that polymer matrices absorb moisture when exposed to a humid environment but that inorganic fibers do not. Moisture absorbed by the matrix results in a volume change of the composite. However, the expansion of unidirectional composites in the longitudinal direction is negligible because of the much higher stiffness of the fibers. Therefore, the longitudinal coefficient of moisture expansion β_L of a unidirectional composite is taken to be zero. The transverse coefficient of moisture expansion β_T of a unidirectional composite is related to the moisture expansion coefficient of the matrix β_m as follows [50]:

$$\beta_T = \frac{\rho_c}{\rho_m}(1 + \nu_m)\beta_m \tag{3.71}$$

where ρ_c and ρ_m are the densities of the composite and matrix material, respectively, and ν_m is the Poisson's ratio of the matrix. Thermal and moisture expansion coefficients of several commercial composites are given in Appendix 4.

3.7.3 Transport Properties

Fiber composites are employed frequently in engineering applications that are influenced by considerations of heat conduction, permeation, electrical con-

3.7 EXPANSION COEFFICIENTS AND TRANSPORT PROPERTIES

duction, or transport of electrical and magnetic fields. Accordingly, it is important to know the relationship between composite structure and such physical constants as thermal conductivity, mass diffusivity, electrical conductivity, dielectric constants, and magnetic permeability. It has been suggested [51–53] that a transport coefficient of a unidirectional composite in the longitudinal direction k_L can be calculated by the rule of mixtures as

$$k_L = V_f k_f + V_m k_m \tag{3.72}$$

The transverse coefficient k_T can be computed by invoking an analogy from classical physics between the in-plane shear field equations and boundary conditions to the transverse transport phenomenon [52,53]. Thus the transverse transport coefficient k_T may be computed by the Halpin–Tsai equation:

$$\frac{k_T}{k_m} = \frac{1 + \xi \eta V_f}{1 - \eta V_f} \tag{3.73}$$

where

$$\eta = \frac{(k_f/k_m) - 1}{(k_f/k_m) + \xi} \tag{3.74}$$

$$\log \xi = \sqrt{3} \log \frac{a}{b} \tag{3.75}$$

where k_f and k_m are the appropriate transfer coefficients for fibers and matrix and a and b are the dimensions of the fiber along and perpendicular to the direction of measurement of the transfer coefficient. For circular cross-sectional fibers, the ratio a/b is 1 if transverse coefficients are to be estimated.

Example 3-4: Find the thermal conductivities of unidirectional glass-fiber- and carbon-fiber-reinforced epoxy composites in the longitudinal and transverse directions. Fiber volume fraction is 60% in both cases. Following are the thermal conductivities for the fibers and the matrix (note that the carbon fiber itself is anisotropic):

Epoxy matrix $K_m = 0.25$ W/m/°C

Glass fibers $K_f = 1.05$ W/m/°C

Carbon fibers $(K_f)_L = 80$ W/m/°C

$(K_f)_T = 12.5$ W/m/°C

Glass–Epoxy Composite

$$K_L = 0.6 \times 1.05 + 0.4 \times 0.25 = 0.73 \text{ W/m/°C}$$

For circular or square cross-sectional fibers

$$\xi = 1$$

$$\eta = \frac{(1.05/0.25) - 1}{(1.05/0.25) + 1} = 0.615$$

$$\frac{K_T}{K_m} = \frac{1 + 0.615 \times 0.6}{1 - 0.615 \times 0.6} = 2.17$$

$$K_T = 0.543 \text{ W/m/°C}$$

Carbon-Fiber–Epoxy Composite

$$K_L = 0.6 \times 80 + 0.4 \times 0.25 = 48.1 \text{ W/m/°C}$$

$$\eta = \frac{(12.5/0.25) - 1}{(12.5/0.25) + 1} = 0.961$$

$$\frac{K_T}{K_m} = \frac{1 + 0.961 \times 0.6}{1 - 0.961 \times 0.6} = 3.72$$

$$K_T = 0.93 \text{ W/m/°C}$$

Longitudinal and transverse thermal conductivities of high-modulus (HMS) and high-strength (HTS) carbon-fiber–epoxy composites are shown as a function of fiber volume fraction in Figs. 3-34 and 3-35 [54]. The linearity of the longitudinal thermal conductivities is consistent with Eq. (3.72). The plots show that the thermal conductivity of the matrix may be neglected for prediction of the longitudinal thermal conductivity. The transverse thermal conductivity of these composites also increases with increasing fiber volume fraction. The experimental data could not be compared with the theoretical predictions because of the absence of pertinent data, that is, transverse thermal conductivities for fibers. Longitudinal and transverse electrical conductivities of these composites are shown in Figs. 3-36 and 3-37. The trends exhibited by the electrical conductivities are similar to those exhibited by the corresponding thermal conductivities.

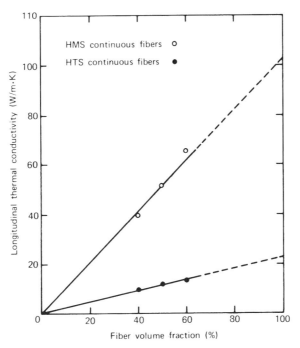

Figure 3-34. Longitudinal thermal conductivity of unidirectional carbon–epoxy composites at 20°C. (From Knibbs et al. [54].)

3.7.4 Mass Diffusion

Polymer matrix composites, when exposed to humid environments or immersed in water, absorb moisture. The moisture content in the composite, besides causing volumetric changes, influences mechanical properties such as modulus and strength. It is of considerable practical interest to know the moisture absorption (and also desorption) characteristics of composite materials. The effects of moisture content on physical dimensions were discussed in Sec. 3.7.2, and the effects on mechanical properties will be discussed in Chap. 9. A brief discussion is presented here on the concepts of mass diffusion and some simplified equations governing the diffusion process are given.

The percent moisture content C in a body is defined as

$$C = \frac{\text{weight of moist material} - \text{weight of dry material}}{\text{weight of dry material}} \times 100 \quad (3.76)$$

In any problem involving moisture absorption or desorption, it is desired to obtain the moisture content at any given time. The moisture content depends on the environmental conditions, the initial moisture content in the

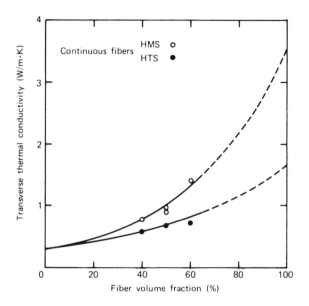

Figure 3-35. Transverse thermal conductivity of unidirectional carbon–epoxy composites at 20°C. (From Knibbs et al. [54].)

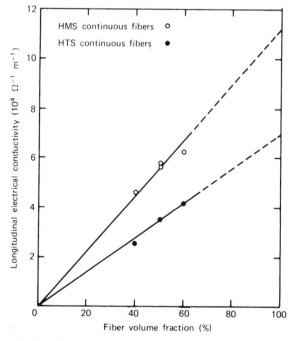

Figure 3-36. Longitudinal electrical conductivity of unidirectional carbon–epoxy composites at 20°C. (From Knibbs et al. [54].)

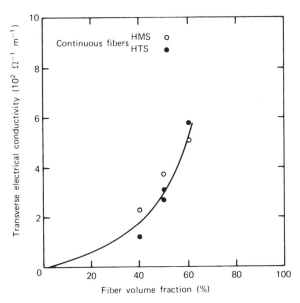

Figure 3-37. Transverse electrical conductivity of unidirectional carbon–epoxy composites at 20°C. (From Knibbs et al. [54].)

body, and the mass diffusivity of the material. For a one-dimensional diffusion, the relationship between different variables is discussed in the following paragraphs.

Consider a plate of thickness h initially with uniform temperature and moisture distributions inside it (Fig. 3-38). The plate is considered infinitely large so that the diffusion takes place in the thickness direction only. The

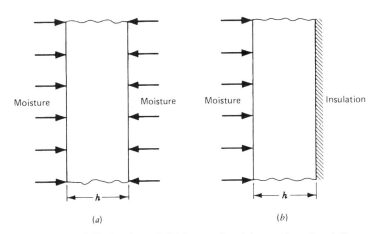

Figure 3-38. Diffusion through thickness of a plate: problem description.

plate is exposed to an environment with constant moisture and temperature on one or both sides. The moisture content C in the plate during absorption or desorption is given by

$$C = G(C_m - C_0) + C_0 \tag{3.77}$$

where C_0 is the initial moisture content of the plate, C_m is the equilibrium or maximum moisture content that can be attained under the given environmental conditions, and G is a time-dependent parameter given by the following equation [50]:

$$G = 1 - \frac{8}{\pi^2} \sum_{j=0}^{\infty} \frac{1}{(2j+1)^2} \exp\left[-\frac{(2j+1)^2 \pi^2 Dt}{S^2}\right] \tag{3.78}$$

where D is the mass diffusivity of the material in the direction of diffusion (thickness direction in this case), t is time, and $S = h$ if the material is exposed on both sides and $S = 2h$ if the material is exposed on one side only. For sufficiently large values of t, Eq. (3.78) can be approximated by the first term of the series:

$$G = 1 - \frac{8}{\pi^2} \exp\left(-\frac{\pi^2 Dt}{S^2}\right) \tag{3.79}$$

Experimental evidence indicates that the equilibrium moisture content C_m is insensitive to the temperature but depends on the moisture content of the environment. For a material exposed to humid air, it is related to the relative humidity ϕ (in percent) by a power law:

$$C_m = a\left(\frac{\phi}{100}\right)^b \tag{3.80}$$

The parameters a and b are determined experimentally [53]. For a typical graphite–epoxy composite, $a = 0.018$ and $b = 1$.

For an isotropic material (e.g., polymeric matrix material), the diffusivity is the same in all directions. For a unidirectional composite, the diffusivity, like other properties, is direction-dependent. There are two principal values of diffusivities, namely, the longitudinal diffusivity D_L and the transverse diffusivity D_T. The diffusivities D_L and D_T may be calculated using Eqs. (3.72) and (3.73). Generally, the diffusivity of the fiber is much smaller than that of the matrix ($D_f \ll D_m$) and frequently is neglected in calculating the diffusivity of the composite. In that case, the rule of mixtures [Eq. (3.72)] for diffusivity reduces to

3.7 EXPANSION COEFFICIENTS AND TRANSPORT PROPERTIES

$$D_L = (1 - V_f)D_m \tag{3.81}$$

It is also suggested [53] that the following simple equation be used for transverse diffusivity if $V_f < 0.78$:

$$D_T = \left(1 - 2\sqrt{\frac{V_f}{\pi}}\right)D_m \tag{3.82}$$

Diffusivities D_L and D_T for a graphite–epoxy composite along with the matrix diffusivity D_m are shown in Fig. 3-39.

Equations (3.77) and (3.79) can be used to calculate the time t_m required for a material to attain 99.9% of its maximum possible moisture content:

$$t_m = \frac{0.678 S^2}{D} \tag{3.83}$$

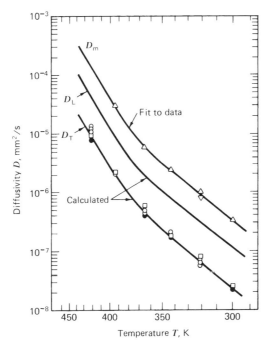

Figure 3-39. Matrix diffusivity and longitudinal and transverse diffusivities of a graphite–epoxy composite.

The time required to reach the maximum moisture content is insensitive to the moisture content of the environment but depends on the temperature because D depends on temperature.

Example 3-5: A 12.5-mm-thick plate of graphite–epoxy composite with an initial moisture content of 0.5% is exposed on both sides to air at 25°C and 90% relative humidity. Estimate the time required to reach 1% moisture content. For the composite, at 25°C, assume $D_T = 2.6 \times 10^{-7}$ mm²/s.

Solution: Substitution of Eq. (3.79) in Eq. (3.77) gives

$$t = \frac{S^2}{D}\left[-\frac{1}{\pi^2}\ln\frac{\pi^2}{8}\left(\frac{C_m - C}{C_m - C_0}\right)\right]$$

In the composite laminate, diffusion is taking place in the thickness direction. Therefore, the transverse diffusivity D_T is to be used for calculation, which is given. Also given are $S = 12.5$ mm, $C = 1.0\%$, and $C_0 = 0.5\%$; C_m may be calculated using Eq. (3.80):

$$C_m = 0.018\left(\frac{90}{100}\right) = 0.0162 \quad \text{or} \quad 1.62\%$$

Therefore,

$$t = \frac{(12.5)^2}{2.6 \times 10^{-7}}\left[-\frac{1}{\pi^2}\ln\left(\frac{\pi^2}{8}\cdot\frac{1.62 - 1.0}{1.62 - 0.5}\right)\right]$$

$$= 2.32 \times 10^7 \text{ s} \quad \text{or} \quad 269 \text{ days}$$

Example 3-6: The plate specified in Example 3-5, having attained the moisture content of 1%, is exposed on both sides to humid air at 15°C and 10% relative humidity. Estimate the moisture content after 10 days. Assume $D_T = 1.13 \times 10^{-7}$ mm²/s at 15°C.

Solution: Given $S = 12.5$ mm, $C_0 = 1.0\%$. At 10% relative humidity,

$$C_m = 0.018\left(\frac{10}{100}\right) = 0.0018 \quad \text{or} \quad 0.18\%$$

From Eq. (3.77),

$$C = \left[1 - \frac{8}{\pi^2}\exp\left(-\frac{\pi^2 \times 10 \times 24 \times 3600 \times 1.13 \times 10^{-7}}{12.5 \times 12.5}\right)\right]$$
$$\times (0.18 - 1.0) + 1.0 = 0.84\%$$

Thus in 10 days the moisture content would be reduced from 1% to 0.84%.

3.8 TYPICAL UNIDIRECTIONAL FIBER COMPOSITE PROPERTIES

The preceding discussions on property-prediction methods and failure modes should be of use in comparing various physical properties of unidirectional composites. It is also valuable to appreciate the difference between different types of fiber composites and their respective properties. From the known values of fiber and matrix properties (see Chap. 2), the interested reader can try to predict the properties shown in Table 3-1. A more detailed insight into mechanical behavior can be gained by reviewing the stress–strain curves presented in Appendix 4 for unidirectional composites.

A few points should be made concerning the data presented in Table 3-1. The measured compression strength in the fiber direction of a unidirectional composite generally is less than the tensile strength. Since compression strength of this type of composite is so difficult to measure, the reported value

Table 3-2 Summary of influence of constituents on properties of unidirectional polymer composites[a]

Composite Property	Fibers	Matrix	Interface
	Tensile Properties		
Longitudinal modulus	S	W	N
Longitudinal strength	S	W	N
Transverse modulus	W	S	N
Transverse strength	W	S	S
	Compression Properties		
Longitudinal modulus	S	W	N
Longitudinal strength	S	S	N
Transverse modulus	W	S	N
Transverse strength	W	S	N
	Shear Properties		
In-plane shear modulus	W	S	N
In-plane shear strength	W	S	S
Interlaminar shear strength	N	S	S

[a] S = strong influence; W = weak influence; N = negligible influence.

often is merely a reflection of the quality of the test technique. Occasionally, one sees values reported in the literature that might exceed the value of the tensile strength, and in such cases, the compression test fixture is often such that it prevents certain failure modes, perhaps producing artificially large values of strength. The very low value of compression strength for the Kevlar composite is caused by the exceptionally low shear and transverse tensile strength of the highly oriented Kevlar fiber, which initiates failure in the composite when subjected to compression.

An overview of the influence of constituent properties on properties of unidirectional polymer composites is presented in Table 3-2. The influences indicated in the table generally are applicable when the ratio E_f/E_m is greater than 10. Since $E_f/E_m < 10$ for metal matrix composites, the influences indicated in Table 3-2 may not be valid.

EXERCISE PROBLEMS

3.1. In a unidirectional composite, cylindrical fibers may be packed in square or hexagonal (sometimes called *triangular*) arrays, as shown in Fig. 3-40. By selecting representative area elements (repeating units), show that the fiber volume fractions are

Square packing: $$V_f = \frac{\pi d^2}{4s^2}$$

Hexagonal packing: $$V_f = \frac{\pi d^2}{2\sqrt{3}s^2}$$

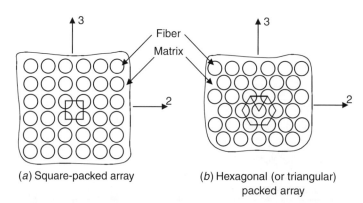

(a) Square-packed array

(b) Hexagonal (or triangular) packed array

Figure 3-40. Fiber packing in unidirectional composites (Exercise Problem 3.1).

where d is the fiber diameter, and s is the distance between centers of two adjacent fibers. Theoretical maximum fiber volume fraction occurs when adjacent fibers touch each other. Calculate theoretical maximum fiber volume fractions in each case.

3.2. A burn-off test was performed to determine the volume fractions of constituents in a glass-fiber-reinforced epoxy composite. The following observations were made:

Weight of empty crucible = 47.6504 g
Weight of crucible and a small piece of composite = 50.1817 g
Weight of crucible and glass after the burn-off = 49.4476 g

Calculate the weight and volume fractions of glass fibers and epoxy resin. Assume that the densities of the fibers and resin are 2.5 and 1.2 g/cm^3, respectively.

3.3. Derive Eq. (3.8). (*Hint:* For a given weight, actual volume of the composite with voids is obtained from the experimentally measured composite density, whereas the volume calculated from theoretical density consists of the volumes of fibers and matrix only.)

3.4. Calculate the density of the composite in Exercise Problem 3.2 using weight fractions and densities of the constituents. If the density was determined experimentally to be 1.86 g/cm^3, calculate the void content in the composite.

3.5. Calculate the ratios of fiber stress to matrix stress and fiber stress to composite stress for unidirectional composites with V_f = 10%, 25%, 50%, and 75%. Assume that the composites are loaded in the fiber direction; E_f = 400 GPa and E_m = 3.2 GPa.

3.6. Estimate E_L, E_T, G_{LT}, and ν_{LT}, of glass–epoxy, graphite–epoxy, Kevlar–epoxy, and boron–aluminum composites with V_f = 25%, 50%, and 75%. Constituent properties are

Material	E (GPa)	ν
Epoxy	3.5	0.35
Glass fibers	70	0.20
Graphite fibers	250	0.20
Kevlar fibers	140	0.20
Boron fibers	350	0.20
Aluminum	70	0.33

For the purpose of calculations, assume all fibers to be isotropic.

3.7. A rod consists of a binder and two types of filamentous reinforcement with the following constituent properties:

Material	Density (g/cm³)	Wt. %	E (GPa)	σ_u (GPa)
Binder	1.3	35	3.5	0.06
Fiber A	2.5	45	70	1.4
Fiber B	1.6	20	6	0.45

(a) What maximum load can this rod carry without rupturing any of the constituents? (Assume that the cross-sectional area of the rod = 10 cm².)
(b) What is the maximum load the rod can carry?
(c) What constituent will rupture last?
(d) Plot the load–elongation curves for the rod to failure in load-maintained and elongation-maintained tests.

3.8. Two composites are fabricated with glass fibers ($V_f = 50\%$) and matrices A and B, whose stress–strain curves are shown in Fig. 3-41. The glass fibers are elastic up to failure and have an elastic modulus of 70 GPa and ultimate tensile strength of 2.8 GPa. Assuming that the composites are stressed parallel to the unidirectional glass fibers, calculate (a) the composite stress at 1% and 4% strains for each composite and (b) the minimum and critical fiber volume fractions for both composites.

3.9. Draw the stress–strain diagrams for the two composites indicated in Problem 3.8. Also calculate the longitudinal strengths of the composites.

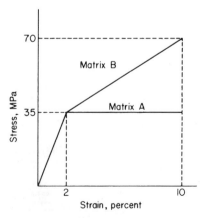

Figure 3-41. Stress–strain curves for matrices A and B (Exercise Problem 3.8).

3.10. The material of a tension link is changed from an aluminum alloy to a unidirectional graphite–epoxy composite. Calculate the volume fraction of graphite fibers required in the composite to match its longitudinal modulus with that of the aluminum alloy. What is the percentage weight saving in this material replacement? Use the properties given in Table 1-1. The elastic modulus of epoxy is 3.5 GPa, and its density is 1.2 g/cm^3.

3.11. Estimate the longitudinal strength of the composite considered in Problem 3.10 by neglecting the load carried by the epoxy.

3.12. Repeat Exercise Problems 3.10 and 3.11 assuming that Kevlar fibers are used in place of graphite fibers in the composite. Can glass fibers be used in this application?

3.13. A uniaxial tension rod is expected to carry a maximum load of 2000 N. Consider a graphite–epoxy composite ($V_f = 0.65$) to replace steel as the rod material. This graphite–epoxy composite costs five times as much as the steel on a weight basis. Would you choose the composite if the criterion depends just on
1. Weight of the rod
2. Cost

Assume the following properties for steel:

Elastic modulus = 210 GPa
Poisson's ratio = 0.3
Tensile strength = 450 MPa
Specific gravity = 7.8

Properties of graphite fibers (PAN) and epoxy resin are given in Tables 2-3 and 2-11, respectively. For calculation of the composite strength, neglect the load carried by the epoxy.

3.14. A unidirectional glass–epoxy composite has $V_m/V_f = 1.5$. What minimum volume fraction of carbon fibers should be added to the glass–epoxy composite, without changing the ratio of volume fractions of epoxy resin and glass fibers, to obtain any strengthening? Following are the constituent properties.

Material	Density (g/cm^3)	E (GPa)	σ_u (MPa)
Epoxy resin	1.2	3.5	52.5
Glass fibers	2.5	70	700
Carbon fibers	1.8	350	700

128 BEHAVIOR OF UNIDIRECTIONAL COMPOSITES

3.15. While maintaining the matrix volume fraction at 30%, addition of what volume fraction of carbon fibers will double the longitudinal modulus of a unidirectional glass–epoxy composite? What is the percent reduction in density by this addition of carbon fibers? Also calculate the longitudinal strengths of the two composites. Take the constituent properties given in Exercise Problem 3.14. Explain why the composite strength decreases even though the strengths of carbon fibers and glass fibers are equal.

3.16. A unidirectional composite shows the following properties in tension:

$$E_L = 40 \text{ GPa} \qquad \epsilon_{LU} = 0.0276$$

$$E_T = 10 \text{ GPa} \qquad \epsilon_{TU} = 0.0036$$

$$\nu_{LT} = 0.25$$

Estimate the longitudinal compressive strength of the composite. Assume the composite moduli to be equal in tension and compression.

3.17. Using Eqs. (3.39)–(3.41), calculate and plot normalized transverse composite strength (σ_{TU}/σ_{mU}) as a function of fiber volume fraction for glass-fiber-reinforced epoxy. Assume that $E_m/E_f = 0.05$. (*Note:* For better appreciation of results, the strengths obtained by using SCF and SMF should be plotted on the same graph.)

3.18. Plot the longitudinal and transverse coefficients of thermal expansion for a unidirectional glass–polyester composite as functions of fiber volume fraction. Assume the following constituent properties:

$$\alpha_f = 0.5 \times 10^{-5}/°C \qquad \alpha_m = 9.0 \times 10^{-5}/°C$$

$$E_f = 70 \text{ GPa} \qquad E_m = 3.5 \text{ GPa}$$

$$\nu_f = 0.2 \qquad \nu_m = 0.35$$

31.9. Plot the longitudinal and transverse thermal conductivities of unidirectional glass fiber and carbon-fiber-reinforced epoxy composites as functions of fiber volume fraction. Assume the same constituent properties as in Example 3-4.

3.20. The coefficient of longitudinal thermal expansion of a unidirectional composite is measured as $-0.61 \times 10^{-6}/°C$. Estimate the coefficient of thermal expansion of the fibers. Given: $E_f = 294$ GPa, $E_m = 3.5$ GPa, $\alpha_m = 54 \times 10^{-6}/°C$, and $V_f = 55\%$.

3.21. The maximum water absorption of a typical epoxy is 6%. What is the maximum amount of water in a graphite–epoxy composite ($V_f = 70\%$)? Assume that graphite fibers do not absorb moisture and that the specific gravities of the epoxy and composite are 1.2 and 1.6, respectively.

REFERENCES

1. A. Kelly and G. J. Davies, "The Principles of the Fiber Reinforcement of Metals," *Metallurg. Rev.,* **10,** 37 (1965).
2. B. W. Rosen, "Strength of Uniaxial Fibrous Composites," in F. W. Wendt, H. Liebowitz, and N. Perrone, eds., *Mechanics of Composite Materials,* Pergamon, New York, 1970.
3. B. H. Jones, "Probabilistic Design and Reliability," in C. C. Chamis, ed., *Structural Design and Analysis, Part II,* Academic, New York, 1975.
4. H. T. Corten, "Micromechanics and Fracture Behavior of Composites," in Lawrence J. Broutman and Richard H. Krock, eds., *Modern Composite Materials,* Addison-Wesley, Reading, MA, 1967, Chap. 2.
5. L. J. Broutman, "Fiber Reinforced Plastics," in Lawrence J. Broutman and Richard H. Krock, eds., *Modern Composite Materials,* Addison-Wesley, Reading, MA, 1967, Chap. 13.
6. L. J. Broutman and R. H. Krock, "Principles of Composites and Composite Reinforcement," in Lawrence J. Broutman and Richard H. Krock, eds., *Modern Composite Materials,* Addison-Wesley, Reading, MA, 1967, Chap. 1.
7. L. J. Broutman, "Mechanical Behavior of Fiber-Reinforced Plastics," in Albert G. H. Dietz, ed., *Composite Engineering Laminates,* MIT Press, Cambridge, MA, 1969, Chap. 6.
8. B. W. Rosen, "Mechanics of Fiber Strengthening," in *Fiber Composite Materials,* ASM Publication, Metals Park, OH, 1964, Chap. 3.
9. N. F. Dow, "Study of Stresses Near a Discontinuity in a Filament-Reinforced Composite Material," General Electric Company Report No. TIS R63 SD61, Barrington, IL, 1963.
10. W. R. Tyson and G. J. Davies, "A Photoelastic Study of the Shear Stresses Associated with the Transfer of Stresses During Fiber Reinforcement," *Br. J. Appl. Phys.,* **16,** 199–205 (1965).
11. D. M. Schuster and E. Scala, "The Mechanical Interaction of Saphire Whiskers with a Birefringent Matrix," *Transact. Metallurg. Soc. AIME,* **230,** 1635 (1964).
12. I. M. Daniel, "Photoelastic Investigations of Composites," in G. P. Sendeckyj, ed., *Mechanics of Composite Materials,* Academic, New York, 1974.
13. T. F. MacLaughlin and R. M. Barker, "Effect of Modulus Ratio on Stress Near a Discontinuous Fiber," *Exp. Mech.,* **12**(4), 178–183 (1972).
14. B. D. Agarwal, J. M. Lifshitz, and L. J. Broutman, "Elastic–Plastic Finite Element Analysis of Short Fiber Composites," *Fibre Sci. Technol.,* **7,** 45–62 (1974).
15. L. J. Broutman and B. D. Agarwal, "A Theoretical Study of the Effect of an Interfacial Layer on the Properties of Composites," *Polym. Eng. Sci.,* **14**(8), 581 (1974).

16. A. G. Metcalfe, *Interfaces in Metal Matrix Composites*, Academic, New York, 1974.
17. E. P. Plueddemann, *Interfaces in Polymer Matrix Composites*, Academic, New York, 1974.
18. C. C. Chamis, "Micromechanics Strength Theories," in Lawrence J. Broutman, ed., *Fracture and Fatigue*, Academic, New York, 1974, Chap. 3.
19. S. W. Tsai and V. D. Azzi, "Strength of Laminated Composite Materials," *AIAA J.*, **4**(2), 296–301 (1966).
20. B. Paul, "Predictions of Elastic Constants of Multiphase Materials," *Transact. Metallurg. Soc. AIME*, 36–41 (February 1960).
21. Z. Hashin and B. W. Rosen, "The Elastic Moduli of Fiber-Reinforced Materials," *J. Appl. Mech., Transact. ASME*, **31**, 233 (June 1964), errata, 219 (March 1965).
22. L. M. Kurshin and L. A. Fil'shtinskii, "Determination of Reduced Elastic Moduli of an Isotropic Plane, Weaker by a Doubly-Periodic Array of Circular Holes" (in Russian), *Izv. Akad. Nauk SSSR, Mekh. Mash.* **6**, 110 (1961).
23. L. A. Fil'shtinskii, "Stresses and Displacements in an Elastic Sheet Weakened by a Doubly-Periodic Set of Equal Circular Holes," *Prikl. Mat. Mekh.*, **28**, 430 (1964).
24. G. A. Van Fo Fy, "On the Equations Connecting the Stresses and Strains in Glass Reinforced Plastics" (in Russian), *Prikladnaia Mekhanika*, **1**(2), 110 (1965).
25. H. B. Wilson and J. L. Hill, "Mathematical Studies of Composite Materials," Rohm and Haas Special Report No. 5-50, AD 468569, 1965.
26. D. F. Adams and D. R. Doner, "Transverse Normal Loading of a Unidirectional Composite," *J. Compos. Mater.*, **1**, 152 (1967).
27. P. E. Chen and J. M. Lin, "Transverse Properties of Fibrous Composites," *Mater. Res. Stand., MTRSA*, **9**(8), 29–33 (1969).
28. R. Hill, "Theory of Mechanical Properties of Fibre-Strengthened Materials: III. Self-Consistent Model," *J. Mech. Phys. Solids*, **13**, 189 (1965).
29. J. M. Whitney and M. B. Riley, "Elastic Properties of Fiber Reinforced Composite Materials," *J. AIAA*, **4**, 1537 (1966).
30. J. J. Hermans, "The Elastic Properties of Fiber Reinforced Materials when the Fibers are Aligned," *Proc. Konigl. Nederl. Akad. van Weteschappen Amsterdam*, **B70**(1), 1 (1967).
31. C. C. Chamis and G. P. Sendeckyj, "Critique on Theories Predicting Thermoelastic Properties of Fibrous Composites," *J. Compos. Mater.*, **2**(3), 332 (1968).
32. J. C. Halpin and S. W. Tsai, "Effects of Environmental Factors on Composite Materials," AFML-TR 67-423, Dayton, OH, June 1969.
33. L. B. Greszczuk, "Theoretical and Experimental Studies on Properties and Behavior of Filamentary Composites," Society of Plastics Industry, 21st Annual Conference, Washington, DC, 1966.
34. J. A. Kies, "Maximum Strains in the Resin of Fiberglass Composites," NRL Report No. 5752, Washington, DC, AD-274560.
35. J. C. Schulz, Society of Plastics Industry, 18th Annual Conference, Washington, D.C., 1963.
36. L. E. Nielsen, *Mechanical Properties of Polymers and Composites*, Vol. 2, Marcel Dekker, New York, 1974, p. 407.

37. B. W. Rosen, "Tensile Failure of Fibrous Composites," *AIAA J.,* **2**(11), 1985 (1964).
38. B. D. Agarwal and J. N. Narang, "Strength and Failure Mechanism of Anisotropic Composites," *Fibre Sci. Technol.,* **10**(1), 37 (1977).
39. L. J. Broutman, "Glass-Resin Joint Strengths and Their Effect on Failure Mechanisms in Reinforced Plastics," *Modern Plast.* (April 1965).
40. L. B. Greszezuk, "Microbuckling Failure of Circular Fiber Reinforced Composites," AIAA/ASME/SAE 15th Structure, Structural Dynamics and Materials Conference, Las Vegas, NV, April 1974.
41. H. Schuerch, "Prediction of Compressive Strength in Uniaxial Boron Fiber–Metal Matrix Composite Materials," *AIAA J.,* **4**(1), 102 (1966).
42. W. C. Dale and F. Baer, "Fiber-Buckling in Composite Systems: A Model for the Ultrastructure of Uncalcified Collagen Tissues," *J. Mater. Sci.,* **9**(3), 369 (1974).
43. N. L. Hancox, "The Compression Strength of Unidirectional Carbon Fibre Reinforced Plastic," *J. Mater. Sci.,* **10**(2), 234–242 (1975).
44. B. W. Rosen, Ed., *Fiber Composite Materials,* American Society for Metals, Metals Park, OH, 1965.
45. N. Fried, in G. Lubin, ed., *Handbook of Fiberglass and Advanced Plastics Composites,* Van Nostrand–Reinhold, Princeton, NJ, 1969, p. 747.
46. C. C. Chamis, M. P. Hanson, and T. T. Serafini, "Designing for Impact Resistance With Unidirectional Fiber Composites," NASA TN D-6463, Cleveland, OH.
47. T. A. Collings, "Transverse Compressive Behavior of Unidirectional Carbon Fibre Reinforced Plastics," *Composites,* **5**(3), 108 (1974).
48. R. A. Schapery, "Thermal Expansion Coefficients of Composite Materials Based on Energy Principles," *J. Compos. Mater.,* **2**(3), 280–404 (1968).
49. *Flightworthy Graphite Reinforced Aircraft Primary Structural Assemblies,* Vol. II, AFML-TR-70-207, Northrop Corporation, October 1970.
50. S. W. Tsai and H. T. Hahn, *Introduction to Composite Materials,* Technomic, Westport, CT, 1980.
51. Z. Hashin, "Assessment of the Self-Consistent Scheme Approximation: Conductivity of Particulate Composites," *J. Compos. Mater.,* **2**(3), 284–300 (1968); S. G. Springer and S. W. Tsai, "Thermal Conductivities of Unidirectional Materials," *J. Compos. Mater.,* **1**(2), 166 (1967).
52. J. C. Halpin, *Primer on Composite Materials: Analysis,* Technomic Westport, CT, 1984.
53. G. S. Springer, *Environmental Effects on Composite Materials,* Technomic, Westport, CT, 1981.
54. R. H. Knibbs, J. D. Baker, and G. Rhodes, "The Thermal and Electrical Properties of Carbon Fibre Unidirectional Reinforced Epoxy Composites," SPI, 26th Annual Technical Conference, Washington, DC, 1971, Sec. 8-F.

4

SHORT-FIBER COMPOSITES

4.1 INTRODUCTION

A distinguishing feature of the unidirectional composites discussed in Chap. 3 is that they have higher strength and modulus in the direction of the fiber axis and generally are very weak in the transverse direction. This is very advantageous when these composites are used in applications where the state of stress can be determined accurately so that laminates can be fabricated from the unidirectional laminae having strengths matched to the design needs. However, in applications where the state of stress may not be predictable or where it is known that the stresses are approximately equal in all directions, unidirectional composites or laminae may not be required or cost-effective. Such applications require composites that have approximately equal strengths in all directions. Multilayered composites can be constructed from layers of unidirectional laminae having different fiber orientations so that the resulting composite is essentially isotropic in a plane. Such composites have the disadvantage that although the overall composite is equally strong in all directions, the surface layers, where the failure is quite often initiated, are still very weak in the transverse direction. In applications where protection from a corrosive environment is an important factor, such as in storage tanks in the chemical industry and in many applications in the automobile industry, laminates of unidirectional laminae do not solve the problem completely. Thus it is advantageous to have each layer or lamina isotropic in some applications. An effective way of producing an isotropic layer is to use randomly oriented short fibers as the reinforcement. Molding compounds consisting of short fibers can produce generally isotropic composites. They can be molded easily by injection or compression molding and also are economical. A summary of these types of composites is given in Table 4-1.

Table 4-1 Short-fiber composites: examples

Type	Molding Method	Fiber Lengths (cm)	Fiber Orientation
Fiber-reinforced thermoplastics	Injection molding	<1.25	Random or dependent on flow in mold
Sheet-molding compound (polyester resin matrix)	Compression molding or sheet stamping	2.5–7.5	Random in compound but dependent on flow in mold
Bulk-molding compound	Compression molding	<2.5	Random in compound but dependent on flow in mold
Nonwoven-mat-reinforced thermoplastic	Sheet stamping	<7.5	Random
Nonwoven-mat-reinforced thermoset	Contact molding or laminating	<7.5	Random

The composites containing short fibers as the reinforcement are called *short-fiber composites*. They are also referred to as *discontinuous-fiber-reinforced composites*. Various aspects of this class of composites are discussed in this chapter.

4.2 THEORIES OF STRESS TRANSFER

In composites, loads are not applied directly on the fibers but are applied to the matrix material and transferred to the fibers through the fiber ends and also through the cylindrical surface of the fiber near the ends. When the length of a fiber is much greater than the length over which the transfer of stress takes place, the end effects can be neglected, and the fiber may be considered to be continuous. In the case of short-fiber composites, the end effects cannot be neglected, and the composite properties are a function of fiber length. The end effects significantly influence the behavior of and reinforcing effects in discontinuous fiber composites. For a good understanding of the behavior of discontinuous-fiber composites, it is necessary to first understand the mechanism of stress transfer.

4.2.1 Approximate Analysis of Stress Transfer

Early studies concerning variation of stresses along the length of a fiber were performed by Cox [1] and Outwater [2]. Probably the most often quoted theory of stress transfer is the shear-lag analysis applied by Rosen [3], who modified an earlier analysis of Dow [4]. The stress distribution along the length of a fiber can be understood in a simple manner by considering the equilibrium of a small element of fiber, as shown in Fig. 4-1. The force equilibrium of an infinitesimal length dz requires

$$(\pi r^2)\sigma_f + (2\pi r\, dz)\tau = (\pi r^2)(\sigma_f + d\sigma_f)$$

or

$$\frac{d\sigma_f}{dz} = \frac{2\tau}{r} \quad (4.1)$$

where σ_f is the fiber stress in the axial direction, τ is the shear stress on the cylindrical fiber–matrix interface, and r is the fiber radius. Equation (4.1) indicates that for a fiber of uniform radius, the rate of increase of fiber stress is proportional to the shear stress at the interface and can be integrated to obtain the fiber stress on a cross section a distance z away from the fiber end:

$$\sigma_f = \sigma_{f0} + \frac{2}{r}\int_0^z \tau\, dz \quad (4.2)$$

where σ_{f0} is the stress on the fiber end. In many analyses, σ_{f0} is neglected because of yielding of the matrix adjacent to the fiber end or separation of the fiber end from the matrix as a result of large stress concentrations. When σ_{f0} is negligible, Eq. (4.2) can be written as

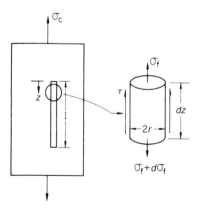

Figure 4-1. Equilibrium of a small length of fiber in a short-fiber composite.

$$\sigma_f = \frac{2}{r} \int_0^z \tau \, dz \tag{4.3}$$

The right-hand side of Eq. (4.3) can be evaluated if the variation of shear stress along the fiber length is known. In practice, the shear stress is not known beforehand and is determined as a part of the complete solution. To obtain analytical solutions such as those obtained in the works just cited [1–4], it is necessary to make assumptions regarding the deformation of material surrounding the fiber and the fiber-end conditions. For example, it may be assumed that the interface shear stress at the midfiber length and the normal stress at the fiber ends are zero. A frequently used approximate method of determining fiber stress is to assume that the matrix material surrounding the fibers is a rigid perfectly plastic material having a shear stress–strain diagram as shown in Fig. 4-2. For this case, the interface shear stress is constant along the fiber length and is equal to the matrix yield stress in shear τ_y. Equation (4.3) then yields

$$\sigma_f = \frac{2\tau_y z}{r} \tag{4.4}$$

For short fibers, the maximum fiber stress occurs at the midfiber length (i.e., $z = l/2$). Therefore,

$$(\sigma_f)_{max} = \frac{\tau_y l}{r} \tag{4.5}$$

where l is the fiber length. As the fiber length increases, the maximum fiber stress also increases. However, for a sufficiently long fiber, the maximum fiber stress is limited by the stress applied to the composite and the fiber volume fraction. The limiting stress value is the stress experienced by the fibers in a

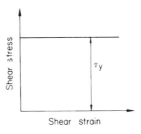

Figure 4-2. Shear-stress–shear-strain curve for an idealized rigid, perfectly plastic matrix material.

unidirectional composite, with infinitely long fibers but equal fiber volume fraction, and subjected to the same composite stress. Using the analysis carried out in Chap. 3 (assuming equal strains in fibers, matrix, and composite), it can be shown that the maximum fiber stress is limited to

$$(\sigma_f)_{max} = \frac{E_f}{E_c} \sigma_c \quad (4.6)$$

where σ_c is the applied composite stress, and the composite modulus E_c can be calculated from the rule of mixtures [Eq. (3.15)]. The minimum fiber length, in which the maximum fiber stress $(\sigma_f)_{max}$ can be achieved, may be defined as a load-transfer length l_t. It is over this length of the fiber that the load is transferred from matrix to fiber. It is given by

$$\frac{l_t}{d} = \frac{(\sigma_f)_{max}}{2\tau_y} = \frac{(E_f/E_c)\sigma_c}{2\tau_y} \quad (4.7)$$

where d ($=2r$) is the fiber diameter, and $(\sigma_f)_{max}$ is given by Eq. (4.6). Since $(\sigma_f)_{max}$ is a function of applied stress, the load-transfer length is also a function of applied stress. A critical fiber length l_c independent of applied stress may be defined as the minimum fiber length in which the maximum allowable fiber stress (or the fiber ultimate strength) σ_{fu} can be achieved. Thus

$$\frac{l_c}{d} = \frac{\sigma_{fu}}{2\tau_y} \quad (4.8)$$

It may be noted that the critical fiber length is the maximum value of load-transfer length. The critical fiber length is an important system property and affects ultimate composite properties.

Sometimes the load-transfer length and critical length are referred to as the *ineffective length* because over this length the fiber supports a stress less than the maximum fiber stress. Stress distribution (fiber stress and interface shear stress) in fibers of different lengths are shown in Fig. 4-3a for a given composite stress. Figure 4-3b shows the variations of fiber stress for increasing composite stress on a fiber longer than the critical length. It may be observed that a small length adjoining the fiber end is stressed at less than the maximum fiber stress. This affects the strength and elastic modulus of the composite, as is discussed in later sections. It may be pointed out here, however, that when the fiber length is much greater than the load-transfer length, the composite behavior approaches the behavior of continuous-fiber-reinforced composites.

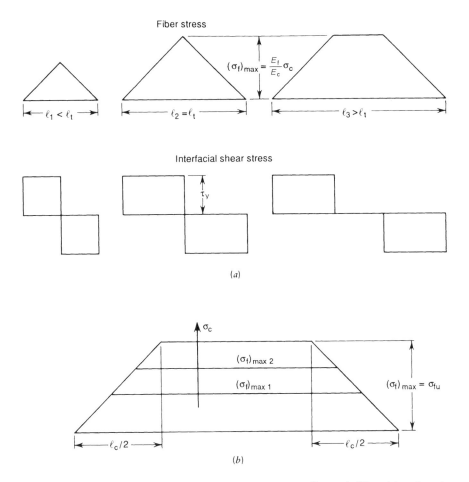

Figure 4-3. (a) Fiber stress and interfacial shear stress on fibers of different lengths when subjected to the same composite stress. (b) Influence of composite stress on variation of stress along fiber length when fiber is longer than its critical length.

4.2.2 Stress Distributions from Finite-Element Analysis

The stress distributions shown in Fig. 4-3 are approximate because they were obtained by assuming that the matrix material is perfectly plastic. In reality, most matrix materials are elastic–plastic in behavior. Accurate stress distributions thus can be obtained only by assuming the matrix to be elastic–plastic. This, however, presents many difficulties in performing a theoretical analysis of the composite. Numerical solutions probably are best for facile analysis of complex problems. In numerical methods, very few simplifying assumptions are required, and an accurate solution can be obtained easily. Finite-element analyses of aligned short-fiber composites have been carried out by many investigators to study the various aspects of these composites. In some of the analyses [5–8] the matrix material has been assumed to be completely elastic, whereas in others [9–14] it is assumed to be elastic–plastic. The analyses provide very useful information regarding stress distributions in the fibers as well as the matrix. Some of the stress distributions obtained by elastic analyses are shown in Figs. 4-4 and 4-5. Variation of fiber stress shows that there is significant stress transfer at the end because it is only an elastic analysis, and perfect adhesion has been assumed. Interfacial shear-stress distribution is consistent with Eq. (4.2) in that the interfacial shear-stress becomes zero when the fiber stress attains its maximum value. Variation of matrix stresses (axial and radial) is shown in Fig. 4-5. It may be noted that there is a stress concentration near the fiber end. It can be shown easily that the ratio of maximum fiber stress (Fig. 4-4) to the maximum matrix stress is equal to the ratio of their respective elastic moduli. Equivalently, the maximum fiber stress is in agreement with Eq. (4.6). It is interesting to note that the matrix radial stress has a compressive value for the case shown. This indicates that even if the interfacial bond between the fiber and the matrix is broken, load transfer still can occur because of friction forces between the two. However, if the fibers are normal to the load direction or the interfiber distances become quite small, the preceding assumption is not always true.

Stress distributions obtained in an elastic–plastic analysis [12] are shown in Fig. 4-6. The elastic–plastic analysis shows that the stress transfer through the fiber end is insignificant. In this case also, the maximum fiber stress was shown to be in agreement with Eq. (4.6). Interfacial shear stress near the fiber end is not a constant, although the matrix deforms plastically. This is so because a three-dimensional stress criterion was used to predict yielding of the matrix. However, the interfacial shear-stress distribution is in agreement with Eq. (4.2).

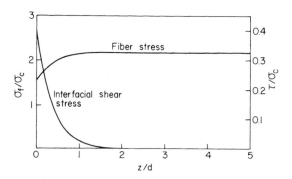

Figure 4-4. FEA results: elastic analysis. Fiber stresses (axial and interfacial shear) along fiber length [$(E_f/E_m) = 29.5$, $(l/d) = 10.4$, $V_f = 42\%$]. (From Broutman and Agarwal [8].)

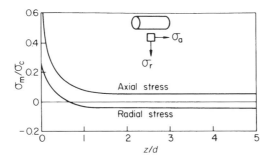

Figure 4-5. FEA results: elastic analysis. Matrix stresses (axial and radial) along fiber length [(E_f/E_m) = 29.5, (l/d) = 10.4, V_f = 42%]. (From Broutman and Agarwal [8].)

4.2.3 Average Fiber Stress

From the preceding discussion it is clear that the ends of finite-length fibers are stressed to less than the maximum fiber stress. The influence of fiber ends is to lower the elastic modulus and strength of short-fiber composites. In an analysis for elastic modulus and strength, average fiber stress is a very useful quantity. It can be evaluated as follows:

$$\bar{\sigma}_f = \frac{1}{l} \int_0^l \sigma_f \, dz \tag{4.9}$$

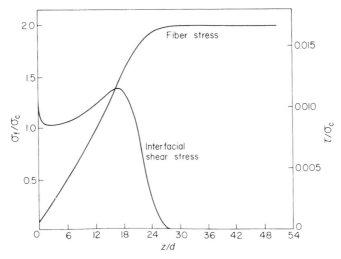

Figure 4-6. FEA results: elastoplastic analysis. Fiber stresses (axial and interfacial shear) along fiber length [(E_f/E_m) = 117, (l/d) = 100, V_f = 49.6%; matrix yield strain = 2.4%]. (From Agarwal and Bansal [13].)

where $\bar{\sigma}_f$ is the average fiber stress. The quantity represented by the integral is given by the area under the curve of fiber stress plotted against length. For the approximate stress distribution shown in Fig. 4-3, the average stress can be found to be

$$\bar{\sigma}_f = \frac{1}{2}(\sigma_f)_{max} = \frac{\tau_y l}{d} \qquad l < l_t$$

$$\bar{\sigma}_f = (\sigma_f)_{max}\left(1 - \frac{l_t}{2l}\right) \qquad l > l_t \qquad (4.10)$$

For more accurate stress analyses, the stress variation at the fiber end will be different from a linear one, although the fiber stress in the middle portion of the length $(l - l_t)$ will be constant at $(\sigma_f)_{max}$. Another difference can occur if the load-transfer length is not defined by Eq. (4.7). In such cases, the average fiber stress can be obtained by considering the actual stress distribution. However, there may be only a small difference between the average stresses obtained from actual and approximate (linear) stress distributions, particularly when fibers are longer than the critical length. The average: maximum fiber-stress ratios as predicted by Eq. (4.10) are given in Table 4-2 for different fiber lengths, where it can be observed that when the fiber length is 50 times the load-transfer length, the average fiber stress is 99% of the maximum fiber stress. Therefore, when the fiber length exceeds 50 times its critical length, the composite behavior approaches that of a continuous-fiber composite for an equivalent fiber orientation.

4.3 MODULUS AND STRENGTH OF SHORT-FIBER COMPOSITES

The stress distributions obtained through finite-element methods, as discussed in the preceding section, also have been used to predict the strength and modulus of short-fiber composites [8,11,13]. The results are available in the form of curves for specific values of system variables such as fiber aspect ratio (l/d), fiber volume fraction, and properties of the constituents. Whenever

Table 4-2 Average-maximum fiber-stress ratios

l/l_t	$\bar{\sigma}_f/(\sigma_f)_{max}$
1	0.50
2	0.75
5	0.90
10	0.95
50	0.99
100	0.995

a change in any of the system variables takes place, a new set of results has to be obtained. Thus the results have limited adaptability to design procedures. For design purposes, it is usually desirable to have simple and rapid computational procedures for estimating composite properties, even though the estimations are only approximate.

4.3.1 Prediction of Modulus

The Halpin–Tsai equations [15], which were discussed in Chap. 3 for predicting the transverse modulus of unidirectional composites, are also very useful in predicting the longitudinal and transverse moduli of aligned short-fiber composites (shown schematically in Fig. 4-7). The Halpin–Tsai equations for longitudinal and transverse moduli can be written as

$$\frac{E_L}{E_m} = \frac{1 + (2l/d)\eta_L V_f}{1 - \eta_L V_f} \qquad (4.11)$$

and

$$\frac{E_T}{E_m} = \frac{1 + 2\eta_T V_f}{1 - \eta_T V_f} \qquad (4.12)$$

where

$$\eta_L = \frac{(E_f/E_m) - 1}{(E_f/E_m) + 2(l/d)} \qquad (4.13)$$

and

$$\eta_T = \frac{(E_f/E_m) - 1}{(E_f/E_m) + 2} \qquad (4.14)$$

It may be pointed out that Eqs. (3.36), (4.11), and (4.12) are only particular cases of a general equation. The form of the general equation coincides with

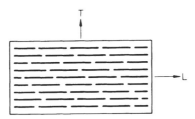

Figure 4-7. Model of an aligned short-fiber composite.

that of Eq. (3.36), in which ξ is a measure of reinforcement and is given by Eq. (3.38). Equations (4.11) and (4.12) can be obtained from Eq. (3.36) by substituting $\xi = 2l/d$ for the case of longitudinal modulus and $\xi = 2$ for the case of transverse modulus. The two values of ξ are consistent with Eq. (3.38). Further, the Halpin–Tsai equations predict that the transverse modulus of an aligned short-fiber composite is not influenced by the fiber aspect ratio (l/d) and that its value is the same as that for the transverse modulus of a continuous-fiber composite. Variations of transverse composite modulus are shown as a function of fiber volume fraction in Fig. 3-11 for different constituent modulus ratios. Variations of longitudinal modulus have been shown as functions of fiber aspect ratios in Figs. 4-8 and 4-9 for different fiber volume fractions for the cases where the modulus ratios are 20 and 100, which approximately represent glass–epoxy and graphite–epoxy systems, respectively.

It was pointed out earlier that randomly oriented short-fiber composites are produced to obtain composites that are essentially isotropic in a plane. Such composites are prepared by injection or compression molding or from nonwoven mats that often form the surface layers of laminated composites to prevent otherwise easy initiation of cracking resulting from poor transverse strength of layers of unidirectional composites. Analysis of such laminates, which have some layers of isotropic composites as well, requires knowledge of the elastic properties of isotropic layers. An effective method of predicting elastic properties of an isotropic layer made from randomly oriented short fibers is to assume them equal to the properties, averaged for angular dependence, of a unidirectional composite. The angular dependence of properties of a unidirectional lamina will be discussed in Chap. 5. A good estimate

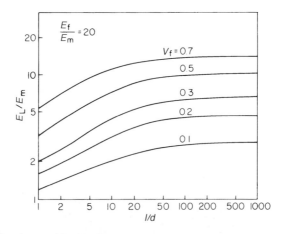

Figure 4-8. Dependence of longitudinal modulus of short-fiber composites on fiber aspect ratio (l/d) and fiber volume fraction ($E_f/E_m = 20$).

4.3 MODULUS AND STRENGTH OF SHORT-FIBER COMPOSITES 143

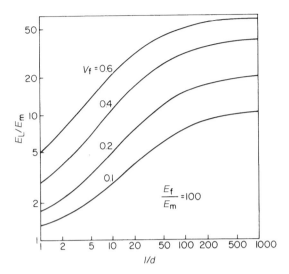

Figure 4-9. Dependence of longitudinal modulus of short-fiber composites on fiber aspect ratio (l/d) and fiber volume fraction ($E_f/E_m = 100$).

of the averaged properties also can be obtained through the analysis of a quasi-isotropic laminate made from unidirectional laminae. Construction of quasi-isotropic laminates and the analysis procedures will be discussed in Chap. 6.

The following empirical equations are often used to predict the elastic modulus and shear modulus of composites containing fibers that are randomly oriented in a plane:

$$E_{\text{random}} = \tfrac{3}{8}E_L + \tfrac{5}{8}E_T$$

$$G_{\text{random}} = \tfrac{1}{8}E_L + \tfrac{1}{4}E_T \tag{4.15}$$

where E_L and E_T are, respectively, the longitudinal and transverse moduli of an aligned short-fiber composite having the same fiber aspect ratio and fiber volume fraction as the composite under consideration. Moduli E_L and E_T either can be determined experimentally or can be calculated using Eqs. (4.11) and (4.12).

Example 4-1: A glass-fiber-reinforced nylon with a fiber volume fraction of 20% is injection-molded to produce a random fiber orientation. The fiber length is 3.2 mm, and the fiber diameter is 10 μm. Calculate the elastic modulus, shear modulus, and Poisson's ratio of the random fiber composite.

Solution: First calculate the modulus E_L assuming that the fibers were oriented. Using Eq. (4.11),

$$E_L = E_m \frac{1 + (2l/d)\eta_L V_f}{1 - \eta_L V_f}$$

For nylon: $E_m = 2.76$ GPa

Glass: $E_f = 72.4$ GPa

$$E_L = 2.76 \frac{1 + (6.4/10^{-2})\eta_L \times 0.2}{1 - 0.2\eta_L}$$

Using Eq. (4.13),

$$\eta_L = \frac{(E_f/E_m) - 1}{(E_f/E_m) + (2l/d)}$$

$$= \frac{(72.4/2.76) - 1}{(72.4/2.76) + (6.4/10^{-2})} = \frac{25.23}{666.23} = 0.03787$$

Thus

$$E_L = 2.76 \frac{1 + 128 \times 0.03787}{1 - 0.03787 \times 0.2}$$

$$= 16.26 \text{ GPa}$$

Now calculate the transverse modulus E_T using Eq. (4.12):

$$E_T = E_m \frac{1 + 2\eta_T V_f}{1 - \eta_T V_f}$$

where

$$\eta_T = \frac{(E_f/E_m) - 1}{(E_f/E_m) + 2} = \frac{26.2 - 1}{26.2 + 2} = 0.89$$

$$E_T = 2.76 \frac{1 + 1.78 \times 0.2}{1 - 0.2 \times 0.89} = 4.53 \text{ GPa}$$

Now the elastic modulus and shear modulus of the random fiber composite can be calculated using Eq. (4.15):

4.3 MODULUS AND STRENGTH OF SHORT-FIBER COMPOSITES

$$E_R = \tfrac{3}{8} \times 16.26 + \tfrac{5}{8} \times 4.53 = 8.93 \text{ GPa}$$

$$G_R = \tfrac{1}{8} \times 16.26 + \tfrac{1}{4} \times 4.53 = 3.17 \text{ GPa}$$

Since a random fiber composite is considered isotropic in its plane, its in-plane Poisson's ratio can be calculated using the following isotropic relationship between E_R, G_R, and Poisson's ratio ν_R:

$$G_R = \frac{E_R}{2(1 + \nu_R)}$$

or

$$\nu_R = \frac{E_R}{2G_R} - 1$$

$$\nu_R = \frac{8.93}{2 \times 3.17} - 1 = 0.41$$

4.3.2 Prediction of Strength

The average longitudinal stress on an aligned short-fiber composite can be calculated by the rule of mixtures:

$$\sigma_c = \overline{\sigma}_f V_f + \sigma_m V_m \tag{4.16}$$

where $\overline{\sigma}_f$ is the average fiber stress given by Eq. (4.9). For a linear stress distribution at the fiber ends as shown in Fig. 4-3, values of $\overline{\sigma}_f$ are given by Eq. (4.10). Thus the average composite stress can be written as

$$\sigma_c = \tfrac{1}{2}(\sigma_f)_{max} V_f + \sigma_m V_m, \qquad l < l_t \tag{4.17}$$

and

$$\sigma_c = (\sigma_f)_{max}\left(1 - \frac{l_t}{2l}\right)V_f + \sigma_m V_m \qquad l > l_t \tag{4.18}$$

If the fiber length is much greater than the load-transfer length (e.g., $l = 100 l_t$), the factor $1 - (l_t/l)$ approaches 1, and Eq. (4.18) can be written as

$$\sigma_c = (\sigma_f)_{max} V_f + \sigma_m V_m \qquad l \gg l_t \tag{4.19}$$

Depending on the fiber length, Eq. (4.17), Eq. (4.18), or Eq. (4.19) can be used for predicting the ultimate strength of the composite. When fibers are smaller than the critical length, the maximum fiber stress is less than the average fiber strength so that the fibers will not fracture regardless of the magnitude of the applied stress. In this case the composite failure occurs when the matrix or interface fails, and the composite ultimate strength σ_{cu} is approximated by

$$\sigma_{cu} = \frac{\tau_y l}{d} V_f + \sigma_{mu} V_m \qquad l < l_c \qquad (4.20)$$

When the fiber length is greater than the critical length, the fibers can be stressed to their average strength. In this case it can be assumed that the fiber failure initiates when the maximum fiber stress is equal to the ultimate strength of the fibers. Thus

$$\sigma_{cu} = \sigma_{fu}\left(1 - \frac{l_c}{2l}\right) V_f + (\sigma_m)_{\epsilon_f^*} V_m \qquad l > l_c \qquad (4.21)$$

and

$$\sigma_{cu} = \sigma_{fu} V_f + (\sigma_m)_{\epsilon_f^*} V_m \qquad l \gg l_c \qquad (4.22)$$

where $(\sigma_m)_{\epsilon_f^*}$ is the matrix stress at the fiber fracture strain ϵ_f^*. A useful approximation in these equations can be made by using the value of matrix ultimate strength σ_{mu} in place of $(\sigma_m)_{\epsilon_f^*}$. In writing Eqs. (4.21) and (4.22), it has been assumed that the fiber volume fraction is above a certain minimum V_{min} so that the matrix will not be able to support the full load when all the fibers break, and composite failure then will take place. The minimum and also critical volume fractions of fibers, which were defined for the case of continuous-fiber-reinforced composites in Chap. 3, can be defined in an analogous manner in this case also. It is left to the reader to define them and to obtain the following expressions for V_{min} and V_{crit}:

$$V_{min} = \frac{\sigma_{mu} - (\sigma_m)_{\epsilon_f^*}}{\sigma_f + \sigma_{mu} - (\sigma_m)_{\epsilon_f^*}} \qquad (4.23)$$

$$V_{crit} = \frac{\sigma_{mu} - (\sigma_m)_{\epsilon_f^*}}{\sigma_f - (\sigma_m)_{\epsilon_f^*}} \qquad (4.24)$$

Comparisons of Eq. (4.23) with Eq. (3.25) and Eq. (4.24) with Eq. (3.27) show that, for identical properties of fibers and matrix material, short-fiber composites require higher values of V_{min} and V_{crit} than do the continuous-fiber-reinforced composites. The reason is obvious in that the short fibers are not fully effective. However, as the fiber length becomes very large compared with the transfer length, the average fiber stress approaches the maximum fiber stress, and the behavior of short-fiber composites approaches that of continuous-fiber composites.

If the fiber volume fraction is less than V_{min}, the composite will not fracture when all fibers break because the remaining cross-sectional area of matrix material can support the full load. The composite fracture will occur when the matrix fails. Thus the ultimate strength of composite with $V_f < V_{min}$ is given by

$$\sigma_{cu} = \sigma_{mu}(1 - V_f) \qquad V_f < V_{min} \qquad (4.25)$$

In the case of discontinuous-fiber composites, an additional factor influences the failure, namely, the large stress concentrations in the matrix produced as a result of the fiber ends. The effect of the stress concentration is to further lower the composite strength.

The problem of off-axis strength of an aligned short-fiber composite takes on added complexity because the failure-mode changes as the angle between the stress and fiber direction increases from 0° to 90°. In the case of continuous fibers, at intermediate angles, matrix shear failure occurs at the interface, and at large angles near 90°, the matrix or interface fails in plain strain. A discussion on the angular dependence of strength is presented in Chap. 5.

It was pointed out earlier that randomly oriented short-fiber composites are of particular significance because they are quasi-isotropic; that is, they have the same properties in all directions. The strength, like modulus, of a randomly oriented short-fiber composite may be obtained by assuming it equal to the strength, averaged for angular dependence, of a unidirectional composite. However, an approach used often to predict the strength of random short-fiber composites utilizes a laminate-analysis procedure. In this approach, the strength of an isotropic laminate constructed from unidirectional plies is used to approximate the strength of random-fiber composites. In practice, the strength of a [0/±45/90] symmetric laminate is close to that of an isotropic laminate made up of many more orientations [16]. Construction of isotropic laminates, along with the analysis of laminates, and the stepwise procedure to calculate ultimate strength of laminates are discussed in Chap. 6. Kardos [16] compared experimental results obtained by Lavengood [17] with the predictions of isotropic laminate analogy calculations as shown in Fig. 4-10. For the purpose of calculating ultimate strength of the laminate, a modified maximum strain theory of failure was applied. Also shown in Fig. 4-10 are the predictions of Chen [18] and Lees [19]. It is clear that the maximum strain criterion in conjunction with the laminate model comes by far the closest to predicting the strength of random-orientation systems at volume fractions of engineering interest. The fact that the predicted strengths are somewhat below experimental findings may be attributed to the interaction

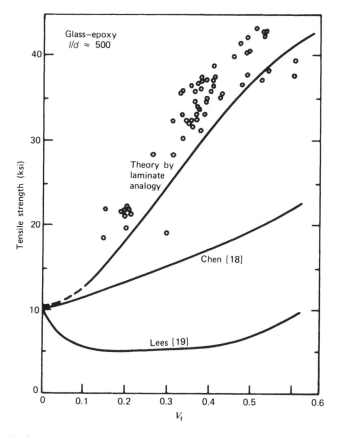

Figure 4-10. Comparison of experimentally measured and theoretically predicted tensile strengths of short-fiber composites. (From Kardos [16].)

between failure modes in the laminate or lack of information on how the allowable ply strains vary with volume fraction of fibers and fabrication conditions.

It would be wise to point out here that in molded short-fiber composites, the fiber orientation throughout the molding varies greatly according to the flow within the mold (Fig. 4-11). Thus the molded-part properties will vary from section to section according to the local fiber orientation. The strength of the molding then will be dependent on both the local stress state and local fiber orientation. Methods to measure fiber-orientation distributions and to account for their effect on mechanical properties are discussed in the literature [20–22]. A recent method of measuring spatial orientation of short-fiber-reinforced thermoplastics is based on image analysis [23].

4.3 MODULUS AND STRENGTH OF SHORT-FIBER COMPOSITES

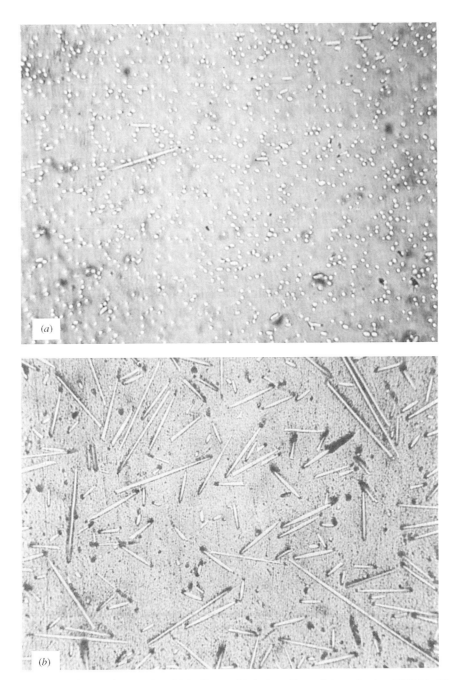

Figure 4-11. Photomicrographs of injection-molded glass-fiber-reinforced nylon (100×) in (a) an area where fibers are aligned due to flow and (b) an area where fibers are randomly oriented.

4.3.3 Effect of Matrix Ductility

The effect of matrix ductility and interface treatment on the properties of short-fiber composites have been investigated by Gaggar and Broutman [24]. Studies were conducted on glass-fiber-mat-reinforced epoxy resins. Ductility of the matrix was varied by mixing a brittle epoxy resin (DER 334) with a flexible epoxy resin (DER 736) in different proportions. The properties of three blends of matrix material are given in Table 4-3. Tensile stress–strain curves for the composites with the three different matrix materials are shown in Fig. 4-12. The ductile resin (designated as blend C) composite shows lower strength and modulus but a slightly larger strain to failure compared with the brittle resin composites. However, the difference between failure strains of the three composites (and also their strengths and moduli) is very insignificant compared with the difference in the corresponding properties of the respective matrix materials. The reason for the low elongation to failure of the ductile-matrix composite is that the matrix is confined by the fibers and cannot deform. In a composite, the matrix is subjected to a triaxial state of stress, even when a uniaxial load is applied to the composite. The effect of triaxial tension causes the ductile matrix to fail at a very low strain. Thus the elongation to failure drops drastically when fibers are added to the matrix. Composites with ductile and brittle resins are observed to fail by the same mechanism. The tensile strength is shown as a function of fiber volume fraction in Fig. 4-13. The maximum tensile strength for the brittle-matrix composite occurs at a fiber volume fraction of approximately 50%, beyond which there is a slight decrease in strength because of fabrication difficulties in obtaining a good-quality composite at these high volume fractions of fibers. The improved strength of the brittle-matrix composite results from the higher matrix strength, which has an observable influence in the case of randomly oriented fiber composites.

Table 4-3 Properties of different blends of matrix material

Material	Tensile Yield Strength, psi[a] (MPa)	Elastic Modulus, psi, (GPa)	Strain to Failure (%)
70% DER 334 30% DER 736 (material A)	10,500 (72.3)	4.75×10^5 (3.27)	4.2
50% DER 334 50% DER 736 (material B)	8,100 (55.8)	4.1×10^5 (2.82)	5.4
30% DER 334 70% DER 736 (material C)	3,300 (22.7)	2.8×10^5 (1.93)	>80

[a]Pounds per square inch.

4.3 MODULUS AND STRENGTH OF SHORT-FIBER COMPOSITES

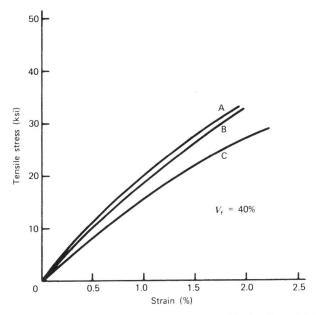

Figure 4-12. Tensile stress–strain curves for composites with ductile and brittle matrices. (From Gaggar and Broutman [24].)

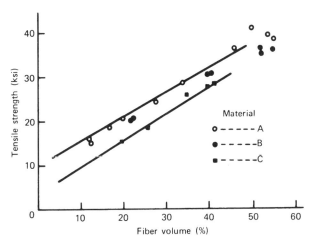

Figure 4-13. Tensile strength of composites with ductile and brittle matrices. (From Gaggar and Broutman [24].)

152 SHORT-FIBER COMPOSITES

Izod impact strength of the composites made with the three blends of matrix materials has been plotted in Fig. 4-14. It is observed that the matrix ductility has little influence on the notched impact strength of such composites. This behavior can be explained by considering how the presence of fibers influences the behavior of brittle and ductile matrices. In the case of ductile matrices, fibers limit the elongation of the matrix between them, and thus addition of rigid fibers greatly reduces the toughness. On the other hand, addition of fibers to a brittle matrix can increase toughness because of crack blunting, branching, and arrest effects. A more detailed discussion on the crack-propagation mechanics in fiber composites is given in Chap. 8 in the section on fracture mechanics, and different energy-absorbing mechanism are described in Chap. 9 in the section on impact.

4.4 RIBBON-REINFORCED COMPOSITES

A *ribbon* can be defined as a filament having a rectangular cross section in which the width is much greater than the thickness. Ribbon (or tape)–reinforced composites may have several possible advantages over the composites containing fibers with a circular cross section. Ribbon-reinforced composites can have high strength and stiffness in two directions: the longitudinal direction and the in-plane transverse direction (i.e., the direction perpendicular to the length of the ribbon in the plane of the sheet). Thus such composites can be nearly isotropic in the plane of a sheet exhibiting nearly equal strength in all directions. This is a big advantage over the aligned-fiber composites, which have poor transverse strength, making them prone to easy

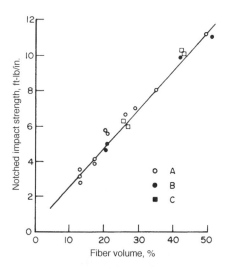

Figure 4-14. Notched Izod impact strength of composites with ductile and brittle matrices. (From Gaggar and Broutman [24].)

crack initiation and hence requiring great care in their handling. Ribbon composites tend to be very resistant to puncture by sharp objects. They exhibit greatly decreased permeability to gases and liquids compared with polymers or other kinds of composites. To permeate through a ribbon composite, a diffusing molecule must follow a long, circuitous path around the impermeable ribbons. Another advantage that the ribbons offer as the reinforcement is that they can be packed in larger volume fractions than can circular fibers. For a typical ribbon-reinforced composite, whose cross section is shown in Fig. 4-15, the volume fraction of ribbons V_r is given by

$$V_r = \frac{W_r t_r}{(W_r + W_m)(t_r + t_m)} = \frac{1}{[1 + (W_m/W_r)][1 + (t_m/t_r)]} \quad (4.26)$$

where W_r and t_r are, respectively, the width and thickness of the ribbons, t_m is the spacing between two layers of the ribbons, and W_m is the spacing between two ribbons in a layer. Equation (4.26) can be written in terms of the amount of overlap B of the ribbons as

$$V_r = \frac{1}{2[1 - (B/W_r)][1 + (t_m/t_r)]} \quad (4.27)$$

It is quite clear from Eq. (4.26) that V_r can be made quite large by reducing spacings between ribbons and the layers of ribbons. Theoretically, the lower limit of W_m or t_m is zero. This, however, puts more stringent requirements on the properties of the matrix for ribbon composites than for fibrous composites. These requirements are discussed in a later paragraph.

The longitudinal modulus of ribbon composites, like that of the continuous-fiber composites, is also given by the rule of mixtures as

$$E_L = E_r V_r + E_m V_m \quad (4.28)$$

where E_r is the elastic modulus of ribbons. The in-plane transverse behavior of ribbon composites is analogous to the longitudinal behavior of aligned

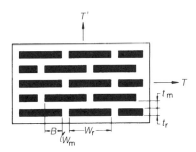

Figure 4-15. Representation of a cross section of ribbon-reinforced composite.

short-fiber composites. Therefore, the in-plane transverse modulus is given by the Halpin–Tsai equation as

$$\frac{E_T}{E_m} = \frac{1 + (2W_r/t_r)\eta_r V_r}{1 - \eta_r V_r} \qquad (4.29)$$

where

$$\eta_r = \frac{(E_r/E_m) - 1}{(E_r/E_m) + 2(W_r/t_r)} \qquad (4.30)$$

However, when the aspect ratio of ribbons (i.e., the ratio of width to thickness of the ribbons W_r/t_r) is large, the in-plane transverse modulus also will be given by the rule of mixtures [Eq. (4.28)]. In that case, the ribbon composites become essentially isotropic in the plane of the sheet. This is in contrast with the aligned-fiber composites, which are transversely isotropic but exhibit a high degree of orthotropy in the plane of the composite. It may be pointed out that the transverse modulus in the thickness direction of ribbon composites may be calculated by replacing W_r/t_r by its reciprocal in Eqs. (4.29) and (4.30) and will have a value much less than that of the longitudinal or in-plane transverse modulus of the composite.

The in-plane transverse tensile strength of a ribbon composite can be nearly as large as the longitudinal strength, provided that the failure occurs by longitudinal splitting of the ribbons and not by interlaminar shear failure. This will require that the ribbon aspect ratio W_r/t_r be quite large to provide enough overlap between the ribbons. An essential condition for the longitudinal splitting of ribbons is that the shear force required for shear failure of the matrix in the region of overlapping ribbons be greater than the force required for tensile splitting. With reference to Fig. 4-15, this condition requires that

$$2B\tau_{mu} \geq t_r \sigma_{ru} \qquad (4.31)$$

where τ_{mu} is the shear strength of the matrix, and σ_{ru} is the ultimate tensile strength of the ribbons.

To attain high ultimate strength of ribbon-reinforced composites, good adhesion between the matrix and the reinforcement is essential. If the interfacial bonding is weak, the strength decreases rapidly with ribbon concentration. However, even with good adhesion, the experimental values may be low because the matrix does not have the required properties. The matrix must be ductile with a high ultimate elongation to minimize the effect of thermal stresses induced by the fabrication process. In addition, there must be enough elongation left over to fully transmit the stresses to the ribbons. The ribbon-reinforced composites also require better control during fabrication processes to maintain a regular arrangement of ribbons to ensure that all areas of overlap are above a critical value. There must be few, if any, voids or areas of poor adhesion.

EXERCISE PROBLEMS

4.1. A composite is fabricated of glass fibers (diameter = 0.03 mm) in an epoxy-resin matrix. All the fibers are aligned parallel to the direction of load application. The volume fraction of fibers is 40%. Assume that the matrix behaves as a rigid–plastic material with a yield strength of 28 MPa and that E_f = 70 GPa and E_m = 3.5 GPa.
   (a) Determine the load-transfer length l_t for composite stresses σ_c = 70 MPa and σ_c = 210 MPa.
   (b) Plot fiber stress and interfacial shear stress along fiber length for fibers with length $l = \frac{1}{2}l_t$, l_t, and $4l_t$. Also calculate average fiber stress in each case.
   (c) Show plots of fiber strain for $l = l_t$ and $l = \frac{1}{2}l_t$.

4.2. Assuming that the ultimate strength of the fibers considered in Exercise Problem 4.1 is 1.4 GPa, calculate critical fiber length l_c. Construct a plot of composite strength versus fiber length for fiber lengths between $0.1 l_c$ and $100 l_c$. Assume matrix strength to be 28 MPa.

4.3. In Sec. 4.2.1 the interfacial shear stress was assumed constant ($= \tau_y$). A better approximation may be to assume the interfacial shear stress varying linearly as $\tau = \tau_y - az$, where a is constant and z is distance from fiber end (see Fig. 4-1). With this assumption, derive an expression for the load-transfer length in terms of the maximum fiber stress, fiber diameter, and τ_y. Compare your result with Eq. (4.7), and explain the difference. Also plot the fiber stress and interfacial shear stress along the fiber length when $l > l_t$.

4.4. The material used in the transmission gears of an automobile is an injection-molded nylon 6,6 containing 20% by weight of randomly oriented chopped-glass fibers. Calculate the length of glass fibers if the material shows a tensile modulus of 7 GPa, given that E_f = 70 GPa, E_m = 3.5 GPa, ρ_f = 2.5 g/cm^3, ρ_m = 1.4 g/cm^3, and d_f = 20 µm.

4.5. In a more demanding application, the modulus of the material is increased to twice the value indicated in Exercise Problem 4.4 by replacing the glass fibers with carbon fibers. Calculate the length of carbon fibers required if the matrix material and fiber weight fraction are the same as in Exercise Problem 4.4. For carbon fibers, take E_f = 210 GPa, ρ_f = 1.8 g/cm^3, and d_f = 15 µm.

4.6. Derive Eqs. (4.23) and (4.24) by following the procedure adopted in Chap. 3 for obtaining expressions for V_{min} and V_{crit} in the case of a continuous-fiber composite.

4.7. Obtain an expression for the ratio of the strengths of an aligned discontinuous-fiber composite and a continuous-fiber composite with equal fiber volume fraction. For a limiting case of $V_f = 1.0$, plot this ratio as a function of l/l_c, and indicate the region where values of the ratio will lie if $V_f < 1.0$.

4.8. In a ribbon-reinforced composite, the spacing between two consecutive layers of ribbons is 10% of the ribbon thickness, whereas the spacing between two ribbons in a layer is 50% of the ribbon thickness. Construct a plot of ribbon volume fraction–ribbon aspect ratio that varies from 1 to 1000.

4.9. Assume that the diameter of long fibers (circular in cross section) is the same as the thickness of ribbons in Exercise Problem 4.8 and that the fiber spacing is the same as the spacing between two layers of ribbons. Calculate the volume fractions of fibers when they are packed in regular square and hexagonal arrays. Compare results with those of Exercise Problem 4.8.

REFERENCES

1. H. L. Cox, "The Elasticity and Strength of Paper and other Fibrous Materials," *Br. J. Appl. Phys.,* **3,** 72–79 (1952).
2. J. O. Outwater, "New Predictions and Interpretations," *Mod. Plast.,* 56 (March 1956).
3. B. W. Rosen, "Mechanics of Composite Strengthening," in *Fiber Composite Materials,* American Society for Metals, Metals Park, OH, 1964, Chap. 3.
4. N. F. Dow, "Study of Stresses near a Discontinuity in a Filament-Reinforced Composite Material," General Electric Company Report No. TISR63SD61, Barrington, IL, August 1963.
5. A. S. Carrara and F. J. McGarry, "Matrix and Interface Stresses in a Discontinuous Fiber Composite Model," *J. Compos. Mater.,* **2**(2), 222–243 (1968).
6. T. F. Maclaughlin and R. M. Barker, "Effect of Modulus Ratio on Stress near a Discontinuous Fiber," *Exp. Mech.,* **12**(4), 178–183 (1972).
7. D. R. Owen and J. F. Lyness, "Investigations of Bond Failure in Fibre Reinforced Materials by the Finite Element Methods," *Fibre Sci. Technol.,* **5,** 129–141 (1972).
8. L. J. Broutman and B. D. Agarwal, "A Theoretical Study of the Effect of an Interfacial Layer on the Properties of Composites," *Polym. Eng. Sci.,* **14**(8), 581–588 (1974).
9. T. H. Lin, D. Salinas, and Y. M. Ito, "Elastic–Plastic Analysis of Unidirectional Composites," *J. Compos. Mater.,* **6**(1), 48–60 (1972).

10. B. D. Agarwal, "Micromechanics Analysis of Composite Materials Using Finite Element Methods," Ph.D. thesis, Illinois Institute of Technology, Chicago, IL, May 1972.
11. B. D. Agarwal, J. M. Lifshitz, and L. J. Broutman, "Elastic–Plastic Finite Element Analysis of Short Fibre Composites," *Fibre Sci. Technol.*, **7**(1), 45–62 (1974).
12. R. K. Bansal, "Finite Element Analysis of Fibre Interaction in Discontinuous Fibre Reinforced Composite," M.tech. thesis, Indian Institute of Technology, Kanpur (India), December 1976.
13. B. D. Agarwal and R. K. Bansal, "Plastic Analysis of Fibre Interactions in Discontinuous Fibre Composites," *Fibre Sci. Technol.*, **10**(4), 281–297 (1977).
14. B. D. Agarwal and R. K. Bansal, "Effect of an Interfacial Layer on the Properties of Fibrous Composites: A Theoretical Analysis," *Fibre Sci. Technol.*, **12**(2), 149–158 (1979).
15. J. C. Halpin and S. W. Tsai, "Effects of Environmental Factors on Composite Materials," AFML-TR 67-423, Dayton, OH, June 1969.
16. J. L. Kardos, "Structure–Property Relations for Short-Fiber Reinforced Plastics," Divisional Technical Meeting, Engineering Properties and Structure Division, SPE, Akron, OH, October 7–8, 1975.
17. R. E. Lavengood, "Strength of Short-fiber Reinforced Composites," *Polymer. Eng. Sci.*, **12**(1), 48–52 (1972).
18. M. Lin, P. E. Chen, A. T. Dibenedetto, "Transverse Properties of Unidirectional Aluminum Matrix Fibrous Composites," *Polymer. Eng. Sci.*, **11**(4), 344–352 (1971).
19. J. K. Lees, "A Study of the Tensile Modulus of Short Fiber Reinforced Plastics," *Polymer. Eng. Sci.*, **8**(3), 186–194 (1968).
20. M. W. Darlington and P. L. McGinley, "Fiber Orientation Distribution in Short Fibre Reinforced Plastics," *J. Mater. Sci.*, **10**, 906–910 (1975).
21. L. Kain, O. Ishai, and M. Narkis, "Oriented Short Glass-Fiber Composites: IV. Dependence of Mechanical Properties on the Distribution of Fiber Orientations," *Polym. Eng. Sci.*, **1**(18), 45–52 (1978).
22. P. T. Curtis, M. G. Bader, and J. E. Bailey, "The Stiffness and Strength of a Polyamide Thermoplastic Reinforced with Glass and Carbon Fibres," *J. Mater. Sci.*, **13**, 377–390 (1978).
23. G. Fischer and P. Iyerer, "Measuring Spatial Orientation of Short Fiber Reinforced Thermoplastics by Image Analysis," *Polym. Composite*, **9**(4), 297–304 (1988).
24. S. K. Gaggar and L. J. Broutman, "Effect of Matrix Ductility and Interface Treatment on Mechanical Properties Glass Fiber Mat Composites," *Polym. Eng. Sci.*, **16**(8), 537–543 (1976).

5

ANALYSIS OF AN ORTHOTROPIC LAMINA

5.1 INTRODUCTION

A single layer of a laminated composite material generally is referred to as a *ply* or *lamina*. It usually contains a single layer of reinforcement, unidirectional or multidirectional. A single lamina generally is too thin to be used directly in any engineering application. Several laminae are bonded together to form a structure termed a *laminate*. Properties and orientation of the laminae in a laminate are chosen to meet the laminate design requirements. Properties of a laminate may be predicted by knowing the properties of its constituent laminae. Behavior of the laminate is governed by the behavior of individual laminae. Thus analysis or design of a laminate requires a complete knowledge of the behavior of the laminae. The analysis of a lamina is discussed in this chapter, and analysis of a laminate, in Chap. 6.

Microscopically, fiber composites are heterogeneous materials. Their properties and behavior are controlled by their microstructure and the properties of their constituents. Properties and behavior of a unidirectional lamina, subjected to loads in the fiber direction and perpendicular to it, have been studied in Chap. 3, and those of short-fiber composite, in Chap. 4. It is observed that their elastic properties and strengths are quite different in these two directions. It is expected that their properties in other directions (off-axis properties) are different from those in the longitudinal and transverse directions but related to them. The off-axis properties may be predicted by carrying out macroscopic analysis of a lamina using the principles of mechanics and assuming a lamina to be macroscopically homogeneous.

5.1.1 Orthotropic Materials

From the mechanics standpoint, fiber composites are among the class of materials called *orthotropic materials,* whose behavior lies between that of isotropic and that of anisotropic materials. Differences between these materials can be best explained through their responses to tensile and shear loads. Consider rectangular specimens made of isotropic, anisotropic, and orthotropic materials. A uniaxial tensile load on the specimen of isotropic material will produce an elongation in the load direction and a shortening in the perpendicular direction, as shown in Fig. 5-1a. However, there will be no change in the angles between two adjacent sides. A pure shear load will produce distortion of the specimen through changes in angles between its adjacent sides but will cause no change in lengths. Further, when the direction of applied load is changed, the material response does not change qualitatively or quantitatively. That is, equal loads applied in different directions produce equal changes in lengths and angles. Thus the deformation behavior of iso-

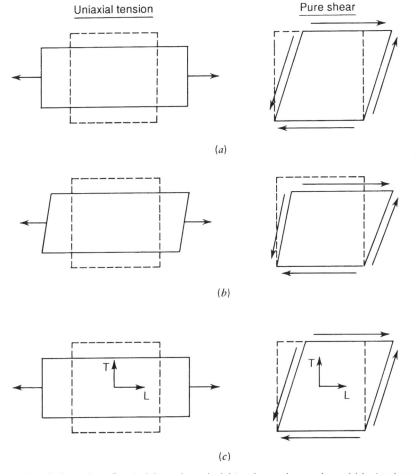

Figure 5-1. Deformation of materials under uniaxial tension and pure shear: (a) isotropic material, (b) anisotropic and generally orthotropic material, and (c) specially orthotropic material. Undeformed plate is shown by broken lines

160 ANALYSIS OF AN ORTHOTROPIC LAMINA

tropic materials is direction-independent and is characterized by "normal stresses produce normal strains only but no shear strain" and "shear stresses produce shear strains only but no normal strains."

In the case of anisotropic materials, uniaxial tension will produce changes in lengths as well as in angles (see Fig. 5-1b). Similarly, pure shear also will produce changes in lengths as well as angles. Further, when the direction of applied load is changed, the material response does not change qualitatively but changes quantitatively. That is, equal loads applied in different directions produce unequal changes in lengths and angles. In other words, the deformation behavior of anisotropic material is direction-dependent.

Deformation response of an orthotropic material, in general, is similar to that of the anisotropic material. That is, it is direction-dependent, and normal stresses and shear stresses alike give rise to normal strains as well as shear strains. However, in special cases, when the loads are applied in some specific directions, the material response is similar to that of isotropic materials in that the normal stresses produce normal strains only and shear stresses produce shear strains only (see Fig. 5-1c). These directions with special behavior are the axes of symmetry of the material. In a unidirectional composite, longitudinal and transverse directions are the axes of symmetry. The existence of the axes of symmetry and their number are governed by the microstructure of the material. A general three-dimensional orthotropic material has three mutually perpendicular axes of symmetry. A unidirectional composite is an orthotropic material but has more than three axes of symmetry—the longitudinal direction and all directions perpendicular to it. Because of these additional axes of symmetry, the unidirectional composites are transversely isotropic.

Analysis procedures for orthotropic materials are developed in this chapter. Developments are limited to small deformations only so that their deformation behavior can be considered linearly elastic. Stress–strain relations for a two-dimensional orthotropic lamina (e.g., a unidirectional composite) in terms of engineering constants are explained in the next section. Stress–strain relations in terms of stiffness and compliance matrices are derived from the generalized Hooke's law in another section. Procedures for transformation of stiffness and compliance matrices are also developed. These developments are essential for the laminate analysis procedures discussed in the next chapter. The last section deals with the strengths of orthotropic laminae.

5.2 STRESS–STRAIN RELATIONS AND ENGINEERING CONSTANTS

Consider a two-dimensional orthotropic lamina (e.g., a unidirectional composite), as shown in Fig. 5-2, with the principal material axes (axes of symmetry) as the longitudinal direction and the transverse direction. Four independent engineering constants are required to relate stresses and strains in this lamina. These constants are the elastic moduli in the longitudinal and transverse directions E_L and E_T, respectively, the shear modulus or the mod-

5.2 STRESS–STRAIN RELATIONS AND ENGINEERING CONSTANTS

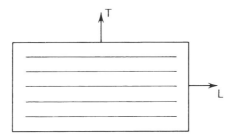

Figure 5-2. Specially orthotropic lamina.

ulus of rigidity associated with the axes of symmetry G_{LT}, and the major Poisson's ratio ν_{LT}, giving transverse strain caused by a longitudinal stress. Minor Poisson's ratio ν_{TL}, which gives longitudinal strain caused by a transverse stress, is also used frequently in the stress–strain relations, but it is related to other independent constants [Eq. (3.60)]. Procedures for predicting these constants for unidirectional composites were discussed in Chap. 3. Analysis presented in this chapter assumes a lamina to be macroscopically homogeneous and uniform in properties.

When reference axes in an analysis coincide with the lamina axes of symmetry, as in Fig. 5-2, the lamina usually is called a *specially orthotropic lamina*. When the reference axes are different from the lamina axes of symmetry, the lamina is called a *generally orthotropic lamina*. Stress–strain relations for specially and generally orthotropic lamina are discussed in this section.

5.2.1 Stress–Strain Relations for Specially Orthotropic Lamina

Deformation behavior of a specially orthotropic lamina is shown in Fig. 5-3. Lamina strains for each stress component can be written as follows:

1. When σ_L is the only nonzero stress ($\sigma_T = \tau_{LT} = 0$), the strains produced are

$$\epsilon_L = \frac{\sigma_L}{E_L} \tag{5.1}$$

$$\epsilon_T = -\nu_{LT}\epsilon_L = -\nu_{LT}\frac{\sigma_L}{E_L} \tag{5.2}$$

$$\gamma_{LT} = 0 \tag{5.3}$$

2. When σ_T is the only nonzero stress ($\sigma_L = \tau_{LT} = 0$), the strains produced are

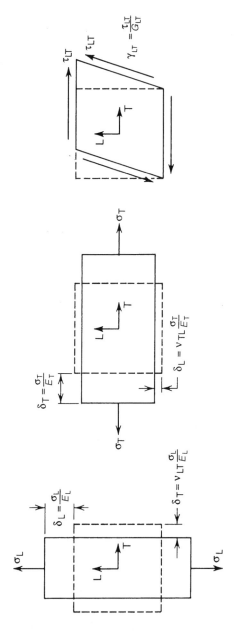

Figure 5-3. Deformation behavior of a specially orthotropic lamina. Undeformed lamina, shown in broken lines, is a square with sides of unit length.

5.2 STRESS–STRAIN RELATIONS AND ENGINEERING CONSTANTS

$$\epsilon_T = \frac{\sigma_T}{E_T} \tag{5.4}$$

$$\epsilon_L = -\nu_{TL}\epsilon_T = -\nu_{TL}\frac{\sigma_T}{E_T} \tag{5.5}$$

$$\gamma_{LT} = 0 \tag{5.6}$$

3. When τ_{LT} is the only nonzero stress ($\sigma_L = \sigma_T = 0$), the strains produced are

$$\epsilon_L = 0 \tag{5.7}$$

$$\epsilon_T = 0 \tag{5.8}$$

$$\gamma_{LT} = \frac{\tau_{LT}}{G_{LT}} \tag{5.9}$$

Superposition of these three states of stresses gives a most general state of stress on the lamina consisting of σ_L, σ_T, and τ_{LT}. In view of the assumption of linearly elastic material, the strains given by Eqs. (5.1)–(5.9) can be superposed to give the following relations:

$$\begin{aligned} \epsilon_L &= \frac{\sigma_L}{E_L} - \nu_{TL}\frac{\sigma_T}{E_T} \\ \epsilon_T &= \frac{\sigma_T}{E_T} - \nu_{LT}\frac{\sigma_L}{E_L} \\ \gamma_{LT} &= \frac{\tau_{LT}}{G_{LT}} \end{aligned} \tag{5.10}$$

Equations (5.10) are the stress–strain relations for a specially orthotropic lamina in terms of engineering constants. It may be noted that Eqs. (5.10) are similar to the stress–strain relations of an isotropic material under plane stress conditions. However, Eqs. (5.10) involve four independent elastic constants, whereas isotropic stress–strain relations under plane stress conditions require only two constants. Stress–strain relations of a generally orthotropic lamina (i.e., an orthotropic lamina referred to arbitrary axes) differ from Eqs. (5.10) and are discussed in the next section.

Example 5-1: A unidirectional composite is subjected to the following stresses:

$$\sigma_L = 3.0 \text{ MPa}, \quad \sigma_T = 0.5 \text{ MPa}, \quad \text{and} \quad \tau_{LT} = 3.5 \text{ MPa}$$

Find the normal and shear strains. Engineering constants are

$$E_L = 14.0 \text{ GPa}, \quad E_T = 3.5 \text{ GPa}, \quad G_{LT} = 4.2 \text{ GPa}$$

$$\nu_{LT} = 0.4 \quad \text{and} \quad \nu_{TL} = 0.1$$

Strains can be obtained by using Eq. (5.10):

$$\varepsilon_L = \frac{3.0}{14 \times 10^3} - 0.1 \times \frac{0.5}{3.5 \times 10^3} = 200 \times 10^{-6}$$

$$\varepsilon_T = \frac{0.5}{3.5 \times 10^3} - 0.4 \times \frac{3.0}{14 \times 10^3} = 57 \times 10^{-6}$$

$$\gamma_{LT} = \frac{3.5}{4.2 \times 10^3} = 833 \times 10^{-6}$$

5.2.2 Stress–Strain Relations for Generally Orthotropic Lamina

A generally orthotropic lamina is shown in Fig. 5-4, which has the principal material axes (L and T) oriented at an angle θ to the reference coordinate axes (x and y). In this case, engineering constants associated with the x and y axes are required to relate stresses and strains. As explained earlier, this lamina responds to the loads like an anisotropic material. That is, normal stresses (σ_x and σ_y) produce shear strains (γ_{xy}) in addition to the normal strains (ϵ_x and ϵ_y), and shear stress (τ_{xy}) produces normal strains (ϵ_x and ϵ_y) along with the shear strain (γ_{xy}). In other words, the shear and normal components of stresses and strains are coupled. Therefore, in this case, in addition to the usual engineering constants associated with the x and y axes (E_x, E_y, G_{xy}, ν_{xy}, and ν_{yx}), cross-coefficients m_x and m_y are also required to relate stresses and strains. These cross-coefficients relate the shear and normal components of stresses and strains. The significance of cross-coefficients will become clear from the stress–strain relations described below. It may be pointed out that

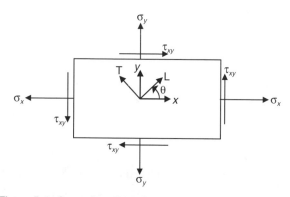

Figure 5-4. Generally orthotropic lamina with applied stresses.

5.2 STRESS–STRAIN RELATIONS AND ENGINEERING CONSTANTS

the engineering constants and cross-coefficients associated with the x and y axes are related to the four independent engineering constants (E_L, E_T, G_{LT}, and ν_{LT}). Transformation equations relating these constants will be derived in Sec. 5.2.3.

Lamina strains for a generally orthotropic lamina (see Fig. 5-4) for each stress component can be written as follows.

1. When σ_x is the only nonzero stress ($\sigma_y = \tau_{xy} = 0$), the strains produced are

$$\epsilon_x = \frac{\sigma_x}{E_x} \tag{5.11}$$

$$\epsilon_y = -\nu_{xy}\epsilon_x = -\nu_{xy}\frac{\sigma_x}{E_x} \tag{5.12}$$

$$\gamma_{xy} = -m_x \frac{\sigma_x}{E_L} \tag{5.13}$$

2. When σ_y is the only nonzero stress ($\sigma_x = \tau_{xy} = 0$), the strains produced are

$$\varepsilon_y = \frac{\sigma_y}{E_y} \tag{5.14}$$

$$\varepsilon_x = -\nu_{yx}\varepsilon_y = -\nu_{yx}\frac{\sigma_y}{E_y} \tag{5.15}$$

$$\gamma_{xy} = -m_y \frac{\sigma_y}{E_L} \tag{5.16}$$

3. When τ_{xy} is the only nonzero stress ($\sigma_x = \sigma_y = 0$), the strains produced are

$$\varepsilon_x = -m_x \frac{\tau_{xy}}{E_L} \tag{5.17}$$

$$\varepsilon_y = -m_y \frac{\tau_{xy}}{E_L} \tag{5.18}$$

$$\gamma_{xy} = \frac{\tau_{xy}}{G_{xy}} \tag{5.19}$$

166 ANALYSIS OF AN ORTHOTROPIC LAMINA

In view of the assumptions of linearly elastic material, the strains given by Eqs. (5.11)–(5.19) can be superposed to obtain the following relations:

$$\varepsilon_x = \frac{\sigma_x}{E_x} - \nu_{yx}\frac{\sigma_y}{E_y} - m_x\frac{\tau_{xy}}{E_L}$$

$$\varepsilon_y = \frac{\sigma_y}{E_y} - \nu_{xy}\frac{\sigma_x}{E_x} - m_y\frac{\tau_{xy}}{E_L} \qquad (5.20)$$

$$\gamma_{xy} = \frac{\tau_{xy}}{G_{xy}} - m_x\frac{\sigma_x}{E_L} - m_y\frac{\sigma_y}{E_L}$$

Equations (5.20) are the stress–strain relations for a generally orthotropic lamina in terms of its engineering constants.

5.2.3 Transformation of Engineering Constants

It was pointed out in the preceding section that the engineering constants in an arbitrary direction, such as the coordinate axis x or y, are related to the four independent engineering constants (E_L, E_T, G_{LT}, and ν_{LT}) for the lamina. To derive transformation equations relating these constants, first consider a generally orthotropic lamina (see Fig. 5-4) subjected to a stress σ_x with $\sigma_y = \tau_{xy} = 0$. Transformation of stresses [Eq. (A2.12) in Appendix 2] gives the normal and shearing stresses along the longitudinal and transverse directions as

$$\sigma_L = \sigma_x \cos^2\theta$$

$$\sigma_T = \sigma_x \sin^2\theta \qquad (5.21)$$

$$\tau_{LT} = -\sigma_x \sin\theta \cos\theta$$

The strains in the longitudinal and transverse directions are given by Eq. (5.10).

$$\epsilon_L = \frac{\sigma_L}{E_L} - \nu_{TL}\frac{\sigma_T}{E_T} = \sigma_x\left(\frac{\cos^2\theta}{E_L} - \nu_{TL}\frac{\sin^2\theta}{E_T}\right)$$

$$\epsilon_T = \frac{\sigma_T}{E_T} - \nu_{LT}\frac{\sigma_L}{E_L} = \sigma_x\left(\frac{\sin^2\theta}{E_T} - \nu_{LT}\frac{\cos^2\theta}{E_L}\right) \qquad (5.22)$$

$$\gamma_{LT} = \frac{\tau_{LT}}{G_{LT}} = -\frac{\sigma_x \sin\theta \cos\theta}{G_{LT}}$$

5.2 STRESS–STRAIN RELATIONS AND ENGINEERING CONSTANTS

The strains in the x and y directions can be obtained by the inverse of the strain-transformation law [Eq. (A.2.2)], which can be written in the expanded form as

$$\epsilon_x = \epsilon_L \cos^2\theta + \epsilon_T \sin^2\theta - \gamma_{LT} \sin\theta \cos\theta$$
$$\epsilon_y = \epsilon_L \sin^2\theta + \epsilon_T \cos^2\theta + \gamma_{LT} \sin\theta \cos\theta \quad (5.23)$$
$$\gamma_{xy} = 2(\epsilon_L - \epsilon_T)\sin\theta \cos\theta + \gamma_{LT}(\cos^2\theta - \sin^2\theta)$$

Substitution of Eq. (5.22) in (5.23) gives the strains

$$\epsilon_x = \sigma_x\left[\frac{\cos^4\theta}{E_L} + \frac{\sin^4\theta}{E_T} + \frac{1}{4}\left(\frac{1}{G_{LT}} - \frac{2\nu_{LT}}{E_L}\right)\sin^2 2\theta\right]$$

$$\epsilon_y = -\sigma_x\left[\frac{\nu_{LT}}{E_L} - \frac{1}{4}\left(\frac{1}{E_L} + \frac{2\nu_{LT}}{E_L} + \frac{1}{E_T} - \frac{1}{G_{LT}}\right)\sin^2 2\theta\right]$$

$$\gamma_{xy} = -\sigma_x \sin 2\theta\left[\frac{\nu_{LT}}{E_L} + \frac{1}{E_T} - \frac{1}{2G_{LT}} - \cos^2\theta\left(\frac{1}{E_L} + \frac{2\nu_{LT}}{E_L} + \frac{1}{E_T} - \frac{1}{G_{LT}}\right)\right]$$

$$(5.24)$$

Now the definitions of elastic constants can be used to calculate them from Eq. (5.24). Since

$$E_x \equiv \frac{\sigma_x}{\epsilon_x}$$

the first relation in Eq. (5.24) gives

$$\frac{1}{E_x} = \frac{\cos^4\theta}{E_L} + \frac{\sin^4\theta}{E_T} + \frac{1}{4}\left(\frac{1}{G_{LT}} - \frac{2\nu_{LT}}{E_L}\right)\sin^2 2\theta \quad (5.25)$$

An expression for E_y can be obtained by substituting $\theta + 90°$ for θ in Eq. (5.25). The result is

$$\frac{1}{E_y} = \frac{\sin^4\theta}{E_L} + \frac{\cos^4\theta}{E_T} + \frac{1}{4}\left(\frac{1}{G_{LT}} - \frac{2\nu_{LT}}{E_L}\right)\sin^2 2\theta \quad (5.26)$$

The Poisson ratio is defined as

$$\nu_{xy} = -\frac{\epsilon_y}{\epsilon_x}$$

when only a uniaxial stress σ_x is applied. Equation (5.25) and the second relation in Eq. (5.24) can be used to obtain

$$\frac{\nu_{xy}}{E_x} = \frac{\nu_{LT}}{E_L} - \frac{1}{4}\left(\frac{1}{E_L} + \frac{2\nu_{LT}}{E_L} + \frac{1}{E_T} - \frac{1}{G_{LT}}\right)\sin^2 2\theta \qquad (5.27)$$

Similarly,

$$\frac{\nu_{yx}}{E_y} = \frac{\nu_{TL}}{E_T} - \frac{1}{4}\left(\frac{1}{E_L} + \frac{2\nu_{LT}}{E_L} + \frac{1}{E_T} - \frac{1}{G_{LT}}\right)\sin^2 2\theta \qquad (5.28)$$

It should be noted that when the normal stress σ_x is applied in a direction other than the longitudinal or transverse direction, it may induce a shearing strain such as that given by Eq. (5.24). Therefore, a cross-coefficient m_x may be defined that relates the shearing strain to the normal stress σ_x in the following manner:

$$\gamma_{xy} = -m_x \frac{\sigma_x}{E_L} \qquad (5.29)$$

Comparing Eq. (5.29) with the third relation in Eq. (5.24) gives

$$m_x = \sin 2\theta \left[\nu_{LT} + \frac{E_L}{E_T} - \frac{E_L}{2G_{LT}} - \cos^2\theta\left(1 + 2\nu_{LT} + \frac{E_L}{E_T} - \frac{E_L}{G_{LT}}\right)\right] \qquad (5.30)$$

A second cross-coefficient m_y that relates the shearing strain to normal stress σ_y is defined as

$$\gamma_{xy} = -m_y \frac{\sigma_y}{E_L} \qquad (5.31)$$

It can be shown in a similar manner that

$$m_y = \sin 2\theta \left[\nu_{LT} + \frac{E_L}{E_T} - \frac{E_L}{2G_{LT}} - \sin^2\theta\left(1 + 2\nu_{LT} + \frac{E_L}{E_T} - \frac{E_L}{G_{LT}}\right)\right] \qquad (5.32)$$

Finally, to obtain an expression for G_{xy}, assume that the only nonzero stress acting on the lamina of Fig. 5-4 is τ_{xy}. This will induce the following stresses in the longitudinal and transverse directions:

$$\sigma_L = -\sigma_T = 2\tau_{xy} \sin\theta \cos\theta$$

$$\tau_{LT} = (\cos^2\theta - \sin^2\theta)\tau_{xy} \qquad (5.33)$$

The strains in longitudinal and transverse directions are given by Eq. (5.10).

5.2 STRESS–STRAIN RELATIONS AND ENGINEERING CONSTANTS

$$\epsilon_L = 2\tau_{xy} \sin\theta \cos\theta \left(\frac{1}{E_L} + \frac{\nu_{TL}}{E_T}\right)$$

$$\epsilon_T = -2\tau_{xy} \sin\theta \cos\theta \left(\frac{1}{E_T} + \frac{\nu_{LT}}{E_L}\right) \quad (5.34)$$

$$\gamma_{LT} = \frac{\tau_{xy}}{G_{LT}}(\cos^2\theta - \sin^2\theta)$$

Now the shearing strain γ_{xy} is obtained by substituting Eq. (5.34) in Eq. (5.23). This results in

$$\gamma_{xy} = \tau_{xy}\left[\frac{1}{E_L} + \frac{2\nu_{LT}}{E_L} + \frac{1}{E_T} - \left(\frac{1}{E_L} + \frac{2\nu_{LT}}{E_L} + \frac{1}{E_T} - \frac{1}{G_{LT}}\right)\cos^2 2\theta\right] \quad (5.35)$$

Now the definition of shear modulus G_{xy}, will give

$$\frac{1}{G_{xy}} = \frac{1}{E_L} + \frac{2\nu_{LT}}{E_L} + \frac{1}{E_T} - \left(\frac{1}{E_L} + \frac{2\nu_{LT}}{E_L} + \frac{1}{E_T} - \frac{1}{G_{LT}}\right)\cos^2 2\theta \quad (5.36)$$

It should be noted that the shearing stress τ_{xy} will cause normal strains ϵ_x and ϵ_y in the x and y directions, respectively, given by

$$\epsilon_x = -m_x \frac{\tau_{xy}}{E_L}$$

$$\epsilon_y = -m_y \frac{\tau_{xy}}{E_L} \quad (5.37)$$

where the cross-coefficients were defined in Eqs. (5.29) and (5.31) and were evaluated in Eqs. (5.30) and (5.32).

To better appreciate the variation of the elastic constants, they can be plotted as functions of orientation θ for specific materials. Variations in E_x, G_{xy}, ν_{xy}, m_x, and m_y are shown in Figs. 5-5 to 5-7 for typical glass–epoxy, graphite–epoxy, and boron–epoxy systems. It may be noted that for the glass–epoxy and graphite–epoxy systems chosen, E_x monotonically decreases from E_L at $\theta = 0°$ to E_T at $\theta = 90°$. It should be pointed out that the variation of E_x will be quite dependent on the assumed value of G_{LT}. It is recommended that this value be obtained experimentally rather than by a predictive technique based on the constituent properties. For the boron–epoxy system, E_x is less than both E_L and E_T for values of θ between 45° and 90°. The value of G_{xy} for all three systems is largest at $\theta = 45°$ and its variation is symmetric about this orientation. The largest value of the Poisson ratio ν_{xy} is nearly $1.2\nu_{LT}$ at $\theta = 30°$ for the glass–epoxy, $1.1\nu_{LT}$ at $\theta = 15°$ for graphite–epoxy, and $1.9\nu_{LT}$ at $\theta = 25°$ for the boron–epoxy systems. The cross-coefficients m_x and m_y are

170 ANALYSIS OF AN ORTHOTROPIC LAMINA

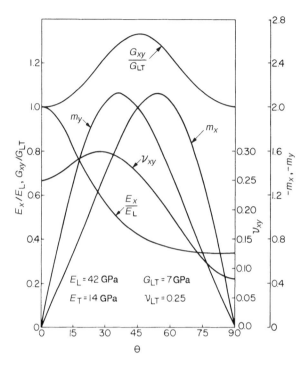

Figure 5-5. Elastic constants of a glass–epoxy lamina: variation with fiber orientation.

zero at $\theta = 0°$ and $\theta = 90°$, but for intermediate values of θ, they achieve large values compared with ν_{xy}. The variations shown in Figs. 5-5 to 5-7 are not entirely typical of all composite materials. The curves shown can be changed considerably by relatively small variations in the properties of the fiber–matrix combination. It should be observed from these curves that the extremum (largest and smallest) material properties do not necessarily occur in principal material directions, which is actually the case for E_x in the boron–epoxy system and for G_{xy}, ν_{xy}, m_x and m_y in all three systems. It was shown by Jones [1] that whenever values of properties in the principal material directions violate certain inequalities, extrema in their values occur in the directions other than the principal material directions. For example, it can be shown by finding maximum and minimum values of E_x as given by Eq. (5.25) that E_x is greater than both E_L and E_T for some values of θ other than $0°$ or $90°$ if

$$G_{LT} > \frac{E_L}{2(1 + \nu_{LT})} \tag{5.38}$$

and that E_x is less than both E_L and E_T for some values of θ other than $0°$ or $90°$ if

5.2 STRESS–STRAIN RELATIONS AND ENGINEERING CONSTANTS

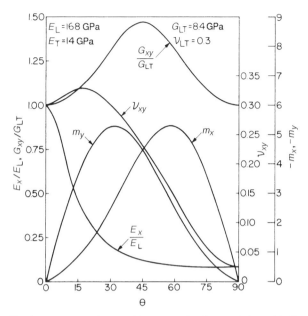

Figure 5-6. Elastic constants of a graphite–epoxy lamina: variation with fiber orientation.

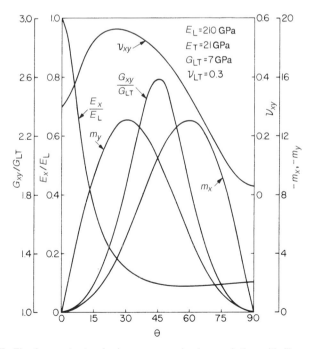

Figure 5-7. Elastic constants of a boron–epoxy lamina: variation with fiber orientation.

$$G_{LT} < \frac{E_L}{2[(E_L/E_T) + \nu_{LT}]} \tag{5.39}$$

The inequality in Eq. (5.38) is not violated by any of the systems chosen earlier, but the inequality in Eq. (5.39) is violated by the boron–epoxy system and hence the behavior shown in Fig. 5-7. Similar conditions can be obtained for other material properties as well.

It is of practical interest to consider a lamina having identical properties in the two principal material directions (i.e., $E_L = E_T$ and $\nu_{LT} = \nu_{TL}$). Such a lamina is called a *balanced orthotropic lamina,* an example of which is a glass-fabric-reinforced material with equal volume fractions of fibers in two mutually perpendicular directions. Typical variations in the elastic constants of a balanced lamina are shown in Fig. 5-8. The elastic constants show symmetry in their variations about an orientation of 45° to the principal material axes. Fabric-reinforced laminae are of practical significance because with them almost any ratio of E_L/E_T can be established through the fabric weave (i.e., by changing the ratio of fiber volume fractions in the two mutually perpendicular directions).

Example 5-2: For the lamina shown in Fig. 5-9, find the strains in the xy directions. Lamina has the same engineering constants in the longitudinal and transverse directions as given in Example 5-1.

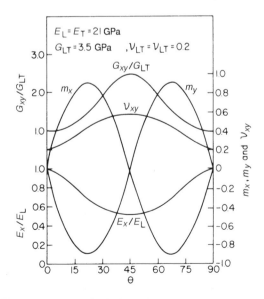

Figure 5-8. Elastic constants of a balanced lamina: variation with fiber orientation.

5.2 STRESS–STRAIN RELATIONS AND ENGINEERING CONSTANTS

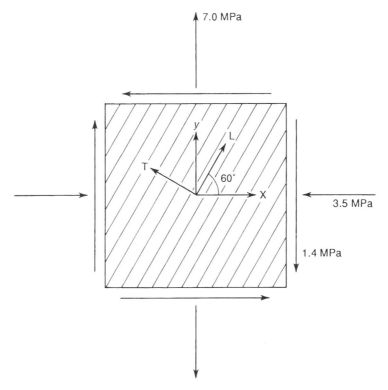

Figure 5-9. State of stress on a lamina for Example 5-2.

From Fig. 5-9, the stresses are

$$\sigma_x = -3.5 \text{ MPa}$$

$$\sigma_y = 7.0 \text{ MPa}$$

$$\tau_{xy} = 1.4 \text{ MPa}$$

Engineering constants in the x and y directions are required to calculate strains. These are obtained by substituting $\theta = 60°$ in Eqs. (5.25)–(5.28) and (5.36). The transformed constants are

$$E_x = 5.02 \text{ GPa}, \quad E_y = 10.87 \text{ GPa}, \quad G_{xy} = 2.70 \text{ GPa}$$

$$\frac{\nu_{xy}}{E_x} = \frac{\nu_{yx}}{E_y} = -0.00446 \text{ GPa}^{-1}$$

$$m_x = 1.833, \quad m_y = 0.765$$

Therefore, strains are

$$\epsilon_x = \frac{-3.5 \times 10^{-3}}{5.02} - (-0.00446)(7.0 \times 10^{-3})$$

$$- 1.833\left(\frac{-1.4 \times 10^{-3}}{14}\right) = -483 \times 10^{-6}$$

$$\epsilon_y = \frac{7.0 \times 10^{-3}}{10.87} - (-0.00446)(-3.5 \times 10^{-3})$$

$$- 0.765\left(\frac{-1.4 \times 10^{-3}}{14}\right) = 705 \times 10^{-6}$$

$$\gamma_{xy} = \frac{-1.4 \times 10^{-3}}{2.70} - 1.833\left(\frac{-3.5 \times 10^{-3}}{14}\right)$$

$$- 0.765\left(\frac{7.0 \times 10^{-3}}{14}\right) = -443 \times 10^{-6}$$

5.3 HOOKE'S LAW AND STIFFNESS AND COMPLIANCE MATRICES

5.3.1 General Anisotropic Material

In general, the state of stress at a point in a body is described by the nine components of the stress tensor σ_{ij}, as shown in Fig. 5-10. Correspondingly, there is a strain tensor, ϵ_{ij}, with nine components.

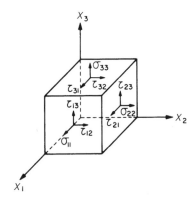

Figure 5-10. Components of stress tensor on a cube element.

5.3 HOOKE'S LAW AND STIFFNESS AND COMPLIANCE MATRICES

The most general linear relationship that connects stress to strain is known as the *generalized Hooke's law* and can be expressed mathematically as

$$\sigma_{ij} = E_{ijkl}\epsilon_{kl} \tag{5.40}$$

where E_{ijkl} is a fourth order tensor. The elements of E_{ijkl} are known as the *elastic constants*. Equation (5.40) also can be written in the matrix form as

$$\begin{bmatrix} \sigma_{11} \\ \sigma_{22} \\ \sigma_{33} \\ \tau_{23} \\ \tau_{31} \\ \tau_{12} \\ \tau_{32} \\ \tau_{13} \\ \tau_{21} \end{bmatrix} = \begin{bmatrix} E_{1111} & E_{1122} & E_{1133} & E_{1123} & E_{1131} & E_{1112} & E_{1132} & E_{1113} & E_{1121} \\ E_{2211} & E_{2222} & E_{2233} & E_{2223} & E_{2231} & E_{2212} & E_{2232} & E_{2213} & E_{2221} \\ E_{3311} & E_{3322} & E_{3333} & E_{3323} & E_{3331} & E_{3312} & E_{3332} & E_{3313} & E_{3321} \\ E_{2311} & E_{2322} & E_{2333} & E_{2323} & E_{2331} & E_{2312} & E_{2332} & E_{2313} & E_{2321} \\ E_{3111} & E_{3122} & E_{3133} & E_{3123} & E_{3131} & E_{3112} & E_{3132} & E_{3113} & E_{3121} \\ E_{1211} & E_{1222} & E_{1233} & E_{1223} & E_{1231} & E_{1212} & E_{1232} & E_{1213} & E_{1221} \\ E_{3211} & E_{3222} & E_{3233} & E_{3223} & E_{3231} & E_{3212} & E_{3232} & E_{3213} & E_{3221} \\ E_{1311} & E_{1322} & E_{1333} & E_{1323} & E_{1331} & E_{1312} & E_{1332} & E_{1313} & E_{1321} \\ E_{2111} & E_{2122} & E_{2133} & E_{2123} & E_{2131} & E_{2112} & E_{2132} & E_{2113} & E_{2121} \end{bmatrix} \begin{bmatrix} \epsilon_{11} \\ \epsilon_{22} \\ \epsilon_{33} \\ \epsilon_{23} \\ \epsilon_{31} \\ \epsilon_{12} \\ \epsilon_{32} \\ \epsilon_{13} \\ \epsilon_{21} \end{bmatrix} \tag{5.41}$$

The first two subscripts on the elastic constants correspond to those of stress, whereas the last two subscripts correspond to those of strain.

It is seen that each stress component is related to all nine components of the strain tensor, and there are 81 elastic constants defining the tensor E_{ijkl}. Fortunately, this tensor exhibits certain symmetry properties that reduce the total number of independent components to 21 for a material that does not have any axes of symmetry. Such a material is called *aeolotropic* or *anisotropic*.

The first set of reductions in elastic constants is obtained by considering the symmetry of strain. It can be shown easily that because of the symmetry of the strain tensor, there is no loss of generality if E_{ijkl} is assumed symmetric with respect to the last two indices; in other words,

$$E_{ijkl} = E_{ijlk} \tag{5.42}$$

which will reduce the number of constants from 81 to 54. A second reduction in constants comes by assuming E_{ijkl} symmetric with respect to the first two indices because of the symmetry of the stress tensor. Thus

$$E_{ijkl} = E_{jikl} \tag{5.43}$$

This causes a further reduction in constants by 18, reducing the number of independent constants to 36.

The further reduction in the number of independent constants to the final total of 21 can be accomplished only by thermodynamic considerations, according to which a strain–energy density function can be assumed to exist as follows:

$$U = U(\epsilon_{ij}) \tag{5.44}$$

with the property

$$\frac{\partial U}{\partial \epsilon_{ij}} = \sigma_{ij} \tag{5.45}$$

Then, by Eq. (5.40),

$$\frac{\partial U}{\partial \epsilon_{ij}} = E_{ijkl}\epsilon_{kl} \tag{5.46}$$

The partial differentiation of Eq. (5.46) with respect to ϵ_{kl} yields

$$\frac{\partial}{\partial \epsilon_{kl}}\left(\frac{\partial U}{\partial \epsilon_{ij}}\right) = E_{ijkl} \tag{5.47}$$

Interchanging the indices in Eq. (5.47) gives

$$\frac{\partial}{\partial \epsilon_{ij}}\left(\frac{\partial U}{\partial \epsilon_{kl}}\right) = E_{klij} \tag{5.48}$$

Since the order of partial differentiation is immaterial, that is,

$$\frac{\partial}{\partial \epsilon_{ij}}\left(\frac{\partial U}{\partial \epsilon_{kl}}\right) = \frac{\partial}{\partial \epsilon_{kl}}\left(\frac{\partial U}{\partial \epsilon_{ij}}\right) \tag{5.49}$$

it is clear that

$$E_{ijkl} = E_{klij} \tag{5.50}$$

Thus the first pair of indices in the elasticity tensor can be interchanged with the second pair without any change in the values. This operation reduces the number of independent elastic constants for an aeolotropic or anisotropic material to only 21.

The response of an anisotropic material to impressed forces can be predicted with the help of these 21 constants, and generally it will be different along each axis. If one pushes along a given direction, changes in length as well as in angle will occur along and between all the axes. It should be noted that since the components both of the stress tensor and of the strain tensor are functions of the orientation of the axis system, the elastic constants (the elements of the elasticity tensor) also will be functions of axis orientation. The elasticity tensor is a fourth-order tensor, and hence its transformation law can be written as

$$E'_{mnrs} = a_{im}a_{jn}a_{kr}a_{ls}E_{ijkl} \quad (5.51)$$

where E'_{mnrs} is the elasticity tensor in the transformed (x') axis system, E_{ijkl} is the elasticity tensor in the original (x) axis system, and a_{im} are the direction cosines of the new axes with respect to the original axes. Once the elastic constants are known in one reference coordinate system, the transformation law [Eq. (5.51)] enables us to calculate the elastic constants in any other reference coordinate system. In general, the elastic constants will change with the transformation, but under some specific transformations, the elastic constants may remain unchanged as a result of additional symmetries existing in the material properties.

5.3.2 Specially Orthotropic Material

Because of their macroscopic structure, many materials exhibit symmetry in their elastic properties with respect to certain planes; that is, the elastic constants do not change when the direction of the axis perpendicular to the plane of symmetry is reversed. The number of elastic constants will reduce when the number of planes of symmetry increases. The transformation law can be used to derive the number of independent elastic constants for various symmetry conditions. Fiber composites come under the category of orthotropic materials that exhibit symmetry of their elastic properties with respect to two orthogonal planes. The number of independent elastic constants for orthotropic materials is now derived.

First, consider that one of the planes of symmetry of orthotropic materials is the x_1x_2 plane (Fig. 5-11). This symmetry requires that the elastic constants do not change under the following coordinate transformation:

$$x'_1 = x_1 \qquad x'_2 = x_2 \qquad x'_3 = -x_3 \quad (5.52)$$

The corresponding direction cosines can be expressed as follows:

	x'_1	x'_2	x'_3
x_1	$a_{11} = 1$	$a_{12} = 0$	$a_{13} = 0$
x_2	$a_{21} = 0$	$a_{22} = 1$	$a_{23} = 0$
x_3	$a_{31} = 0$	$a_{32} = 0$	$a_{33} = -1$

(5.53)

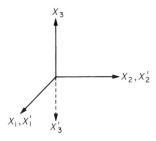

Figure 5-11. Transformation of coordinate axes when x_1x_2 is a plane of symmetry.

The invariance of elastic properties under the preceding coordinate transformation imposes certain restrictions on the elasticity tensor. These restrictions are actually the conditions necessary to satisfy the invariance condition (i.e., $E'_{ijkl} = E_{ijkl}$) and are obtained by applying the transformation law [Eq. (5.51)]. To this end, examine the dependence of components of E'_{ijkl} on E_{ijkl}:

$$E'_{1111} = E_{ijkl}a_{i1}a_{j1}a_{k1}a_{l1} = E_{1111}$$
$$E'_{1112} = E_{ijkl}a_{i1}a_{j1}a_{k1}a_{l2} = E_{1112} \tag{5.54}$$
$$E'_{1113} = E_{ijkl}a_{i1}a_{j1}a_{k1}a_{l3} = -E_{1113}$$

Since there are only three nonzero direction cosines [Eq. (5.53)], the expansion of the transformation law is simplified. The result given in Eq. (5.54) states that the invariance conditions are satisfied for the first two components examined, but not for the third one. To satisfy the invariance condition for the third one, it is necessary to set E_{1113} equal to zero. In a similar manner, it can be verified easily that the condition of no change in the elastic constants under the coordinate transformation [Eq. (5.52)] would require that 8 of the 21 elastic constants should be zero. These 8 components that must be set equal to zero are

$$E_{1113}, \quad E_{2223}, \quad E_{1123}, \quad E_{2213}, \quad E_{1213}, \quad E_{1223}, \quad E_{1333}, \quad E_{2333} \tag{5.55}$$

Now to complete the symmetry requirements for an orthotropic material, consider that the second plane of symmetry is the x_2x_3 plane (Fig. 5-12). This means that the elastic constants do not change under the following coordinate transformation:

$$x'_1 = -x_1 \qquad x'_2 = x_2 \qquad x'_3 = x_3 \tag{5.56}$$

The direction cosines corresponding to the preceding transformations are

5.3 HOOKE'S LAW AND STIFFNESS AND COMPLIANCE MATRICES

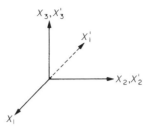

Figure 5-12. Transformation of coordinate axes when x_2x_3 is a plane of symmetry.

	x_1'	x_2'	x_3'	
x_1	$a_{11} = -1$	$a_{12} = 0$	$a_{13} = 0$	(5.57)
x_2	$a_{21} = 0$	$a_{22} = 1$	$a_{23} = 0$	
x_3	$a_{31} = 0$	$a_{32} = 0$	$a_{33} = 1$	

The application of the transformation law [Eq. (5.51)] subject to the direction cosines given by Eq. (5.57) will lead to contradictions of the type shown in Eq. (5.54). These are resolved by setting some of the elastic constants equal to zero. It can be verified easily that the following additional constants must be zero:

$$E_{1233}, \; E_{1323}, \; E_{1222}, \; E_{1112} \tag{5.58}$$

It may be pointed out that the symmetry of elastic properties with respect to two mutually perpendicular planes implies the symmetry with respect to the third orthogonal plane. An interested reader may verify that the third plane of symmetry (the x_3x_1 plane) does not yield any additional reduction in the elastic constants in the present case. Thus elastic constants of an orthotropic material can be described by the following array of nine constants:

$$(E_{ijkl}) = \begin{bmatrix} E_{1111} & E_{1122} & E_{1133} & 0 & 0 & 0 \\ E_{1122} & E_{2222} & E_{2233} & 0 & 0 & 0 \\ E_{1133} & E_{2233} & E_{3333} & 0 & 0 & 0 \\ 0 & 0 & 0 & E_{2323} & 0 & 0 \\ 0 & 0 & 0 & 0 & E_{1313} & 0 \\ 0 & 0 & 0 & 0 & 0 & F_{1212} \end{bmatrix} \tag{5.59}$$

By a careful examination of Eq. (5.59), it is quickly realized that it is now unnecessary to use four subscripts of the original elasticity tensor to describe the nine nonzero elastic constants of orthotropic materials. It is more convenient to write Hooke's law for an orthotropic material in the contracted notation as

180 ANALYSIS OF AN ORTHOTROPIC LAMINA

$$\sigma_i = C_{ij}\varepsilon_j \qquad i,j = 1, 2, 3, 4, 5, 6 \qquad (5.60)$$

where σ_i are the stress components, C_{ij} is the stiffness matrix, and ε_j are the engineering strain components.

The engineering strains are different from the tensor strains used in Eq. (5.40). The two types of strain are described in Appendix 2. The difference between the two arises only in the shearing-strain components. An engineering shear strain is twice the corresponding tensorial shear strain. Equation (5.60), which uses engineering strains, can be written in the matrix form as

$$\begin{Bmatrix} \sigma_1 \\ \sigma_2 \\ \sigma_3 \\ \tau_{23} \\ \tau_{13} \\ \tau_{12} \end{Bmatrix} = \begin{bmatrix} C_{11} & C_{12} & C_{13} & 0 & 0 & 0 \\ C_{12} & C_{22} & C_{23} & 0 & 0 & 0 \\ C_{13} & C_{23} & C_{33} & 0 & 0 & 0 \\ 0 & 0 & 0 & C_{44} & 0 & 0 \\ 0 & 0 & 0 & 0 & C_{55} & 0 \\ 0 & 0 & 0 & 0 & 0 & C_{66} \end{bmatrix} \begin{Bmatrix} \varepsilon_1 \\ \varepsilon_2 \\ \varepsilon_3 \\ \gamma_{23} \\ \gamma_{13} \\ \gamma_{12} \end{Bmatrix} \qquad (5.61)$$

Equation (5.61) represents three-dimensional stress–strain relations for an orthotropic material when the reference coordinate axes coincide with the material axes. Therefore, these are called stress–strain relations for a *specially orthotropic material*. The stiffness matrix contains 12 nonzero elements, with only 9 of those being independent (C_{11}, C_{22}, C_{33}, C_{12}, C_{13}, C_{23}, C_{44}, C_{55}, and C_{66}). When coordinate axes do not coincide with the material axes, the same orthotropic material is called a *generally orthotopic material*. Stress–strain relations for a generally orthotropic material can be obtained by tensor transformation, as used in the preceding sections. The stiffness matrix for a generally orthotropic material is usually fully populated; that is, it has no zero elements. However, all 36 elements are obtained only from the 9 independent elements mentioned above.

5.3.3 Transversely Isotropic Material

It was pointed out earlier that for unidirectional composites, mechanical properties in all directions perpendicular to the longitudinal direction generally are assumed to be equal. Thus, for a unidirectional composite, the transverse plane (plane perpendicular to the longitudinal axis) is a plane of isotropy, and a unidirectional composite is an example of a *transversely isotropic material*. In general, an orthotropic material is called *transversely isotropic* when one of its planes of symmetry is isotropic. Since transversely isotropic material has more axes of symmetry than a specially orthotropic material, its stiffness matrix has a smaller number of independent elements. If it is assumed that plane 23 is the plane of isotropy, the following relations between the elements of stiffness matrix can be shown to exist:

5.3 HOOKE'S LAW AND STIFFNESS AND COMPLIANCE MATRICES

$$C_{22} = C_{33}, \quad C_{12} = C_{13}$$

$$C_{55} = C_{66}$$

$$C_{44} = \frac{(C_{22} - C_{23})}{2} \tag{5.62}$$

The stiffness matrix therefore can be written as

$$\begin{Bmatrix} \sigma_1 \\ \sigma_2 \\ \sigma_3 \\ \tau_{23} \\ \tau_{13} \\ \tau_{12} \end{Bmatrix} = \begin{bmatrix} C_{11} & C_{12} & C_{12} & 0 & 0 & 0 \\ C_{12} & C_{22} & C_{23} & 0 & 0 & 0 \\ C_{12} & C_{23} & C_{22} & 0 & 0 & 0 \\ 0 & 0 & 0 & \frac{C_{22} - C_{23}}{2} & 0 & 0 \\ 0 & 0 & 0 & 0 & C_{66} & 0 \\ 0 & 0 & 0 & 0 & 0 & C_{66} \end{bmatrix} \begin{Bmatrix} \epsilon_1 \\ \epsilon_2 \\ \epsilon_3 \\ \gamma_{23} \\ \gamma_{13} \\ \gamma_{12} \end{Bmatrix} \tag{5.63}$$

A transversely isotropic material thus has only five independent elastic constants (C_{11}, C_{22}, C_{12}, C_{23}, and C_{66}).

5.3.4 Isotropic Material

A material is called *isotropic* when its properties are independent of direction. As a result, every coordinate axis is an axis of symmetry. It can be shown that the following relations exist between the elements of the stiffness matrix [Eq. (5.63)]:

$$C_{11} = C_{22}, \quad C_{12} = C_{23}$$

$$C_{66} = \frac{(C_{22} - C_{23})}{2} = \frac{(C_{11} - C_{12})}{2} \tag{5.64}$$

Therefore, the stiffness matrix for an isotropic material becomes

$$\begin{Bmatrix} \sigma_1 \\ \sigma_2 \\ \sigma_3 \\ \tau_{23} \\ \tau_{13} \\ \tau_{12} \end{Bmatrix} = \begin{bmatrix} C_{11} & C_{12} & C_{12} & 0 & 0 & 0 \\ C_{12} & C_{11} & C_{12} & 0 & 0 & 0 \\ C_{12} & C_{12} & C_{11} & 0 & 0 & 0 \\ 0 & 0 & 0 & \dfrac{C_{11}-C_{12}}{2} & 0 & 0 \\ 0 & 0 & 0 & 0 & \dfrac{C_{11}-C_{12}}{2} & 0 \\ 0 & 0 & 0 & 0 & 0 & \dfrac{C_{11}-C_{12}}{2} \end{bmatrix} \begin{Bmatrix} \epsilon_1 \\ \epsilon_2 \\ \epsilon_3 \\ \gamma_{23} \\ \gamma_{13} \\ \gamma_{12} \end{Bmatrix}$$

(5.65)

There are only two independent elastic constants for isotropic materials (C_{11} and C_{12}).

A summary of independent and nonzero elastic constants for different materials is given in Table 5-1 for both three- and two-dimensional cases.

5.3.5 Specially Orthotropic Material under Plane Stress

In most structural applications, composite laminates are loaded in the plane of the laminate. Such a loading is called a *plane-stress condition* in which all out-of-plane stress components are zero. If axis 3 is the out-of-plane direction, a plane-stress condition gives

$$\sigma_3 = \tau_{23} = \tau_{13} = 0 \tag{5.66}$$

By substituting Eq. (5.66) in Eq. (5.61) and writing it in expanded form, we get

Table 5-1 Number of elastic constants

Material	Three-Dimensional		Two-Dimensional	
	Number of Nonzero Constants	Number of Independent Constants	Number of Nonzero Constants	Number of Independent Constants
Anisotropic	36	21	9	6
Specially orthotropic	12	9	5	4
Generally orthotropic	36	9	9	4
Transversely isotropic	12	5	5	4
Isotropic	12	2	5	2

5.3 HOOKE'S LAW AND STIFFNESS AND COMPLIANCE MATRICES

$$\sigma_1 = C_{11}\varepsilon_1 + C_{12}\varepsilon_2 + C_{13}\varepsilon_3$$
$$\sigma_2 = C_{12}\varepsilon_1 + C_{22}\varepsilon_2 + C_{23}\varepsilon_3$$
$$0 = C_{13}\varepsilon_1 + C_{23}\varepsilon_2 + C_{33}\varepsilon_3 \quad (5.67)$$
$$\gamma_{23} = \gamma_{13} = 0$$
$$\tau_{12} = C_{66}\gamma_{12}$$

Elimination of strain ε_3 from Eq. (5.67) gives

$$\sigma_1 = \left(C_{11} - \frac{C_{13}^2}{C_{33}}\right)\varepsilon_1 + \left(C_{12} - \frac{C_{13}C_{23}}{C_{33}}\right)\varepsilon_2$$
$$\sigma_2 = \left(C_{12} - \frac{C_{23}C_{13}}{C_{33}}\right)\varepsilon_1 + \left(C_{22} - \frac{C_{23}^2}{C_{33}}\right)\varepsilon_2 \quad (5.68)$$
$$\tau_{12} = C_{66}\gamma_{12}$$

For simplicity, new stiffness coefficients are defined for a specially orthotropic material under plane stress conditions as follows:

$$Q_{11} = C_{11} - \frac{C_{13}^2}{C_{33}}$$
$$Q_{22} = C_{22} - \frac{C_{23}^2}{C_{33}} \quad (5.69)$$
$$Q_{12} = C_{12} - \frac{C_{13}C_{23}}{C_{33}}$$
$$Q_{66} = C_{66}$$

Now the stress–strain relations for a specially orthotropic lamina under plane-stress conditions can be written as

$$\begin{Bmatrix} \sigma_1 \\ \sigma_2 \\ \tau_{12} \end{Bmatrix} - \begin{bmatrix} Q_{11} & Q_{12} & 0 \\ Q_{12} & Q_{22} & 0 \\ 0 & 0 & Q_{66} \end{bmatrix} \begin{Bmatrix} \varepsilon_1 \\ \varepsilon_2 \\ \gamma_{12} \end{Bmatrix} \quad (5.70)$$

The stress–strain relations written as Eq. (5.70) are used in this text to develop laminae and laminate analysis procedures.

5.3.6 Compliance Tensor and Compliance Matrix

The stress–strain relations given by Eq. (5.40) can be expressed in the inverted form as

$$\epsilon_{ij} = S_{ijkl}\sigma_{kl} \tag{5.71}$$

where S_{ijkl} is known as the *compliance tensor*. It should be evident that the compliance tensor has the same symmetry properties as the elasticity tensor and the same type of transformation law. The number of independent components of the compliance tensor may be reduced in a manner similar to the one used for the elasticity tensor. Finally, the stress–strain relations for an orthotropic material for the two-dimensional case can be written in terms of a compliance matrix as follows:

$$\begin{Bmatrix} \epsilon_1 \\ \epsilon_2 \\ \gamma_{12} \end{Bmatrix} = \begin{bmatrix} S_{11} & S_{12} & 0 \\ S_{12} & S_{22} & 0 \\ 0 & 0 & S_{66} \end{bmatrix} \begin{Bmatrix} \sigma_1 \\ \sigma_2 \\ \tau_{12} \end{Bmatrix} \tag{5.72}$$

where directions 1 and 2 coincide with the natural axes of the material. It may be noted that the Eqs. (5.70) and (5.72) are inversions of each other. Therefore, the following relations between the elements of the stiffness matrix and compliance matrix may be obtained by the matrix inversion (see Appendix 1):

$$\begin{aligned} Q_{11} &= \frac{S_{22}}{S_{11}S_{22} - S_{12}^2} \\ Q_{22} &= \frac{S_{11}}{S_{11}S_{22} - S_{12}^2} \\ Q_{12} &= \frac{S_{12}}{S_{11}S_{22} - S_{12}^2} \\ Q_{66} &= \frac{1}{S_{66}} \end{aligned} \tag{5.73}$$

It may be pointed out that whereas three-dimensional orthotropy requires nine independent elastic constants, as shown in Eq. (5.61), only four constants are needed for two-dimensional orthotropy. The number of elastic constants required for an isotropic material is only two for both two- and three-dimensional stress states. The increased number of elastic constants indicates the additional complexity of orthotropic problems. The two-dimensional orthotropic problem is of primary concern in the remainder of this book.

5.3.7 Relations between Engineering Constants and Elements of Stiffness and Compliance Matrices

Relations between the five engineering constants of Eq. (5.10) and four independent elastic constants of Eq. (5.70) and Eq. (5.72) can be established easily by considering a specially orthotropic lamina with the longitudinal and transverse directions as the material axes of symmetry. For such a lamina, stress–strain relations in terms of stiffness and compliance matrices can be written as follows by changing subscripts of stresses and strains from 1 and 2 to L and T, respectively:

$$\begin{Bmatrix} \sigma_L \\ \sigma_T \\ \tau_{LT} \end{Bmatrix} = \begin{bmatrix} Q_{11} & Q_{12} & 0 \\ Q_{12} & Q_{22} & 0 \\ 0 & 0 & Q_{66} \end{bmatrix} \begin{Bmatrix} \varepsilon_L \\ \varepsilon_T \\ \gamma_{LT} \end{Bmatrix} \quad (5.74)$$

$$\begin{Bmatrix} \varepsilon_L \\ \varepsilon_T \\ \gamma_{LT} \end{Bmatrix} = \begin{bmatrix} S_{11} & S_{12} & 0 \\ S_{12} & S_{22} & 0 \\ 0 & 0 & S_{66} \end{bmatrix} \begin{Bmatrix} \sigma_L \\ \sigma_T \\ \tau_{LT} \end{Bmatrix} \quad (5.75)$$

Now consider that the lamina is subjected to a general state of stress consisting of σ_L, σ_T, and τ_{LT}. The resulting strains are given by Eq. (5.10) in terms of engineering constants. The strains in terms of elements of the compliance matrix are given by Eq. (5.75), which can be written in the expanded form as

$$\varepsilon_L = S_{11}\sigma_L + S_{12}\sigma_T$$
$$\varepsilon_T = S_{12}\sigma_L + S_{22}\sigma_T \quad (5.76)$$
$$\gamma_{LT} = S_{66}\tau_{LT}$$

Comparison of Eq. (5.10) with Eq. (5.76) gives the desired relations between engineering constants and elements of the compliance matrix:

$$S_{11} = \frac{1}{E_L}$$
$$S_{22} = \frac{1}{E_T} \quad (5.77)$$
$$S_{12} = -\frac{\nu_{LT}}{E_L} = -\frac{\nu_{TL}}{E_T}$$
$$S_{66} = \frac{1}{G_{LT}}$$

Substitution of Eqs. (5.77) in Eqs. (5.73) gives the relations between engineering constants and elements of the stiffness matrix:

$$Q_{11} = \frac{E_L}{1 - \nu_{LT}\nu_{TL}}$$

$$Q_{22} = \frac{E_T}{1 - \nu_{LT}\nu_{TL}}$$ (5.78)

$$Q_{12} = \frac{\nu_{LT}E_T}{1 - \nu_{LT}\nu_{TL}} = \frac{\nu_{TL}E_L}{1 - \nu_{LT}\nu_{TL}}$$

$$Q_{66} = G_{LT}$$

It may be pointed out that although five engineering constants have been mentioned, only four of them are independent. The following functional relationship, which is evident from Eq. (5.78), exists between four of the five constants:

$$\nu_{LT}E_T = \nu_{TL}E_L$$

or

$$\frac{\nu_{LT}}{E_L} = \frac{\nu_{TL}}{E_T}$$ (5.79)

Engineering constants for a number of commercially available composites are given in Appendix 4.

Example 5-3: Determine the stiffness and compliance matrices for a unidirectional AS4/3501-6 graphite–epoxy lamina that has the following engineering constants:

$$E_L = 148.0 \text{ GPa}, \quad E_T = 10.5 \text{ GPa}$$

$$G_{LT} = 5.61 \text{ GPa}, \quad \nu_{LT} = 0.3$$

From Eq. (5.79),

$$\nu_{TL} = \frac{0.3 \times 10.5}{148.0} = 0.021$$

Elements of the stiffness matrix are obtained by using Eq. (5.78):

$$Q_{11} = \frac{148.0}{1 - 0.3 \times 0.021} = 148.95 \text{ GPa}$$

$$Q_{22} = \frac{10.5}{1 - 0.3 \times 0.021} = 10.57 \text{ GPa}$$

$$Q_{12} = \frac{0.3 \times 10.5}{1 - 0.3 \times 0.021} = 3.17 \text{ GPa}$$

$$Q_{66} = 5.61 \text{ GPa}$$

Thus

$$[Q] = \begin{bmatrix} 148.95 & 3.17 & 0 \\ 3.17 & 10.57 & 0 \\ 0 & 0 & 5.61 \end{bmatrix} \text{ GPa}$$

Elements of the compliance matrix are obtained by using Eq. (5.77):

$$S_{11} = \frac{1}{148.0} = 0.0068 \text{ GPa}^{-1}$$

$$S_{22} = \frac{1}{10.5} = 0.0952 \text{ GPa}^{-1}$$

$$S_{12} = -\frac{0.3}{148.0} = -0.0020 \text{ GPa}^{-1}$$

$$S_{66} = \frac{1}{5.61} = 0.1783 \text{ GPa}^{-1}$$

Thus

$$[S] = \begin{bmatrix} 0.0068 & -0.0020 & 0 \\ -0.0020 & 0.0952 & 0 \\ 0 & 0 & 0.1783 \end{bmatrix} \text{ GPa}^{-1}$$

5.3.8 Restrictions on Elastic Constants

It was pointed out in an earlier section that for an orthotropic material, three-dimensional stress–strain relations require nine independent elastic constants and two-dimensional relations require four constants. For isotropic materials, the number of independent elastic constants is only two for both the two- and three-dimensional stress–strain relations. Consequently, for characterization purposes, more measurements have to be made for an orthotropic material than for an isotropic material. However, additional measurements could be made for an isotropic material to calculate different engineering constants, but it will be found that the new constants could be calculated using known relations between them. For example, the Young modulus (E) and Poisson ratio (ν) of an isotropic material can be determined in a uniaxial tension test, and shear modulus (G), in a torsion test. It is well known that the value of G calculated from the values of E and ν using the relation

$$G = \frac{E}{2(1 + \nu)} \qquad (5.80)$$

is in good agreement with the experimentally observed value. For orthotropic materials, Eq. (5.80) is not, in general, valid. The elastic constants E_L, E_T, G_{LT}, and ν_{LT} therefore should be determined independently (methods for their experimental evaluation are discussed in a later chapter). For a specific case of a transversely isotropic composite (e.g., a unidirectional composite), a relation similar to Eq. (5.80), as well as other relations among different constants, can be derived from the symmetry conditions. If the axis perpendicular to the longitudinal and transverse axes is denoted by T', the following relations among the properties can be shown to exist:

$$E_T = E_{T'}$$
$$G_{LT} = G_{LT'}$$
$$\nu_{LT} = \nu_{LT'} \qquad (5.81)$$
$$G_{TT'} = \frac{E_T}{2(1 + \nu_{TT'})}$$

It may be noted that for a transversely isotropic material there are only five independent elastic constants, namely E_L, E_T, G_{LT}, ν_{LT}, and $\nu_{TT'}$. An interested reader may verify this statement by following a procedure based on the invariance of elastic constants for an arbitrary rotation of the T − T' axes about the longitudinal axis [i.e, the procedure similar to the one adopted in deriving Eq. (5.59)].

The stress–strain relations represent a mathematical formulation that describes the behavior of a mathematical model of a real physical problem. Therefore, the elastic constants involved in the formulation should have values that will not violate certain basic physical principles. For example, a tensile force on a real solid body should produce an extension in the direction of the force, or a hydrostatic pressure should not cause an expansion of material. Constraints on the values of the elastic constants of isotropic materials imposed by such conditions are simple. They are as follows: (1) the elastic modulus E, shear modulus G, and bulk modulus K, all should be positive, and (2) the Poisson ratio should have a value between -1 and 0.5. The constraints on elastic constants of an orthotropic material obtained by Lempriere [2] can be stated as follows:

$$E_L, \ E_T, \ E_{T'}, \ G_{LT}, \ G_{LT'}, \ G_{TT'} > 0 \tag{5.82}$$

$$(1 - \nu_{LT}\nu_{TL}), \ (1 - \nu_{LT'}\nu_{T'L}), \ (1 - \nu_{TT'}\nu_{T'T}) > 0 \tag{5.83}$$

$$1 - \nu_{LT}\nu_{TL} - \nu_{LT'}\nu_{T'L} - \nu_{TT'}\nu_{T'T} - 2\nu_{LT}\nu_{TT'}\nu_{T'L} > 0 \tag{5.84}$$

In view of Eq. (5.79), the conditions of Eq. (5.83) can be written as

$$
\begin{aligned}
|\nu_{LT}| &< \left(\frac{E_L}{E_T}\right)^{1/2} & |\nu_{TL}| &< \left(\frac{E_T}{E_L}\right)^{1/2} \\
|\nu_{LT'}| &< \left(\frac{E_L}{E_{T'}}\right)^{1/2} & |\nu_{T'L}| &< \left(\frac{E_{T'}}{E_L}\right)^{1/2} \\
|\nu_{TT'}| &< \left(\frac{E_T}{E_{T'}}\right)^{1/2} & |\nu_{T'T}| &< \left(\frac{E_{T'}}{E_T}\right)^{1/2}
\end{aligned}
\tag{5.85}
$$

A more detailed discussion of the constraints on the elastic constants can be found in Jones [3].

The preceding restrictions on engineering constants can be used to examine experimental data to determine whether they are physically consistent with the mathematical model. If the measured material properties satisfy the constraints, one can proceed with confidence to design structures using these material properties. Otherwise, one has reasons to doubt the experimental techniques.

The restrictions on engineering constants can be helpful in arriving at the physically admissible solution to a practical engineering analysis problem. For example, a governing differential equation may have several solutions depending on the relative values of coefficients in a differential equation. These coefficients in a problem of deformation of a body generally involve the elastic constants. The solutions corresponding to the elastic constants that violate the constraints can be rejected.

5.3.9 Transformation of Stiffness and Compliance Matrices

It was mentioned earlier that a composite material structure is constructed by stacking several unidirectional laminae in a specified sequence of orientation. That is, the principal material directions of each lamina make a different angle with a common set of reference axes. Each lamina is orthotropic and obeys the previously described stress–strain relations referred to its principal material axes. However, for the purpose of analysis and synthesis of laminated structures, it is convenient, in fact necessary, to refer the stress–strain relation to a common-reference coordinate system. Therefore, the stiffness and compliance matrices for an orthotropic lamina referred to arbitrary axes are derived in this section.

Stresses and strains can be transformed from one set of axes to another by following the procedures described in Appendix 2. The stress-transformation equation [Eq. (A.2.12)] can be written as

$$\begin{Bmatrix} \sigma_L \\ \sigma_T \\ \tau_{LT} \end{Bmatrix} = [T_1] \begin{Bmatrix} \sigma_x \\ \sigma_y \\ \tau_{xy} \end{Bmatrix} \tag{5.86}$$

where the stress-transformation matrix $[T_1]$ is

$$[T_1] = \begin{bmatrix} \cos^2 \theta & \sin^2 \theta & 2\sin \theta \cos \theta \\ \sin^2 \theta & \cos^2 \theta & -2\sin \theta \cos \theta \\ -\sin \theta \cos \theta & \sin \theta \cos \theta & \cos^2 \theta - \sin^2 \theta \end{bmatrix} \tag{5.87}$$

The strain-transformation equations [Eq. (A.2.2)] can written as

$$\begin{Bmatrix} \varepsilon_L \\ \varepsilon_T \\ \gamma_{LT} \end{Bmatrix} = [T_2] \begin{Bmatrix} \varepsilon_x \\ \varepsilon_y \\ \gamma_{xy} \end{Bmatrix} \tag{5.88}$$

where the strain-transformation matrix $[T_2]$ is

$$[T_2] = \begin{bmatrix} \cos^2 \theta & \sin^2 \theta & \sin \theta \cos \theta \\ \sin^2 \theta & \cos^2 \theta & -\sin \theta \cos \theta \\ -2\sin \theta \cos \theta & 2\sin \theta \cos \theta & \cos^2 \theta - \sin^2 \theta \end{bmatrix} \tag{5.89}$$

It should be noted that the angle θ is taken positive when the angle of the LT axes measured from xy axes is in the counterclockwise direction, as shown in Fig. 5-4. It may be pointed out that the strain-transformation matrix $[T_2]$ is different from the stress-transformation matrix $[T_1]$ because engineering strains are being transformed rather than the tensorial strains.

Now, to obtain a transformed stiffness matrix, we first write Eq. (5.86) in the inverted form by premultiplying its both sides by the inverse of the $[T_1]$ matrix:

$$\begin{Bmatrix} \sigma_x \\ \sigma_y \\ \tau_{xy} \end{Bmatrix} = [T_1]^{-1} \begin{Bmatrix} \sigma_L \\ \sigma_T \\ \tau_{LT} \end{Bmatrix} \tag{5.90}$$

The procedure for obtaining the $[T_1]^{-1}$ matrix is illustrated in Appendix 1. However, in the present case, $[T_1]^{-1}$ may be obtained from $[T_1]$ by replacing the angle θ by $-\theta$ so that

5.3 HOOKE'S LAW AND STIFFNESS AND COMPLIANCE MATRICES

$$[T_1]^{-1} = \begin{bmatrix} \cos^2\theta & \sin^2\theta & -2\sin\theta\cos\theta \\ \sin^2\theta & \cos^2\theta & 2\sin\theta\cos\theta \\ \sin\theta\cos\theta & -\sin\theta\cos\theta & \cos^2\theta - \sin^2\theta \end{bmatrix} \quad (5.91)$$

Substitution of Eq. (5.74) into Eq. (5.90) gives

$$\begin{Bmatrix} \sigma_x \\ \sigma_y \\ \tau_{xy}^y \end{Bmatrix} = [T_1]^{-1} \begin{bmatrix} Q_{11} & Q_{12} & 0 \\ Q_{12} & Q_{22} & 0 \\ 0 & 0 & Q_{66} \end{bmatrix} \begin{Bmatrix} \varepsilon_L \\ \varepsilon_T \\ \gamma_{LT} \end{Bmatrix} \quad (5.92)$$

Now, substitution of Eq. (5.88) into Eq. (5.92) gives

$$\begin{Bmatrix} \sigma_x \\ \sigma_y \\ \tau_{xy} \end{Bmatrix} = [T_1]^{-1} \begin{bmatrix} Q_{11} & Q_{12} & 0 \\ Q_{12} & Q_{22} & 0 \\ 0 & 0 & Q_{66} \end{bmatrix} [T_2] \begin{Bmatrix} \varepsilon_x \\ \varepsilon_y \\ \gamma_{xy} \end{Bmatrix} \quad (5.93)$$

Thus Eq. (5.93) gives stress–strain relation for an orthotropic lamina referred to arbitrary axes. For the purpose of uniformity, a $[\overline{Q}]$ matrix similar to the $[Q]$ matrix of Eq. (5.74) is defined that relates engineering strains to the stresses referred to arbitrary axes. Then the $[\overline{Q}]$ matrix is defined by the equation

$$\begin{Bmatrix} \sigma_x \\ \sigma_y \\ \tau_{xy} \end{Bmatrix} = \begin{bmatrix} \overline{Q}_{11} & \overline{Q}_{12} & \overline{Q}_{16} \\ \overline{Q}_{12} & \overline{Q}_{22} & \overline{Q}_{26} \\ \overline{Q}_{16} & \overline{Q}_{26} & \overline{Q}_{66} \end{bmatrix} \begin{Bmatrix} \epsilon_x \\ \epsilon_y \\ \gamma_{xy} \end{Bmatrix}$$

$$(5.94)$$

A careful comparison between Eqs. (5.93) and (5.94), along with some algebraic manipulations, will yield the relations between the elements of the $[\overline{Q}]$ matrix and $[Q]$ matrix. The result is as follows:

$$\overline{Q}_{11} = Q_{11}\cos^4\theta + Q_{22}\sin^4\theta + 2(Q_{12} + 2Q_{66})\sin^2\theta\cos^2\theta$$
$$\overline{Q}_{22} = Q_{11}\sin^4\theta + Q_{22}\cos^4\theta + 2(Q_{12} + 2Q_{66})\sin^2\theta\cos^2\theta$$
$$\overline{Q}_{12} = (Q_{11} + Q_{22} - 4Q_{66})\sin^2\theta\cos^2\theta + Q_{12}(\cos^4\theta + \sin^4\theta)$$
$$\overline{Q}_{66} = (Q_{11} + Q_{22} - 2Q_{12} - 2Q_{66})\sin^2\theta\cos^2\theta + Q_{66}(\sin^4\theta + \cos^4\theta)$$
$$\overline{Q}_{16} = (Q_{11} - Q_{12} - 2Q_{66})\cos^3\theta\sin\theta - (Q_{22} - Q_{12} - 2Q_{66})\cos\theta\sin^3\theta$$
$$\overline{Q}_{26} = (Q_{11} - Q_{12} - 2Q_{66})\cos\theta\sin^3\theta - (Q_{22} - Q_{12} - 2Q_{66})\cos^3\theta\sin\theta$$

$$(5.95)$$

192 ANALYSIS OF AN ORTHOTROPIC LAMINA

The $[\overline{Q}]$ matrix is now fully populated and similar in appearance to the $[Q]$ matrix for a fully anisotropic lamina ($\overline{Q}_{16} \neq 0$, $\overline{Q}_{26} \neq 0$). It would seem that there are now six elastic constants that govern the behavior of the lamina. However, $\overline{Q}_{16}$ and $\overline{Q}_{26}$ are not independent but merely linear combinations of the four basic elastic constants. Sometimes Eq. (5.74) is referred to as the constitutive equation for "specially" orthotropic lamina because $Q_{16} = Q_{26} = 0$, and Eq. (5.94), as the constitutive equation for a generally" orthotropic lamina, although both of them apply to the same lamina.

The inverse stress–strain relations referred to arbitrary axes now can be written as

$$\begin{Bmatrix} \epsilon_x \\ \epsilon_y \\ \gamma_{xy} \end{Bmatrix} = \begin{bmatrix} \overline{S}_{11} & \overline{S}_{12} & \overline{S}_{16} \\ \overline{S}_{12} & \overline{S}_{22} & \overline{S}_{26} \\ \overline{S}_{16} & \overline{S}_{26} & \overline{S}_{66} \end{bmatrix} \begin{Bmatrix} \sigma_x \\ \sigma_y \\ \tau_{xy} \end{Bmatrix} \quad (5.96)$$

In a manner similar to the one adopted for obtaining the elements of the $[\overline{Q}]$ matrix in terms of the elements of the $[Q]$ matrix, the elements of the $[\overline{S}]$ matrix also can be obtained in terms of the elements of the compliance matrix. The result is as follows:

$$\overline{S}_{11} = S_{11} \cos^4 \theta + S_{22} \sin^4 \theta + (2S_{12} + S_{66}) \sin^2 \theta \cos^2 \theta$$

$$\overline{S}_{22} = S_{11} \sin^4 \theta + S_{22} \cos^4 \theta + (2S_{12} + S_{66}) \sin^2 \theta \cos^2 \theta$$

$$\overline{S}_{12} = (S_{11} + S_{22} - S_{66}) \cos^2 \theta \sin^2 \theta + S_{12} (\cos^4 \theta + \sin^4 \theta)$$

$$\overline{S}_{66} = 2(2S_{11} + 2S_{22} - 4S_{12} - S_{66}) \cos^2 \theta \sin^2 \theta + S_{66} (\cos^4 \theta + \sin^4 \theta)$$

$$\overline{S}_{16} = (2S_{11} - 2S_{12} - S_{66}) \cos^3 \theta \sin \theta - (2S_{22} - 2S_{12} - S_{66}) \cos \theta \sin^3 \theta$$

$$\overline{S}_{26} = (2S_{11} - 2S_{12} - S_{66}) \cos \theta \sin^3 \theta - (2S_{22} - 2S_{12} - S_{66}) \cos^3 \theta \sin \theta$$

(5.97)

Example 5-4: For the lamina considered in Example 5-2 (Fig. 5-9), first find the stresses and strains in the longitudinal and transverse directions, and then transform the strains to the x and y directions. Compare results with those of Example 5-2.

Stresses in the longitudinal and transverse directions may be obtained from Eq. (5.86) by substituting $\theta = 60°$:

$$\begin{Bmatrix} \sigma_L \\ \sigma_T \\ \tau_{LT} \end{Bmatrix} = \begin{bmatrix} 0.25 & 0.75 & 0.866 \\ 0.75 & 0.25 & -0.866 \\ -0.433 & 0.433 & -0.5 \end{bmatrix} \begin{Bmatrix} -3.5 \\ 7.0 \\ -1.4 \end{Bmatrix} = \begin{Bmatrix} 3.16 \\ 0.34 \\ 5.24 \end{Bmatrix}$$

5.3 HOOKE'S LAW AND STIFFNESS AND COMPLIANCE MATRICES

$$\sigma_L = 3.16 \text{ MPa}$$

$$\sigma_T = 0.34 \text{ MPa}$$

$$\tau_{LT} = 5.24 \text{ MPa}$$

Strains in the longitudinal and transverse directions may be obtained from Eq. (5.10) as follows:

$$\varepsilon_L = \frac{3.16 \times 10^6}{14 \times 10^9} - 0.1 \left(\frac{0.34 \times 10^6}{3.5 \times 10^9} \right) = 216 \times 10^{-6}$$

$$\varepsilon_T = \frac{0.34 \times 10^6}{3.5 \times 10^9} - 0.4 \left(\frac{3.16 \times 10^6}{14 \times 10^9} \right) = 6.9 \times 10^{-6}$$

$$\gamma_{LT} = \frac{5.24 \times 10^6}{4.2 \times 10^9} = 1248 \times 10^{-6}$$

The preceding strains now can be transformed to obtain strains in the x and y directions using the inverse of Eq. (5.88) as follows:

$$\begin{Bmatrix} \varepsilon_x \\ \varepsilon_y \\ \gamma_{xy} \end{Bmatrix} = \begin{bmatrix} 0.25 & 0.75 & -0.433 \\ 0.75 & 0.25 & 0.433 \\ 0.866 & -0.866 & -0.5 \end{bmatrix} \begin{Bmatrix} 216 \times 10^{-6} \\ 6.9 \times 10^{-6} \\ 1248 \times 10^{-6} \end{Bmatrix}$$

$$\varepsilon_x = -481 \times 10^{-6}$$

$$\varepsilon_y = 704 \times 10^{-6}$$

$$\gamma_{xy} = -442 \times 10^{-6}$$

These strains differ slightly from those calculated in Example 5-2 only because of rounding-off errors.

Example 5-5: If longitudinal and transverse axes of the lamina considered in Example 5-3 make a counterclockwise angle of 30° with the reference axes, determine $\overline{Q}$ and $\overline{S}$ matrices.

Substitution of $\theta = 30°$ in Eqs. (5.87) and (5.89) gives stress- and strain-transformation matrices as

$$[T_1] = \begin{bmatrix} 0.750 & 0.250 & 0.866 \\ 0.250 & 0.750 & -0.866 \\ -0.433 & 0.433 & 0.500 \end{bmatrix}$$

$$[T_2] = \begin{bmatrix} 0.750 & 0.250 & 0.433 \\ 0.250 & 0.750 & -0.433 \\ -0.866 & 0.866 & 0.500 \end{bmatrix}$$

Inversion of $[T_1]$ and $[T_2]$ gives

$$[T_1]^{-1} = \begin{bmatrix} 0.750 & 0.250 & -0.866 \\ 0.250 & 0.750 & 0.866 \\ 0.433 & -0.433 & 0.500 \end{bmatrix}$$

$$[T_2]^{-1} = \begin{bmatrix} 0.750 & 0.250 & -0.433 \\ 0.250 & 0.750 & 0.433 \\ 0.866 & -0.866 & 0.500 \end{bmatrix}$$

From Eq. (5.93),

$$[\overline{Q}] = [T_1]^{-1} [Q] [T_2]$$

Substitution of $[T_1]^{-1}$, $[Q]$, and $[T_2]$ gives

$$[\overline{Q}] = \begin{bmatrix} 89.84 & 27.68 & 44.11 \\ 27.68 & 20.65 & 15.81 \\ 44.11 & 15.81 & 30.12 \end{bmatrix} \text{ GPa}$$

Similarly,

$$[\overline{S}] = [T_2]^{-1} [S][T_1]$$

Substitution of $[T_2]^{-1}$, $[S]$, and $[T_1]$ gives

$$[\overline{S}] = \begin{bmatrix} 0.0424 & -0.0156 & -0.0539 \\ -0.0156 & 0.0867 & -0.0227 \\ -0.0539 & -0.0227 & 0.1241 \end{bmatrix} \text{ GPa}^{-1}$$

5.3.10 Invariant Forms of Stiffness and Compliance Matrices

Design of laminates invariably requires a decision on constituent laminae orientations to meet stiffness and strength requirements in different directions.

5.3 HOOKE'S LAW AND STIFFNESS AND COMPLIANCE MATRICES

Contribution of a lamina to the laminate stiffness in any direction can be obtained through the stiffness transformation equations (5.95). However, that form of equations is not very convenient to visualize the consequences of changing orientation of a lamina in a laminate. Tsai and Pagano [4] have recast transformation equations in an invariant form, that makes it easier to visualize the effects of changing lamina orientation, and thereby, simplifying the laminate design process. The invariant form of Eq. (5.95) is

$$\overline{Q}_{11} = U_1 + U_2 \cos 2\theta + U_3 \cos 4\theta$$
$$\overline{Q}_{22} = U_1 - U_2 \cos 2\theta + U_3 \cos 4\theta$$
$$\overline{Q}_{12} = U_4 - U_3 \cos 4\theta \qquad (5.98)$$
$$\overline{Q}_{16} = \frac{1}{2} U_2 \sin 2\theta - U_3 \sin 4\theta$$
$$\overline{Q}_{26} = \frac{1}{2} U_2 \sin 2\theta - U_3 \sin 4\theta$$
$$\overline{Q}_{66} = U_5 - U_3 \cos 4\theta$$

where

$$U_1 = \frac{3Q_{11} + 3Q_{22} + 2Q_{12} + 4Q_{66}}{8}$$
$$U_2 = \frac{Q_{11} - Q_{22}}{2}$$
$$U_3 = \frac{Q_{11} + Q_{22} - 2Q_{12} - 4Q_{66}}{8} \qquad (5.99)$$
$$U_4 = \frac{Q_{11} + Q_{22} + 6Q_{12} - 4Q_{66}}{8}$$
$$U_5 = \frac{Q_{11} + Q_{22} - 2Q_{12} + 4Q_{66}}{8}$$

In the above equations, expressions for $\overline{Q}_{11}$, $\overline{Q}_{22}$, $\overline{Q}_{12}$ and $\overline{Q}_{66}$, are composed of an invariant or constant part (U_1, U_4, or U_5), that does not change with the lamina orientation, θ, and another part that changes with θ. This form of expression is useful when examining the consequences of changing lamina orientation to achieve a certain stiffness profile. This concept of invariance will be more useful in the study of laminates (Chap. 6) because laminates are made of a collection of laminae at different orientations to achieve certain mechanical properties.

The invariant form of the compliance matrix [Eq. (5.97)] is

$$\overline{S}_{11} = V_1 + V_2 \cos 2\theta + V_3 \cos 4\theta$$
$$\overline{S}_{22} = V_1 - V_2 \cos 2\theta + V_3 \cos 4\theta$$
$$\overline{S}_{12} = V_4 - V_3 \cos 4\theta \quad\quad (5.100)$$
$$\overline{S}_{16} = V_2 \sin 2\theta + 2V_3 \sin 4\theta$$
$$\overline{S}_{26} = V_2 \sin 2\theta - 2V_3 \sin 4\theta$$
$$\overline{S}_{66} = V_5 - V_3 \cos 4\theta$$

where

$$V_1 = \frac{3S_{11} + 3S_{22} + 2S_{12} + 4S_{66}}{8}$$
$$V_2 = \frac{S_{11} - S_{22}}{2}$$
$$V_3 = \frac{S_{11} + S_{22} - 2S_{12} - S_{66}}{8} \quad\quad (5.101)$$
$$V_4 = \frac{S_{11} + S_{22} + 6S_{12} - S_{66}}{8}$$
$$V_5 = \frac{S_{11} + S_{22} + 2S_{12} - S_{66}}{2}$$

5.4 STRENGTHS OF AN ORTHOTROPIC LAMINA

Strengths of a material are obtained experimentally by subjecting suitable specimens to loads that produce simple stress fields in the test specimen and by determining the load at which failure occurs. For example, ultimate tensile and compressive strengths of isotropic materials are obtained through tests that produce uniaxial tensile and compressive stresses, respectively, in the test section of the specimen.

A design engineer estimates the load-carrying capacity of a structure or a component through the procedures that involve stress analysis of the structure and a comparison of the actual stress field with the strengths of the material. When the actual stress field is simple, such as the one produced in the specimens during strength-determination tests, a direct comparison can be made and the load-carrying capacity of the structure obtained. Otherwise, a direct comparison may not be valid. In reality, structures may be subjected to biaxial stress states, and it is impractical to establish the strength characteristics of materials for every possible biaxial stress state. A "failure criterion" or "failure theory," if valid, can predict strength of materials under biaxial stress states using strength data obtained from uniaxial tests. For isotropic materials, the failure criteria are written in terms of principal stresses and ultimate tensile, compressive, and shear strengths. Thus the load-carrying capacity of a structure made of an isotropic material can be predicted from knowledge of these three strengths.

In the case of orthotropic materials, the situation is considerably more complex. The most important complexity arises from the fact that their strengths, like their elastic constants, are direction-dependent. Thus, for an orthotropic material, an infinite number of strength values can be obtained even through uniaxial tests, depending on the direction of load application. For prediction purposes, they can be limited to five strengths in the principal material directions. These strengths are the longitudinal tensile strength (σ_{LU}), transverse tensile strength (σ_{TU}), shear strength (τ_{LTU}), longitudinal compressive strength (σ'_{LU}), and transverse compressive strength (σ'_{TU}). As a consequence of the fact that the basic uniaxial strength values for an orthotropic material are known only along the principal material axes, the first step in all calculations related to their strengths has to be the transformation of the actual stress field to the principal material axes so that a comparison can be made with the appropriate strengths. Further, whenever the actual stress field referred to the principal material axes is multiaxial, a direct comparison of the actual stress field with the preceding uniaxial strengths may not be valid, and a suitable failure criterion must be used. It may be pointed out here that a uniaxial stress applied in any direction other than the principal material axes produces multiaxial stresses along the principal material axes. Therefore, off-axis uniaxial strengths of orthotropic materials must be predicted, like their strengths under complex stress states, through an appropriate failure criterion. All the failure criteria for orthotropic materials are, quite obviously, written in terms of stresses along principal material axes rather than the principal stresses.

There are a number of theories for predicting the failure of isotropic materials and orthotropic materials subjected to a complex stress state. Nahas [5] has presented a useful survey of failure and postfailure theories of laminated-fiber-reinforced composites, giving a large number of references. Many failure theories for orthotropic materials have been developed from the failure theories of isotropic materials. Many failure theories are not general but are applicable only to some specific types of composites. In this section, three of the strength theories used widely for fiber composites, based on maximum stress, maximum strain, and maximum work, are discussed.

5.4.1 Maximum-Stress Theory

This theory states that failure will occur if any of the stresses in the principal material axes exceed the corresponding allowable stress. Thus the following inequalities must be satisfied to avoid failure:

ANALYSIS OF AN ORTHOTROPIC LAMINA

$$\sigma_L < \sigma_{LU}$$
$$\sigma_T < \sigma_{TU} \quad (5.102)$$
$$\tau_{LT} < \tau_{LTU}$$

If the normal stresses are compressive, σ_{LU} and σ_{TU} in Eq. (5.102) must be replaced by the allowable compressive stresses:

$$\sigma_L < \sigma'_{LU}$$
$$\sigma_T < \sigma'_{TU} \quad (5.103)$$

According to this theory, when one of the inequalities indicated by Eqs. (5.102) and (5.103) is violated, the material is considered to have failed by a failure mode associated with the allowable stress. There is no interaction between the modes of failure in this criterion. Thus this is actually not one criterion but five subcriteria.

To illustrate the application of this theory, consider an orthotropic lamina subjected to a stress σ_x making an angle θ with the longitudinal direction. The stresses in the principal material direction are obtained by transformation as

$$\sigma_L = \sigma_x \cos^2 \theta$$
$$\sigma_T = \sigma_x \sin^2 \theta \quad (5.104)$$
$$\tau_{LT} = -\sigma_x \sin \theta \cos \theta$$

Now, from Eq. (5.102), failure will occur if σ_x exceeds the smallest of $(\sigma_{LU}/\cos^2 \theta)$, $(\sigma_{TU}/\sin^2 \theta)$, or $(\tau_{LTU}/\sin \theta \cos \theta)$. The maximum-stress theory is applied to a typical glass–epoxy composite with the following normalized material properties:

$$\frac{\sigma_{TU}}{\sigma_{LU}} = 0.025 \qquad \frac{\tau_{LTU}}{\sigma_{LU}} = 0.05$$

$$\frac{\sigma'_{LU}}{\sigma_{LU}} = 1 \qquad \frac{\sigma'_{TU}}{\sigma_{LU}} = 0.125$$

$$\nu_{LT} = 0.25 \qquad \nu_{TL} = 0.08$$

The same material properties are used for discussing two other strength theories. Predictions of the maximum-stress theory are shown in Fig. 5-13. Solid lines represent the variation of allowable stress σ_x with the orientation θ.

5.4 STRENGTHS OF AN ORTHOTROPIC LAMINA

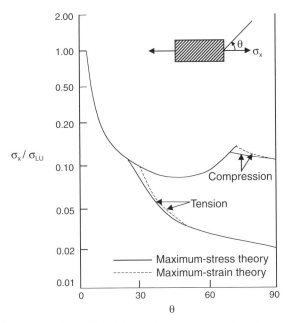

Figure 5-13. Off-axis strength predicted by maximum-stress and maximum-strain theories of failure.

Example 5-6: A unidirectional glass–epoxy lamina, shown in Fig. 5-14, has the following allowable stresses:

$$\sigma_{LU} = 1062 \text{ MPa}$$

$$\sigma'_{LU} = 610 \text{ MPa}$$

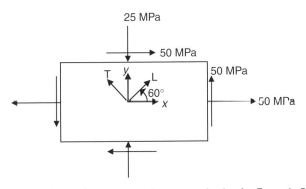

Figure 5-14. State of stress on a glass–epoxy lamina for Example 5-6.

$$\sigma_{TU} = 31 \text{ MPa}$$

$$\sigma'_{TU} = 118 \text{ MPa}$$

$$\tau_{LTU} = 72 \text{ MPa}$$

Determine if, according to the maximum-stress theory, the lamina will fail under the applied stresses.

The off-axis stresses applied to the lamina are

$$\begin{Bmatrix} \sigma_x \\ \sigma_y \\ \tau_{xy} \end{Bmatrix} = \begin{Bmatrix} 50 \\ -25 \\ 50 \end{Bmatrix} \text{ MPa}$$

Since the fibers are oriented at 60° to the x axis, the stress-transformation matrix is

$$[T_1] = \begin{bmatrix} 0.250 & 0.750 & 0.866 \\ 0.750 & 0.250 & -0.866 \\ -0.433 & 0.433 & -0.500 \end{bmatrix}$$

Stress transformation gives

$$\begin{Bmatrix} \sigma_L \\ \sigma_T \\ \tau_{LT} \end{Bmatrix} = \begin{bmatrix} 0.250 & 0.750 & 0.866 \\ 0.750 & 0.250 & -0.866 \\ -0.433 & 0.433 & -0.500 \end{bmatrix} \begin{Bmatrix} 50 \\ -25 \\ 50 \end{Bmatrix} = \begin{Bmatrix} 37.05 \\ -12.05 \\ -57.48 \end{Bmatrix} \text{ MPa}$$

Comparison of allowable stresses σ_{LU}, σ'_{TU}, and τ_{LTU} with the calculated stresses σ_L, σ_T, and τ_{LT} shows that, according to the maximum-stress theory, the lamina will not fail under the applied stresses.

5.4.2 Maximum-Strain Theory

This theory states that failure will occur if any of the strains in the principal material axes exceed the corresponding allowable strain. Thus the following inequalities must be satisfied for "no failure":

$$\epsilon_L < \epsilon_{LU}$$
$$\epsilon_T < \epsilon_{TU} \qquad (5.105)$$
$$\gamma_{LT} < \gamma_{LTU}$$

If normal strains are compressive, ϵ_{LU} and ϵ_{TU} in Eq. (5.105) must be replaced by the allowable compressive strains:

5.4 STRENGTHS OF AN ORTHOTROPIC LAMINA

$$\epsilon_L < \epsilon'_{LU}$$
$$\epsilon_T < \epsilon'_{TU} \quad (5.106)$$

The maximum-strain theory is similar to the maximum-stress theory. All the stresses are replaced by the corresponding strains first to apply the maximum-strain theory. If the material is assumed to be linearly elastic up to ultimate failure, the ultimate strains (allowable strains) in Eqs. (5.105) and (5.106) can be related directly to the strengths:

$$\epsilon_{LU} = \frac{\sigma_{LU}}{E_L}$$

$$\epsilon_{TU} = \frac{\sigma_{TU}}{E_T} \quad (5.107)$$

$$\gamma_{LTU} = \frac{\tau_{LTU}}{G_{LT}}$$

Consider an orthotropic lamina subjected to a stress σ_x making an angle θ with the longitudinal direction, as considered earlier to illustrate the maximum-stress theory. The stresses in the principal material directions can be obtained by Eq. (5.104), and the strains in the principal material direction are calculated by using Eq. (5.10):

$$\epsilon_L = \frac{1}{E_L}(\cos^2\theta - \nu_{LT}\sin^2\theta)\sigma_x$$

$$\epsilon_T = \frac{1}{E_T}(\sin^2\theta - \nu_{TL}\cos^2\theta)\sigma_x \quad (5.108)$$

$$\gamma_{LT} = -\frac{1}{G_{LT}}(\sin\theta\cos\theta)\sigma_x$$

Now Eqs. (5.105) and (5.107) predict that failure will occur if σ_x exceeds the smallest of $\sigma_{LU}/(\cos^2\theta - \nu_{LT}\sin^2\theta)$, $\sigma_{TU}/(\sin^2\theta - \nu_{TL}\cos^2\theta)$, or $\tau_{LTU}/(\sin\theta\cos\theta)$. Predictions of the maximum-strain theory for the glass–epoxy composite considered earlier also have been shown in Fig. 5-13 along with the predictions of the maximum-stress theory. Predictions of the two theories are quite close to each other. This is so because the material has been assumed to be linearly elastic up to ultimate failure. The differences are a result of the effect of the Poisson ratio. When the material does not remain linearly elastic up to failure, the two theories are completely independent and have to be applied separately. In that case, larger differences in their predictions should be expected.

ANALYSIS OF AN ORTHOTROPIC LAMINA

Example 5-7: The lamina considered in Example 5-6 has the following elastic constants:

$$E_L = 38.6 \text{ GPa}$$

$$E_T = 8.27 \text{ GPa}$$

$$\nu_{LT} = 0.26$$

$$G_{LT} = 4.14 \text{ GPa}$$

Determine if, according to the maximum-strain theory, the lamina will fail. Assume that the lamina deforms linearly up to failure.

Lamina strains can be obtained using Eq. (5.10) and the stresses σ_L, σ_T, and τ_{LT} calculated in Example 5-6:

$$\varepsilon_L = \frac{37.05}{38.6 \times 10^3} - \frac{0.26}{38.6 \times 10^3} \times (-12.05) = 0.00104$$

$$\varepsilon_T = -\frac{12.05}{8.27 \times 10^3} - \left(\frac{0.26}{38.6 \times 10^3}\right) \times (37.05) = -0.00170$$

$$\gamma_{LT} = -\frac{57.48}{4.14 \times 10^3} = -0.01388$$

The lamina allowable strains can be obtained from allowable stresses and elastic constants as follows:

$$\varepsilon_{LU} = \frac{\sigma_{LU}}{E_L} = 0.0275$$

$$\varepsilon'_{LU} = \frac{\sigma'_{LU}}{E_L} = 0.0158$$

$$\varepsilon_{TU} = \frac{\sigma_{TU}}{E_T} = 0.0037$$

$$\varepsilon'_{TU} = \frac{\sigma'_{TU}}{E_T} = 0.0143$$

$$\gamma_{LTU} = \frac{\tau_{LTU}}{G_{LT}} = 0.0174$$

Comparison of allowable strains ε_{LU}, ε'_{TU}, and γ_{LTU} with the calculated strains ε_L, ε_T, and γ_{LT} shows that, according to the maximum-strain theory, the lamina will not fail.

5.4.3 Maximum-Work Theory

This theory states that in plane-stress states the failure initiates when the following inequality is violated:

$$\left(\frac{\sigma_L}{\sigma_{LU}}\right)^2 - \left(\frac{\sigma_L}{\sigma_{LU}}\right)\left(\frac{\sigma_T}{\sigma_{LU}}\right) + \left(\frac{\sigma_T}{\sigma_{TU}}\right)^2 + \left(\frac{\tau_{LT}}{\tau_{LTU}}\right)^2 < 1 \qquad (5.109)$$

When normal stresses are compressive, the corresponding compressive strengths are to be used in Eq. (5.109).

The theory was derived in this form by Tsai [6] from a yield criterion for anisotropic materials proposed by Hill [7]. Therefore, it is sometimes referred to as the *Tsai–Hill theory*. Application of this theory can be illustrated by the example used in the previous cases where an off-axis stress σ_x acts on an orthotropic lamina. The stresses in the principal material directions given by Eq. (5.104) can be substituted directly into Eq. (5.109) to obtain the failure criterion as

$$\frac{\cos^4 \theta}{\sigma_{LU}^2} - \frac{\cos^2 \theta \sin^2 \theta}{\sigma_{LU}^2} + \frac{\sin^4 \theta}{\sigma_{TU}^2} + \frac{\sin^2 \theta \cos^2 \theta}{\tau_{LTU}^2} < \frac{1}{\sigma_x^2} \qquad (5.110)$$

Thus the maximum-work theory provides a single function to predict strength. This criterion does take into consideration the interaction between strengths, which was not considered in the maximum-stress or maximum-strain theory. Predictions of the maximum-work theory for the glass–epoxy composite considered earlier are shown in Fig. 5-15. Also shown in the figure are the predictions of the maximum-stress theory. The maximum-work theory predicts slightly lower strength compared with that predicted by the maximum-stress theory. The largest differences occur at the points where the maximum-stress theory predicts a change in the failure mode, that is, from the shear mode to the longitudinal or transverse tensile failure mode. The maximum-work theory has found wider acceptability compared with the other two theories primarily because of the smooth variation of strength according to a single equation [Eq. (5.109)]. Experimental support for this theory has been reported by many investigators [8–12].

The theories discussed in the preceding paragraphs are applicable to cases where the state of stress is biaxial. When all three stresses σ_x, σ_y, and τ_{xy} are acting at a point, the theories may be applied in a manner similar to the one used earlier for a uniaxial off-axis stress. The first step in the application of the theories is to transform the stresses σ_x, σ_y, and τ_{xy} to the stresses σ_L, σ_T,

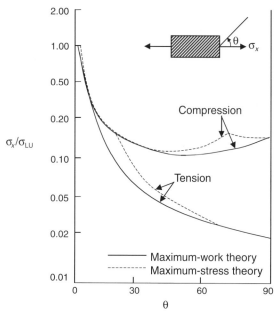

Figure 5-15. Off-axis strength predicted by maximum-work (Tsai–Hill) and maximum-stress theories of failure.

and τ_{LT} in the longitudinal and transverse directions by means of the transformation law [Eq. (5.86)]. Now failure may be predicted by comparing σ_L, σ_T, and τ_{LT} with the corresponding allowable stresses in the case of the maximum-stress theory or by examining the failure criterion [Eq. (5.109)] for the maximum-work theory. Thus, for one set of reference axes (xy axes), there will be infinite combinations of σ_x, σ_y, and τ_{xy} that will cause failure. A graphic representation of the failure conditions is quite difficult in this case because (1) one graphic representation is valid for only one set of reference axes (a new representation is needed for any change in the orientation of reference axes), and (2) a two-dimensional representation of the type used for isotropic materials is not possible because the direction of principal stresses does not, in general, coincide with either the reference axes or the longitudinal and transverse directions. Thus, in the case of multiaxial stress states, it is best to apply the failure conditions separately for each case.

Although it is not possible to construct a graphic representation of a failure criterion for a general case of biaxial stresses in which all three stresses σ_L, σ_T, and τ_{LT} are nonzero, it is interesting to consider some specific cases. For a case in which τ_{LT} vanishes (i.e., when principal stress directions coincide with the longitudinal and transverse directions), the Tsai–Hill criterion can be represented on normalized stress axes σ_L/σ_{LU} and σ_T/σ_{TU}. The failure envelope in the normalized stress plane will depend on the ratio of strengths in the longitudinal and transverse directions (σ_{LU}/σ_{TU}). Failure envelopes for three different values of (σ_{LU}/σ_{TU}) are shown in Fig. 5-16. When strengths in the longitudinal and transverse directions are equal ($\sigma_{LU}/\sigma_{TU} = 1$), the failure ellipse represents a case of an isotropic material. In the other extreme case as the ratio (σ_{LU}/σ_{TU}) approaches an infinite value (e.g., when the transverse strength is negligibly small), the failure envelope becomes a circle. For a practical case of an orthotropic material in which the ratio of strengths has a value between one and infinity, the failure envelope will be an ellipse lying between the ellipse for an isotropic case and the circle. Influence of a nonzero shear stress τ_{LT} is shown in Fig. 5-17. For a fixed value of σ_{LU}/σ_{TU}), the effect of shear stress is to reduce the size of the failure envelope. As the value of τ_{LT}/τ_{LTU} increases, the major and minor axes of the failure ellipse become smaller with no change in orientations.

Example 5-8: Determine if, according to the maximum-work theory, the lamina in Example 5-6 will fail under the applied stresses.

To apply the maximum-work theory, the left-hand side of Eq. (5.109) can be evaluated as follows:

$$\left(\frac{37.05}{1062}\right)^2 - \left(\frac{37.05}{1062}\right)\left(\frac{12.05}{610}\right) + \left(\frac{12.05}{118}\right)^2 + \left(\frac{57.48}{72}\right)^2 = 0.65 < 1$$

Therefore, according to the maximum-work theory, the lamina will not fail under the applied stresses.

5.4.4 Importance of Sign of Shear Stress on Strength of Composites

A sign convention for stresses that is almost universally accepted is discussed in Appendix 2. It can be stated as "on a plane where the outward normal is in the positive direction of a coordinate axis, all the stress components acting in the positive directions of the axes are positive." According to this conven-

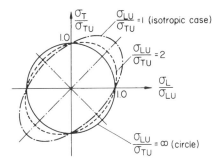

Figure 5-16. Failure envelopes on a normalized stress plane for zero shear stress.

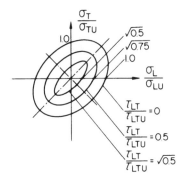

Figure 5-17. Influence of shear stress on failure envelopes.

tion, a stress component in the negative direction of a coordinate axis is negative when the outward normal to the surface on which it is acting is in the positive direction of the coordinate axis parallel to the normal. Positive and negative stress components on a two-dimensional element are shown in Fig. 5-18a and Fig. 5-18b, respectively. This sign convention is in agreement with the understanding that tensile stress is positive and compressive stress is negative.

It is generally recognized that the tensile and compressive strengths of materials are different. It is not generally appreciated, however, that the shear strength of an anisotropic material is dependent on the direction of the shear stress, particularly when the reference axes are different from the principal material axes. This statement can be illustrated by considering positive and negative shear stresses applied to a unidirectionally reinforced lamina. When the fibers are aligned parallel to a reference axis (as in Fig. 5-19), there is no difference between the stress fields labeled "positive" and "negative" shear stress. The two stress fields are mirror images of each other even when the principal stresses are examined as in the lower half of Fig. 5-19. Thus the shear strength is the same in both cases and is independent of the longitudinal and transverse strengths of the lamina.

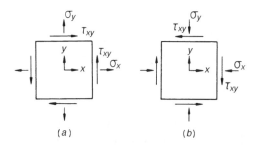

Figure 5-18. Sign conventions for stress components: (a) positive and (b) negative.

5.4 STRENGTHS OF AN ORTHOTROPIC LAMINA

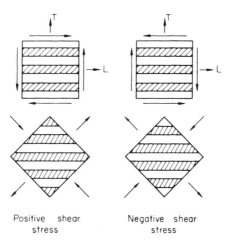

Figure 5-19. Stress fields for positive and negative shear stress with fibers aligned parallel to reference axis.

Consider that the lamina now has fibers oriented at 45° to one of the reference axes and is subjected to positive and negative shear stresses, as shown in Fig. 5-20. In this case, positive and negative shear stresses result in normal stresses of opposite signs in the longitudinal and transverse directions. For positive shear stress, tensile stresses result in the fiber direction and compressive stresses in the transverse direction. For negative shear stress, compressive stresses exist in the fiber direction and tensile stresses in the

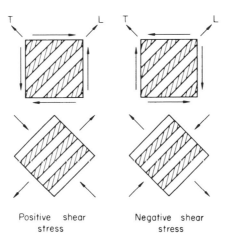

Figure 5-20. Stress fields for positive and negative shear stress with fibers oriented at 45° to reference axis.

transverse direction. Since the stresses in the fiber direction and perpendicular to it are of equal magnitude in both cases, it is reasonable to assume that the shear strength is largely controlled by the transverse strength of the lamina (longitudinal strength generally is much greater). Thus, for this 45° lamina, the apparent shear strength for a negative shear stress will be lower than that for a positive shear stress because transverse tensile strength of a lamina generally is smaller than the compressive strength.

Similar arguments can be extended to other fiber orientations. Thus off-axis shear strength of a lamina depends not only on the fiber orientation but also on the sign of applied shear stress. The following numerical example illustrates the point further.

Example 5-9: A unidirectional lamina of glass–epoxy composite shows the following strength properties:

$$\sigma_{LU} = 500 \text{ MPa}$$

$$\sigma'_{LU} = 350 \text{ MPa}$$

$$\sigma_{TU} = 5 \text{ MPa}$$

$$\sigma'_{TU} = 75 \text{ MPa}$$

$$\tau_{LTU} = 35 \text{ MPa}$$

Estimate off-axis shear strength of the lamina for fiber orientations of 15°, 45° and 60°. Failure may be predicted by using the Tsai–Hill criterion.

When τ_{xy} is the only nonzero stress, the stress in the longitudinal and transverse directions may be obtained as

$$\sigma_L = \tau_{xy} \sin 2\theta$$

$$\sigma_T = -\tau_{xy} \sin 2\theta$$

$$\tau_{LT} = \tau_{xy} \cos 2\theta$$

When the sign of applied shear stress is positive, σ_L will be tensile and σ_T compressive. In that case, the Tsai–Hill failure criterion may be written as

$$\frac{\tau_{xy}^2 \sin^2 2\theta}{500 \times 500} - \frac{\tau_{xy}^2 \sin^2 2\theta}{500 \times 350} + \frac{\tau_{xy}^2 \sin^2 2\theta}{75 \times 75} + \frac{\tau_{xy}^2 \cos^2 2\theta}{35 \times 35} = 1$$

When the sign of applied stress is negative, σ_L will be compressive and σ_T tensile. In that case, the failure criterion may be written as

$$\frac{\tau_{xy}^2 \sin^2 2\theta}{350 \times 350} - \frac{\tau_{xy}^2 \sin^2 2\theta}{500 \times 350} + \frac{\tau_{xy}^2 \sin^2 2\theta}{5 \times 5} + \frac{\tau_{xy}^2 \cos^2 2\theta}{35 \times 35} = 1$$

By substituting different values of θ, the following shear strengths are obtained:

$\theta =$	15°	45°	60°
$+\tau_{xy} =$	39.03	75.36	54.54
$-\tau_{xy} =$	9.71	5.00	5.75

EXERCISE PROBLEMS

5.1. Derive expressions for E_y, ν_{yx}, and m_y by assuming that σ_y, is the only nonzero stress acting on the lamina shown in Fig. 5-4. Compare your results with Eqs. (5.26), (5.28), and (5.32).

5.2. Derive expressions for m_x and m_y by assuming that τ_{xy} is the only non-zero stress acting on the lamina shown in Fig. 5-4. Compare your results with Eqs. (5.30) and (5.32).

5.3. Plot the variation of E_x, G_{xy}, ν_{xy}, m_x, and m_y for a lamina with the following properties:

$$E_L = 35 \text{ GPa} \quad E_T = 3.5 \text{ GPa}$$

$$G_{LT} = 4 \text{ GPa} \quad \nu_{LT} = 0.45$$

5.4. Plot the variations of E_x, G_{xy}, ν_{xy}, m_x, and m_y for a balanced lamina with the following properties:

$$E_L = E_T = 15 \text{ GPa}$$

$$G_{LT} = 2.5 \text{ GPa}$$

$$\nu_{LT} = \nu_{TL} = 0.20$$

5.5. Calculate E_x, G_{xy}, ν_{xy}, m_x, and m_y at 30°, 45°, and 60° for an orthotropic lamina having the following properties:

$$E_L = 14 \text{ GPa} \quad E_T = 3.5 \text{ GPa}$$

$$G_{LT} = 4.2 \text{ GPa} \quad \nu_{LT} = 0.4$$

5.6. Verify that the elastic constants given in Eq. (5.55) are actually zero for a material that has the $x_1 x_2$ plane as a plane of symmetry.

5.7. Verify that the elastic constants given in Eq. (5.58) are actually zero for a material that has the $x_2 x_3$ plane as a plane of symmetry.

5.8. Obtain the elastic constants that must be zero for a material having $x_1 x_3$ plane as a plane of symmetry. Indicate which of these constants are common to Eqs. (5.55) and (5.58).

5.9. Obtain the stiffness and compliance matrices for a unidirectional lamina that has the following elastic constants:

$$E_L = 20 \text{ GPa} \qquad E_T = 2 \text{ GPa}$$

$$G_{LT} = 0.7 \text{ GPa} \qquad \nu_{LT} = 0.35$$

5.10. Following are the experimentally observed elastic constants of a boron–epoxy composite:

$$E_L = 81.7 \text{ GPa} \qquad E_T = 9.1 \text{ GPa}$$

$$\nu_{LT} = 1.97 \qquad \nu_{TL} = 0.22$$

Determine whether the data satisfy the constraints on the elastic constants of an orthotropic material. Comment on the large value of ν_{LT}. Will it be an admissible number for the Poisson ratio of an isotropic material?

5.11. Show that the inequalities in Eq. (5.83) reduce to $-1 < \nu < 1$ for isotropic materials.

5.12. Show that the inequality in Eq. (5.84) correctly reduces to $\nu < \frac{1}{2}$ for isotropic materials.

5.13. Calculate the $\overline{Q}$ matrix at 30°, 45° and 60° for a lamina whose Q matrix is given by

$$[Q] = \begin{bmatrix} 20 & 0.7 & 0 \\ 0.7 & 2 & 0 \\ 0 & 0 & 0.7 \end{bmatrix}$$

5.14. A tensile specimen of a unidirectional composite with a rectangular cross section of dimensions 12.5 mm × 4 mm has the fibers oriented at 45° to a longitudinal edge of the specimen. It is subjected to an axial force of 500 N.

(a) Calculate normal strains in the axial and perpendicular directions and shear strain on the specimen. Properties of the composite in

the longitudinal and transverse directions are the same as those given in Exercise Problem 5.5.
(b) Calculate off-axis modulus, the Poisson ratio, and cross-coupling coefficient for this specimen. Compare these values with the ones obtained in Exercise Problem 5.5 by direct transformation.

5.15. Repeat Exericse Problem 5.14 for a fiber orientation of 30°.

5.16. Longitudinal axis of an orthotropic lamina makes an angle of 45° with the x axis. It is subjected to the following stresses:

$$\sigma_x = 20 \text{ MPa}$$

$$\sigma_y = 0$$

$$\tau_{xy} = 20 \text{ MPa}$$

(a) Draw a neat sketch of the lamina indicating x, y, L, and T axes. Also show the applied stresses.
(b) Calculate stresses along longitudinal and transverse directions.
(c) Does the lamina fail under these stresses? Use the maximum work theory to predict failure and take the following strength values for the lamina:

$$\sigma_{LU} = 500 \text{ MPa} \qquad \sigma_{TU} = 10 \text{ MPa}$$

$$\sigma'_{LU} = 350 \text{ MPa} \qquad \sigma'_{TU} = 75 \text{ MPa}$$

$$\tau_{LTU} = 35 \text{ MPa}$$

(d) Will your answer change if the direction of applied shear stress is reversed?

5.17. Repeat Exercise Problem 5.16 assuming that the angle between the longitudinal and x axes is 30°.

5.18. A graphite–epoxy lamina shows the following strength properties:

$$\sigma_{LU} = 1725 \text{ MPa} \qquad \sigma_{TU} - 40 \text{ MPa}$$

$$\sigma'_{LU} = 1350 \text{ MPa} \qquad \sigma'_{TU} = 275 \text{ MPa}$$

$$\tau_{LTU} = 95 \text{ MPa}$$

Using the maximum-work theory of failure, estimate off-axis shear strength of the lamina for orientations of 30° and 45°.

REFERENCES

1. R. M. Jones, "Stiffness of Orthotropic Materials and Laminated-Fiber-Reinforced Composites," *AIAA J.*, **12**(1), 112–114 (1974).
2. B. M. Lempriere, "Poisson's Ratio in Orthotropic Materials," *AIAA J.*, **6**(11), 2226–2277 (1968).
3. R. M. Jones, *Mechanics of Composite Materials*, 2d ed., Taylor & Francis, Philadelphia, 1999.
4. S. W. Tsai and N. J. Pagano, "Invariant Properties of Composite Materials," in Composite Materials Workshop, S. W. Tsai, J. C. Halpin, and N. J. Pagano, eds., St. Louis, July 13–21, 1967, Technomic, Westport, CT, 1968; also AFML-TR-67-349, March 1968.
5. M. N. Nahas, "Survey of Failure and Post-Failure Theories of Laminated Fiber-Reinforced Composites," *J. Comp. Tech. Res.*, **8**(4), 138–153 (1986).
6. S. W. Tsai, "Strength Theories of Filamentary Structures," in R. T. Schwartz and H. S. Schwartz, eds., *Fundamental Aspects of Fiber Reinforced Plastic Composites*, Interscience, New York, 1968, Chap. 1.
7. R. Hill, *The Mathematical Theory of Plasticity*, Oxford University Press, Oxford, U.K., 1950.
8. B. D. Agarwal and J. N. Narang, "Strength and Failure Mechanism of Anisotropic Composites," *Fibre Sci. Technol.*, **10**(1), 37–52 (1977).
9. B. H. Jones, "Determination of Design Allowables for Composite Materials," in *Composite Materials: Testing and Design*, STP 460, American Society for Testing and Materials, Philadelphia, PA, 1969, pp. 307–320.
10. H. E. Brandmaier, "Optimum Filament Orientation Criterion," *J. Compos. Mater.*, **4**(3), 422–425 (1970).
11. B. Harris, "The Strength of Fibre Composites," *Composites*, **3**, 152–167 (1972).
12. A. F. Johnson, "Engineering Design Properties of GRP," The British Plastics Federation, London, 1978.

6

ANALYSIS OF LAMINATED COMPOSITES

6.1 INTRODUCTION

One of the most important advantages of fibrous composites is that their anisotropy or properties can be controlled very effectively; that is, desired property values in different directions can be obtained easily by altering the material and manufacturing variables. For example, in a unidirectional composite, the longitudinal-strength (or stiffness) to transverse-strength (or stiffness) ratio can be altered by changing the volume fraction of fibers. The longitudinal properties of unidirectional composites are controlled by fiber properties, whereas the transverse properties are matrix-dominated. In most engineering applications, the transverse properties of unidirectional composites are found to be unsatisfactory. This apparent limitation on the use of purely unidirectional composites is overcome by forming laminates from the unidirectional layers. A laminate is formed from two or more laminae bonded together to act as an integral structural element. The laminae principal material directions are oriented to produce a structural element with the desired properties in all directions. A laminate made up of four laminae with different fiber orientations is shown in Fig. 6-1. In this chapter, procedures are discussed for laminate analysis from the known properties of the constituent laminae.

6.2 LAMINATE STRAINS

Laminates are fabricated such that they act as single-layer materials. The bond between two laminae in a laminate is assumed to be perfect, that is, infinitesimally thin and not shear deformable. Thus the laminae cannot slip over

214 ANALYSIS OF LAMINATED COMPOSITES

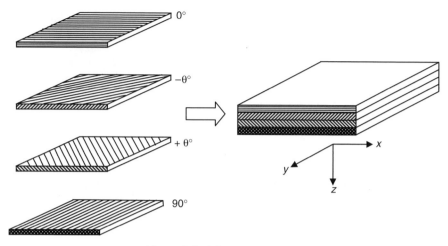

Figure 6-1. A four-ply laminate.

each other, and the displacements remain continuous across the bond. In this section, equations are developed that relate the strain at any point in a thin laminate undergoing deformation to the displacements and curvatures of its geometric midplane. Then, recognizing the fact that the laminate consists of several laminae with different directional properties, the variations of stress across the thickness of the laminate are discussed. Theoretical developments presented in this chapter are often referred to as the *classical lamination theory*.

Consider the deformation of a section of a laminate in the *xz* plane, as shown in Fig. 6-2. Assume that a line *ABCD* originally straight and perpendicular to the midplane of the laminate also remains straight and perpendicular to the midplane in the deformed state. This assumption is equivalent to neglecting shearing deformations γ_{xz} and γ_{yz} and is also equivalent to assuming that the laminae that make up the cross section do not slip over each other. Further assume that the point *B* at the geometric midplane undergoes displacements u_0, v_0, and w_0 along *x*, *y*, and *z* directions, respectively. The displacement *u* in the *x* direction of a point *C* that is located on the normal *ABCD* at a distance *z* from the midplane is given by

$$u = u_0 - z\alpha \tag{6.1}$$

where α is the slope of the laminate midplane in the *x* direction, that is,

$$\alpha = \frac{\partial w_0}{\partial x} \tag{6.2}$$

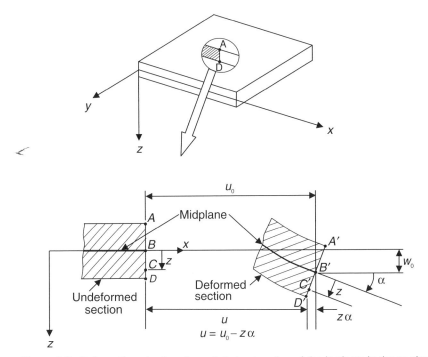

Figure 6-2. Deformation of a line element during bending of the laminate in the xz plane.

By combining Eqs. (6.1) and (6.2), an expression for the displacement u in the x direction of an arbitrary point at a distance z from the midplane is obtained:

$$u = u_0 - z \frac{\partial w_0}{\partial x} \qquad (6.3)$$

By similar reasoning, the displacement v in the y direction of an arbitrary point at a distance z from the geometric midplane is

$$v = v_0 - z \frac{\partial w_0}{\partial y} \qquad (6.4)$$

The displacement w in the z direction of any point on *ABCD* is the displacement w_0 of the midplane plus the stretching of the normal. It is assumed that the stretching (or shortening) of the normal (*ABCD*) is insignificant compared with the displacement w_0, and thus the normal displacement of any point in the laminate is taken to be equal to the displacement w_0 of the corresponding point at the midplane. Thus the normal strains ϵ_z are neglected. This reduces the laminate strains to ϵ_x, ϵ_y, and γ_{xy}. These strains can be obtained for the derived displacements u and v as follows:

$$\epsilon_x = \frac{\partial u}{\partial x} = \frac{\partial u_0}{\partial x} - z\frac{\partial^2 w_0}{\partial x^2}$$

$$\epsilon_y = \frac{\partial v}{\partial y} = \frac{\partial v_0}{\partial y} - z\frac{\partial^2 w_0}{\partial y^2} \qquad (6.5)$$

$$\gamma_{xy} = \frac{\partial u}{\partial y} + \frac{\partial v}{\partial x} = \frac{\partial u_0}{\partial y} + \frac{\partial v_0}{\partial x} - 2z\frac{\partial^2 w_0}{\partial x\, \partial y}$$

The preceding strain–displacement relation can be written in terms of the midplane strains and the plate curvatures as follows:

$$\begin{Bmatrix} \epsilon_x \\ \epsilon_y \\ \gamma_{xy} \end{Bmatrix} = \begin{Bmatrix} \epsilon_x^0 \\ \epsilon_y^0 \\ \gamma_{xy}^0 \end{Bmatrix} + z \begin{Bmatrix} k_x \\ k_y \\ k_{xy} \end{Bmatrix} \qquad (6.6)$$

where the midplane strains are

$$\begin{Bmatrix} \epsilon_x^0 \\ \epsilon_y^0 \\ \gamma_{xy}^0 \end{Bmatrix} = \begin{Bmatrix} \dfrac{\partial u_0}{\partial x} \\ \dfrac{\partial v_0}{\partial y} \\ \dfrac{\partial u_0}{\partial y} + \dfrac{\partial v_0}{\partial x} \end{Bmatrix} \qquad (6.7)$$

and the plate curvatures are

$$\begin{Bmatrix} k_x \\ k_y \\ k_{xy} \end{Bmatrix} = -\begin{Bmatrix} \dfrac{\partial^2 w_0}{\partial x^2} \\ \dfrac{\partial^2 w_0}{\partial y^2} \\ 2\dfrac{\partial^2 w_0}{\partial x\, \partial y} \end{Bmatrix} \qquad (6.8)$$

6.3 VARIATION OF STRESSES IN A LAMINATE

Strains at any point in a laminate can be calculated, using Eq. 6.6, from midplane strains (ϵ_x^0, ϵ_y^0, and γ_{xy}^0), plate curvatures (k_x, k_y, and k_{xz}), and its

6.3 VARIATION OF STRESSES IN A LAMINATE

distance from the midplane. Strains vary linearly across the thickness even though the laminate is composed of laminae with different directional properties. This linear strain variation is a result of the assumption that the adjacent laminae do not slip over each other. A linear strain variation will produce a linear stress variation for a material with identical elastic properties across the thickness. However, in a laminate, laminae usually have different elastic properties because of different fiber orientations. Therefore, while the stress variation across a single lamina thickness will be linear, stress variation across the total laminate thickness will be composed of several linear segments. Stresses at any point in a lamina (say, k) can be obtained by substituting Eq. (6.6) in the stress–strain relation [Eq. (5.94)] for the lamina as follows:

$$\begin{Bmatrix} \sigma_x \\ \sigma_y \\ \tau_{xy} \end{Bmatrix} = \begin{bmatrix} \overline{Q}_{11} & \overline{Q}_{12} & \overline{Q}_{16} \\ \overline{Q}_{12} & \overline{Q}_{22} & \overline{Q}_{26} \\ \overline{Q}_{16} & \overline{Q}_{26} & \overline{Q}_{66} \end{bmatrix}_k \begin{Bmatrix} \epsilon_x^0 \\ \epsilon_y^0 \\ \gamma_{xy}^0 \end{Bmatrix} + z \begin{bmatrix} \overline{Q}_{11} & \overline{Q}_{12} & \overline{Q}_{16} \\ \overline{Q}_{12} & \overline{Q}_{22} & \overline{Q}_{26} \\ \overline{Q}_{16} & \overline{Q}_{26} & \overline{Q}_{66} \end{bmatrix}_k \begin{Bmatrix} k_x \\ k_y \\ k_{xy} \end{Bmatrix} \quad (6.9)$$

or

$$\{\sigma\}_k = [\overline{Q}]_k \{\epsilon^0 + zk\}$$

Since the lamina stiffness matrix $[\overline{Q}]$ is a constant for each lamina, Eq. (6.9) gives a straight-line stress variation across lamina thickness. Combination of these straight-line variations for all the laminae gives the stress variation for the entire laminate. In general, the stress variation across the laminate thickness, is not a single straight line, like the strain variation, but is composed of several line segments with one line for each lamina. Whenever adjacent laminae have different stiffnesses, there is a stress discontinuity or jump in the stress value across the laminae interface, and the stress gradients in the two laminae (represented by the slopes of the line segments) are also different. Stress and strain variations in an imaginary three-ply laminate are shown in Fig. 6-3.

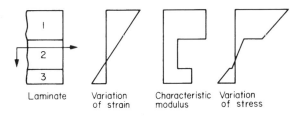

Figure 6-3. Variations of strain and stress in a hypothetical three-ply laminate.

6.4 RESULTANT FORCES AND MOMENTS: SYNTHESIS OF STIFFNESS MATRIX

The stresses in a laminate vary from layer to layer. Hence it is convenient to deal with a simpler but equivalent system of forces and moments acting on a laminate cross section. Therefore, the resultant forces and moments acting on a laminate cross section are defined as follows. Resultant force is obtained by integrating the corresponding stress through the laminate thickness h:

$$N_x = \int_{-h/2}^{h/2} \sigma_x \, dz$$

$$N_y = \int_{-h/2}^{h/2} \sigma_y \, dz \tag{6.10}$$

$$N_{xy} = \int_{-h/2}^{h/2} \tau_{xy} \, dz$$

Similarly, the resultant moment is obtained by integration through the thickness of the corresponding stress times the moment arm with respect to the midplane:

$$M_x = \int_{-h/2}^{h/2} \sigma_x z \, dz$$

$$M_y = \int_{-h/2}^{h/2} \sigma_y z \, dz \tag{6.11}$$

$$M_{xy} = \int_{-h/2}^{h/2} \tau_{xy} z \, dz$$

In the preceding equations, N_x, N_y, and N_{xy} have the units of force per unit length (e.g., width of the beam), and M_x, M_y, and M_{xy} have the units of moment per unit length. The positive sense of the resultant forces and moments are consistent with the sign convention for stresses, as shown in Fig. 6-4. Together the six force and moment resultants form a system that is statically equivalent to the stress system on the laminate but that is applied at the geometric midplane. By defining these six forces and moments, the loading has been reduced to a system that does not contain the laminate thickness or z coordinate explicitly.

Consider a laminate consisting of n orthotropic laminae, as shown in Fig. 6-5. The force–moment system acting at the midplane of this laminate can be obtained by replacing the continuous integral in Eqs. (6.10) and (6.11) with the summation of integrals representing the contribution of each layer as follows:

6.4 RESULTANT FORCES AND MOMENTS: SYNTHESIS OF STIFFNESS MATRIX 219

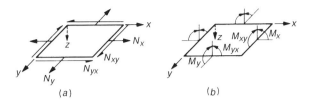

Figure 6-4. Sign convention for resultant forces and moments (all forces and moments shown are positive).

$$\begin{Bmatrix} N_x \\ N_y \\ N_{xy} \end{Bmatrix} = \int_{-h/2}^{h/2} \begin{Bmatrix} \sigma_x \\ \sigma_y \\ \tau_{xy} \end{Bmatrix} dz = \sum_{k=1}^{n} \int_{h_{k-1}}^{h_k} \begin{Bmatrix} \sigma_x \\ \sigma_y \\ \tau_{xy} \end{Bmatrix}_k dz \quad (6.12)$$

and

$$\begin{Bmatrix} M_x \\ M_y \\ M_{xy} \end{Bmatrix} = \int_{-h/2}^{h/2} \begin{Bmatrix} \sigma_x \\ \sigma_y \\ \tau_{xy} \end{Bmatrix} z \, dz = \sum_{k=1}^{n} \int_{h_{k-1}}^{h_k} \begin{Bmatrix} \sigma_x \\ \sigma_y \\ \tau_{xy} \end{Bmatrix}_k z \, dz \quad (6.13)$$

The stresses in Eqs. (6.12) and (6.13) can be written in terms of midplane strains and plate curvatures from Eq. (6.9), and thus the resultant forces and

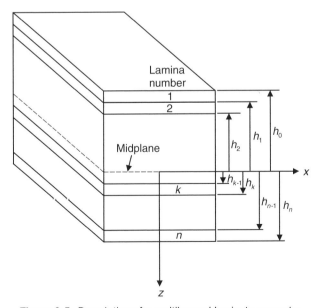

Figure 6-5. Description of a multilayered laminate geometry.

moments can be related directly to the midplane strains and plate curvatures. Substitution of Eq. (6.9) in Eqs. (6.12) and (6.13) gives

$$\begin{Bmatrix} N_x \\ N_y \\ N_{xy} \end{Bmatrix} = \sum_{k=1}^{n} \left\{ \int_{h_{k-1}}^{h_k} \begin{bmatrix} \overline{Q}_{11} & \overline{Q}_{12} & \overline{Q}_{16} \\ \overline{Q}_{12} & \overline{Q}_{22} & \overline{Q}_{26} \\ \overline{Q}_{16} & \overline{Q}_{26} & \overline{Q}_{66} \end{bmatrix}_k \begin{Bmatrix} \epsilon_x^0 \\ \epsilon_y^0 \\ \gamma_{xy}^0 \end{Bmatrix} dz + \int_{h_{k-1}}^{h_k} \right.$$

$$\left. \cdot \begin{bmatrix} \overline{Q}_{11} & \overline{Q}_{12} & \overline{Q}_{16} \\ \overline{Q}_{12} & \overline{Q}_{22} & \overline{Q}_{26} \\ \overline{Q}_{16} & \overline{Q}_{26} & \overline{Q}_{66} \end{bmatrix}_k \begin{Bmatrix} k_x \\ k_y \\ k_{xy} \end{Bmatrix} z \, dz \right\} \qquad (6.14)$$

$$\begin{Bmatrix} M_x \\ M_y \\ M_{xy} \end{Bmatrix} = \sum_{k=1}^{n} \left\{ \int_{h_{k-1}}^{h_k} \begin{bmatrix} \overline{Q}_{11} & \overline{Q}_{12} & \overline{Q}_{16} \\ \overline{Q}_{12} & \overline{Q}_{22} & \overline{Q}_{26} \\ \overline{Q}_{16} & \overline{Q}_{26} & \overline{Q}_{66} \end{bmatrix}_k \begin{Bmatrix} \epsilon_x^0 \\ \epsilon_y^0 \\ \gamma_{xy}^0 \end{Bmatrix} z \, dz + \int_{h_{k-1}}^{h_k} \right.$$

$$\left. \cdot \begin{bmatrix} \overline{Q}_{11} & \overline{Q}_{12} & \overline{Q}_{16} \\ \overline{Q}_{12} & \overline{Q}_{22} & \overline{Q}_{26} \\ \overline{Q}_{16} & \overline{Q}_{26} & \overline{Q}_{66} \end{bmatrix}_k \begin{Bmatrix} k_x \\ k_y \\ k_{xy} \end{Bmatrix} z^2 \, dz \right\} \qquad (6.15)$$

Evaluation of Eqs. (6.14) and (6.15) can be simplified by noting that the midplane strains and plate curvatures remain constant not only within a lamina but also for all the laminae and hence can be taken outside the summation sign. Second, the stiffness matrix $[\overline{Q}]$ remains constant within a lamina and hence can be taken outside the integration sign. Thus Eqs. (6.14) and (6.15) become

$$\begin{Bmatrix} N_x \\ N_y \\ N_{xy} \end{Bmatrix} = \left[\sum_{k=1}^{n} \begin{bmatrix} \overline{Q}_{11} & \overline{Q}_{12} & \overline{Q}_{16} \\ \overline{Q}_{12} & \overline{Q}_{22} & \overline{Q}_{26} \\ \overline{Q}_{16} & \overline{Q}_{26} & \overline{Q}_{66} \end{bmatrix}_k \int_{h_{k-1}}^{h_k} dz \right] \begin{Bmatrix} \epsilon_x^0 \\ \epsilon_y^0 \\ \gamma_{xy}^0 \end{Bmatrix}$$

$$+ \left[\sum_{k=1}^{n} \begin{bmatrix} \overline{Q}_{11} & \overline{Q}_{12} & \overline{Q}_{16} \\ \overline{Q}_{12} & \overline{Q}_{22} & \overline{Q}_{26} \\ \overline{Q}_{16} & \overline{Q}_{26} & \overline{Q}_{66} \end{bmatrix}_k \int_{h_{k-1}}^{h_k} z \, dz \right] \begin{Bmatrix} k_x \\ k_y \\ k_{xy} \end{Bmatrix} \qquad (6.16)$$

$$\begin{Bmatrix} M_x \\ M_y \\ M_{xy} \end{Bmatrix} = \left[\sum_{k=1}^{n} \begin{bmatrix} \overline{Q}_{11} & \overline{Q}_{12} & \overline{Q}_{16} \\ \overline{Q}_{12} & \overline{Q}_{22} & \overline{Q}_{26} \\ \overline{Q}_{16} & \overline{Q}_{26} & \overline{Q}_{66} \end{bmatrix}_k \int_{h_{k-1}}^{h_k} z \, dz \right] \begin{Bmatrix} \epsilon_x^0 \\ \epsilon_y^0 \\ \gamma_{xy}^0 \end{Bmatrix}$$

$$+ \left[\sum_{k=1}^{n} \begin{bmatrix} \overline{Q}_{11} & \overline{Q}_{12} & \overline{Q}_{16} \\ \overline{Q}_{12} & \overline{Q}_{22} & \overline{Q}_{26} \\ \overline{Q}_{16} & \overline{Q}_{26} & \overline{Q}_{66} \end{bmatrix}_k \int_{h_{k-1}}^{h_k} z^2 \, dz \right] \begin{Bmatrix} k_x \\ k_y \\ k_{xy} \end{Bmatrix} \qquad (6.17)$$

6.4 RESULTANT FORCES AND MOMENTS: SYNTHESIS OF STIFFNESS MATRIX

By introducing definitions of three new matrices, Eqs. (6.16) and (6.17) can be rewritten in a relatively simple form as follows:

$$\begin{Bmatrix} N_x \\ N_y \\ N_{xy} \end{Bmatrix} = \begin{bmatrix} A_{11} & A_{12} & A_{16} \\ A_{12} & A_{22} & A_{26} \\ A_{16} & A_{26} & A_{66} \end{bmatrix} \begin{Bmatrix} \epsilon_x^0 \\ \epsilon_y^0 \\ \gamma_{xy}^0 \end{Bmatrix} + \begin{bmatrix} B_{11} & B_{12} & B_{16} \\ B_{12} & B_{22} & B_{26} \\ B_{16} & B_{26} & B_{66} \end{bmatrix} \begin{Bmatrix} k_x \\ k_y \\ k_{xy} \end{Bmatrix} \quad (6.18)$$

$$\begin{Bmatrix} M_x \\ M_y \\ M_{xy} \end{Bmatrix} = \begin{bmatrix} B_{11} & B_{12} & B_{16} \\ B_{12} & B_{22} & B_{26} \\ B_{16} & B_{26} & B_{66} \end{bmatrix} \begin{Bmatrix} \epsilon_x^0 \\ \epsilon_y^0 \\ \gamma_{xy}^0 \end{Bmatrix} + \begin{bmatrix} D_{11} & D_{12} & D_{16} \\ D_{12} & D_{22} & D_{26} \\ D_{16} & D_{26} & D_{66} \end{bmatrix} \begin{Bmatrix} k_x \\ k_y \\ k_{xy} \end{Bmatrix} \quad (6.19)$$

where

$$A_{ij} = \sum_{k=1}^{n} (\overline{Q}_{ij})_k (h_k - h_{k-1})$$

$$B_{ij} = \tfrac{1}{2} \sum_{k=1}^{n} (\overline{Q}_{ij})_k (h_k^2 - h_{k-1}^2) \quad (6.20)$$

$$D_{ij} = \tfrac{1}{3} \sum_{k=1}^{n} (\overline{Q}_{ij})_k (h_k^3 - h_{k-1}^3)$$

Combining Eqs. (6.18) and (6.19), the total plate constitutive equation can be written as follows:

$$\begin{Bmatrix} N \\ \overline{M} \end{Bmatrix} = \begin{bmatrix} A & \vdots & B \\ \cdots & & \cdots \\ B & \vdots & D \end{bmatrix} \begin{Bmatrix} \epsilon^0 \\ \overline{k} \end{Bmatrix} \quad (6.21)$$

In Eqs. (6.18)–(6.21), the matrices A, B, and D are called the *extensional stiffness matrix*, *coupling stiffness matrix*, and *bending stiffness matrix*, respectively. The extensional stiffness matrix relates the resultant forces to the midplane strains, and the bending stiffness matrix relates the resultant moments to the plate curvatures.

The presence of the coupling matrix $[B]$ in the plate constitutive equation implies coupling between bending and extension of a laminated plate. That is, normal and shear forces acting at the midplane of the plate result in not only the in-plane deformations, leading to the midplane strains, but also twisting and bending, producing plate curvatures. Similarly, bending and twisting moments are accompanied by midplane strains. Thus stretching a laminate that has nonzero B_{ij} terms will produce bending and/or twisting of the laminate in addition to the extensional and shear deformation.

Example 6-1: Consider a two-ply laminate with the ply orientations of 0° and 45° with the laminate axes as shown in Fig. 6-6. The bottom lamina is a 0° layer with a thickness of 5 mm, whereas the 45° top lamina is 3

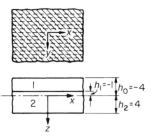

Figure 6-6. Two-ply laminate for Example 6-1.

mm thick. Evaluate A, B, and D matrices for the laminate if both the laminae have identical stiffness matrix Q as follows:

$$[Q] = \begin{bmatrix} 20 & 0.7 & 0 \\ 0.7 & 2.0 & 0 \\ 0 & 0 & 0.7 \end{bmatrix} \text{GPa}$$

It may be pointed out that the units of the elements in the Q matrix are the same as the units of stress (e.g., gigapascals, as indicated in this example). The units of elements in the A matrix are those of stress times length; they are stress times length squared in the B matrix and stress times length cubed in the D matrix. In the examples considered here, laminae thicknesses are given in millimeters. Therefore, the units of the elements in the A, B, and D matrices will be GPa·mm, GPa·mm², and GPa·mm³, respectively. These units are not being indicated in Examples 6-1–6-5 because these examples are used only to illustrate calculations that are not affected by the units. However, when laminate stresses and strains are to be calculated (as in Example 6-7), consistent units must be used.

Evaluation of matrices A, B, and D requires finding the $\overline{Q}$ matrices for the two layers. For the 0° lamina the Q and $\overline{Q}$ matrices are the same; that is,

$$[\overline{Q}]_{0°} = [Q]_{0°} = \begin{bmatrix} 20 & 0.7 & 0 \\ 0.7 & 2.0 & 0 \\ 0 & 0 & 0.7 \end{bmatrix}$$

The $\overline{Q}_{ij}$ terms for the 45° lamina are found by using the transformation [Eq. (5.61)]:

6.4 RESULTANT FORCES AND MOMENTS: SYNTHESIS OF STIFFNESS MATRIX

$$\overline{Q}_{11} = 20(\cos 45)^4 + 2(\sin 45)^4 + 2(0.7 + 2 \times 0.7)(\sin 45)^2(\cos 45)^2$$

$$= 6.55$$

$$\overline{Q}_{22} = 20(\sin 45)^4 + 2(\cos 45)^4 + 2(0.7 + 2 \times 0.7)(\sin 45)^2(\cos 45)^2$$

$$= 6.55$$

$$\overline{Q}_{12} = \left(\frac{1}{\sqrt{2}}\right)^4 [(20 + 2 - 4 \times 0.7) + 2 \times 0.7] = 5.15$$

$$\overline{Q}_{66} = \left(\frac{1}{\sqrt{2}}\right)^4 [(20 + 2 - 2 \times 0.7 - 2 \times 0.7) + 2 \times 0.7] = 5.15$$

$$\overline{Q}_{16} = \left(\frac{1}{\sqrt{2}}\right)^4 [(20 - 0.7 - 2 \times 0.7) - (2 - 0.7 - 2 \times 0.7)] = 4.50$$

$$\overline{Q}_{26} = \left(\frac{1}{\sqrt{2}}\right)^4 [(20 - 0.7 - 2 \times 0.7) - (2 - 0.7 - 2 \times 0.7)] = 4.50$$

Therefore,

$$[\overline{Q}]_{45°} = \begin{bmatrix} 6.55 & 5.15 & 4.50 \\ 5.15 & 6.55 & 4.50 \\ 4.50 & 4.50 & 5.15 \end{bmatrix}$$

Now the basic terms in Eq. (6.20) are known ($h_0 = -4.0$, $h_1 = -1.0$, and $h_2 = 4.0$, as shown in Fig. 6-6). Thus the terms of the laminate stiffness matrices A, B, and D can be calculated as follows:

$$A_{ij} = \sum_{k=1}^{n} (\overline{Q}_{ij})_k (h_k - h_{k-1}) = \overset{\text{Top}}{(\overline{Q}_{ij})_{45°}[(-1) - (-4)]} + \overset{\text{Bottom}}{(\overline{Q}_{ij})_{0°}[4 - (-1)]}$$
$$\qquad\qquad\qquad\qquad\qquad\quad h_1 - h_0 \qquad\qquad h_2 - h_1$$

or

$$A_{ij} = 3(\overline{Q}_{ij})_{45°} + 5(\overline{Q}_{ij})_{0°}$$

Thus

$$\begin{bmatrix} A_{11} & A_{12} & A_{16} \\ A_{12} & A_{22} & A_{26} \\ A_{16} & A_{26} & A_{66} \end{bmatrix} = 3 \begin{bmatrix} 6.55 & 5.15 & 4.50 \\ 5.15 & 6.55 & 4.50 \\ 4.50 & 4.50 & 5.15 \end{bmatrix} + 5 \begin{bmatrix} 20 & 0.7 & 0 \\ 0.7 & 2.0 & 0 \\ 0 & 0 & 0.7 \end{bmatrix}$$

$$[A] = \begin{bmatrix} 119.65 & 18.95 & 13.50 \\ 18.95 & 29.65 & 13.50 \\ 13.50 & 13.50 & 18.95 \end{bmatrix}$$

$$B_{ij} = \frac{1}{2} \sum_{k=1}^{n} (\overline{Q}_{ij})_k (h_k^2 - h_{k-1}^2) = \frac{1}{2}(\overline{Q}_{ij})_{45°}[(-1)^2 - (-4)^2]$$
$$+ \frac{1}{2}(\overline{Q}_{ij})_{0°}[(4)^2 - (-1)^2]$$
$$= 7.5[-(\overline{Q}_{ij})_{45°} + (\overline{Q}_{ij})_{0°}]$$

$$\begin{bmatrix} B_{11} & B_{12} & B_{16} \\ B_{12} & B_{22} & B_{26} \\ B_{16} & B_{26} & B_{66} \end{bmatrix} = 7.5 \left[\begin{bmatrix} 20 & 0.7 & 0 \\ 0.7 & 2.0 & 0 \\ 0 & 0 & 0.7 \end{bmatrix} - \begin{bmatrix} 6.55 & 5.15 & 4.50 \\ 5.15 & 6.55 & 4.50 \\ 4.50 & 4.50 & 5.15 \end{bmatrix} \right]$$

$$[B] = \begin{bmatrix} 100.9 & -33.4 & -33.75 \\ -33.4 & -34.1 & -33.75 \\ -33.75 & -33.75 & -33.40 \end{bmatrix}$$

$$D_{ij} = \frac{1}{3} \sum_{k=1}^{n} (\overline{Q}_{ij})_k (h_k^3 - h_{k-1}^3) = \frac{1}{3}(\overline{Q}_{ij})_{45°}[(-1)^3 - (-4)^3]$$
$$+ \frac{1}{3}(\overline{Q}_{ij})_{0°}[(4)^3 - (-1)^3]$$
$$= 21(\overline{Q}_{ij})_{45°} + 21.67(\overline{Q}_{ij})_{0°}$$

Thus

$$\begin{bmatrix} D_{11} & D_{12} & D_{16} \\ D_{12} & D_{22} & D_{26} \\ D_{16} & D_{26} & D_{66} \end{bmatrix} = 21 \begin{bmatrix} 6.55 & 5.15 & 4.50 \\ 5.15 & 6.55 & 4.50 \\ 4.50 & 4.50 & 5.15 \end{bmatrix} + 21.67 \begin{bmatrix} 20 & 0.7 & 0 \\ 0.7 & 2.0 & 0 \\ 0 & 0 & 0.7 \end{bmatrix}$$

$$[D] = \begin{bmatrix} 571 & 123 & 94.5 \\ 123 & 181 & 94.5 \\ 94.5 & 94.5 & 123 \end{bmatrix}$$

Combining the preceding results, the total set of constitutive equations for this particular two-ply laminate can be written as

$$\begin{Bmatrix} N_x \\ N_y \\ N_{xy} \\ \hline M_x \\ M_y \\ M_{xy} \end{Bmatrix} = \begin{bmatrix} 119.6 & 18.9 & 13.5 & | & 100.9 & -33.4 & -33.8 \\ 18.9 & 29.6 & 13.5 & | & -33.4 & -34.1 & -33.8 \\ 13.5 & 13.5 & 18.9 & | & -33.8 & -33.8 & -33.4 \\ \hline 100.9 & -33.4 & -33.8 & | & 571 & 123 & 94.5 \\ -33.4 & -34.1 & -33.8 & | & 123 & 181 & 94.5 \\ -33.8 & -33.8 & -33.4 & | & 94.5 & 94.5 & 123 \end{bmatrix} \begin{Bmatrix} \epsilon_x^0 \\ \epsilon_y^0 \\ \gamma_{xy}^0 \\ \hline k_x \\ k_y \\ k_{xy} \end{Bmatrix}$$

None of the elements of the A, B, and D matrices is zero for the laminate considered in this example. Such laminates are the most general laminates.

6.5 LAMINATE DESCRIPTION SYSTEM

Laminate properties and characteristics are influenced directly by the laminate makeup. It is therefore necessary to adopt a laminate description system that will provide a positive identification of the laminate makeup. A positive identification of a laminate requires the following:

1. Orientation of each lamina relative to a reference axis (the x axis in this text)
2. Number of laminae at each orientation
3. The exact geometric sequence of laminae

A laminate orientation code that provides a positive and concise identification of the laminates is described in Appendix 3. Basic features of the code are discussed here.

In the standard laminate code, it is assumed that all laminae are identical in thickness and properties. Following are the elements of the code:

1. Each lamina is denoted by a number representing the angle in degrees between its fiber direction and the x axis.
2. Individual adjacent laminae are separated in the code by a slash if their angles are different.
3. The laminae are listed in sequence from one laminate face to the other, starting with the first lamina laid up, with brackets indicating the beginning and end of the code.
4. Adjacent laminae of the same orientation are denoted by a numerical subscript.
5. The laminate possessing symmetry of laminae orientations about the geometric midplane requires specifying only one-half the laminate stacking sequence. A subscript s to the code signifies that only one-half

of the laminate is described, with the other half being symmetric about the midplane. The following two examples illustrate use of the code:

Laminate	Code
45°	
0°	
45°	
90°	[45/0/45/90$_2$/30]
90°	
30°	
90°	
0°	
0°	
45°	
45°	[90/0$_2$/45]$_s$
0°	
0°	
90°	

A more detailed description of the laminate orientation code is given in Appendix 3.

6.6 CONSTRUCTION AND PROPERTIES OF LAMINATES

As explained in an earlier section, a nonzero coupling matrix [B] implies coupling between bending and extension of the laminate. That is, in-plane normal and shear forces produce plate curvatures (bending and twisting) in addition to the in-plane deformations. Similarly, bending and twisting moments produce midplane strains in addition to the plate curvatures. A typical coupling can be illustrated by considering deformation of a two-ply laminate with the plies oriented at $+\theta$ and $-\theta$ angles. When such a laminate is subjected to an axial force, with the ends free to rotate, it will exhibit twisting, as shown schematically in Fig. 6-7. Ashton et al. [1] demonstrated this coupling experimentally through a specimen fabricated with two layers of nylon-fabric-reinforced rubber.

6.6 CONSTRUCTION AND PROPERTIES OF LAMINATES

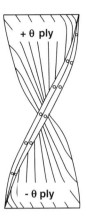

Figure 6-7. Twisting of an asymmetric ($\pm \theta$) laminate by an in-plane axial force: schematic representation.

The coupling between the extension and bending of the laminate, as evident mathematically by the presence of a nonzero [B] matrix, is not attributable to the orthotropy or anisotropy of the layers but rather to the nonsymmetric stacking of laminae. In fact, this coupling exists for a laminate made of layers of isotropic materials with different elastic moduli. Bimetallic strips are used as temperature-controlling devices because they exhibit this behavior. A bimetallic strip subjected to a temperature change causes its bending owing to unequal thermal strains in the two materials (see Exercise Problem 6.1 at the end of this chapter).

Couplings such as the ones just discussed usually are undesirable because they may induce unwanted plate curvatures owing to temperature changes and produce unwanted assembly stresses. Therefore, laminates often are constructed such that some of the couplings are minimized or eliminated. Construction and properties of important classes of laminates are discussed in this section. These special laminate constructions reduce many critical elements of the stiffness matrix to zero, which usually simplifies calculations in the laminate analysis.

6.6.1 Symmetric Laminates

Equations (6.18) and (6.19) show that the bending–stretching coupling occurs owing to a nonzero coupling stiffness matrix [B]. This coupling can be eliminated if the laminate construction reduces each element of the [B] matrix to zero. To that end, consider the composition of the B_{ij} terms in Eq. (6.20). The contribution of a lamina to a particular term of the B matrix is given by the product of the corresponding term in the $\overline{Q}$ matrix and the difference of the squares of z coordinates of the top and bottom of each ply. The contribution of a lamina above the geometric midplane can be nullified by placing an identical (in properties and orientation) lamina an equal distance below the midplane. Thus the matrix $[B]$ is identically zero for laminates in which for each ply above the midplane there is an identical ply placed an equal distance below the midplane. The laminates, which are constructed by placing the laminae symmetrically with respect to the midplane, are called *symmetric laminates* and represent an important class of laminates. For symmetric laminates, since the $[B]$ matrix is zero, the constitutive equations [Eqs. (6.18) and (6.19)] simplify to the following uncoupled equations:

$$\begin{Bmatrix} N_x \\ N_y \\ N_{xy} \end{Bmatrix} = \begin{bmatrix} A_{11} & A_{12} & A_{16} \\ A_{12} & A_{22} & A_{26} \\ A_{16} & A_{26} & A_{66} \end{bmatrix} \begin{Bmatrix} \varepsilon_x^0 \\ \varepsilon_y^0 \\ \gamma_{xy}^0 \end{Bmatrix} \quad (6.22)$$

$$\begin{Bmatrix} M_x \\ M_y \\ M_{xy} \end{Bmatrix} = \begin{bmatrix} D_{11} & D_{12} & D_{16} \\ D_{12} & D_{22} & D_{26} \\ D_{16} & D_{26} & D_{66} \end{bmatrix} \begin{Bmatrix} k_x \\ k_y \\ k_{xy} \end{Bmatrix} \quad (6.23)$$

Symmetric laminates are commonly constructed because the bending–stretching coupling is eliminated, which in nonsymmetric laminates causes an undesirable warping owing to in-plane loads. Also, temperature changes will cause warping, and thus, in the fabrication of a laminate at an elevated temperature, warpage will result when the laminate is cooled to room temperature. This effect is discussed in greater detail in the section on thermal stresses.

6.6.2 Unidirectional, Cross-Ply, and Angle-Ply Laminates

Another possibility is the fabrication of a laminate that behaves as an orthotropic layer with respect to in-plane forces and strains, that is, a laminate in which there is no coupling between the normal stresses (or forces) and shear strain. This is possible when $A_{16} = A_{26} = 0$. The contribution of a lamina to a particular term of the A matrix is given by the corresponding term in the $[Q]$ matrix times the lamina thickness. Thus the contribution of one lamina to a term A_{ij} can be nullified by another lamina of the same thickness but whose corresponding $\overline{Q}_{ij}$ term is opposite in sign to the $\overline{Q}_{ij}$ term of the first lamina. The $\overline{Q}_{ij}$ terms are obtained from the Q_{ij} terms using the transformation equations [Eqs. (5.95)]. It is apparent from these equations that the $\overline{Q}_{11}$, $\overline{Q}_{22}$, $\overline{Q}_{12}$, and $\overline{Q}_{66}$, are always positive in sign and greater than zero in magnitude. Consequently, A_{11}, A_{22}, A_{12}, and A_{66} cannot be made equal to zero. On the other hand, $\overline{Q}_{16}$ and $\overline{Q}_{26}$, are zero for orientations of 0° or 90° and can be positive or negative for intermediate orientations. Since $\overline{Q}_{16}$ and $\overline{Q}_{26}$ are odd functions of θ, for equal positive and negative orientations they are equal in magnitude but opposite in sign. Therefore, the terms A_{16} and A_{26} can be made equal to zero if for every lamina oriented at a positive angle θ in the laminate there exists another lamina of equal thickness and the same orthotropic properties but oriented at the equal negative angle θ. Relative positions of the two laminae are immaterial. Hence it is possible to design a laminate that will be symmetric (the coupling matrix $[B]$ identically zero) and at the same time specially orthotropic with respect to in-plane forces and strains ($A_{16} = A_{26} = 0$). Three different types of orthotropic laminates can be constructed. These are (1) unidirectional laminate with all the laminae oriented in the same direction, (2) cross-ply laminate with laminae oriented at 0° or 90° only, and (3) angle-ply laminate with equal number of laminae oriented at $\pm\theta$ angle. All these laminae can be symmetric also.

The simplification of the bending matrix $[D]$ also can be considered. The contribution of a lamina to a particular term of the $[D]$ matrix is given by the product of the corresponding term in the $[\overline{Q}]$ matrix and the difference of the cubes of the z coordinates of the top and the bottom of the lamina. Since the geometric contribution $(h_k^3 - h_{k-1}^3)$ is always positive, it follows from the preceding discussion that D_{11}, D_{22}, D_{12}, and D_{66} are always positive. On the other hand, since $\overline{Q}_{16}$ and $\overline{Q}_{26}$ are odd functions of θ, D_{16} and D_{26} can be made equal to zero if all the laminae are oriented at 0° or 90° or if for every lamina oriented at a positive angle θ above the midplane there exists an identical lamina placed at an equal distance below the midplane but oriented at a negative angle θ. However, laminates of the latter type then will not possess midplane symmetry, and $B_{ij} \neq 0$. Thus D_{16} and D_{26} are not zero for any midplane symmetric laminate except for a cross-ply laminate where all the laminae are oriented at 0° or 90°. However, if the laminate is constructed by stacking alternate laminae at equal positive and negative angles, the D_{16} and D_{26} terms do become small, particularly if the number of laminae is large. This is so because the contribution of the $+\theta$ laminae is opposite in sign to the contribution of the $-\theta$ laminae, and hence they partially cancel each other even though they are located at different distances from the midplane.

6.6.3 Quasi-isotropic Laminates

A laminate construction of considerable practical importance is called *quasi-isotropic*. In a quasi-isotropic laminate, the extensional stiffness matrix $[A]$ is isotropic; that is, it has elastic coefficients that are independent of orientation in the plane. This will require that there be only two independent elastic constants, like those in the stiffness matrix of an isotropic material. Extensional stiffness matrix $[A]$ for an isotropic plate is as follows:

$$[A] = \begin{bmatrix} \dfrac{Et}{1-\nu^2} & \dfrac{\nu Et}{1-\nu^2} & 0 \\ \dfrac{\nu Et}{1-\nu^2} & \dfrac{Et}{1-\nu^2} & 0 \\ 0 & 0 & \dfrac{Et}{2(1+\nu)} \end{bmatrix} \qquad (6.24)$$

where E is the elastic modulus of the material, ν is the Poisson ratio, and t is the plate thickness. It may be noted from Eq. (6.24) that for the isotropic plates,

$$A_{11} = A_{22}$$
$$A_{11} - A_{12} = 2A_{66} \quad (6.25)$$
$$A_{16} = A_{26} = 0$$

Therefore, the [A] matrix for a quasi-isotropic material also should satisfy Eq. (6.25). It can be shown that a laminate constructed by meeting the following conditions will be quasi-isotropic:

1. The total number of layers must be three or more.
2. The individual layers must have identical stiffness matrices [Q] and thicknesses.
3. The layers must be oriented at equal angles. For example, if the total number of layers is n, the angle between two adjacent layers should be π/n. If a laminate is constructed from identical sets of three or more layers each, the condition on orientation must be satisfied by the layers in each set.

Since a laminate constructed according to the preceding design is isotropic with regard to extensional stiffness matrix [A] and not, in general, with regard to coupling and bending stiffness matrices ([B] and [D]), this design is called *quasi-isotropic*.

The concept of a quasi-isotropic laminate is very helpful in predicting the properties of randomly oriented short-fiber composites, as discussed in Chap. 4. A randomly oriented short-fiber composite may be modeled as a laminate having an infinite number of plies with continuously varying orientations. In practice, the properties are calculated from a quasi-isotropic laminate consisting of three or four plies only. It is left to the reader to show that [0/±60] and [0/±45/90] laminates are quasi-isotropic (the modulus and strength of randomly oriented short-fiber composites can be analyzed using these types of laminates).

The reader is advised to attempt relevant exercise problems at the end of this chapter to further appreciate properties of quasiisotropic laminates.

Example 6-2: Consider a three-ply laminate as shown in Fig. 6-8. The top and bottom layers are each 3 mm thick and oriented at 45° to the laminate reference axis, whereas the 6-mm-thick middle layer is oriented at 0°. Obtain the A, B, and D matrices if each lamina has the same properties as the lamina considered in Example 6-1.

6.6 CONSTRUCTION AND PROPERTIES OF LAMINATES

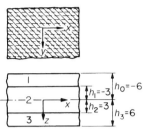

Figure 6-8. Three-ply laminate for Example 6-2.

Stiffness matrices for each lamina are

$$[\overline{Q}]_2 = \begin{bmatrix} 20 & 0.7 & 0 \\ 0.7 & 2.0 & 0 \\ 0 & 0 & 0.7 \end{bmatrix}$$

$$[\overline{Q}]_1 = [\overline{Q}]_3 = \begin{bmatrix} 6.55 & 5.15 & 4.50 \\ 5.15 & 6.55 & 4.50 \\ 4.50 & 4.50 & 5.15 \end{bmatrix}$$

For the laminate under consideration. $h_0 = -6$, $h_1 = -3$, $h_2 = 3$, and $h_3 = 6$, as shown in Fig. 6-8. Therefore,

$$A_{ij} = \sum_{k=1}^{3} (\overline{Q}_{ij})_k (h_k - h_{k-1}) = (\overline{Q}_{ij})_1(3) + (\overline{Q}_{ij})_2(6) + (\overline{Q}_{ij})_3(3)$$

$$= (\overline{Q}_{ij})_1(6)^* + (\overline{Q}_{ij})_2(6)$$

$$= [(\overline{Q}_{ij})_1 + (\overline{Q}_{ij})_2](6)$$

Substitution of the values of $(\overline{Q}_{ij})_1$ and $(\overline{Q}_{ij})_2$ will give

$$[A] = \begin{bmatrix} 159.3 & 35.1 & 27.0 \\ 35.1 & 51.3 & 27.0 \\ 27.0 & 27.0 & 35.1 \end{bmatrix}$$

$$2B_{ij} = \sum_{k=1}^{3} (\overline{Q}_{ij})_k (h_k^2 - h_{k-1}^2)$$

$$= (\overline{Q}_{ij})_1[(-3)^2 - (-6)^2] + (\overline{Q}_{ij})_2[(3)^2 - (-3)^2] + (\overline{Q}_{ij})_3[(6)^2 - (3)^2]$$

$$= [(\overline{Q}_{ij})_3 - (\overline{Q}_{ij})_1](27)$$

*Since $(\overline{Q}_{ij})_1 = (\overline{Q}_{ij})_3$.

But $(\overline{Q}_{ij})_1 = (\overline{Q}_{ij})_3$; therefore,

$$B_{ij} = 0$$

$$\begin{aligned}
D_{ij} &= \tfrac{1}{3}\sum_{k=1}^{3}(\overline{Q}_{ij})_k(h_k^3 - h_{k-1}^3) \\
&= \tfrac{1}{3}(\overline{Q}_{ij})_1[(-3)^3 - (-6)^3] + \tfrac{1}{3}(\overline{Q}_{ij})_2[(3)^3 - (-3)^3] \\
&\quad + \tfrac{1}{3}(\overline{Q}_{ij})_3[(6)^3 - (3)^3] \\
&= 126(\overline{Q}_{ij})_1 + 18(\overline{Q}_{ij})_2
\end{aligned}$$

Substitution of values of $(\overline{Q}_{ij})_1$ and $(\overline{Q}_{ij})_2$ will give

$$[D] = \begin{bmatrix} 1185.3 & 661.5 & 567.0 \\ 661.5 & 861.3 & 567.0 \\ 567.0 & 567.0 & 661.5 \end{bmatrix}$$

This three-ply laminate does not exhibit bending–stretching coupling because of the midplane symmetry of the laminate but does exhibit both in-plane and bending anisotropy because the A_{16}, A_{26}, D_{16}, and D_{26} terms are all nonzero.

Example 6-3: Consider a four-ply laminate $[\pm 45]_S$. Each layer is assumed to have a thickness of 3 mm and the same orthotropic properties as in Example 6-1.

In this laminate, the stiffness matrices are as follows:

$$[\overline{Q}]_1 = [\overline{Q}]_4 = \begin{bmatrix} 6.55 & 5.15 & 4.50 \\ 5.15 & 6.55 & 4.50 \\ 4.50 & 4.50 & 5.15 \end{bmatrix}$$

$$[\overline{Q}]_2 = [\overline{Q}]_3 = \begin{bmatrix} 6.55 & 5.15 & -4.50 \\ 5.15 & 6.55 & -4.50 \\ -4.50 & -4.50 & 5.15 \end{bmatrix}$$

$$\begin{aligned}
A_{ij} &= 3[(\overline{Q}_{ij})_1 + (\overline{Q}_{ij})_2 + (\overline{Q}_{ij})_3 + (\overline{Q}_{ij})_4] \\
&= 6[(\overline{Q}_{ij})_1 + (\overline{Q}_{ij})_2]
\end{aligned}$$

Thus

6.6 CONSTRUCTION AND PROPERTIES OF LAMINATES

$$[A] = \begin{bmatrix} 78.6 & 61.8 & 0 \\ 61.8 & 78.6 & 0 \\ 0 & 0 & 61.8 \end{bmatrix}$$

Since the laminate possess midplane symmetry, $B_{ij} = 0$.

$$D_{ij} = \tfrac{1}{3}\{(\overline{Q}_{ij})_1[(-3)^3 - (-6)^3] + (\overline{Q}_{ij})_2[(0)^3 - (-3)^3] \\ + (\overline{Q}_{ij})_3[(3)^3 - (0)^3] + (\overline{Q}_{ij})_4[(6)^3 - (3)^3]\}$$
$$= 126(\overline{Q}_{ij})_1 + 18(\overline{Q}_{ij})_2$$

Substitution of $(\overline{Q}_{ij})_1$ and $(\overline{Q}_{ij})_2$ gives

$$[D] = \begin{bmatrix} 943.2 & 741.6 & 486.0 \\ 741.6 & 943.2 & 486.0 \\ 486.0 & 486.0 & 741.6 \end{bmatrix}$$

Owing to symmetry, this laminate does not show bending–stretching coupling ($B_{ij} = 0$). Further, since $A_{16} = A_{26} = 0$, it is also free from normal stress–shear strain coupling.

Example 6-4: The effect of alternating-angle lamination can be illustrated by considering the following eight-ply laminates, with each lamina having a thickness of 3 mm and the properties the same as considered in Example 6-1: **(a)** all laminae at +45, **(b)** $[(45)_2/(-45)_2]_S$, **(c)** $[(\pm 45)_2]_S$, and **(d)** $[\pm \mp 45]_S$.

(a) This laminate is equivalent to a single lamina of thickness 24 mm. The stiffness matrices can be easily found to be

$$A = \begin{bmatrix} 157.2 & 123.6 & 108.0 \\ 123.6 & 157.2 & 108.0 \\ 108.0 & 108.0 & 123.6 \end{bmatrix}$$

$$B_{ij} = 0$$

$$D_{ij} = \tfrac{1}{3}(\overline{Q}_{ij})[(12)^3 - (-12)^3] = 1152(\overline{Q}_{ij})$$

$$[D] = 10^3 \begin{bmatrix} 7.55 & 5.93 & 5.18 \\ 5.93 & 7.55 & 5.18 \\ 5.18 & 5.18 & 5.93 \end{bmatrix}$$

$$\frac{D_{16}}{D_{11}} = 0.686$$

(b) The terms A_{11}, A_{22}, A_{12}, and A_{66} are the same as calculated in case (a), and A_{16} and A_{26} are zero. Thus

$$[A] = \begin{bmatrix} 157.2 & 123.6 & 0 \\ 123.6 & 157.2 & 0 \\ 0 & 0 & 123.6 \end{bmatrix}$$

$$B_{ij} = 0$$

Calculation of the D_{ij} terms is simplified by noting that because of the symmetry of orientations, the contribution of layers above the midplane is equal to the contribution of the layers below the midplane. Further, since $\overline{Q}_{11}$, $\overline{Q}_{22}$, $\overline{Q}_{12}$, and $\overline{Q}_{66}$ are the same for all layers, D_{11}, D_{22}, D_{12}, and D_{66} remain the same as in case (a). Now

$$D_{16} = D_{26} = 2 \times \tfrac{1}{3}\{(4.50)[(12)^3 - (6)^3] - (4.50)[(6)^3 - (0)^3]\}$$
$$= 3.89 \times 10^3$$

$$[D] = 10^3 \begin{bmatrix} 7.55 & 5.93 & 3.89 \\ 5.93 & 7.55 & 3.89 \\ 3.89 & 3.89 & 5.93 \end{bmatrix}$$

$$\frac{D_{16}}{D_{11}} = 0.515$$

(c) In this case, the $[A]$ and $[B]$ matrices and the D_{11}, D_{22}, D_{12}, and D_{66} terms of the $[D]$ matrix remain the same as in case (b). Now, D_{16} and D_{26} are calculated as follows:

$$D_{16} = D_{26} = 2 \times \tfrac{1}{3}[(4.5)(12^3 - 9^3) - (4.5)(9^3 - 6^3)$$
$$+ (4.5)(6^3 - 3^3) - (4.5)(3^3 - 0^3)]$$
$$= 1.94 \times 10^3$$

Thus

$$[D] = 10^3 \times \begin{bmatrix} 7.55 & 5.93 & 1.94 \\ 5.93 & 7.55 & 1.94 \\ 1.94 & 1.94 & 5.93 \end{bmatrix}$$

$$\frac{D_{16}}{D_{11}} = 0.257$$

(d) In this case, D_{16} and D_{26} are also the only terms affected by changing the stacking sequence and can be calculated as follows:

$$D_{16} = D_{26} = 2 \times \tfrac{1}{3}[(4.5)(12^3 - 9^3) - (4.5)(9^3 - 3^3) + (4.5)(3^3 - 0^3)]$$
$$= 0.97 \times 10^3$$

Thus

$$[D] = 10^3 \begin{bmatrix} 7.55 & 5.93 & 0.97 \\ 5.93 & 7.55 & 0.97 \\ 0.97 & 0.97 & 5.93 \end{bmatrix}$$

$$\frac{D_{16}}{D_{11}} = 0.129$$

This simple example has demonstrated the effect of stacking sequence on the [A], [B], and [D] matrices. Stacking sequence has no effect on the [A] matrix. If all laminae are oriented at equal positive and negative angles, the D_{11}, D_{22}, D_{12}, and D_{66} terms of the [D] matrix also remain unaffected by stacking sequence. Stacking sequence has no effect on the [B] matrix as long as the symmetry about the midplane is maintained. However, if the number of laminae forming the laminate is large, it is possible, by selection of the proper stacking sequence, to minimize the D_{16} and D_{26} terms of the [D] matrix without disturbing the laminate symmetry.

Example 6-5: Using an analysis for a quasi-isotropic laminate [0/±45/90] made up of the composite considered in Example 4-1, predict elastic modulus, shear modulus, and Poisson's ratio for a randomly oriented fiber composite. Compare results with those obtained earlier.

The following moduli values were obtained in Example 4-1:

$$E_L = 16.26 \text{ GPa}$$

$$E_T = 4.53 \text{ GPa}$$

Shear modulus for the aligned-fibers composite can be calculated using Halpin–Tsai equation [Eq. (3.53)]. For this purpose, we may assume the Poisson ratios of glass fibers and nylon matrix as 0.2 and 0.35, respectively, so that their shear moduli may be obtained as follows:

$$G_f = \frac{72.4}{2(1 + 0.2)} = 30.17 \text{ GPa}$$

$$G_m = \frac{2.76}{2(1 + 0.35)} = 1.02 \text{ GPa}$$

From Eq. (3.54),

$$\eta = \frac{(30.17/1.02) - 1}{(30.17/1.02) + 1} = 0.935$$

From Eq. (3.53),

$$\frac{G_{LT}}{G_m} = \frac{1 + \xi\eta V_f}{1 - \eta V_f}$$

$$G_{LT} = 1.02\left(\frac{1 + 0.935 \times 0.2}{1 - 0.935 \times 0.2}\right) = 1.49 \text{ GPa}$$

Poisson's ratio ν_{LT} may be obtained from the rule of mixtures [Eq. (3.59)]:

$$\nu_{LT} = 0.8 \times 0.35 + 0.2 \times 0.2 = 0.32$$

Minor Poisson's ratio ν_{TL} is obtained through Eq. (3.60):

$$\nu_{TL} = 0.32(4.53/16.26) = 0.089$$

Now the stiffness matrix of the laminae may be obtained from Eq. (5.78):

$$Q_{11} = \frac{16.26}{1 - 0.32 \times 0.089} = 16.74 \text{ GPa}$$

$$Q_{22} = \frac{4.53}{1 - 0.32 \times 0.089} = 4.66 \text{ GPa}$$

$$Q_{12} = \frac{0.32 \times 4.53}{1 - 0.32 \times 0.089} = 1.49 \text{ GPa}$$

$$Q_{66} = 1.49 \text{ GPa}$$

6.6 CONSTRUCTION AND PROPERTIES OF LAMINATES

$$[\bar{Q}_{ij}]_{0°} = \begin{bmatrix} 16.74 & 1.49 & 0 \\ 1.49 & 4.66 & 0 \\ 0 & 0 & 1.49 \end{bmatrix}$$

$$[\bar{Q}_{ij}]_{90°} = \begin{bmatrix} 4.66 & 1.49 & 0 \\ 1.49 & 16.74 & 0 \\ 0 & 0 & 1.49 \end{bmatrix}$$

Transformation equations [Eq. (5.61)] give

$$[\bar{Q}_{ij}]_{45°} = \begin{bmatrix} 7.585 & 4.605 & 3.02 \\ 4.605 & 7.585 & 3.02 \\ 3.02 & 3.02 & 4.605 \end{bmatrix}$$

$$[\bar{Q}_{ij}]_{-45°} = \begin{bmatrix} 7.585 & 4.605 & -3.02 \\ 4.605 & 7.585 & -3.02 \\ -3.02 & -3.02 & 4.605 \end{bmatrix}$$

Assuming unit thickness of the laminate, the [A] matrix is obtained as follows:

$$A_{ij} = \tfrac{1}{4}[(\bar{Q}_{ij})_{0°} + (\bar{Q}_{ij})_{90°} + (\bar{Q}_{ij})_{45°} + (\bar{Q}_{ij})_{-45°}]$$

$$[A] = \begin{bmatrix} 0.1425 & 3.0475 & 0 \\ 3.04751 & 9.1425 & 0 \\ 0 & 0 & 3.0475 \end{bmatrix}$$

Using the results of Exercise 6-7, the following elastic constants can be obtained easily:

$$E_R = \frac{9.1425^2 - 3.0475^2}{9.1425} = 8.13 \text{ GPa}$$

$$G_R = 3.0475 \text{ GPa}$$

$$\nu_R = \frac{3.0475}{9.1425} = 0.33$$

The values of E_R and G_R obtained here compare well with those obtained in Example 4-1, but the value of ν_R is lower than that obtained earlier.

6.7 DETERMINATION OF LAMINAE STRESSES AND STRAINS

The aim of the analysis of laminated composites is to determine the stresses and strains in each of the laminae forming the laminate. These stresses and strains can be used to predict the load at which failure initiates, that is, the load at which the first lamina fails. A step-by-step analysis will be needed to predict the loads at which the subsequent laminae will fail. This procedure is discussed in Sec. 6.8. In this section only the method of determining the laminae stresses and strains for known loads and a given stacking sequence in a laminate is discussed.

The strains in a lamina caused by external loading are functions of the laminate midplane strains, plate curvatures, and distance from the geometric midplane of the laminate. Equation (6.6) gives the relations among these quantities. The lamina stresses can be determined from either the calculated lamina strains using stress–strain relations of Eq. (5.94) or directly from the midplane strains and curvatures using Eq. (6.9). Thus the first step in determining the stress and strain is to calculate the midplane strains and curvatures.

Relations between the applied loads and the midplane strains and plate curvatures are provided by Eqs. (6.18) and (6.19). The two matrix equations represent six simultaneous algebraic equations involving six unknowns that are three midplane strains and three plate curvatures. These equations can be solved for the unknowns. For a general laminate in which the coupling matrix $[B]$ is also nonzero. the solution of these equations requires inverting a 6×6 matrix of Eq. (6.21). However, this inversion can be carried out in steps, and complete inversion can be subdivided into the inversion of smaller matrices and matrix multiplications. Under certain loading conditions, it will be found quite satisfactory to use the intermediate equations with partially inverted matrices for the purpose of calculation. In the following paragraphs, the strains and curvatures are derived as explicit functions of applied loads, that is, the stress resultants and moments.

The general constitutive equations for a laminate have been derived [Eq. (6.21)] to be

$$\left\{\frac{N}{M}\right\} = \left[\begin{array}{c|c} A & B \\ \hline B & D \end{array}\right]\left\{\frac{\epsilon^0}{k}\right\}$$

Consider the equations for N and M separately

$$\{N\} = [A]\{\epsilon^0\} + [B]\{k\}$$
$$\{M\} = [B]\{\epsilon^0\} + [D]\{k\}$$
(6.26)

Solving the first of Eqs. (6.26) for midplane strains gives

6.7 DETERMINATION OF LAMINAE STRESSES AND STRAINS

$$\{\epsilon^0\} = [A^{-1}]\{N\} - [A^{-1}][B]\{k\} \tag{6.27}$$

Substituting Eq. (6.27) in the second relation in Eq. (6.26) yields

$$\{M\} = [B][A^{-1}]\{N\} - [[B][A^{-1}][B] - [D]]\{k\} \tag{6.28}$$

Equations (6.27) and (6.28) can be combined to obtain a partially inverted form of the laminate constitutive equation as follows:

$$\begin{Bmatrix} \epsilon^0 \\ M \end{Bmatrix} = \begin{bmatrix} A^* & B^* \\ C^* & D^* \end{bmatrix} \begin{Bmatrix} N \\ k \end{Bmatrix} \tag{6.29}$$

where (see Appendix 1 for discussion of matrix algebra):

$$[A^*] = [A^{-1}]$$

$$[B^*] = -[A^{-1}][B]$$

$$[C^*] = [B][A^{-1}] = -[B^*]^T$$

$$[D^*] = [D] - [B][A^{-1}][B]$$

It may be noted that to obtain this partially inverted form of the laminate constitutive equation, only one 3 × 3 matrix needs to be inverted and two matrix multiplications carried out.

Now, following the definitions of starred matrices, Eqs. (6.27) and (6.28) can be rewritten as

$$\{\epsilon^0\} = [A^*]\{N\} + [B^*]\{k\}$$
$$\{M\} = [C^*]\{N\} + [D^*]\{k\} \tag{6.30}$$

Solving the second of these equations for plate curvatures gives

$$\{k\} = [D^{*-1}]\{M\} - [D^{*-1}][C^*]\{N\} \tag{6.31}$$

Substituting Eq. (6.31) into the first relation in Eqs. (6.30) yields

$$\{\epsilon^0\} = [[A^*] - [B^*][D^{*-1}][C^*]]\{N\} + [B^*][D^{*-1}]\{M\} \tag{6.32}$$

Equations (6.31) and (6.32) can be combined to obtain a fully inverted form of the laminated constitutive equations as follows:

$$\left\{\begin{array}{c}\epsilon^0\\k\end{array}\right\} = \left[\begin{array}{c|c}A' & B'\\ \hline C' & D'\end{array}\right]\left\{\begin{array}{c}N\\M\end{array}\right\} = \left[\begin{array}{c|c}A' & B'\\ \hline B' & D'\end{array}\right]\left\{\begin{array}{c}N\\M\end{array}\right\} \qquad (6.33)$$

where

$$[A'] = [A^*] - [B^*][D^{*-1}][C^*] = [A^*] + [B^*][D^{*-1}][B^*]^T$$

$$[B'] = [B^*][D^{*-1}]$$

$$[C'] = -[D^{*-1}][C^*] = [B']^T = [B']$$

$$[D'] = [D^{*-1}]$$

Thus a fully inverted form of the laminate constitutive equations is obtained by inverting an additional 3×3 matrix and carrying out two more matrix multiplications.

The laminate constitutive equations in one of the three forms discussed earlier can be used to calculate the laminae stresses and strains. The choice of a particular form depends on the loading condition of the laminate. It is important to note that each form can be obtained through the basic elastic properties of the laminae (i.e., $\overline{Q}_{ij}$ matrices) and their stacking sequence.

For symmetric laminates, the constitutive equations [Eqs. (6.22) and (6.23)] can be written as

$$\{N\} = [A]\{\varepsilon^0\} \qquad (6.34)$$

$$\{M\} = [D]\{k\} \qquad (6.35)$$

Each of the Eqs. (6.34) and (6.35) is a set of three algebraic equations with three unknowns. Solution of these equations is relatively simple. Equation (6.34) is solved by premultiplying both sides by $[A^{-1}]$, and Eq. (6.35), by premultiplying by $[D^{-1}]$. The solutions are

$$\{\varepsilon^0\} = [A^{-1}]\{N\} \qquad (6.36)$$

$$\{k\} = [D^{-1}]\{M\} \qquad (6.37)$$

Thus the inverted form of the constitutive equations for symmetric laminates can be obtained by inversion of only two 3×3 matrices.

Example 6-6: Obtain the partially inverted and fully inverted forms of the laminate constitutive equations for the laminate considered in Example 6-1. $[A]$, $[B]$, and $[D]$ matrices for the laminate are

6.7 DETERMINATION OF LAMINAE STRESSES AND STRAINS

$$[A] = \begin{bmatrix} 119.6 & 18.9 & 13.5 \\ 18.9 & 29.6 & 13.5 \\ 13.5 & 13.5 & 18.9 \end{bmatrix}$$

$$[B] = \begin{bmatrix} 100.9 & -33.4 & -33.8 \\ -33.4 & -34.1 & -33.8 \\ -33.8 & -33.8 & -33.4 \end{bmatrix}$$

$$[D] = \begin{bmatrix} 571.0 & 123.0 & 94.5 \\ 123.0 & 181.0 & 94.5 \\ 94.5 & 94.5 & 123.0 \end{bmatrix}$$

First, $[A^{-1}]$ can be found to be (see Appendix 1 for the procedure):

$$[A^{-1}] = [A^*] = 10^{-2} \begin{bmatrix} 0.95 & -0.44 & -0.36 \\ -0.44 & 5.21 & -3.41 \\ -0.36 & -3.41 & 7.99 \end{bmatrix}$$

Now the other matrices in the semi-inverted form of the constitutive equations can be obtained easily by matrix multiplications and subtraction as follows:

$$[B^*] = -10^{-2} \begin{bmatrix} 0.95 & -0.44 & -0.36 \\ -0.44 & 5.21 & -3.41 \\ -0.36 & -3.41 & 7.99 \end{bmatrix} \begin{bmatrix} 100.9 & -33.4 & -33.8 \\ -33.4 & -34.1 & -33.8 \\ -33.8 & -33.8 & -33.4 \end{bmatrix}$$

$$[B^*] = \begin{bmatrix} -1.224 & 0.044 & 0.050 \\ 1.032 & 0.479 & 0.475 \\ 1.926 & 1.415 & 1.392 \end{bmatrix}$$

$$[C^*] = -[B^*]^T = \begin{bmatrix} 1.224 & -1.032 & -1.926 \\ -0.044 & -0.479 & -1.415 \\ -0.050 & -0.475 & -1.392 \end{bmatrix}$$

$[D^*] = [D] - [B][A^{-1}][B]$

$\quad = [D] + [B][B^*]$

$$= \begin{bmatrix} 571.0 & 123.0 & 94.5 \\ 123.0 & 181.0 & 94.5 \\ 94.5 & 94.5 & 123.0 \end{bmatrix}$$

$$+ \begin{bmatrix} 100.9 & -33.4 & -33.8 \\ -33.4 & -34.1 & -33.8 \\ -33.8 & -33.8 & -33.4 \end{bmatrix} \begin{bmatrix} -1.224 & 0.044 & 0.500 \\ 1.032 & 0.479 & 0.475 \\ 1.926 & 1.415 & 1.392 \end{bmatrix}$$

$$[D^*] = \begin{bmatrix} 347.95 & 63.61 & 36.68 \\ 63.61 & 115.38 & 29.57 \\ 36.68 & 29.57 & 58.75 \end{bmatrix}$$

Thus the partially inverted form of the constitutive equations becomes

$$\begin{Bmatrix} \epsilon_x^0 \\ \epsilon_y^0 \\ \gamma_{xy}^0 \\ \hline M_x \\ M_y \\ M_{xy} \end{Bmatrix} = \left[\begin{array}{ccc:ccc} 0.0095 & -0.0044 & -0.0036 & -1.224 & 0.044 & 0.050 \\ -0.0044 & 0.0521 & -0.0341 & 1.032 & 0.479 & 0.475 \\ -0.0036 & -0.0341 & 0.0799 & 1.926 & 1.415 & 1.392 \\ \hdashline 1.224 & -1.032 & -1.926 & 347.95 & 63.61 & 36.68 \\ -0.044 & -0.479 & -1.415 & 63.61 & 115.38 & 29.57 \\ -0.050 & -0.475 & -1.392 & 36.68 & 29.57 & 58.75 \end{array} \right] \cdot \begin{Bmatrix} N_x \\ N_y \\ N_{xy} \\ \hline k_x \\ k_y \\ k_{xy} \end{Bmatrix}$$

Now, to find the fully inverted form of the constitutive equations, begin with finding the inverse of matrix $[D^*]$.

$$[D^{*-1}] = [D'] = \begin{bmatrix} 0.0033 & -0.0015 & -0.0013 \\ -0.0015 & 0.0106 & -0.0044 \\ -0.0013 & -0.0044 & 0.0201 \end{bmatrix}$$

The other matrices can be obtained by matrix multiplications and subtractions:

$[B'] = [B^*][D^{*-1}]$

$$= \begin{bmatrix} -1.224 & 0.044 & 0.050 \\ 1.032 & 0.479 & 0.475 \\ 1.926 & 1.415 & 1.392 \end{bmatrix} \begin{bmatrix} 0.0033 & -0.0015 & -0.0013 \\ -0.0015 & 0.0106 & -0.0044 \\ -0.0013 & -0.0044 & 0.0201 \end{bmatrix}$$

$$[B'] = \begin{bmatrix} -0.0041 & 0.0021 & 0.0024 \\ 0.0021 & 0.0015 & 0.0060 \\ 0.0024 & 0.0060 & 0.0192 \end{bmatrix}$$

6.7 DETERMINATION OF LAMINAE STRESSES AND STRAINS

$$[C'] = [B']^T = \begin{bmatrix} -0.0041 & 0.0021 & 0.0024 \\ 0.0021 & 0.0015 & 0.0060 \\ 0.0024 & 0.0060 & 0.0192 \end{bmatrix}$$

$$[A'] = [A^*] - [B^*][D^{*-1}][C^*]$$

$$= [A^*] - [B'][C^*]$$

$$= \begin{bmatrix} 0.0095 & -0.0044 & -0.0036 \\ -0.0044 & 0.0521 & -0.0341 \\ -0.0036 & -0.0341 & 0.0799 \end{bmatrix}$$

$$+ \begin{bmatrix} -0.0041 & 0.0021 & 0.0024 \\ 0.0021 & 0.0015 & 0.0060 \\ 0.0024 & 0.0060 & 0.0192 \end{bmatrix} \begin{bmatrix} -1.224 & 1.032 & 1.926 \\ 0.044 & 0.479 & 1.415 \\ 0.050 & 0.475 & 1.392 \end{bmatrix}$$

$$[A'] = \begin{bmatrix} 0.0148 & -0.0065 & -0.0053 \\ -0.0065 & 0.0578 & -0.0196 \\ -0.0053 & -0.0196 & 0.1197 \end{bmatrix}$$

Thus the fully inverted form of the constitutive equation is

$$\begin{Bmatrix} \epsilon_x^0 \\ \epsilon_y^0 \\ \gamma_{xy}^0 \\ \hline k_x \\ k_y \\ k_{xy} \end{Bmatrix} = \begin{bmatrix} 0.0148 & -0.0065 & -0.0053 & \vdots & -0.0041 & 0.0021 & 0.0024 \\ -0.0065 & 0.0578 & -0.0196 & \vdots & 0.0021 & 0.0015 & 0.0060 \\ -0.0053 & -0.0196 & 0.1197 & \vdots & 0.0024 & 0.0060 & 0.0192 \\ \hline -0.0041 & 0.0021 & 0.0024 & \vdots & 0.0033 & -0.0015 & -0.0013 \\ 0.0021 & 0.0015 & 0.0060 & \vdots & -0.0015 & 0.0106 & -0.0044 \\ 0.0024 & 0.0060 & 0.0192 & \vdots & -0.0013 & -0.0044 & 0.0201 \end{bmatrix} \cdot \begin{Bmatrix} N_x \\ N_y \\ N_{xy} \\ \hline M_x \\ M_y \\ M_{xy} \end{Bmatrix}$$

It should be noted that the preceding compliance matrix in the fully inverted form of the constitutive equation is a symmetric matrix, as should be

244 ANALYSIS OF LAMINATED COMPOSITES

expected because of the symmetry of the original stiffness matrix. Small errors caused by the rounding off of the numbers in the intermediate steps should be ignored. It may be pointed out that this compliance matrix could be obtained by directly inverting the original 6 × 6 stiffness matrix.

Example 6-7: Let the three-ply laminate considered in Example 6-2 be subjected to the forces $N_x = 1000$ N/mm, $N_y = 200$ N/mm, and $N_{xy} = 0$, as shown in Fig. 6-9. Calculate the stresses and strains in the individual plies.

The extensional stiffness matrix for the laminate was found to be

$$[A] = \begin{bmatrix} 159.3 & 35.1 & 27.0 \\ 35.1 & 51.3 & 27.0 \\ 27.0 & 27.0 & 35.1 \end{bmatrix}$$

The coupling matrix [B] for this laminate is zero, as shown in Example 6-2. Therefore, the given loading would produce only in-plane normal and shear strains, and no plate curvatures would be produced. This also implies, from Eq. (6.6), that the midplane strains are also the strains for individual plies because there is no strain gradient through the thickness. However, the stresses in each ply will be different and have to be evaluated by taking into consideration the corresponding stiffness matrix.

To obtain the midplane strains, first, $[A^{-1}]$ can be found to be

$$[A^{-1}] = \begin{bmatrix} 0.00759 & -0.00356 & -0.00309 \\ -0.00356 & 0.03441 & -0.02373 \\ -0.00309 & -0.02373 & 0.04911 \end{bmatrix}$$

Since the coupling matrix [B] is equal to zero, Eq. (6.36) can be used to calculate the in-plane strains as

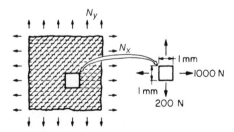

Figure 6-9. Definition of applied forces on laminate for Example 6-7.

6.7 DETERMINATION OF LAMINAE STRESSES AND STRAINS

$$\left\{\begin{array}{c} \epsilon_x^0 \\ \epsilon_y^0 \\ \gamma_{xy}^0 \end{array}\right\} = 10^{-3} \begin{bmatrix} 0.00759 & -0.00356 & -0.00309 \\ -0.00356 & 0.03441 & -0.02373 \\ -0.00309 & -0.02373 & 0.04911 \end{bmatrix} \left\{\begin{array}{c} 1000 \\ 200 \\ 0 \end{array}\right\}$$

$$= \left\{\begin{array}{c} 0.00685 \\ 0.00332 \\ -0.00784 \end{array}\right\}$$

It may be noted that the factor 10^{-3} has been placed before the $[A^{-1}]$ matrix to make its units consistent with those of N_x, N_y, and N_{xy}. The reader is advised to verify this. The preceding midplane strains are also the lamina strains in the xy reference coordinates. The reference coordinates for each lamina are explained in Fig. 6-10. The lamina stresses in the xy coordinates can be obtained from the stress–strain relation [Eq. (594)]. Using the $[\overline{Q}]$ matrices obtained in Example 6-2, stresses are found as

$$\left\{\begin{array}{c} \sigma_x \\ \sigma_y \\ \tau_{xy} \end{array}\right\}_{0° \text{ ply}} = \begin{bmatrix} 20 & 0.7 & 0 \\ 0.7 & 2.0 & 0 \\ 0 & 0 & 0.7 \end{bmatrix} \left\{\begin{array}{c} 0.00685 \\ 0.00332 \\ -0.00784 \end{array}\right\} = \left\{\begin{array}{c} 139.3 \\ 11.4 \\ -5.5 \end{array}\right\} 10^{-3} \text{ GPa}$$

$$\left\{\begin{array}{c} \sigma_x \\ \sigma_y \\ \tau_{xy} \end{array}\right\}_{45° \text{ ply}} = \begin{bmatrix} 6.55 & 5.15 & 4.50 \\ 5.15 & 6.55 & 4.50 \\ 4.50 & 4.50 & 5.15 \end{bmatrix} \left\{\begin{array}{c} 0.00685 \\ 0.00332 \\ -0.00784 \end{array}\right\} = \left\{\begin{array}{c} 26.7 \\ 21.7 \\ 5.4 \end{array}\right\} 10^{-3} \text{ GPa}$$

The laminae stresses and strains in the xy reference coordinates are represented graphically in Fig. 6-11.

For purposes of the laminate strength analysis, it is desirable that the laminae stresses and strains be obtained along their natural (longitudinal and transverse) axes. These now can be obtained easily by using the transformation equations [Eqs. (5.86) and (5.88)]. For the 0° ply, the lamina natural axes coincide with the laminate xy coordinates. Therefore, the stresses and strains obtained in the preceding paragraphs for the 0° ply are also the stresses and strains in the longitudinal and transverse directions; in other words,

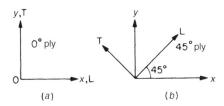

Figure 6-10. Reference coordinate axes for (a) 0° lamina and (b) 45° lamina.

246 ANALYSIS OF LAMINATED COMPOSITES

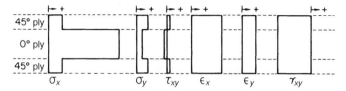

Figure 6-11. Lamina stresses and strains along the reference axes (Example 6-7).

$$\left\{\begin{array}{c}\epsilon_L \\ \epsilon_T \\ \gamma_{LT}\end{array}\right\}_{0° \text{ ply}} = \left\{\begin{array}{c}\epsilon_x \\ \epsilon_y \\ \gamma_{xy}\end{array}\right\}_{0° \text{ ply}} = \left\{\begin{array}{c}0.00685 \\ 0.00332 \\ -0.00784\end{array}\right\}$$

and

$$\left\{\begin{array}{c}\sigma_L \\ \sigma_T \\ \tau_{LT}\end{array}\right\}_{0° \text{ ply}} = \left\{\begin{array}{c}\sigma_x \\ \sigma_y \\ \tau_{xy}\end{array}\right\}_{0° \text{ ply}} = \left\{\begin{array}{c}139.3 \\ 11.4 \\ -5.5\end{array}\right\} \text{ MPa}$$

The lamina stress and strains for the 45° plies in the longitudinal and transverse axes may be obtained using one of two procedures. In the first procedure, both the stresses and strains are transformed using Eqs. (5.86) and (5.88), whereas in the second procedure, only strains are transformed, and then the stresses are calculated from the stress–strain relations [Eq. (5.74)]. Resulting stresses in the two cases will be the same. In problems where the stresses in the xy coordinates are not needed, the second procedure should be preferred because it would save an intermediate step in the calculations.

The following is the transformation matrix for the 45° orientation:

$$\begin{bmatrix} 0.5 & 0.5 & 0.5 \\ 0.5 & 0.5 & -0.5 \\ -1.0 & 1.0 & 0 \end{bmatrix}$$

The strains are obtained by using Eq. (5.88):

$$\left\{\begin{array}{c}\epsilon_L \\ \epsilon_T \\ \gamma_{LT}\end{array}\right\}_{45° \text{ ply}} = \begin{bmatrix} 0.5 & 0.5 & 0.5 \\ 0.5 & 0.5 & -0.5 \\ -1.0 & 1.0 & 0 \end{bmatrix} \left\{\begin{array}{c}0.00685 \\ 0.00332 \\ -0.00784\end{array}\right\} = \left\{\begin{array}{c}0.00116 \\ 0.00900 \\ -0.00352\end{array}\right\}$$

Similarly, the stresses are obtained by using Eq. (5.86):

$$\left\{\begin{array}{c}\sigma_L \\ \sigma_T \\ \tau_{LT}\end{array}\right\} = \left[\begin{array}{ccc} 0.5 & 0.5 & 1.0 \\ 0.5 & 0.5 & -1.0 \\ -0.5 & 0.5 & 0 \end{array}\right]\left\{\begin{array}{c} 26.7 \\ 21.7 \\ 5.4 \end{array}\right\} = \left\{\begin{array}{c} 29.6 \\ 18.8 \\ -2.5 \end{array}\right\} \text{MPa}$$

The laminae stresses and strains in the longitudinal and transverse reference coordinates are represented graphically in Fig. 6-12. These stress–strain variations can be compared with the allowable stresses and strains in each lamina, and thus the load at which the failure initiates in one of the laminae may be calculated. The procedure for calculating the load at failure initiation and the laminae stresses and strains after failure initiation is discussed in the next section.

6.8 ANALYSIS OF LAMINATES AFTER INITIAL FAILURE

Procedures for calculating stresses and strains in individual laminae owing to external loads on the laminate were discussed in the preceding section. The stresses and strains in each lamina may be compared with the corresponding allowable values to predict failure. Commonly employed failure theories were discussed in Chap. 5. Thus, for a given load, it may be determined easily whether any of the plies in the laminate will fail. Conversely, the load at which the first ply failure (FPF) will occur may be calculated. Since the strength of a ply is a function of its orientation, it is expected that all plies will not fail at the same load. Plies will fail successively in the increasing order of strength in the direction of loading. Moreover, the transverse strength of unidirectional laminae is known to be much smaller than the longitudinal strength, so the plies with fibers perpendicular to the load will fail first. Thus the FPF may occur at relatively small loads at which the laminate is in no real danger of fracture. Sometimes the effect of the FPF may not be evident from the macroscopic response of the laminate, but as the number of ply failures increases, the loss of laminate stiffness becomes evident, and the overall response of the laminate deviates from its original straight-line behavior. However, the laminate is still able to carry additional loads, although

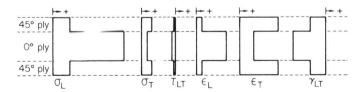

Figure 6-12. Lamina stresses and strains along the longitudinal and transverse axes (Example 6-7).

the additional loads produce larger deflections than those produced by the same loads prior to the FPF. Thus the analysis procedures discussed so far are no longer directly applicable. The procedures may be modified to calculate the maximum load-carrying capacity of the laminate whenever required. In this section an analysis procedure is developed that is applicable over the entire range of load. The procedure is developed for a general laminate, and its application for a special case (namely, the cross-ply laminate) is illustrated.

The analysis procedures discussed so far in this chapter are valid when all the laminae are intact. As the load is increased, the stresses in a lamina may become high enough to cause failure. After the FPF, the laminate response will deviate from that predicted by Eq. (6.21) and show a discontinuity in its behavior. As the load is increased further, more ply failures will occur, showing more discontinuities in the laminate behavior. Thus the complete stress–strain behavior of a laminate up to fracture may be expected to be as shown in Fig. 6-13, where each corner or knee in the curve represents a ply failure. This change in slope may be difficult to detect if the ply that fails carries only a small fraction of the total load.

As shown in Fig. 6-13, the response between two discontinuities may be assumed to be linear because laminae are assumed to show a linear behavior up to fracture. Therefore, the load–strain relationship for each segment (e.g., ith) of the curve may be written for incremental load and incremental strain as follows:

$$\left\{ \begin{array}{c} \Delta N \\ \Delta M \end{array} \right\}_i = \left[\begin{array}{c|c} \overline{A} & \overline{B} \\ \hline \overline{B} & \overline{D} \end{array} \right]_i \left\{ \begin{array}{c} \Delta \epsilon^0 \\ \Delta k \end{array} \right\}_i \tag{6.38}$$

where matrices $[\overline{A}]$, $[\overline{B}]$, and $[\overline{D}]$ are not the same as the matrices $[A]$, $[B]$, and $[D]$. The new matrices $[\overline{A}]$, $[\overline{B}]$, and $[\overline{D}]$ have been modified to take into account the fact that some of the plies already have fractured. They are obtained from Eq. (6.20), in which the stiffness matrices $[\overline{Q}]$ of the fractured plies are appropriately corrected. The exact nature of the correction depends

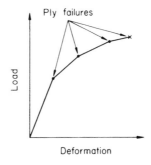

Figure 6-13. Load-deformation behavior of a hypothetical laminate.

on the mode of fracture of the ply in consideration. For example, when a ply fails because of a transverse tensile stress exceeding the transverse tensile strength, it cannot support any additional transverse load, and hence its transverse modulus should be set to zero. The longitudinal modulus may remain unaffected by the transverse failure. However, different failure modes interact in many cases so that all the elastic properties of the ply are affected. The extent by which the properties are influenced is very difficult to ascertain. As a conservative approach, it is sometimes suggested that when a ply fails, all its elastic properties should be set to zero.

The incremental loads and strains as determined from Eq. (6.38) may be added to the loads and strains at the previous ply failure (or those at the end of the previous segment) to obtain their absolute values as follows:

$$\begin{Bmatrix} N_x \\ N_y \\ N_{xy} \end{Bmatrix} = \begin{Bmatrix} N_x \\ N_y \\ N_{xy} \end{Bmatrix}_{i-1} + \begin{Bmatrix} \Delta N_x \\ \Delta N_y \\ \Delta N_{xy} \end{Bmatrix}_i \qquad (6.39)$$

$$\begin{Bmatrix} M_x \\ M_y \\ M_{xy} \end{Bmatrix} = \begin{Bmatrix} M_x \\ M_y \\ M_{xy} \end{Bmatrix}_{i-1} + \begin{Bmatrix} \Delta M_x \\ \Delta M_y \\ \Delta M_{xy} \end{Bmatrix}_i \qquad (6.40)$$

$$\begin{Bmatrix} \epsilon_x^0 \\ \epsilon_y^0 \\ \gamma_{xy}^0 \end{Bmatrix} = \begin{Bmatrix} \epsilon_x^0 \\ \epsilon_y^0 \\ \gamma_{xy}^0 \end{Bmatrix}_{i-1} + \begin{Bmatrix} \Delta \epsilon_x^0 \\ \Delta \epsilon_y^0 \\ \Delta \gamma_{xy}^0 \end{Bmatrix}_i \qquad (6.41)$$

$$\begin{Bmatrix} k_x \\ k_y \\ k_{xy} \end{Bmatrix}_i = \begin{Bmatrix} k_x \\ k_y \\ k_{xy} \end{Bmatrix}_{i-1} + \begin{Bmatrix} \Delta k_x \\ \Delta k_y \\ \Delta k_{xy} \end{Bmatrix}_i \qquad (6.42)$$

The stresses in the laminae now may be determined from Eq. (6.9). Thus the load at which the next ply failure occurs can be calculated. This stepwise procedure can be employed until all plies have failed. The load at which the final fracture of the laminate occurs also can be calculated.

The preceding analysis procedure may be illustrated by application of the analysis to a cross-ply laminate. Consider a cross-ply laminate made up of n identical laminae of which l have fibers along the direction of load and m perpendicular to the load ($l + m = n$). Let the longitudinal and transverse moduli of the laminae be, respectively, E_L and E_T, and their longitudinal and transverse fracture strain, ϵ_{LU} and ϵ_{TU}. The composite modulus in the load direction can be determined by the rule of mixtures as

$$E = \frac{l}{n} E_L + \frac{m}{n} E_T \qquad (6.43)$$

250   ANALYSIS OF LAMINATED COMPOSITES

The initial composite modulus E is sometimes referred to as the *primary modulus*. Equation (6.43) is valid when all the plies are intact. Failure in the 90° plies occurs when the composite strain is equal to ϵ_{TU}. The composite stress at the failure of the 90° plies is

$$\sigma_A = E\epsilon_{TU} \tag{6.44}$$

When the gross stress on the composite exceeds σ_A, the entire load is supported by the 0° plies, and the modulus of the composite is to be calculated by considering the 0° plies only. The composite modulus after the failure of 90° plies, sometimes referred to as the *secondary modulus* E_S, can be calculated as

$$E_S = \frac{l}{n} E_L \tag{6.45}$$

Thus the secondary modulus is equal to the modulus of 0° plies corrected for the area reduction. The composite fracture occurs when the composite strain is equal to ϵ_{LU}. Thus the composite fracture stress is

$$\sigma_F = \sigma_A + E_S(\epsilon_{LU} - \epsilon_{TU}) \tag{6.46}$$

The complete stress–strain diagram of the composite is shown in Fig. 6-14. The point A, where the slope of the curve changes from E to E_S, is often referred to as the *knee* of the curve. The experimental stress–strain curves of the cross-ply materials, a few of which are given later, do show such behavior. The experimental curves usually are nonlinear but can be very closely approximated by two straight lines having slopes equal to the primary modulus (E) and the secondary modulus (E_S).

In the preceding analysis, the maximum strain has been assumed to be the lamina failure criterion for simplicity. Any other criterion, such as those discussed in Chap. 5, also could be used. However, the main difference among these criteria appears only in combined stress and in compression. In the

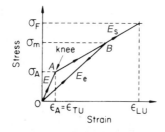

Figure 6-14. Stress–strain diagram of a cross-ply laminate.

6.8 ANALYSIS OF LAMINATES AFTER INITIAL FAILURE

present case, when the material is assumed to be linear, the criteria based on stress can be converted easily into strain criteria. Moreover, for the uniaxial stress considered here, all the criteria yield the same results.

The preceding analysis has been carried out by assuming that the laminate is composed of identical laminae. The analysis can be modified easily if the laminae differ in properties and thickness. When the thicknesses are different, the relative number of plies in Eqs. (6.43) and (6.45) should be replaced by relative cross-sectional areas. When the elastic moduli and strengths of the plies are different, the knee occurs at the transverse fracture strain of the 90° plies, whereas the composite fracture occurs at the longitudinal fracture strain of the 0° plies. It is left to the reader to work out the details of modified equations.

Based on the stress–strain curve shown in Fig. 6-14, the composite strain may be written as

$$\epsilon = \frac{\sigma}{E} \qquad \sigma \leq \sigma_A \qquad (6.47)$$

$$\epsilon = \frac{\sigma_A}{E} + \frac{\sigma - \sigma_A}{E_S} \qquad \sigma \geq \sigma_A \qquad (6.48)$$

It is also desirable to write Eq. (6.48) in a simple form, such as

$$\epsilon = \frac{\sigma}{E_e} \qquad \sigma \geq \sigma_A \qquad (6.49)$$

where E_e may be called the *effective modulus*. Comparison of Eq. (6.49) with Eq. (6.48) will give

$$E_e = \frac{E}{1 + [(E/E_S) - 1][1 - (\sigma_A/\sigma)]} \qquad (6.50)$$

Thus the effective modulus is a function of instantaneous stress σ, and it relates the instantaneous stress to total strain.

Equations (6.47)–(6.50) have been written for monotonically increasing load. The stress during unloading may not correspond to the loading-path stress. The unloading path will depend on whether the maximum stress during loading is more or less than the stress at the knee σ_A. When the maximum stress is less than σ_A, unloading retraces the loading path, and Eq. (6.47) holds for unloading also. When the maximum stress is more than σ_A (e.g., point B in Fig. 6-14), unloading does not trace the preceding loading path [2]. Unloading takes place along a straight line with a slope different from the initial slope and results in a residual strain. On reloading, a hysteresis loop forms, and eventually the unloading point (B) is recovered. At this point the slope changes to that of the original loading curve. Although this behavior resembles that of common metals in the plastic range, there is one major difference, specifically that the elastic modulus decreases with degradation (i.e., failure) of the 90° plies almost to the extent that the residual strain can be neglected. Thus the unloading takes place along paths such as $B0$ in Fig. 6-14. During the unloading, the stress and strain are related through Eq. (6.49) with the modification that the effective modulus is not a function of instantaneous stress but the maximum stress (σ_m), that is, the stress at the point of the start of unloading. The effective modulus during unloading and reloading thus becomes

$$E_e = \frac{E}{1 + [(E/E_S) - 1][1 - (\sigma_A/\sigma_m)]} \tag{6.51}$$

In the preceding discussion it has been assumed that the stress–strain curve is continuous. The failure (or degradation) of 90° plies is assumed to cause a change in the slope of the curve but not a sudden jump in the magnitude of stress or strain. This means that it has been implicitly assumed that the failure does not produce stress relaxation in the 90° plies. If stress relaxation occurs, the stress–strain curve will show a sudden change in the magnitude of stress or strain or both, as shown in Fig. 6-15. Thus, instead of path A, the material may follow a path such as B or C. The actual path will depend on the type of loading. For example, in a load-controlled test, path B will be traversed, whereas a displacement-controlled test will trace path C. An intermediate path D is also possible.

Whether complete stress relaxation in the 90° plies occurs is an important question and needs careful consideration. It should be expected that if the plies are independent of one another in the laminate (e.g., when interlaminar bonds are weak), a significant or complete stress relaxation will occur. On the other hand, when interlaminar bonds are strong, the adjoining 0° plies restrain the failure of 90° plies. As a result, the effect of failure of a 90° ply is localized, and only a partial stress relaxation will occur. The restraint is

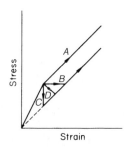

Figure 6-15. Possible stress–strain diagrams for cross-ply laminates.

maximum when the interaction between plies of different orientations is maximum (e.g., with alternate layers 0° and 90°). This situation is analogous to the breaking of a fiber surrounded by the matrix. The stress at the broken end of the fiber reduces to zero but builds up, because of shear-stress transfer, away from the end in the presence of strong interfacial bonds (see Chap. 4). Thus the influence of fiber breaks in a unidirectional composite loaded in the fiber direction becomes evident only gradually. Similarly, when there is good interaction between plies, degradation of the 90° plies is gradual. That is, when the composite strain is just equal to the transverse fracture strain of the 90° plies, cracks develop in the 90° plies, but their effect is confined to only small lengths, and the composite modulus does not drop instantly to the level predicted by Eq. (6.45). More cracks develop as the load is increased, and when the cracks are very close to each other, the contribution of the 90° plies to the composite stiffness reduces to near zero, and the composite modulus eventually drops to the level predicted by Eq. (6.45). The experimental results of Hahn and Tsai [2] are quite relevant in the present context. They investigated the stress–strain behavior of glass–epoxy cross-ply laminates with two different stacking sequences. The $[0/90]_{2S}$ laminate consists of eight plies with alternate layers being 0° and 90°, whereas the $[0/90_2]_S$ laminate has six plies with two surface plies 0° and the four inner plies 90°. The stress–strain curves for the two laminates are shown in Figs. 6-16 and 6-17 along with theoretical predictions [i.e., Eqs. (6.47) and (6.48). The stress–strain curve for the $[0/90]_{2S}$ laminate does not show a distinct knee, whereas the curve for the $[0/90_2]_S$ laminate shows a distinct knee or obvious change in slope. This is what is expected in view of the preceding discussion. Although the interlaminar bonds are equally strong in the two laminates, the $[0/90]_{2S}$ lam-

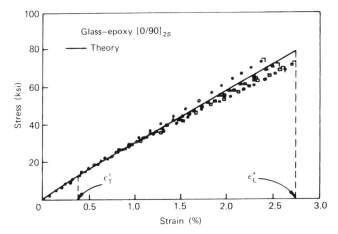

Figure 6-16. Experimental and theoretical stress–strain curves for a $[0/90]_{2s}$ laminate. (From Hahn and Tsai [2].)

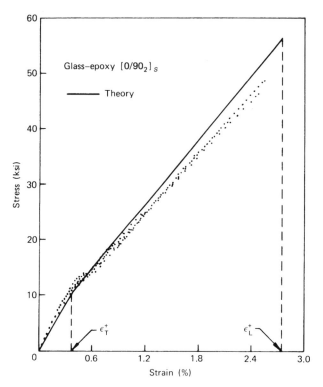

Figure 6-17. Experimental and theoretical stress–strain curves for a $[0/90_2]_s$ laminate. (From Hahn and Tsai [2].)

inate has interspersed plies, except for the middle plies, so that failure of the 90° plies is effectively restrained by the 0° layers. In the $[0/90_2]_s$ laminate, all the 90° plies are bonded together so that the restraint exerted by the 0° plies is not so effective; consequently, the knee is observed.

The preceding discussion illustrates how a cross-ply laminate subjected to uniaxial stress is analyzed. In this analysis it is sufficient to assume that on failure of the 90° plies, their transverse modulus (E_T) reduces to zero. Other properties, such as the shear modulus (G_{LT}), have not been discussed with regard to how they are affected by the failure when the transverse strain in the lamina exceeds the allowable strain in that direction. In real situations, when angle-ply laminates are subjected to complex stresses, it is important to know how the lamina elastic constants are affected when failure in the lamina takes place by a single failure mode. Modifications of the laminate stiffness matrix [e.g., as in (Eq. 6.38)] can be carried out only when the influence of failure on lamina properties is known. A common practice is to set all lamina properties equal to zero when failure occurs. This type of analysis gives a conservative estimate of the load-carrying capacity of the laminate because lamina failure in one direction does not, in general, result in complete stress relaxation in all directions. For example, when failure occurs in the transverse direction, the lamina stresses in the longitudinal direction may not be affected significantly. Some studies [3,4] have been carried out to establish experimentally the quantitative influence on lamina properties resulting from failure in one direction. These studies are inconclusive, and the rationale in reducing elastic constants to specific values has not been established. Thus, in the present circumstances, the practice of setting all lamina properties equal to zero may be continued.

Example 6-8: A 5-mm-thick symmetric cross-ply laminate is constructed from 15 identical laminae having the following stiffness matrix and strengths:

$$[Q] = \begin{bmatrix} 56 & 4.6 & 0 \\ 4.6 & 18.7 & 0 \\ 0 & 0 & 8.9 \end{bmatrix} \text{GPa}$$

$$\sigma_{LU} = 1050 \text{ MPa}$$

$$\sigma_{TU} = 28 \text{ MPa}$$

$$\tau_{LTU} = 42 \text{ MPa}$$

A uniaxial load is applied, and the laminate construction is such that nine laminae are in the load direction. Calculate the load at which the 90° plies fail and the load-carrying capacity of the laminate.

Solution: $[\overline{Q}]$ matrices for the 0° and 90° laminae can be written as

$$[\overline{Q}]_{0°} = \begin{bmatrix} 56.0 & 4.6 & 0 \\ 4.6 & 18.7 & 0 \\ 0 & 0 & 8.9 \end{bmatrix} \text{GPa}$$

$$[\overline{Q}]_{90°} = \begin{bmatrix} 18.7 & 4.6 & 0 \\ 4.6 & 56.0 & 0 \\ 0 & 0 & 8.9 \end{bmatrix} \text{GPa}$$

The [A] matrix for the laminate can be obtained if the thicknesses of all 0° and 90° plies are known. Since all plies have the same thickness, thicknesses of all 0° and 90° plies are proportional to their number. Therefore,

Thickness of 0° plies (9 in number) = $\frac{5}{15} \times 9 = 3$ mm

Thickness of 90° plies (6 in number) = $\frac{5}{15} \times 6 = 2$ mm

The terms of the [A] matrix are given by

$$A_{ij} = 3(\overline{Q}_{ij})_{0°} + 2(\overline{Q}_{ij})_{90°}$$

Therefore the [A] matrix is obtained as

$$[A] = \begin{bmatrix} 205.4 & 23 & 0 \\ 23 & 168.1 & 0 \\ 0 & 0 & 44.5 \end{bmatrix} \text{GPa} \cdot \text{mm}$$

For analysis of this cross-ply laminate, we shall use the maximum-strain theory to predict failure of the laminae. Maximum allowable strains can be obtained from the given strength values, and the moduli values can be calculated from the given stiffness matrix. The moduli values are

$$E_L = 54.87 \text{ GPa}$$

$$E_T = 18.32 \text{ GPa}$$

Fracture strains in the longitudinal and transverse directions become

$$\varepsilon_{LU} = \frac{1050 \times 10^{-3}}{54.87} = 0.01914$$

$$\varepsilon_{TU} = \frac{28 \times 10^{-3}}{18.32} = 0.00153$$

Therefore, the 90° plies will fail when $\varepsilon_x = 0.00153$. The load N_x at failure of the 90° plies can be obtained as follows:

$$\begin{Bmatrix} N_x \\ 0 \\ 0 \end{Bmatrix} = \begin{bmatrix} 205.4 & 23 & 0 \\ 23 & 168.1 & 0 \\ 0 & 0 & 44.5 \end{bmatrix} \begin{Bmatrix} 0.00153 \\ \varepsilon_y \\ \gamma_{xy} \end{Bmatrix}$$

Solution of this matrix equation gives

$$N_x = 0.3094 \text{ GPa} \cdot \text{mm} = 309.4 \text{ MPa} \cdot \text{mm}$$

The 90° plies will fail when $N_x = 309.4$ MPa·mm. After failure of the 90° plies, the $[A]$ matrix is modified according to Eq. (6.38). The modified $[A]$ matrix is obtained by substituting $(Q_{ij})_{90°} = 0$. Therefore,

$$[A] = \begin{bmatrix} 168 & 13.8 & 0 \\ 13.8 & 56.1 & 0 \\ 0 & 0 & 26.7 \end{bmatrix} \text{GPa·mm}$$

The laminate will fail at a strain of $\varepsilon_x = 0.01914$, that is, an additional strain $\Delta\varepsilon_x = 0.01914 - 0.00153 = 0.01761$. Now, additional load at fracture can be obtained from the following:

$$\begin{Bmatrix} \Delta N_x \\ 0 \\ 0 \end{Bmatrix} = \begin{bmatrix} 168 & 13.8 & 0 \\ 13.8 & 56.1 & 0 \\ 0 & 0 & 26.7 \end{bmatrix} \begin{Bmatrix} 0.01761 \\ \Delta\varepsilon_y \\ \Delta\gamma_{xy} \end{Bmatrix}$$

Solution of this equation gives

$$\Delta N_x = 2.8987 \text{ GPa·mm} = 2898.7 \text{ MPa·mm}$$

Therefore, the total load at fracture or the load-carrying capacity of the laminate is

$$N_x = 2898.7 + 309.4 = 3208.1 \text{ MPa·mm}$$

Example 6-9: Following are the elastic constants and strengths of laminae in a quasi-isotropic laminate $[0/\pm 45/90]_S$:

Elastic constants

$$E_L = 40 \text{ GPa} \qquad E_T = 10 \text{ GPa}$$

$$G_{LT} = 4 \text{ GPa} \qquad \nu_{LT} = 0.285$$

Strengths

$$\sigma_{LU} = 1050 \text{ MPa} \qquad \sigma'_{LU} = 650 \text{ MPa}$$

$$\sigma_{TU} = 20 \text{ MPa} \qquad \sigma'_{TU} = 140 \text{ MPa}$$

$$\tau_{LTU} = 65 \text{ MPa}$$

From the laminate, a rectangular specimen with the dimensions 250 mm × 20 mm × 2 mm is tested in uniaxial tension. Predict the load elongation curve for the specimen if the grips are initially 200 mm apart. Assume that the laminae fail according to the maximum-stress theory and that all the elastic constants of a lamina become zero when it fails. Calculate the fracture load of the specimen.

Solution: The $[Q]$ matrix for the laminae can be obtained using Eq. (5.78):

$$[Q] = \begin{bmatrix} 40.83 & 2.91 & 0 \\ 2.91 & 10.21 & 0 \\ 0 & 0 & 4 \end{bmatrix} \text{GPa}$$

The $[\overline{Q}]$ matrices for different ply orientations can be obtained using the transformation equations [Eq. (5.95)]:

$$[\overline{Q}]_{0°} = \begin{bmatrix} 40.83 & 2.91 & 0 \\ 2.91 & 10.21 & 0 \\ 0 & 0 & 4 \end{bmatrix} \text{GPa}$$

$$[\overline{Q}]_{90°} = \begin{bmatrix} 10.21 & 2.91 & 0 \\ 2.91 & 40.83 & 0 \\ 0 & 0 & 4 \end{bmatrix} \text{GPa}$$

$$[\overline{Q}]_{45°} = \begin{bmatrix} 18.215 & 10.215 & 7.655 \\ 10.215 & 18.215 & 7.655 \\ 7.655 & 7.655 & 11.305 \end{bmatrix} \text{GPa}$$

$$[\overline{Q}]_{-45°} = \begin{bmatrix} 18.215 & 10.215 & -7.655 \\ 10.215 & 18.215 & -7.655 \\ -7.655 & -7.655 & 11.305 \end{bmatrix} \text{GPa}$$

Analysis of Initial Behavior The $[A]$ matrix is obtained by using Eq. (6.20) and noting that the thickness of each ply is $\frac{2}{8} = 0.25$ mm:

$$A_{ij} = 2 \times \tfrac{2}{8}[(\overline{Q}_{ij})_{0°} + (\overline{Q}_{ij})_{90°} + (\overline{Q}_{ij})_{45°} + (\overline{Q}_{ij})_{-45°}]$$

Therefore,

$$[A] = \begin{bmatrix} 43.735 & 13.125 & 0 \\ 13.125 & 43.735 & 0 \\ 0 & 0 & 15.305 \end{bmatrix} \text{GPa} \cdot \text{mm}$$

Owing to the symmetry of the laminate, the $[B]$ matrix vanishes, initial stress–strain relation for uniaxial tension ($N_y = N_{xy} = 0$) may be written as

6.8 ANALYSIS OF LAMINATES AFTER INITIAL FAILURE

$$\begin{Bmatrix} N_x \\ 0 \\ 0 \end{Bmatrix} = \begin{bmatrix} 43.735 & 13.125 & 0 \\ 13.125 & 43.735 & 0 \\ 0 & 0 & 15.305 \end{bmatrix} \begin{Bmatrix} \varepsilon_x^0 \\ \varepsilon_y^0 \\ \gamma_{xy}^0 \end{Bmatrix}$$

Solution of this matrix equation yields

$$\varepsilon_y^0 = -0.3\varepsilon_x^0 \qquad \gamma_{xy}^0 = 0$$

$$N_x = \varepsilon_x^0 / 0.0254 \text{ GPa} \cdot \text{mm}$$

For $\varepsilon_y^0 = -0.3\varepsilon_x^0$, and $\gamma_{xy}^0 = 0$, stresses in the laminae in the longitudinal and transverse directions may be obtained in terms of ε_x^0 using Eqs. (5.94) and (5.74):

$$\begin{Bmatrix} \sigma_L \\ \sigma_T \\ \tau_{LT} \end{Bmatrix}_{0°} = \begin{Bmatrix} 39.96 & \varepsilon_x^0 \\ -0.153 & \varepsilon_x^0 \\ 0 \end{Bmatrix} \text{GPa}$$

$$\begin{Bmatrix} \sigma_L \\ \sigma_T \\ \tau_{LT} \end{Bmatrix}_{90°} = \begin{Bmatrix} -9.339 & \varepsilon_x^0 \\ 9.337 & \varepsilon_x^0 \\ 0 \end{Bmatrix} \text{GPa}$$

$$\begin{Bmatrix} \sigma_L \\ \sigma_T \\ \tau_{LT} \end{Bmatrix}_{45°} = \begin{Bmatrix} 15.31 & \varepsilon_x^0 \\ 4.59 & \varepsilon_x^0 \\ -5.20 & \varepsilon_x^0 \end{Bmatrix} \text{GPa}$$

$$\begin{Bmatrix} \sigma_L \\ \sigma_T \\ \tau_{LT} \end{Bmatrix}_{-45°} = \begin{Bmatrix} 15.31 & \varepsilon_x^0 \\ 4.59 & \varepsilon_x^0 \\ -5.20 & \varepsilon_x^0 \end{Bmatrix} \text{GPa}$$

(Notice that the uniaxial stress on the quasi-isotropic laminate produces complex stresses and strains in the constituent laminae.)

It can be shown easily that for the preceding state of stress, 90° plies will fail first when $\sigma_T = \sigma_{TU}$. That is,

$$9.337\varepsilon_x^0 = 20 \times 10^{-3}$$

$$\varepsilon_x^0 = 0.002142$$

Following are the load and elongation at FPF:

Elongation $\delta = 200 \times 0.002142 = 0.4284$ mm

$$N_x = \frac{0.002142}{0.0254} \text{ GPa} \cdot \text{mm} = 0.08433 \text{ kN/mm}$$

For the 20-mm-wide specimen, the load is

$$P = 20 \times 0.08433 = 1.687 \text{ kN}$$

Analysis during Second Segment After FPF, the [A] matrix is modified by neglecting the contributions of 90° plies. The modified [A] matrix is calculated as follows:

$$\overline{A}_{ij} = 2 \times \tfrac{2}{8}[(\overline{Q}_{ij})_{0°} + (\overline{Q}_{ij})_{45°} + (\overline{Q}_{ij})_{-45°}]$$

Thus

$$[\overline{A}] = \begin{bmatrix} 38.63 & 11.67 & 0 \\ 11.67 & 23.32 & 0 \\ 0 & 0 & 13.305 \end{bmatrix} \text{ GPa} \cdot \text{mm}$$

The load–strain relationship for the second segment can be written as

$$\begin{Bmatrix} \Delta N_x \\ 0 \\ 0 \end{Bmatrix}_2 = \begin{bmatrix} 38.63 & 11.67 & 0 \\ 11.67 & 23.32 & 0 \\ 0 & 0 & 13.305 \end{bmatrix}_2 \begin{Bmatrix} \Delta\varepsilon_x^0 \\ \Delta\varepsilon_y^0 \\ \Delta\gamma_{xy}^0 \end{Bmatrix}_2$$

Solution of the matrix equation gives

$$\Delta\varepsilon_y^0 = -0.5\,\Delta\varepsilon_x^0 \qquad \Delta\gamma_{xy}^0 = 0$$

$$\Delta N_x = 32.795\,\Delta\varepsilon_x^0 \text{ GPa} \cdot \text{mm}$$

Incremental stresses in the laminae now may be obtained in terms of incremental strain $\Delta\varepsilon_x^0$ using Eqs. (5.94) and (5.74):

$$\begin{Bmatrix} \Delta\sigma_L \\ \Delta\sigma_T \\ \Delta\tau_{LT} \end{Bmatrix}_{0°} = \begin{Bmatrix} 39.375\,\Delta\varepsilon_x^0 \\ -2.195\,\Delta\varepsilon_x^0 \\ 0 \end{Bmatrix} \text{GPa}$$

$$\begin{Bmatrix} \Delta\sigma_L \\ \Delta\sigma_T \\ \Delta\tau_{LT} \end{Bmatrix}_{\pm 45°} = \begin{Bmatrix} 10.94\,\Delta\varepsilon_x^0 \\ 3.28\,\Delta\varepsilon_x^0 \\ \mp 6.00\,\Delta\varepsilon_x^0 \end{Bmatrix} \text{GPa}$$

6.8 ANALYSIS OF LAMINATES AFTER INITIAL FAILURE

Instantaneous stresses may be obtained by adding these incremental stresses to the stresses at FPF. Thus

$$\left\{\begin{array}{c}\sigma_L \\ \sigma_T \\ \tau_{LT}\end{array}\right\}_{0°} = \left\{\begin{array}{c}0.0856 + 39.3751\,\Delta\varepsilon_x^0 \\ -0.00033 - 2.195\,\Delta\varepsilon_x^0 \\ 0\end{array}\right\} \text{GPa}$$

$$\left\{\begin{array}{c}\sigma_L \\ \sigma_T \\ \tau_{LT}\end{array}\right\}_{\pm 45°} = \left\{\begin{array}{c}0.0328 + 10.94\,\Delta\varepsilon_x^0 \\ 0.0098 + 3.28\,\Delta\varepsilon_x^0 \\ \mp 0.0111 \mp 6.00\,\Delta\varepsilon_x^0\end{array}\right\} \text{GPa}$$

It can be shown easily that $\pm 45°$ plies will fail next when $(\sigma_T)_{45°} = \sigma_{TU}$. That is

$$0.0098 + 3.28\,\Delta\varepsilon_x^0 = 20 \times 10^{-3} \quad \text{or} \quad \Delta\varepsilon_x^0 = 0.00311$$

The incremental elongation and load at the second ply failure are

$$\Delta\delta = 200 \times 0.00311 = 0.622 \text{ mm}$$

$$\Delta P = 20 \times 0.00311 \times 32.795 \text{ GPa} \cdot \text{mm}^2 = 2.04 \text{ kN}$$

The total load and elongation at the second ply failure are

$$P = 1.687 + 2.04 = 3.721 \text{ kN}$$

$$\delta = 0.4284 + 0.622 = 1.054 \text{ mm}$$

Analysis during Third Segment During the third segment, only $0°$ plies are acting, and therefore, the analysis can be carried out by assuming it to be a unidirectional composite. Failure will occur when

$$(\sigma_L)_{0°} = \sigma_{LU} = 1050 \text{ MPa}$$

Therefore, strain at the laminate failure will be

$$\varepsilon_x = \frac{1050 \times 10^{-3}}{40} = 0.02625$$

Incremental strain at laminate failure will be

$$\Delta\varepsilon_x = 0.02625 - 0.002142 - 0.00311 = 0.0210$$

Incremental elongation and load are

$$\Delta\delta = 200 \times 0.021 = 4.2 \text{ mm}$$

$$\Delta P = 20 \times 2 \times \tfrac{2}{8} \times 0.021 \times 40 \text{ GPa} \cdot \text{mm}^2$$

$$\Delta P = 8.4 \text{ kN}$$

Therefore,

$$\text{Load at fracture} = 8.4 + 3.721 = 12.121 \text{ kN}$$

$$\text{Elongation at fracture} = 4.2 + 1.054 = 5.254 \text{ mm}$$

The complete load–elongation curve for the laminate is shown in Fig. 6-18. Each ply failure causes a change in the slope of—that is, produces a knee in—the stress–strain curve. Load levels at the ply failure have been marked. It may be noted that when 90° plies fail, the change in slope of load–elongation curve is hardly noticeable.

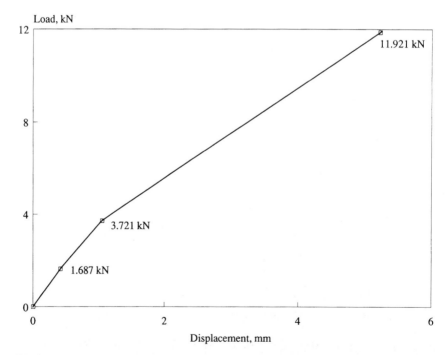

Figure 6-18. Predicted load–elongation curve for the quasi-isotropic laminate in Example 6-9.

6.9 HYGROTHERMAL STRESSES IN LAMINATES

6.9.1 Concepts of Thermal Stresses

Concepts of hygrothermal stresses are relatively simple. However, mathematical formulations for hygrothermal stresses in composite materials may appear quite cumbersome and involved owing to a large number of variables coming from orthotropy of laminae and their lamination. It is easy to lose sight of simple physical concepts as a result of the complicated mathematical formulations. Therefore, before we proceed with the mathematical developments for this problem, let us discuss the concept through a simple example of thermal stresses in a uniaxial case.

A temperature change in a body causes a change in its dimensions proportional to the temperature change. In other words, a temperature change produces thermal strains in the body. These thermal strains are not caused by any external force applied to the body and are not accompanied by internal stresses. However, if the body is, in any manner, restrained from undergoing thermal strains, internal stresses will be produced in the body. Concepts of producing these stresses can be explained by considering a one-dimensional model of a laminate made up of two layers of aluminum and one layer of steel, as shown in Fig. 6-19a. The layers are equal in thickness, bonded, and stacked such that the laminate is symmetric. A temperature change in this laminate, owing to symmetry, produces only extension or shortening of layers but no bending. Initially, the laminate is at temperature T_0. Now consider that the temperature is increased to T. If it is assumed that the three layers are not bonded together, the strains in the aluminum and steel layers are different, as shown in Fig. 6-19b. These strains are called the *free thermal strains* in the layers and are described by a superscript T. However, the aluminum and steel layers are bonded together in the laminate and act as a single unit. Therefore, the actual strains in the aluminum and steel layers are equal. This common strain (ϵ) is somewhere in-between the free thermal strains in steel (ϵ_S^T) and aluminum (ϵ_{Al}^T), and will depend on the elastic moduli of steel and aluminum. Because of the bonding between the aluminum and steel layers, the aluminum layers apply a tensile force to the steel layer and the steel layer applies compressive forces to the aluminum layers such that the aluminum and steel layers have equal strain in the final state. Further, since there is no external force applied to the laminate, the internal forces applied by the layers on each other balance themselves. That is, the net internal force is zero. These internal forces on individual layers are responsible for the difference between the final strain (ε) and free thermal strains (ϵ_{Al}^T and ϵ_S^T), as indicated in Fig. 6-19c. The differential strains (ϵ_{Al}^M and ϵ_S^M) are often called *mechanical strains* that are produced by internal forces and cause internal stresses. These internal stresses are called *thermal stresses* and are proportional to the mechanical strains and not the free thermal strains. Note that the thermal stresses in steel are tensile, whereas those in aluminum are compressive, such that the resultant force on the laminate is zero (see Fig. 6-19d).

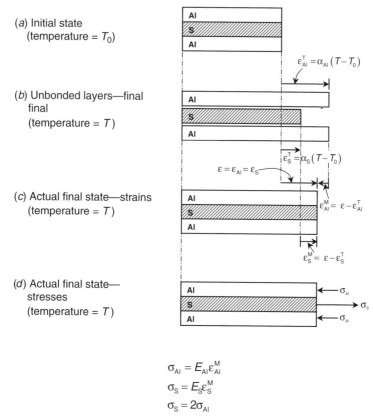

Figure 6-19. Concepts of thermal strains and stresses in a three-ply symmetric laminate.

These concepts of stresses and strains owing to temperature changes can be generalized to develop procedures for calculating hygrothermal stresses and strains in composite laminates. There are two important concepts to be kept in mind: (1) Since no external forces are applied, the resultant forces and moments are zero, and (2) the laminate stresses are caused by the difference in actual strain and strain for free expansion of laminae. Mathematical formulations for hygrothermal stresses are described in the next subsection.

6.9.2 Hygrothermal Stress Calculations

A change in temperature or moisture content of a body causes a change in its dimensions proportional to the change in temperature or moisture content and its initial dimensions. Thus thermal and hygroscopic strains develop in the body as a result of temperature and hygroscopic changes. The thermal strain ε^T is equal to the product of the coefficient of thermal expansion α of

6.9 HYGROTHERMAL STRESSES IN LAMINATES

the body and the change in temperature ΔT, and similarly, the hygroscopic strains ε^H is equal to the product of coefficient of moisture expansion β of the body and the change in moisture content ΔC:

$$\varepsilon^T = \alpha \, \Delta T \qquad (6.52)$$

$$\varepsilon^H = \beta \, \Delta C \qquad (6.53)$$

In the case of an orthotropic material, such as a unidirectional lamina, the coefficients of Thermal and moisture expansion, like its other properties, change with direction. Thus the hygrothermal changes result in unequal strains in the longitudinal and transverse directions given by the following equations:

$$\begin{aligned} \varepsilon_L^T &= \alpha_L \, \Delta T \\ \varepsilon_T^T &= \alpha_T \, \Delta T \end{aligned} \qquad (6.54)$$

$$\begin{aligned} \varepsilon_L^H &= \beta_L \, \Delta C \\ \varepsilon_T^H &= \beta_T \, \Delta C \end{aligned} \qquad (6.55)$$

where α_L, α_T, β_L, and β_T are coefficients of thermal and moisture expansion in the longitudinal and transverse directions, respectively. The hygrothermal strains also can be transformed to an arbitrary coordinate system, such as the x and y axes, by means of the transformation equations [Eq. (5.88)]. Therefore, the same transformation law is applicable to the coefficients of hygrothermal expansions, so

$$\begin{Bmatrix} \alpha_x \\ \alpha_y \\ \alpha_{xy} \end{Bmatrix} = [T_2]^{-1} \begin{Bmatrix} \alpha_L \\ \alpha_T \\ 0 \end{Bmatrix} \qquad (6.56)$$

and

$$\begin{Bmatrix} \beta_x \\ \beta_y \\ \beta_{xy} \end{Bmatrix} = [T_2]^{-1} \begin{Bmatrix} \beta_L \\ \beta_T \\ 0 \end{Bmatrix} \qquad (6.57)$$

where α_{xy} and β_{xy} are apparent coefficients of thermal and moisture shear, and the $[T_2]$ matrix is given by Eq. (5.89). Thus the hygrothermal strains may be written directly in terms of the transformed coefficients as

266 ANALYSIS OF LAMINATED COMPOSITES

$$\begin{Bmatrix} \varepsilon_x^T \\ \varepsilon_y^T \\ \gamma_{xy}^T \end{Bmatrix} = \begin{Bmatrix} \alpha_x \Delta T \\ \alpha_y \Delta T \\ \alpha_{xy} \Delta T \end{Bmatrix} \quad (6.58)$$

and

$$\begin{Bmatrix} \varepsilon_x^H \\ \varepsilon_y^H \\ \gamma_{xy}^H \end{Bmatrix} = \begin{Bmatrix} \beta_x \Delta C \\ \beta_y \Delta C \\ \beta_{xy} \Delta C \end{Bmatrix} \quad (6.59)$$

Hygrothermal strains do not produce a resultant force or moment when the body is completely free to expand, bend, and twist. Thus, when a laminate as a whole is considered, a hygrothermal change does not affect the resultant force or moment. However, an individual lamina in a laminate is not completely free to deform. Its deformation is influenced by other laminae. Lamina strains can be obtained from the laminate midplane strains and plate curvatures through Eq. (6.6). The lamina stresses are induced by the constraints placed on its deformation by adjacent laminae. The stresses in a lamina are produced only by the strains that are in excess of the hydrothermal strains for its free expansion given by Eqs. (6.58) and (6.59). The strains that cause stresses may be referred to as the *mechanical strains* and may be denoted by $\{\epsilon^M\}$. The mechanical strains then are given as

$$\begin{Bmatrix} \varepsilon_x^M \\ \varepsilon_y^M \\ \gamma_{xy}^M \end{Bmatrix} = \begin{Bmatrix} \varepsilon_x \\ \varepsilon_y \\ \gamma_{xy} \end{Bmatrix} - \begin{Bmatrix} \varepsilon_x^T \\ \varepsilon_y^T \\ \gamma_{xy}^T \end{Bmatrix} - \begin{Bmatrix} \varepsilon_x^H \\ \varepsilon_y^H \\ \gamma_{xy}^H \end{Bmatrix} \quad (6.60)$$

where ε are the total lamina strains given by Eq. (6.6). With substitution of Eqs. (6.6), (6.58), and (6.59) into Eq. (6.60), the mechanical strains are obtained as

$$\begin{Bmatrix} \varepsilon_x^M \\ \varepsilon_y^M \\ \gamma_{xy}^M \end{Bmatrix} = \begin{Bmatrix} \varepsilon_x^0 + zk_x \\ \varepsilon_y^0 + zk_y \\ \gamma_{xy}^0 + zk_{xy} \end{Bmatrix} - \begin{Bmatrix} \alpha_x \Delta T \\ \alpha_y \Delta T \\ \alpha_{xy} \Delta T \end{Bmatrix} - \begin{Bmatrix} \beta_x \Delta C \\ \beta_y \Delta C \\ \beta_{xy} \Delta C \end{Bmatrix} \quad (6.61)$$

The lamina hygrothermal stresses can be obtained by substituting Eq. (6.61) into Eq. (5.94).

$$\begin{Bmatrix} \sigma_x^T \\ \sigma_y^T \\ \tau_{xy}^T \end{Bmatrix} = \begin{bmatrix} \overline{Q}_{11} & \overline{Q}_{12} & \overline{Q}_{16} \\ \overline{Q}_{12} & \overline{Q}_{22} & \overline{Q}_{26} \\ \overline{Q}_{16} & \overline{Q}_{26} & \overline{Q}_{66} \end{bmatrix} \begin{Bmatrix} \varepsilon_x^0 + zk_x - \alpha_x \Delta T - \beta_x \Delta C \\ \varepsilon_y^0 + zk_y - \alpha_y \Delta T - \beta_y \Delta C \\ \gamma_{xy}^0 + zk_{xy} - \alpha_{xy} \Delta T - \beta_{xy} \Delta C \end{Bmatrix} \quad (6.62)$$

6.9 HYGROTHERMAL STRESSES IN LAMINATES

In Eq. (6.62), the midplane strains $\{\varepsilon^0\}$ and the plate curvatures $\{k\}$ are unknown but can be evaluated by employing the fact that no external force is responsible for these stresses. The resultant forces and moments are obtained from Eqs. (6.12) and (6.13). Therefore, by substituting Eq. (6.62) into Eqs. (6.12) *and* (6.13) *and* integrating *and* then equating the resulting expression to zero, the following equations are obtained:

$$\begin{bmatrix} A_{11} & A_{12} & A_{16} \\ A_{12} & A_{22} & A_{26} \\ A_{16} & A_{26} & A_{66} \end{bmatrix} \begin{Bmatrix} \varepsilon_x^0 \\ \varepsilon_y^0 \\ \gamma_{xy}^0 \end{Bmatrix} + \begin{bmatrix} B_{11} & B_{12} & B_{16} \\ B_{12} & B_{22} & B_{26} \\ B_{16} & B_{26} & B_{66} \end{bmatrix} \begin{Bmatrix} k_x \\ k_y \\ k_{xy} \end{Bmatrix} = \begin{Bmatrix} N_x^T \\ N_y^T \\ N_{xy}^T \end{Bmatrix} + \begin{Bmatrix} N_x^H \\ N_y^H \\ N_{xy}^H \end{Bmatrix} \tag{6.63}$$

$$\begin{bmatrix} B_{11} & B_{12} & B_{16} \\ B_{12} & B_{22} & B_{26} \\ B_{16} & B_{26} & B_{66} \end{bmatrix} \begin{Bmatrix} \varepsilon_x^0 \\ \varepsilon_y^0 \\ \gamma_{xy}^0 \end{Bmatrix} + \begin{bmatrix} D_{11} & D_{12} & D_{16} \\ D_{12} & D_{22} & D_{26} \\ D_{16} & D_{26} & D_{66} \end{bmatrix} \begin{Bmatrix} k_x \\ k_y \\ k_{xy} \end{Bmatrix} = \begin{Bmatrix} M_x^T \\ M_y^T \\ M_{xy}^T \end{Bmatrix} + \begin{Bmatrix} M_x^H \\ M_y^H \\ M_{xy}^H \end{Bmatrix} \tag{6.64}$$

where $\{N^T\}$, $\{M^T\}$, $\{N^H\}$, and $\{M^H\}$ are

$$\begin{Bmatrix} N_x^T \\ N_y^T \\ N_{xy}^T \end{Bmatrix} = \Delta T \sum_{k=1}^{n} \begin{bmatrix} \bar{Q}_{11} & \bar{Q}_{12} & \bar{Q}_{16} \\ \bar{Q}_{12} & \bar{Q}_{22} & \bar{Q}_{26} \\ \bar{Q}_{16} & \bar{Q}_{26} & \bar{Q}_{66} \end{bmatrix}_k \begin{Bmatrix} \alpha_x \\ \alpha_y \\ \alpha_{xy} \end{Bmatrix}_k (h_k - h_{k-1}) \tag{6.65}$$

$$\begin{Bmatrix} M_x^T \\ M_y^T \\ M_{xy}^T \end{Bmatrix} = \frac{1}{2} \Delta T \sum_{k=1}^{n} \begin{bmatrix} \bar{Q}_{11} & \bar{Q}_{12} & \bar{Q}_{16} \\ \bar{Q}_{12} & \bar{Q}_{22} & \bar{Q}_{26} \\ \bar{Q}_{16} & \bar{Q}_{26} & \bar{Q}_{66} \end{bmatrix}_k \begin{Bmatrix} \alpha_x \\ \alpha_y \\ \alpha_{xy} \end{Bmatrix}_k (h_k^2 - h_{k-1}^2) \tag{6.66}$$

$$\begin{Bmatrix} N_x^H \\ N_y^H \\ N_{xy}^H \end{Bmatrix} = \Delta C \sum_{k=1}^{n} \begin{bmatrix} \bar{Q}_{11} & \bar{Q}_{12} & \bar{Q}_{16} \\ \bar{Q}_{12} & \bar{Q}_{22} & \bar{Q}_{26} \\ \bar{Q}_{16} & \bar{Q}_{26} & \bar{Q}_{66} \end{bmatrix}_k \begin{Bmatrix} \beta_x \\ \beta_y \\ \beta_{xy} \end{Bmatrix}_k (h_k - h_{k-1}) \tag{6.67}$$

$$\begin{Bmatrix} M_x^H \\ M_y^H \\ M_{xy}^H \end{Bmatrix} = \frac{1}{2} \Delta C \sum_{k=1}^{n} \begin{bmatrix} \bar{Q}_{11} & \bar{Q}_{12} & \bar{Q}_{16} \\ \bar{Q}_{12} & \bar{Q}_{22} & \bar{Q}_{26} \\ \bar{Q}_{16} & \bar{Q}_{26} & \bar{Q}_{66} \end{bmatrix}_k \begin{Bmatrix} \beta_x \\ \beta_y \\ \beta_{xy} \end{Bmatrix}_k (h_k^2 - h_{k-1}^2) \tag{6.68}$$

where h_k and h_{k-1} define the position of a lamina in a laminate and are shown in Fig. 6-5. The forces $\{N^T\}$ and $\{N^H\}$ and moments $\{M^T\}$ and $\{M^H\}$ are apparent forces that produce the midplane strains $\{\epsilon^0\}$ and plate curvatures $\{k\}$ given by Eqs. (6.63) and (6.64). These fictitious forces and moments are sometimes called the *hygrothermal forces and moments*. They are subjected to the same rules as the externally applied forces. Thus a laminate subjected to a hygrothermal change as well as to external forces and moments may be analyzed in two ways. First, the stresses induced by the hygrothermal change and those induced by the external forces and moments may be evaluated separately and then added to obtain the resulting stresses. Second, the hygrothermal forces and moments may be evaluated by Eqs. (6.65)–(6.68) and added to the external forces and moments. Now the resulting stresses may be obtained directly from the analysis of the laminate, as discussed in the preceding section.

The hygrothermal stresses given by Eq. (6.62) are induced in the laminae whenever the hygrothermal state of a laminate differs from its stress-free state. The thermal stresses are invariably unavoidable as a result of the fabrication of composite laminates caused by temperature changes of several hundred degrees between fabrication temperatures and room temperature. The thermal stresses produced while cooling the laminate after fabrication at elevated temperature are called *residual stresses* or *curing stresses.* In some cases, such residual stresses may be sufficiently large to influence the failure of the laminate and thus should not be neglected in a design analysis. It may be pointed out again that the hygrothermal stresses are induced not because laminae expand or contract because of hygrothermal changes, but because they are not free to expand or contract. The laminates are fabricated such that they act as single-layer materials. Thus each lamina influences the expansion or contraction of the other because their coefficients of expansion are different. It may be noted that residual stresses caused by fabrication are created even in a lamina if the matrix and fiber have different expansion coefficients. In practically all cases the matrix has a greater expansion than the fiber, which subjects the fiber to compressive stress. For most practical volume fractions of fibers, the matrix generally will be subjected to a radial compression at the fiber–matrix interface and a tangential tensile stress. This radial compression acting against the interface is significant in aiding shear-stress transfer into the fiber by friction forces even in the absence of good bonding. Because of the presence of these internal stresses, which can be calculated by micromechanics analyses, the total internal residual stresses in a laminate would have to be obtained by superimposing these stresses on those induced by lamina restraints, as discussed earlier.

It may be noted from Eqs. (6.63) and (6.64) that in a general unsymmetric laminate, the hygrothermal change will induce not only extensional strains but also warping of the laminate represented by the plate curvatures $\{k\}$. However, in a symmetric laminate, the coupling matrix $[B]$ is zero, and the hygrothermal moments given by Eqs. (6.66) and (6.68) are also zero. Therefore, Eq. (6.64) predicts that the plate curvature $\{k\}$ will vanish. Thus the

6.9 HYGROTHERMAL STRESSES IN LAMINATES

warpage due to hygrothermal change during fabrication is avoided by the use of symmetric laminates. A more specific discussion on the residual stresses is given in refs. 5–7.

Example 6-10: Calculate the residual stresses in the laminate considered in Example 6-1 that is fabricated at 125°C and cooled to room temperature of 25°C, given

$$\alpha_L = 7.0 \times 10^{-6}/°C \quad \text{and} \quad \alpha_T = 23 \times 10^{-6}/°C$$

First, transform the coefficients of thermal expansion in the xy coordinate axes:

$$\left\{\begin{array}{c}\alpha_x\\ \alpha_y\\ \alpha_{xy}\end{array}\right\}_{0°} = \left\{\begin{array}{c}\alpha_L\\ \alpha_T\\ 0\end{array}\right\} = 10^{-6}\left\{\begin{array}{c}7\\ 23\\ 0\end{array}\right\}$$

$$\left\{\begin{array}{c}\alpha_x\\ \alpha_y\\ \alpha_{xy}\end{array}\right\}_{45°} = \left[\begin{array}{ccc}0.5 & 0.5 & -0.5\\ 0.5 & 0.5 & 0.5\\ 1 & -1 & 0\end{array}\right]\left\{\begin{array}{c}7 \times 10^{-6}\\ 23 \times 10^{-6}\\ 0\end{array}\right\}$$

$$\left\{\begin{array}{c}\alpha_x\\ \alpha_y\\ \alpha_{xy}\end{array}\right\}_{45°} = \left\{\begin{array}{c}15\\ 15\\ -16\end{array}\right\}10^{-6}$$

Now thermal forces and moments may be calculated by means of Eqs. (6.65) and (6.66), where the $[\overline{Q}]$ matrices for the two plies were obtained in Example 6-1. The calculations may be carried out in the following sequence:

$$\Delta T = 25 - 125 = -100°C$$

$$\Delta T \left[\begin{array}{ccc}\overline{Q}_{11} & \overline{Q}_{12} & \overline{Q}_{16}\\ \overline{Q}_{12} & \overline{Q}_{22} & \overline{Q}_{26}\\ \overline{Q}_{16} & \overline{Q}_{26} & \overline{Q}_{66}\end{array}\right]_{0°}\left\{\begin{array}{c}\alpha_x\\ \alpha_y\\ \alpha_{xy}\end{array}\right\}_{0°} = 10^{-3}\left\{\begin{array}{c}-15.61\\ -5.09\\ 0\end{array}\right\}$$

$$\Delta T \left[\begin{array}{ccc}\overline{Q}_{11} & \overline{Q}_{12} & \overline{Q}_{16}\\ \overline{Q}_{12} & \overline{Q}_{22} & \overline{Q}_{26}\\ \overline{Q}_{16} & \overline{Q}_{26} & \overline{Q}_{66}\end{array}\right]_{45°}\left\{\begin{array}{c}\alpha_x\\ \alpha_y\\ \alpha_{xy}\end{array}\right\}_{45°} = 10^{-3}\left\{\begin{array}{c}-10.35\\ -10.35\\ -5.26\end{array}\right\}$$

$$\begin{Bmatrix} N_x^T \\ N_y^T \\ N_{xy}^T \end{Bmatrix} = [(4) - (-1)]10^{-3} \begin{Bmatrix} -15.61 \\ -5.09 \\ 0 \end{Bmatrix} + [(-1) - (-4)]10^{-3} \begin{Bmatrix} -10.35 \\ -10.35 \\ -5.26 \end{Bmatrix}$$

$$\begin{Bmatrix} N_x^T \\ N_y^T \\ N_{xy}^T \end{Bmatrix} = 10^{-3} \begin{Bmatrix} -109.10 \\ -56.50 \\ -15.78 \end{Bmatrix} \text{ GPa} \cdot \text{mm}$$

$$\begin{Bmatrix} M_x^T \\ M_y^T \\ M_{xy}^T \end{Bmatrix} = \tfrac{1}{2}[(4)^2 - (-1)^2]10^{-3} \begin{Bmatrix} -15.61 \\ -5.09 \\ 0 \end{Bmatrix}$$

$$+ \tfrac{1}{2}[(-1)^2 - (-4)^2]10^{-3} \begin{Bmatrix} -10.35 \\ -10.35 \\ -5.26 \end{Bmatrix}$$

$$\begin{Bmatrix} M_x^T \\ M_y^T \\ M_{xy}^T \end{Bmatrix} = 10^{-3} \begin{Bmatrix} -39.45 \\ 39.45 \\ 39.45 \end{Bmatrix} \text{ GPa} \cdot \text{mm}$$

The midplane strains and plate curvatures may be obtained from Eqs. (6.63) and (6.64). It may be noted, however, that Eqs. (6.63) and (6.64) may be written in an inverted form similar to Eq. (6.33) as follows:

$$\begin{Bmatrix} \epsilon^0 \\ \hline k \end{Bmatrix} = \begin{bmatrix} A' & \vdots & B' \\ \hline B' & \vdots & D' \end{bmatrix} \begin{Bmatrix} N^T \\ \hline M^T \end{Bmatrix}$$

Further, the matrices $[A']$, $[B']$, and $[D']$ have the same meaning as in Eq. (6.33), and for the laminate under consideration, these were evaluated in Example 6-6. Therefore, the midplane strains and plate curvatures can be found to be

$$\begin{Bmatrix} \epsilon_x^0 \\ \epsilon_y^0 \\ \gamma_{xy}^0 \end{Bmatrix} = 10^{-4} \begin{Bmatrix} -8.14 \\ -20.20 \\ 6.99 \end{Bmatrix}$$

and

$$\begin{Bmatrix} k_x \\ k_y \\ k_{xy} \end{Bmatrix} = 10^{-4} \begin{Bmatrix} 0.58 \\ -1.00 \\ -2.35 \end{Bmatrix}$$

The nonzero values of plate curvatures $\{k\}$ in the preceding calculations show that warping of the laminate will occur when the laminate is cooled

6.9 HYGROTHERMAL STRESSES IN LAMINATES

from the curing temperature (125°C) to room temperature (25°C). Mechanical strains that cause the residual stresses are calculated by Eq. (6.61):

$$\left\{\begin{matrix}\epsilon_x^M \\ \epsilon_y^M \\ \gamma_{xy}^M\end{matrix}\right\}_{0°} = 10^{-4}\left\{\begin{matrix}-8.14 + 0.58z + 7.0 \\ -20.20 - 1.00z + 23.0 \\ 6.99 - 2.35z + 0\end{matrix}\right\} = 10^{-4}\left\{\begin{matrix}-1.14 + 0.58z \\ 2.80 - 1.00z \\ 6.99 - 2.35z\end{matrix}\right\}$$

$$\left\{\begin{matrix}\epsilon_x^M \\ \epsilon_y^M \\ \gamma_{xy}^M\end{matrix}\right\}_{45°} = 10^{-4}\left\{\begin{matrix}-8.14 + 0.58z + 15.0 \\ -20.20 - 1.00z - 15.0 \\ 6.99 - 2.35z - 16.0\end{matrix}\right\} = 10^{-4}\left\{\begin{matrix}6.86 + 0.58z \\ -5.20 - 1.00z \\ -9.01 - 2.35z\end{matrix}\right\}$$

The residual stress distribution may be obtained by substituting the preceding strains into Eq. (6.62). Because the strain variation and hence the stress variation are linear across the thickness of a ply, it is sufficient to calculate the stresses only at the ply surfaces to complete the residual stress distribution. The required stresses are calculated as follows:

$0°$ ply, $z = 4$

$$\left\{\begin{matrix}\epsilon_x^M \\ \epsilon_y^M \\ \gamma_{xy}^M\end{matrix}\right\} = 10^{-4}\left\{\begin{matrix}1.18 \\ -1.20 \\ -2.41\end{matrix}\right\}$$

$$\left\{\begin{matrix}\sigma_x^T \\ \sigma_y^T \\ \tau_{xy}^T\end{matrix}\right\} = 10^{-4}\left[\begin{matrix}20 & 0.7 & 0 \\ 0.7 & 2 & 0 \\ 0 & 0 & 0.7\end{matrix}\right]\left\{\begin{matrix}1.18 \\ -1.20 \\ -2.41\end{matrix}\right\} \text{GPa}$$

$$\left\{\begin{matrix}\sigma_x^T \\ \sigma_y^T \\ \tau_{xy}^T\end{matrix}\right\} = \left\{\begin{matrix}\sigma_L^T \\ \sigma_T^T \\ \tau_{LT}^T\end{matrix}\right\} = \left\{\begin{matrix}2.28 \\ -0.16 \\ -0.17\end{matrix}\right\} \text{MPa}$$

$0°$ ply, $z = -1$

$$\left\{\begin{matrix}\epsilon_x^M \\ \epsilon_y^M \\ \gamma_{xy}^M\end{matrix}\right\} = 10^{-4}\left\{\begin{matrix}-1.72 \\ 3.80 \\ 9.34\end{matrix}\right\}$$

$$\left\{\begin{matrix}\sigma_x^T \\ \sigma_y^T \\ \tau_{xy}^T\end{matrix}\right\} = 10^{-4}\left[\begin{matrix}20 & 0.7 & 0 \\ 0.7 & 2 & 0 \\ 0 & 0 & 0.7\end{matrix}\right]\left\{\begin{matrix}-1.72 \\ 3.80 \\ 9.34\end{matrix}\right\} \text{GPa}$$

$$\left\{\begin{matrix}\sigma_x^T \\ \sigma_y^T \\ \tau_{xy}^T\end{matrix}\right\} = \left\{\begin{matrix}\sigma_L^T \\ \sigma_T^T \\ \tau_{LT}^T\end{matrix}\right\} = \left\{\begin{matrix}-3.17 \\ 0.64 \\ 0.65\end{matrix}\right\} \text{MPa}$$

$45°$ ply, $z = -1$

$$\begin{Bmatrix} \epsilon_x^M \\ \epsilon_y^M \\ \gamma_{xy}^M \end{Bmatrix} = 10^{-4} \begin{Bmatrix} 6.28 \\ -4.20 \\ -6.66 \end{Bmatrix}$$

$$\begin{Bmatrix} \sigma_x^T \\ \sigma_y^T \\ \tau_{xy}^T \end{Bmatrix} = 10^{-4} \begin{bmatrix} 6.55 & 5.15 & 4.50 \\ 5.15 & 6.55 & 4.50 \\ 4.50 & 4.50 & 5.15 \end{bmatrix} \begin{Bmatrix} 6.28 \\ -4.20 \\ -6.66 \end{Bmatrix} = 10^{-3} \begin{Bmatrix} -1.05 \\ -2.51 \\ -2.49 \end{Bmatrix} \text{ GPa}$$

$$\begin{Bmatrix} \sigma_L^T \\ \sigma_T^T \\ \tau_{LT}^T \end{Bmatrix} = \begin{bmatrix} 0.5 & 0.5 & 1 \\ 0.5 & 0.5 & -1 \\ -0.5 & 0.5 & 0 \end{bmatrix} \begin{Bmatrix} -1.05 \\ -2.51 \\ -2.49 \end{Bmatrix} = \begin{Bmatrix} -4.27 \\ 0.71 \\ -0.73 \end{Bmatrix} \text{ MPa}$$

$45°$ ply, $z = -4$

$$\begin{Bmatrix} \epsilon_x^M \\ \epsilon_y^M \\ \gamma_{xy}^M \end{Bmatrix} = 10^{-4} \begin{Bmatrix} 4.54 \\ -1.20 \\ 0.39 \end{Bmatrix}$$

$$\begin{Bmatrix} \sigma_x^T \\ \sigma_y^T \\ \tau_{xy}^T \end{Bmatrix} = 10^{-4} \begin{bmatrix} 6.55 & 5.15 & 4.50 \\ 5.15 & 6.55 & 4.50 \\ 4.50 & 4.50 & 5.15 \end{bmatrix} \begin{Bmatrix} 4.54 \\ -1.20 \\ 0.39 \end{Bmatrix} = 10^{-3} \begin{Bmatrix} 2.53 \\ 1.73 \\ 1.70 \end{Bmatrix} \text{ GPa}$$

$$\begin{Bmatrix} \sigma_L^T \\ \sigma_T^T \\ \tau_{LT}^T \end{Bmatrix} = \begin{bmatrix} 0.5 & 0.5 & 1 \\ 0.5 & 0.5 & -1 \\ -0.5 & 0.5 & 0 \end{bmatrix} \begin{Bmatrix} 2.53 \\ 1.73 \\ 1.70 \end{Bmatrix} = \begin{Bmatrix} 3.83 \\ 0.43 \\ -0.40 \end{Bmatrix} \text{ MPa}$$

The variations of the residual stresses across the laminate thickness are shown in Fig. 6-20 for the xy reference axes, as well as the longitudinal and transverse axes. It may be noted from the variations of σ_x, σ_y, and τ_{xy} that the resultant forces N_x, N_y, and N_{xy} and resultant moments, M_x, M_y, and M_{xy} are zero; that is, the net area in each plot and the moment of the area about any point are zero. This shows the self-equilibrating nature of the residual stresses.

6.10 LAMINATE ANALYSIS THROUGH COMPUTERS

Laminate analysis procedures were discussed in this chapter. It is assumed that the laminae elastic properties are known either through the prediction

6.10 LAMINATE ANALYSIS THROUGH COMPUTERS

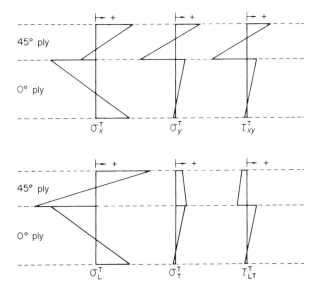

Figure 6-20. Residual stresses (Example 6-10).

techniques discussed in Chap. 3 or determined by the experimental characterization procedures discussed in Chap. 10. The laminate stress-analysis procedure may be summarized as follows:

1. Calculate the laminae stiffness matrix from the laminae elastic properties [Eq. (5.78)].
2. Transform the stiffness matrix to different ply orientations [Eq. (5.95)].
3. Calculate laminate stiffness matrices $[A]$, $[B]$, and $[D]$ [Eq. (6.20)].
4. Calculate midplane strains and plate curvatures for the given loads [Eq. (6.21) or Eq. (6.33)].
5. Calculate laminae strains [Eq. (6.6)].
6. Transform laminae strains from arbitrary directions to the longitudinal and transverse directions [Eq. (5.88)].
7. Calculate laminae stresses [Eq. (5.74)].

This procedure is used to obtain complete states of stress and strain in all laminae for the applied loads. A flowchart for the laminate stress analysis is given in Fig. 6-21, which can be used to streamline calculations or develop software.

The laminate stress analysis may be extended to obtain laminate failure load (Sec. 6.8) if the laminae strengths are known through the prediction techniques (see Chap. 3) or experimental measurements (see Chap. 10). The

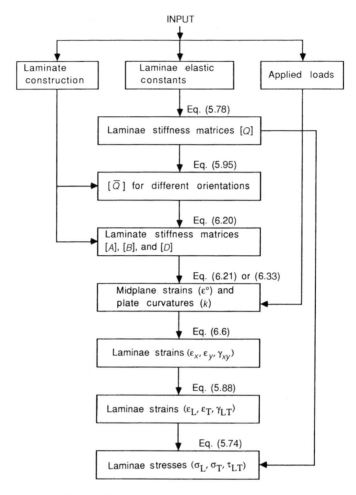

Figure 6-21. Flowchart for laminate stress analysis.

first step in the laminate strength analysis is to carry out laminate stress analysis for an assumed load (e.g., a unit load). The complete laminate strength analysis procedure may be summarized as follows:

1. Carry out laminate stress analysis for a unit load on the laminate.
2. Select an appropriate failure theory (Sec. 5.4).
3. Through the failure theory selected in step 2, compare the laminae stresses and strains with the allowable values, and predict the minimum load at which one of the plies will fail. Obtain stresses and strains at FPF load by multiplying stresses and strains obtained in step 1 by a suitable factor.

4. Modify the laminate stiffness matrices [A], [B], and [D] by assuming the stiffness matrix of the failed lamina to be zero [Eqs. (6.38) and (6.20)].
5. Through laminate stress analysis with modified [A], [B], and [D] matrices, calculate incremental stresses and strains in the laminae for an arbitrary (or unit) increment in load on the laminate.
6. Add the incremental stresses and strains to those at the previous ply failure.
7. Through the failure theory selected in step 2, compare the new laminae stresses and strains with the allowable values, and predict the minimum load increment at which the next ply fails.
8. Calculate the laminate load at which the next ply fails by adding the load increment obtained in step 7 to the load at which the previous ply failed. Also obtain stresses and strains at this ply failure.
9. Repeat steps 4–8 until the last ply fails, and obtain the laminate failure load.

A flowchart for the laminate strength analysis is given in Fig. 6-22.

The laminate stress and strength analyses have been illustrated through several numerical example problems in this chapter. It is realized that the calculations are tedious and time-consuming. In some cases, even rounding off of numbers in the intermediate steps can lead to considerable errors in the final results. Such will be the case in the calculation of hygrothermal stresses. Therefore, great care should be exercised in such calculations for accurate results.

It is apparent from the preceding discussion that laminate analysis calculations should be carried out with the help of computers for accurate results. There are two types of software programs available for the calculations. The first type of software consists of essentially mathematical tools to carry out general calculations. MATLAB and Mathcad are two such software programs. Calculations such as matrix inversions and multiplications and obtaining numerical values using given formulas can be carried out easily using these tools. These software programs are easy to use and relatively inexpensive. Their use improves accuracy of results and minimizes the probability of an error.

The second type of software consists of commercial programs/codes to carry out complete structural analysis. A list of commercial software programs is given in Appendix 5. These software programs also have several other features such as elastic properties and strengths of many commercial materials stored in memory, flexibility regarding the use of a failure theory, iterative procedure for design analysis and strength calculations, hygrothermal stress calculations, finite-element analysis, and graphics. These programs require greater training in their application

276 ANALYSIS OF LAMINATED COMPOSITES

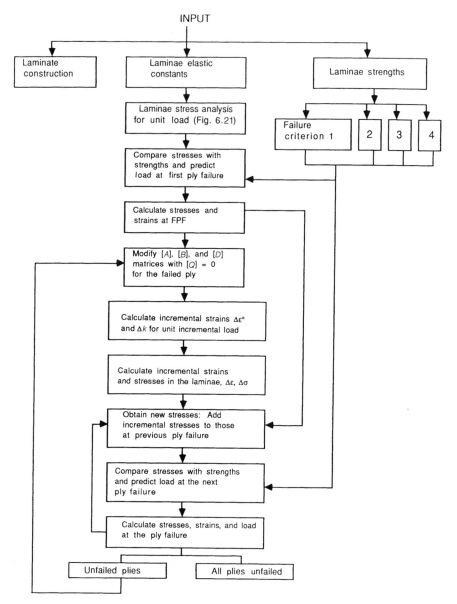

Figure 6-22. Flowchart for laminate strength analysis.

For learning purpose, it is recommended that the calculations be carried out by interactively using mathematical tools (e.g., MATLAB and Mathcad). This provides greater appreciation and understanding of the analysis procedure, properties and behavior of laminae in different directions, failure mechanisms, etc. Several exercise problems included at the end of this chapter require lengthy calculations and are candidates for the use of the mathematical

EXERCISE PROBLEMS

6.1. Determine the extensional, coupling, and bending stiffness matrices of a bimetallic strip made of 5-mm-thick layers of steel and aluminum. Discuss the significance of the coupling matrix.

6.2. An angle-ply laminate $[\pm\theta]_S$ is subjected to a uniaxial stress σ_{x0}. Show that

(a) The laminae stresses along longitudinal and transverse directions are

$$\sigma_L(\theta) = \sigma_L(-\theta) = \sigma_{x0} \cos^2\theta + 2\tau_{xy} \sin\theta \cos\theta$$

$$\sigma_T(\theta) = \sigma_T(-\theta) = \sigma_{x0} \sin^2\theta - 2\tau_{xy} \sin\theta \cos\theta$$

$$\tau_{LT}(\theta) = -\tau_{LT}(-\theta) = -\sigma_{x0} \sin\theta \cos\theta + \tau_{xy}(\cos^2\theta - \sin^2\theta)$$

where τ_{xy} is the shear stress induced in the laminae. (Notice that τ_{xy} is statically indeterminant.)

(b) The laminae strains are

$$\varepsilon_L(\theta) = \varepsilon_L(-\theta) = \varepsilon_x^0 \cos^2\theta + \varepsilon_y^0 \sin^2\theta$$

$$\varepsilon_T(\theta) = \varepsilon_T(-\theta) = \varepsilon_x^0 \sin^2\theta + \varepsilon_y^0 \cos^2\theta$$

$$\gamma_{LT}(\theta) = -\gamma_{LT}(-\theta) = 2(\varepsilon_y^0 - \varepsilon_x^0) \sin\theta \cos\theta$$

where ε_x^0 and ε_y^0 are the laminate midplane strains.

6.3. Obtain stresses and strains in the longitudinal and transverse directions in Exercise Problem 6.2 for $\theta = 45°$, Discuss the importance of these results with regard to the possibility of using the $[\pm 45]_S$ laminate to evaluate the lamina in-plane shear properties. (*Note:* Results of this problem will be used in Chap. 10.)

6.4. Show that a balanced cross-ply laminate (with equal number of identical plies in the 0° and 90° directions) is *not* a quasi-isotropic laminate.

6.5. Show that a laminate constructed by placing an equal number of identical plies at 0°, +60°, and −60° is quasi-isotropic.

6.6. Show that a laminate constructed by placing an equal number of identical plies at 0°, 45°, −45°, and 90° is quasi-isotropic.

6.7. **(a)** Show that the apparent elastic moduli, shear modulus, and Poisson ratios of an orthotropic symmetric laminate ($A_{16} = A_{26} = 0$) are

$$E_x = \frac{A_{11}A_{22} - A_{12}^2}{A_{22}t}$$

$$E_y = \frac{A_{11}A_{22} - A_{12}^2}{A_{11}t}$$

$$\nu_{xy} = \frac{A_{12}}{A_{22}}$$

$$\nu_{yx} = \frac{A_{12}}{A_{11}}$$

$$G_{xy} = \frac{A_{66}}{t}$$

where t is the laminate thickness. (*Hint:* Follow the procedure adopted in Sec. 5.3.7.)

(b) From the results obtained in part (a), derive the expressions for elastic modulus, shear modulus, and Poisson ratio of a quasi-isotropic laminate.

6.8. Repeat the analysis in Example 6-5 for quasi-isotropic laminate [0/±60].

6.9. Repeat the analysis in Example 6-5 for quasi-isotropic laminate [0/±30/±60/90].

*6.10. A balanced cross-ply laminate possessing midplane symmetry is made up of laminae having the following properties:

$$E_L = 15 \text{ GPa} \qquad G_{LT} = 3 \text{ GPa}$$

$$E_T = 6 \text{ GPa} \qquad \nu_{LT} = 0.5$$

The laminate is subjected to a normal axial stress of 15 MPa and a shear stress of 1.0 MPa. Calculate the normal and shear stresses in the 0° and 90° plies.

*6.11. The cross-ply laminate considered in Exercise Problem 6.10 is subjected to a normal stress of 30 MPa at 45° to the fibers in the plies.

---

*The problems marked with an asterisk require lengthy calculations and should be assigned selectively only to students who have access to personal computers and the appropriate software.

Calculate the normal and shear stresses in the plies in the directions parallel and perpendicular to the applied stress.

*6.12. Two laminates have ply orientations $[45/\bar{0}]_S$ and $[45/0/-45]$, where each ply is 4 mm thick and has the following stiffness matrix referred to the longitudinal and transverse axes:

$$[Q] = \begin{bmatrix} 30 & 1 & 0 \\ 1 & 3 & 0 \\ 0 & 0 & 1 \end{bmatrix} \text{GPa}$$

If $N_x = N_y = 4000$ N/mm, $N_{xy} = 0$, $M_x = 25{,}000$ N·mm/mm, and $M_y = M_{xy} = 0$, calculate the midplane strains, plate curvatures, and stresses in the laminae.

6.13. Derive Eq. (6.50) by comparing Eqs. (6.48) and (6.49).

6.14. A symmetric cross-ply laminate has seven plies, of which four have fibers parallel to the applied load. All plies have the following identical properties:

$$E_L = 40 \text{ GPa} \qquad E_T = 8.5 \text{ GPa}$$

$$G_{LT} = 4.2 \text{ GPa} \qquad \nu_{LT} = 0.26$$

$$\epsilon_{LU} = 2.75\% \qquad \epsilon_{TU} = 0.38\%$$

Calculate the primary and secondary moduli, composite stress at the knee, and effective modulus at 1% composite strain. Plot the stress–strain curve by assuming (a) laminate construction $[0/90/0/\bar{90}]_S$, (b) laminate construction $[(0)_2/(90)_3/(0)_2]$ and the testing machine maintaining a constant loading rate, and (c) laminate construction as in (b) but the test is performed under conditions of controlled strain rate.

*6.15. Calculate the residual stresses in the cross-ply laminate considered in Exercise Problem 6.10 that is fabricated at 125°C and cooled to 25°C. Given

$$\alpha_L - 7.0 \times 10^{-6}/°C \qquad \text{and} \qquad \alpha_T = 23.0 \times 10^{-6}/°C$$

how will the residual stresses be affected by interchanging locations of the 0° and 90° plies?

*6.16. Repeat Exercise Problem 6.15 for a balanced cross-ply laminate that does not possess midplane symmetry but has all the 0° plies placed above the midplane and all 90° plies below it. Properties of the constituent laminae are same as given in Exercise Problems 6.10 and 6.15.

*6.17. Calculate residual stresses in a three-ply quasiisotropic laminate [60/0/−60]. The laminae properties, fabrication temperatures, and so on are the same as those given in Exercise Problems 6.10 and 6.15.

6.18. The stress analysis of a filament-wound cylindrical pressure vessel can be carried out using laminate analysis. An angle-ply laminate assumption is quite appropriate for the analysis of a symmetric helically wound cylinder that has fibers oriented at equal angles on either side of the cylinder axis. The analysis procedure will be similar to that used for Exercise 6.2. However, design analyses of filament-wound pressure vessels and pipes often are carried out by assuming that each layer carries load in the longitudinal direction only. That is

$$\sigma_L \neq 0$$

$$\sigma_T = \tau_{LT} = 0$$

This greatly simplifies the analysis. Such an analysis is called the *netting analysis*.

(a) A filament-wound cylinder is represented by a $[\pm\theta]_S$ laminate, where θ is the angle the fibers make with the cylinder axis. Using the netting analysis, show that the hoop stress in the cylinder is $\sigma_L \sin^2\theta$ and that the axial stress is $\sigma_L \cos^2\theta$.

(b) Calculate the resultant shear stress in the axial and circumferential direction.

(c) If this cylinder is a thin-walled, closed-end pressure vessel, what is the optimal winding angle θ?

*6.19. An unsymmetric cross-ply laminate $[0_4/90_4]_2$, which is originally in the form of a flat plate, is bent and glued at the seam to form a tube of 5 cm radius such that the outermost ply has fibers in the hoop direction. Calculate the ply stresses and strains when the tube is subjected to a torque of 0.5 N·m. Assume that each lamina is 0.125 mm thick and has the following properties:

$$E_L = 138 \text{ GPa} \quad G_{LT} = 7.1 \text{ GPa}$$

$$E_T = 8.96 \text{ GPa} \quad \nu_{LT} = 0.3$$

(*Hint:* Use the partially inverted form of the stiffness matrix to obtain midplane strains.)

6.20. A [0/90] asymmetric laminate of material AS/3501 graphite–epoxy is cured at 125°C. Assume that both plies are equal in thickness and have the following properties:

$$E_L = 138.0 \text{ GPa}$$

$$E_T = 8.96 \text{ GPa}$$

$$\nu_{LT} = 0.30$$

$$G_{LT} = 7.10 \text{ GPa}$$

$$\alpha_L = -0.30 \times 10^{-6}/°C$$

$$\alpha_T = 28.10 \times 10^{-6}/°C$$

Calculate the laminate curvatures when it is cooled to room temperature (25°C). Sketch the curved shape of the laminate. Also discuss how the shape would appear if the thickness-to-width ratio is small (i.e., the laminate is thin).

REFERENCES

1. J. C. Halpin, *Primer on Composite Materials: Analysis,* Technomic, Lancaster, PA, 1984.
2. H. T. Hahn and S. W. Tsai, "On the Behavior of Composite Laminates after Initial Failure," *J. Compos. Mater.,* **8**(3), 288 (1974).
3. S. C. Chou, O. Oringer, and J. H. Rainey, "Post-Failure Behavior of Laminates: I—No Stress Concentration," *J. Compos. Mater.,* **10**(4), 371 (1976).
4. S. C. Chou, O. Oringer, and J. H. Rainey, "Post-Failure Behavior of Laminates: II—Stress Concentration," *J. Compos. Mater.,* **11**(1), 71 (1977).
5. H. T. Hahn and N. J. Pagano, "Curing Stresses in Composite Laminates," *J. Compos. Mater.,* **9**(1), 91 (1975).
6. H. T. Hahn, "Residual Stresses in Polymer Matrix Composite Laminates," *J. Compos. Mater.,* **10**(4), 266 (1976).
7. R. B. Pipes, J. R. Vinson, and T. W. Chou, "On the Hygrothermal Response of Laminated Composite Systems," *J. Compos. Mater.,* **10,** 129 (1976).

7

ANALYSIS OF LAMINATED PLATES AND BEAMS

7.1 INTRODUCTION

The classical lamination theory (CLT) was developed in Chap. 6 for the analysis of laminated composites. Equations were derived to obtain strains and stresses at every point in the laminate from the knowledge of midplane strains and plate curvatures. Constitutive equations for the laminated composites relate resultant forces and moments to the laminate midplane strains and plate curvatures through the synthesized stiffness matrix. Analysis with CLT can be used to calculate stresses and strains for known resultant forces and moments. For real structures, this analysis can be used only when the laminate is subjected to constant in-plane forces and moments in the laminate plane. However, practical laminates are subjected to transverse loads (loads perpendicular to the laminate plane) that produce variation in moments. The analysis procedures developed in Chap. 6 cannot consider this variation in moments. Developments in this chapter extend the analysis capabilities using plate theory.

The plate theory for isotropic materials and its application to various problems are well established [1,2]. The basic governing equations (i.e., the equilibrium equations) for the laminate plate theory are the same as those for the isotropic plate theory when written in terms of the resultant forces and moments. However, differences arise when the equations are written in terms of midplane displacements. Owing to the complex constitutive equations for the laminated composites, the laminated-plate governing equations also become complex. Consequently, methods for solving practical problems of laminated composites become more involved [3–10].

In this chapter, equilibrium equations for plates (isotropic or anisotropic material) will be derived first in terms of force and moment resultants. These equations then will be expressed in terms of midplane displacements for a laminated plate. Application of this laminated-plate theory will be illustrated through bending, buckling, and free-vibration problems. Shear deformation theories for moderately thick plates are discussed in one section. The plate theory is simplified in the last section for application to laminated-beam problems. Example problems are provided throughout the chapter.

7.2 GOVERNING EQUATIONS FOR PLATES

7.2.1 Equilibrium Equations

Consider a flat plate of thickness h subjected to a distributed load $p(x, y)$, as shown in Fig. 7-1. The coordinate plane xy is located at the midplane of the plate. The equilibrium equations of the plate are derived by considering equilibrium of a differential element shown in Fig. 7-2. The in-plane force resultants (N_x, N_y, and N_{xy}) on the element are shown in Fig 7-2a, and the moment resultants (M_x, M_y, and M_{xy}) and the shear force resultants (R_{xz}, R_{yz}) are shown in Fig. 7-2b. The in-plane force and moment resultants were defined by Eqs. (6.12) and (6.13), respectively.

It may be recalled that out-of-plane shear stresses τ_{xz} and τ_{yz} were neglected in the analysis developed in Chap. 6. However, τ_{xz} and τ_{yz} must be considered here to balance the applied transverse load $p(x, y)$ on the differential element. Therefore, the shear-force resultants R_{xz} and R_{yz} are defined in a manner similar to the definition of force resultants as follows:

$$R_{xz} = \int_{-h/2}^{h/2} \tau_{xz}\, dz \tag{7.1}$$

$$R_{yz} = \int_{-h/2}^{h/2} \tau_{yz}\, dz \tag{7.2}$$

For equilibrium, resultant forces and moments are zero. Therefore, for equilibrium in the x direction,

$$-N_x\, dy + \left(N_x + \frac{\partial N_x}{\partial x} dx\right) dy - N_{xy}\, dx + \left(N_{xy} + \frac{\partial N_{xy}}{\partial y} dy\right) dx = 0$$

which reduces to

284 ANALYSIS OF LAMINATED PLATES AND BEAMS

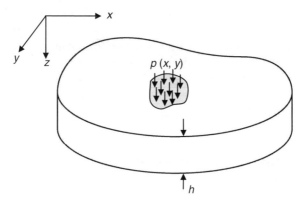

Figure 7-1. A flat plate subjected to transverse load.

$$\frac{\partial N_x}{\partial x} + \frac{\partial N_{xy}}{\partial y} = 0 \qquad (7.3)$$

Equation (7.3) is the first equilibrium equation. The second equilibrium equation can be obtained similarly by considering equilibrium in the y direction. It is given by

$$\frac{\partial N_{xy}}{\partial x} + \frac{\partial N_y}{\partial y} = 0 \qquad (7.4)$$

For force equilibrium in the z direction,

$$-R_{xz}\, dy + \left(R_{xz} + \frac{\partial R_{xz}}{\partial x} dx\right) dy - R_{yz}\, dx + \left(R_{yz} + \frac{\partial R_{yz}}{\partial y} dy\right) dx + p\, dx\, dy = 0$$

which reduces to

$$\frac{\partial R_{xz}}{\partial x} + \frac{\partial R_{yz}}{\partial y} + p = 0 \qquad (7.5)$$

Equations (7.3)–(7.5) are the equations for force equilibrium in the x, y, and z directions.

For equilibrium, resultant moment about any axis is also zero. By summing the moments about the x axis, we get

7.2 GOVERNING EQUATIONS FOR PLATES   285

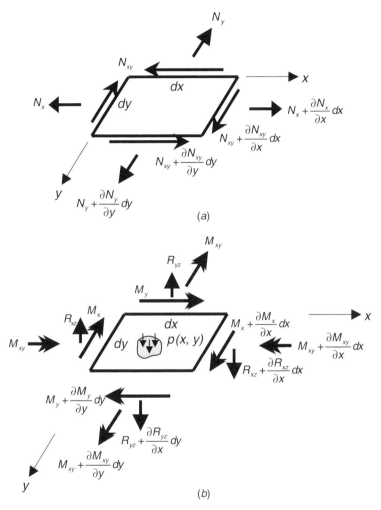

Figure 7-2. A differential element with (a) in-plane force resultants and (b) moment resultants, shear force resultants, and applied transverse forces.

$$M_y \, dx - \left(M_y + \frac{\partial M_y}{\partial y} dy\right) dx + M_{xy} \, dy - \left(M_{xy} + \frac{\partial M_{xy}}{\partial x} dx\right) dy$$

$$+ \left(R_{yz} + \frac{\partial R_{yz}}{\partial y} dy\right) dx \, dy + \left(R_{xz} + \frac{\partial R_{xz}}{\partial x} dx\right) dy \frac{dy}{2} - R_{xz} \, dy \frac{dy}{2}$$

$$+ p \, dx \, dy \frac{dy}{2} = 0$$

When higher-order terms are neglected, this equation simplifies to

$$\frac{\partial M_{xy}}{\partial x} + \frac{\partial M_y}{\partial y} - R_{yz} = 0 \tag{7.6}$$

Similarly, the summation of moments about the y axis gives

$$\frac{\partial M_x}{\partial x} + \frac{\partial M_{xy}}{\partial y} - R_{xz} = 0 \tag{7.7}$$

Substitution of Eqs. (7.6) and (7.7) in Eq. (7.5) gives

$$\frac{\partial^2 M_x}{\partial x^2} + 2 \frac{\partial^2 M_{xy}}{\partial x \partial y} + \frac{\partial^2 M_y}{\partial y^2} + p = 0 \tag{7.8}$$

Equations (7.3), (7.4), and (7.8) represent three equations of equilibrium for a plate subjected to transverse loading. These equations are valid for isotropic as well as anisotropic or laminated composite plates.

7.2.2 Equilibrium Equations in Terms of Displacements

The equilibrium equations [Eqs. (7.3), (7.4) and (7.8)] also can be written in terms of three midplane displacements u_0, v_0, and w_0, using constitutive relations. Since the constitutive relations for plates are material-dependent, the equilibrium equations in terms of displacements will be different for different materials. Equations for the laminated composites are derived in the following paragraphs.

The constitutive equations for a general laminated composite were derived in Chap. 6. Substitution of the constitutive equations [Eqs. (6.18) and (6.19)] into Eqs. (7.3), (7.4), and (7.8) and rearranging the terms give the following equilibrium equations in terms of the midplane displacements u_0, v_0, and w_0:

$$A_{11} \frac{\partial^2 u_0}{\partial x^2} + 2A_{16} \frac{\partial^2 u_0}{\partial x \partial y} + A_{66} \frac{\partial^2 u_0}{\partial y^2} + A_{16} \frac{\partial^2 v_0}{\partial x^2} + (A_{12} + A_{66}) \frac{\partial^2 v_0}{\partial x \partial y}$$
$$+ A_{26} \frac{\partial^2 v_0}{\partial y^2} - B_{11} \frac{\partial^3 w_0}{\partial x^3} - 3B_{16} \frac{\partial^3 w_0}{\partial x^2 \partial y} - (B_{12} + 2B_{66}) \frac{\partial^3 w_0}{\partial x \partial y^2} \tag{7.9}$$
$$- B_{26} \frac{\partial^3 w_0}{\partial y^3} = 0$$

$$A_{16} \frac{\partial^2 u_0}{\partial x^2} + (A_{12} + A_{66}) \frac{\partial^2 u_0}{\partial x \partial y} + A_{26} \frac{\partial^2 u_0}{\partial y^2} + A_{66} \frac{\partial^2 v_0}{\partial x^2} + 2A_{26} \frac{\partial^2 v_0}{\partial x \partial y}$$
$$+ A_{22} \frac{\partial^2 v_0}{\partial y^2} - B_{16} \frac{\partial^3 w_0}{\partial x^3} - (B_{12} + 2B_{66}) \frac{\partial^3 w_0}{\partial x^2 \partial y} - 3B_{26} \frac{\partial^3 w_0}{\partial x \partial y^2} \tag{7.10}$$
$$- B_{22} \frac{\partial^3 w_0}{\partial y^3} = 0$$

7.2 GOVERNING EQUATIONS FOR PLATES

$$D_{11}\frac{\partial^4 w_0}{\partial x^4} + 4D_{16}\frac{\partial^4 w_0}{\partial x^3 \partial y} + 2(D_{12} + 2D_{66})\frac{\partial^4 w_0}{\partial x^2 \partial y^2} + 4D_{26}\frac{\partial^4 w_0}{\partial x \partial y^3}$$
$$+ D_{22}\frac{\partial^4 w_0}{\partial y^4} - B_{11}\frac{\partial^3 u_0}{\partial x^3} - 3B_{16}\frac{\partial^3 u_0}{\partial x^2 \partial y} - (B_{12} + 2B_{66})\frac{\partial^3 u_0}{\partial x \partial y^2}$$
$$- B_{26}\frac{\partial^3 u_0}{\partial y^3} - B_{16}\frac{\partial^3 v_0}{\partial x^3} - (B_{12} + 2B_{66})\frac{\partial^3 v_0}{\partial x^2 \partial y}$$
$$- 3B_{26}\frac{\partial^3 v_0}{\partial x \partial y^2} - B_{22}\frac{\partial^3 v_0}{\partial y^3} = p \quad (7.11)$$

Equations (7.9)–(7.11) are the equilibrium equations for a laminated plate with any stacking sequence. Solution to these differential equations has to be obtained for problems involving a general laminate. A complete solution of the problem also will satisfy the boundary conditions. Closed-form solutions for these equations with arbitrary boundary conditions are not possible. However, closed-form solutions can be obtained for certain laminates whose constitutive relations exhibit symmetry conditions that permit simplification of the equilibrium equations. Two such special cases are discussed below.

Specially orthotropic laminates These are the laminates that behave like a single layer of an orthotropic material. Their constitutive equations satisfy the following conditions:

$$A_{16} = A_{26} = 0$$
$$B_{ij} = 0 \quad (7.12)$$
$$D_{16} = D_{26} = 0$$

Examples of specially orthotropic laminates are the unidirectional laminates with all the plies in 0° or 90° direction and symmetric cross-ply laminates. Incorporation of Eq. (7.12) into Eqs. (7.9)–(7.11) simplifies the equilibrium equations for specially orthotropic laminates as follows:

$$A_{11}\frac{\partial^2 u_0}{\partial x^2} + A_{66}\frac{\partial^2 u_0}{\partial y^2} + (A_{12} + A_{66})\frac{\partial^2 v_0}{\partial x \partial y} = 0 \quad (7.13)$$

$$(A_{12} + A_{66})\frac{\partial^2 u_0}{\partial x \partial y} + A_{66}\frac{\partial^2 v_0}{\partial x^2} + A_{22}\frac{\partial^2 v_0}{\partial y^2} = 0 \quad (7.14)$$

$$D_{11}\frac{\partial^4 w_0}{\partial x^4} + 2(D_{12} + 2D_{66})\frac{\partial^4 w_0}{\partial x^2 \partial y^2} + D_{22}\frac{\partial^4 w_0}{\partial y^4} = p \quad (7.15)$$

Isotropic plates Constitutive relations of an isotropic plate satisfy the following conditions and can be written in terms of the modulus (E), Poisson's ratio (ν), and the plate thickness (h) as:

$$A_{11} = A_{22} = \frac{Eh}{1 - \nu^2} = A$$

$$A_{12} = \nu A \qquad A_{66} = \frac{(1 - \nu)A}{2}$$

$$A_{16} = A_{26} = 0$$

$$B_{ij} = 0 \qquad (7.16)$$

$$D_{11} = D_{22} = \frac{Eh^3}{12(1 - \nu^2)} = D$$

$$D_{12} = \nu D \qquad D_{66} = \frac{(1 - \nu)D}{2}$$

$$D_{16} = D_{26} = 0$$

Incorporation of these conditions into Eqs. (7.13)–(7.15) simplifies the equilibrium equations for isotropic plates as follows:

$$\frac{\partial^2 u_0}{\partial x^2} + \frac{(1 - \nu)}{2} \frac{\partial^2 u_0}{\partial y^2} + \frac{(1 + \nu)}{2} \frac{\partial^2 v_0}{\partial x \partial y} = 0 \qquad (7.17)$$

$$\frac{(1 + \nu)}{2} \frac{\partial^2 u_0}{\partial x \partial y} + \frac{(1 - \nu)}{2} \frac{\partial^2 v_0}{\partial x^2} + \frac{\partial^2 v_0}{\partial y^2} = 0 \qquad (7.18)$$

$$\frac{\partial^4 w_0}{\partial x^4} + 2 \frac{\partial^4 w_0}{\partial x^2 \partial y^2} + \frac{\partial^4 w_0}{\partial y^4} = \frac{p}{D} \qquad (7.19)$$

It may be pointed out here that in both the special cases just considered, there is no coupling between bending and stretching because $B_{ij} = 0$. Therefore, in these cases, bending problems and stretching problems (problems involving in-plane loads only, sometimes called *membrane problems*) can be solved separately and the solutions superimposed.

7.3 APPLICATION OF PLATE THEORY

7.3.1 Bending

It was pointed out earlier that for a general laminate, bending and stretching are coupled as indicated by the governing equations. Thus bending problems cannot be isolated for general laminates. However, bending of specially orthotropic laminates is governed by Eq. (7.15). Methods have been developed to obtain solutions of practical bending problems of such laminates. One such problem is discussed in the following paragraphs, namely, bending of a specially orthotropic plate with simply supported edges.

Consider a specially orthotropic, rectangular plate subjected to a transverse load $p(x, y)$, as shown in Fig. 7-3. The plate edges are simply supported so that the transverse displacements at the edges and resultant moments about each edge are zero. These edge conditions, mathematically expressed as follows, are the boundary conditions:

On edge AD ($x = 0$)

$$w_0(0, y) = M_x(0, y) = 0 \qquad (7.20)$$

On edge BC ($x = a$)

$$w_0(a, y) = M_x(a, y) = 0 \qquad (7.21)$$

On edge AB ($y = 0$)

$$w_0(x, 0) = M_y(x, 0) = 0 \qquad (7.22)$$

On edge CD ($y = b$)

$$w_0(x, b) = M_y(x, b) = 0 \qquad (7.23)$$

In order to solve this problem, it will be convenient to write the resultant moments in terms of the displacement $w_0(x, y)$, using Eq. (6.19), as follows:

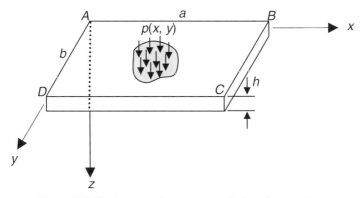

Figure 7-3. Rectangular plate geometry for bending problem.

$$M_x = -D_{11}\frac{\partial^2 w_0}{\partial x^2} - D_{12}\frac{\partial^2 w_0}{\partial y^2} \tag{7.24}$$

$$M_y = -D_{12}\frac{\partial^2 w_0}{\partial x^2} - D_{22}\frac{\partial^2 w_0}{\partial y^2} \tag{7.25}$$

A solution to the problem requires obtaining a function $w_0(x, y)$ such that it satisfies the governing Eq. (7.15) and the preceding boundary conditions [Eqs. (7.20)–(7.23)]. A closed-bound solution to the problem is not possible. A series solution can be obtained, however, using the well-known Navier's approach [1,2]. For this purpose, the displacement $w_0(x, y)$ and the applied load $p(x, y)$ are represented by double Fourier sine series. The coefficients for the series representing load (called *load coefficients*) are related to the applied load, whereas the displacement coefficients are obtained such that the series satisfies the governing equation [Eq. (7.15)]. A solution of the governing equation that also satisfies the preceding boundary conditions is given by

$$w_0(x, y) = \sum_{m=1}^{\infty}\sum_{n=1}^{\infty} w_{mn}\sin\left(\frac{m\pi x}{a}\right)\sin\left(\frac{n\pi y}{b}\right) \tag{7.26}$$

where w_{mn} are the displacement coefficients to be determined, and m and n are positive integers. The transverse load is expressed by a similar series as

$$p(x, y) = \sum_{m=1}^{\infty}\sum_{n=1}^{\infty} p_{mn}\sin\left(\frac{m\pi x}{a}\right)\sin\left(\frac{n\pi y}{b}\right) \tag{7.27}$$

where

$$p_{mn} = \frac{4}{ab}\int_0^a\int_0^b p(x, y)\sin\left(\frac{m\pi x}{a}\right)\sin\left(\frac{n\pi y}{b}\right)dx\,dy \tag{7.28}$$

Substitution of Eqs. (7.26) and (7.27) in Eq. (7.15) and comparison of coefficients on the two sides give the following displacement coefficients:

$$w_{mn} = \frac{p_{mn}}{\pi^4\left[D_{11}\left(\frac{m}{a}\right)^4 + 2(D_{12} + 2D_{66})\left(\frac{m}{a}\right)^2\left(\frac{n}{b}\right)^2 + D_{22}\left(\frac{n}{b}\right)^4\right]} \tag{7.29}$$

Substitution of Eq. (7.29) in Eq. (7.26) gives the displacement function:

$$w_0(x, y) = \sum_{m=1}^{\infty} \sum_{n=1}^{\infty} \frac{p_{mn} \sin\left(\frac{m\pi x}{a}\right) \sin\left(\frac{n\pi y}{b}\right)}{\pi^4 \left[D_{11}\left(\frac{m}{a}\right)^4 + 2(D_{12} + 2D_{66})\left(\frac{m}{a}\right)^2 \left(\frac{n}{b}\right)^2 + D_{22}\left(\frac{n}{b}\right)^4 \right]} \quad (7.30)$$

The transverse deflection w_0 at any point (x, y) can be calculated numerically by evaluating a sufficient number of terms of the series [Eq. (7.30)].

The load coefficients p_{mn} can be found for a given load distribution $p(x, y)$ using Eq. (7.28). It can be shown easily that for a uniformly distributed load $[p(x, y) = p_0]$, the load coefficients are given by

$$p_{mn} = \begin{cases} \dfrac{16 p_0}{\pi^2 mn} & \text{where } m \text{ and } n \text{ are odd} \\ 0 & \text{other cases} \end{cases} \quad (7.31)$$

Substitution of Eq. (7.31) in Eq. (7.30) gives the transverse deflection of a specially orthotropic, simply supported rectangular plate subjected to a uniformly distributed load as follows:

$$w_0(x, y) = \frac{16 p_0}{\pi^6} \sum_{m=1,3,\ldots}^{\infty} \sum_{n=1,3,\ldots}^{\infty} \frac{\sin\left(\frac{m\pi x}{a}\right) \sin\left(\frac{n\pi y}{b}\right)}{mn \left[D_{11}\left(\frac{m}{a}\right)^4 + 2(D_{12} + 2D_{66})\left(\frac{m}{a}\right)^2 \left(\frac{n}{b}\right)^2 + D_{22}\left(\frac{n}{b}\right)^4 \right]} \quad (7.32)$$

For an isotropic plate, Eq. (7.32) simplifies to

$$w_0(x, y) = \frac{16 p_0}{\pi^6 D} \sum_{m=1,3,\ldots}^{\infty} \sum_{n=1,3,\ldots}^{\infty} \frac{\sin\left(\frac{m\pi x}{a}\right) \sin\left(\frac{n\pi y}{b}\right)}{mn \left[\left(\frac{m}{a}\right)^2 + \left(\frac{n}{b}\right)^2 \right]^2} \quad (7.33)$$

Example 7-1: A rectangular plate of length 0.5 m, width 0.25 m, and thickness 0.005 m, simply supported at all edges, carries a uniformly distributed load $p_0 = 10$ N/m². Determine the maximum displacement and stresses in the plate assuming that the plate has a symmetric cross-ply

[0/90]$_s$ lay-up, and is made of the AS4/3501-6 graphite–epoxy laminae considered in Example 5-3.

Elements of the lamina stiffness matrix were calculated in Example 5-3 as

$$Q_{11} = 148.95 \text{ GPa}$$

$$Q_{22} = 10.57 \text{ GPa}$$

$$Q_{12} = 3.17 \text{ GPa}$$

$$Q_{66} = 5.61 \text{ GPa}$$

The bending stiffness coefficients are calculated using Eq. (6.20) as follows:

$$D_{11} = 1371.4 \text{ N} \cdot \text{m}$$

$$D_{22} = 290.26 \text{ N} \cdot \text{m}$$

$$D_{12} = 33.023 \text{ N} \cdot \text{m}$$

$$D_{66} = 58.44 \text{ N} \cdot \text{m}$$

Plate deflections are calculated from Eq. (7.32) by substituting appropriate values of x and y and evaluating a sufficient number of terms in the series. Since the maximum deflection occurs at the center of the plate, where $x = a/2$ and $y = b/2$, the maximum deflection is given by the following series:

$$(w_0)_{max} = 8.67 \times 10^{-6} \sum_{m=1}^{\infty} \sum_{n=1}^{\infty} \frac{(-1)^{1-[(m+n)/2]}}{mn(1.14m^4 + 3.87n^4 + m^2n^2)}$$

Terms of the series are evaluated by substituting values of m and n in increasing order. It will be noted that the numerical value of terms decreases as the values of m and n increase. Deflection is obtained by taking the sum of a sufficient number of terms in the series such that considering additional terms does not change the numerical result for the desired accuracy. In other words, the result converges to a number. To illustrate this point, deflection values have been obtained by taking the sum of different numbers of terms. The results are given in Table 7-1. Deflection is the summation of terms, with the ending term corresponding to the values of m and n shown. For example deflection predicted by only the first term

7.3 APPLICATION OF PLATE THEORY

Table 7-1 Maximum deflection (w_0, 10^{-6} m) predicted by the Navier series solution

m/n	1	3	5	7	9	11
1	1.4402	1.4313	1.4320	1.4319	1.4319	1.4319
3	1.4130	1.4060	1.4065	1.4064	1.4065	1.4065
5	1.4153	1.4079	1.4085	1.4084	1.4084	1.4084
7	1.4148	1.4076	1.4081	1.4080	1.4080	1.4080
9	1.4150	1.4077	1.4082	1.4081	1.4081	1.4081
11	1.4149	1.4076	1.4082	1.4081	1.4081	1.4081

($m = 1$, $n = 1$) is 1.4402×10^{-6} m, and that predicted by the series containing terms up to $m = 5$ and $n = 5$ is 1.4085×10^{-6} m. This table shows that the deflection converges to 1.4081×10^{-6} when all the terms up to $m = 11$ and $n = 11$ are considered. It may be pointed out that a term vanishes if either m or n is an even number. The plate deflections on the plane $y = b/2$ are plotted in Fig. 7-4, against distance from the plate edge. The maximum deflection occurs at the center of the plate.

Strains are obtained by substituting Eq. (7.32) in Eq. (6.5). It may be noted that the strains are also given as a series whose values can be obtained by following the procedure just explained. Stresses then can be obtained using Eq. (6.9). Bending stresses σ_x and σ_y have been calculated on the plane $y = b/2$. The maximum stress across the laminate thickness occurs at the top and bottom of the laminate. The maximum stresses σ_x and σ_y are plotted against the distance from the edge in Fig. 7-5. The absolute maximum stresses σ_x and σ_y occur at the center of the plate on the top surface, where $x = a/2$, $y = b/2$, and $z = -h/2$. The values are

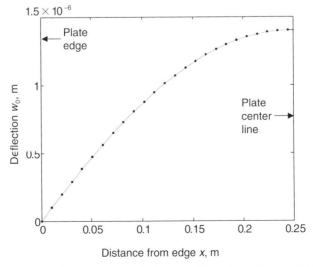

Figure 7-4. Deflection of a simply supported plate subjected to uniformly distributed load.

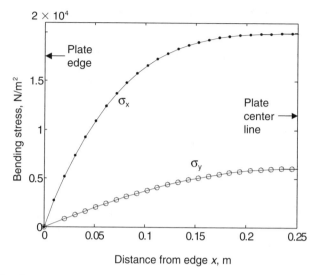

Figure 7-5. Bending stresses (σ_x and σ_y) in a simply supported rectangular plate subjected to uniformly distributed load.

$$\sigma_x = 19{,}915 \text{ Pa}$$

$$\sigma_y = 6{,}058 \text{ Pa}$$

In-plane shear stress τ_{xy} at this point is zero.

7.3.1.1 Bending of General Laminates Plate-governing equations [Eqs. (7.9)–(7.11)], derived in Sec. 7.2.2, are applicable to laminated plates with any stacking sequence. Their application, however, has been illustrated in this section for specially orthotropic laminates to keep the mathematical development and discussion simple. It was pointed out that a closed-form solution of Eqs. (7.9)–(7.11) for general laminates is not possible. Therefore, problems involving general laminates are solved by approximate numerical methods such as finite-element analysis (FEA). A detailed discussion on this technique is beyond the scope of this book. Numerical results for general laminates obtained through FEA are presented in the following paragraphs for a comparative discussion.

Analyses have been carried out on square plates of sides 0.5 m, thickness 5 mm, and subjected to a uniformly distributed load $p_0 = 10 \text{ N/m}^2$. Three laminates of AS4/3501-6 graphite–epoxy laminae (Example 7-1), with layups $[0/90]_s$, $[\pm 45]_s$, and $[30/50]$ have been analyzed. Displacement and stresses σ_x, σ_y, and τ_{xy} at the center of the plate have been obtained. For comparison, results on an aluminum plate ($E = 70$ GPa and $\nu = 0.33$) with the same dimensions also have been obtained. In each case, three support conditions have been considered. These are (1) all four edges simply supported (S), (2) all four edges clamped (C), and (3) two opposite edges simply supported and the other two clamped. Analyses were performed using a commercial FEA software called ABAQUS.

Numerical results of the analyses are given in Table 7-2. For each material, deflection and stresses are the highest when all edges are simply supported. They are the lowest when all edges are clamped. As the plate support conditions change, the stresses and deflection change in different proportions for different materials. Thus analysis of one lay-up cannot be used to predict results of a different lay-up. Similarly, analysis of an isotropic plate should not be used to understand the behavior of composite laminates. It also may be pointed out that shear stress (τ_{xy}) develops in the unsymmetric laminate [30/50] owing to coupling for all three boundary conditions. No shear stress develops under any of the boundary conditions in the aluminum plate or in the cross-ply laminate when fibers are parallel to the edges. However, in a [±45]$_s$ laminate, which is a cross-ply laminate with fibers oriented at ±45° to the plate edges, a shear stress develops when two opposite edges are simply supported and the other two are clamped. However, no shear stress develops for the other two support conditions.

7.3.2 Buckling

In-plane loads on a flat symmetric laminate cause in-plane displacements (u_0 and v_0) but no out-of-plane displacement (w_0). However, it has been known that in-plane compressive loads, when high enough, cause out-of-plane de-

Table 7-2 Center deflection and stresses in square plates with different support conditions

Support Conditions	Material/ Laminate Lay-up	Center Deflection w_0 (10^{-6} m)	Stress, Pa		
			σ_x	σ_y	τ_{xy}
S S S S	Aluminum	3.11	29,420	29,420	0
	[0/90]$_s$	5.18	74,910	6,140	0
	[±45]$_s$	3.74	42,210	5,230	0
	[30/50]	5.26	60,830	8,850	−350
C C C C	Aluminum	0.97	14,080	14,080	0
	[0/90]$_s$	1.17	27,580	1,980	0
	[±45]$_s$	1.41	23,120	2,970	0
	[30/50]	1.89	27,610	4,700	490
S C C S	Aluminum	1.47	20,260	15,170	0
	[0/90]$_s$	1.30	30,970	1,340	0
	[±45]$_s$	2.04	28,770	3,600	−730
	[30/50]	2.48	35,300	4,040	−380

flections that may be excessive and lead to failure. This is called *buckling*. The load at which excessive out-of-plane deflections occur is called the *buckling load*. The displacements at the onset of buckling exhibit a characteristic pattern called the *buckling mode*. Buckling of plates is of great interest in the design of structures. It is discussed in this section.

In buckling, out-of-plane displacements are caused by in-plane loads. Buckling cannot be predicted through the governing equation [Eq. (7.11)] because interaction between in-plane loads and out-of-plane displacements was suppressed in its derivation. It was assumed that the out-of-plane displacements are small such that the resultant forces N_x, N_y, and N_{xy} act in their original direction of action in the xy plane and do not have a component in the z direction. Thus the possibility of these forces producing out-of-plane deflections is not considered. Therefore, to study buckling, the governing equation should be modified by taking into account the out-of-plane component (in the z direction) of the resultant forces (N_x, N_y, and N_{xy}). Such an equation can be derived by assuming the differential element of Fig. 7-2a to be oriented in a general out-of-plane position such that the in-plane forces N_x, N_y, and N_{xy} contribute a component in the z direction. It can be shown that equilibrium of such a differential element in the z direction gives the following equation (details of the derivation can be found in ref. 1):

$$\frac{\partial R_{xz}}{\partial x} + \frac{\partial R_{yz}}{\partial y} - N_x \frac{\partial^2 w_0}{\partial x^2} - 2N_{xy} \frac{\partial^2 w_0}{\partial x \partial y} - N_y \frac{\partial^2 w_0}{\partial y^2} + p = 0 \quad (7.34)$$

Substitution of Eqs. (7.6) and (7.7) into Eq. (7.34) gives

$$\frac{\partial^2 M_x}{\partial x^2} + 2\frac{\partial^2 M_{xy}}{\partial x \partial y} + \frac{\partial^2 M_y}{\partial y^2} - N_x \frac{\partial^2 w_0}{\partial x^2} - 2N_{xy} \frac{\partial^2 w_0}{\partial x \partial y} - N_y \frac{\partial^2 w_0}{\partial y^2} + p = 0 \quad (7.35)$$

Resultant moments M_x, M_y, and M_{xy} in Eq. (7.35) can be written in terms of transverse displacements using constitutive equation [Eq. (6.19)] to obtain

$$D_{11} \frac{\partial^4 w_0}{\partial x^4} + 2(D_{12} + 2D_{66}) \frac{\partial^4 w_0}{\partial x^2 \partial y^2} + D_{22} \frac{\partial^4 w_0}{\partial y^4} - N_x \frac{\partial^2 w_0}{\partial x^2}$$
$$- 2N_{xy} \frac{\partial^2 w_0}{\partial x \partial y} - N_y \frac{\partial^2 w_0}{\partial y^2} + p = 0 \quad (7.36)$$

Equation (7.36) permits coupling between in-plane loads and the transverse displacements and is therefore an appropriate governing equation to predict buckling of specially orthotropic laminates.

Consider a specially orthotropic, rectangular plate subjected to uniformly distributed normal compressive loads N_{x0} and N_{y0} along its edges, as shown in Fig. 7-6. No shear force (N_{xy}) or transverse load (p) is applied to the plate. For this case, resultant forces and transverse load become

7.3 APPLICATION OF PLATE THEORY

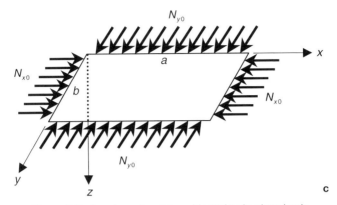

Figure 7-6. A rectangular plate subjected to in-plane loads.

$$N_x = -N_{x0}, \quad N_y = -N_{y0}, \quad N_{xy} = 0, \quad p = 0 \tag{7.37}$$

Substitution of Eq. (7.37) into Eq. (7.36) gives

$$D_{11}\frac{\partial^4 w_0}{\partial x^4} + 2(D_{12} + 2D_{66})\frac{\partial^4 w_0}{\partial x^2 \partial y^2} + D_{22}\frac{\partial^4 w_0}{\partial y^4} = -N_{x0}\frac{\partial^2 w_0}{\partial x^2} - N_{y0}\frac{\partial^2 w_0}{\partial y^2} \tag{7.38}$$

If the plate is considered simply supported along its four edges, the boundary conditions are the same as in bending considered earlier [Eqs. (7.20)–(7.23)]. Solution to this problem is also obtained by following the method used for solving the bending problem. To that end, assume that the displacement $w_0(x, y)$ is given by the following double Fourier sine series:

$$w_0(x, y) = \sum_{m=1}^{\infty}\sum_{n=1}^{\infty} w_{mn} \sin\left(\frac{m\pi x}{a}\right) \sin\left(\frac{n\pi y}{b}\right) \tag{7.39}$$

where m and n are the number of half sine waves along the x and y directions, respectively.

Substitution of Eq. (7.39) into Eq. (7.38) gives

$$\pi^4 w_{mn}\left[D_{11}\left(\frac{m}{a}\right)^4 + 2(D_{12} + 2D_{66})\left(\frac{mn}{ab}\right)^2 + D_{22}\left(\frac{n}{b}\right)^4\right]$$
$$= w_{mn}\pi^2\left[N_{x0}\left(\frac{m}{a}\right)^2 + N_{y0}\left(\frac{n}{b}\right)^2\right] \tag{7.40}$$

For a nontrivial solution of Eq. (7.40), we have

$$\pi^2\left[D_{11}\left(\frac{m}{a}\right)^4 + 2(D_{12} + 2D_{66})\left(\frac{mn}{ab}\right)^2 + D_{22}\left(\frac{n}{b}\right)^4\right]$$
$$= \left[N_{x0}\left(\frac{m}{a}\right)^2 + N_{y0}\left(\frac{n}{b}\right)^2\right] \quad (7.41)$$

Equation (7.41) can be used to predict buckling behavior of specially orthotropic rectangular plates for different loading conditions. Two special cases are discussed in the following paragraphs.

Case I: Buckling under Uniaxial Compression If $N_{x0} = N_0$ and $N_{y0} = 0$, Eq. (7.41) simplifies to

$$N_0(m, n) = \pi^2\left[D_{11}\left(\frac{m}{a}\right)^2 + 2(D_{12} + 2D_{66})\left(\frac{n}{b}\right)^2 + D_{22}\left(\frac{n}{b}\right)^4\left(\frac{a}{m}\right)^2\right] \quad (7.42)$$

Buckling loads can be obtained for any combination of m and n. The critical buckling load is the lowest of all the values. It can be seen by inspection that for any value of m, the smallest value of N_0 occurs when $n = 1$. Thus

$$N_0(m, 1) = \pi^2\left[D_{11}\left(\frac{m}{a}\right)^2 + 2(D_{12} + 2D_{66})\frac{1}{b^2} + D_{22}\frac{1}{b^4}\left(\frac{a}{m}\right)^2\right] \quad (7.43)$$

The value of N_0 for any value of m depends on the bending stiffness and plate aspect ratio a/b. The critical load is obtained by numerical calculations. For a laminated plate with material properties $D_{11}/D_{22} = 10$ and $(D_{12} + 2D_{66})/D_{22} = 1$, Eq. (7.43) reduces to

$$N_0(m, 1) = \frac{\pi^2 D_{22}}{b^2}\left[10m^2\left(\frac{b}{a}\right)^2 + 2 + \frac{1}{m^2}\left(\frac{a}{b}\right)^2\right] \quad (7.44)$$

A nondimensional buckling load $\overline{N}$ may be defined as

$$\overline{N} = \frac{N_0 b^2}{\pi^2 D_{22}} = \left[10m^2\left(\frac{b}{a}\right)^2 + 2 + \frac{1}{m^2}\left(\frac{a}{b}\right)^2\right] \quad (7.45)$$

The nondimensional buckling load $\overline{N}$ is a function of m and the plate aspect ratio a/b. $\overline{N}$ is plotted against a/b for different values of m in Fig. 7-7. Each value of m corresponds to a particular buckling mode. For example, $m = 1$ corresponds to the plate buckling into a one-half sine wave in the x direction, and $m = 2$ corresponds to a full sine wave in the x direction. For the aspect ratio corresponding to the intersection of two buckling load curves for two consecutive m values, the plate can buckle in two different shapes. For example, the buckling curves for $m = 1$ and 2 intersect at $a/b = 2.5$ (see Fig.

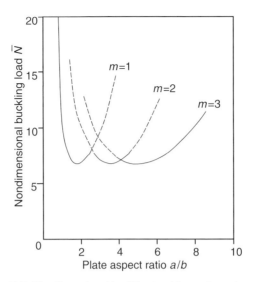

Figure 7-7. Nondimensional buckling load for rectangular plates.

7-7). Therefore, a plate with $a/b = 2.5$ can buckle as a half or a full sine wave. The plate with $a/b < 2.5$ can buckle only as a half sine wave, and that with $a/b > 2.5$ as a full sine wave.

The critical buckling load for a square plate is obtained by substituting the conditions that $m = 1$ and $a = b$ in Eq. (7.45):

$$(N_0)_{\text{crit}} = \frac{13\pi^2 D_{22}}{a^2} \tag{7.46}$$

It can be shown that the lowest value of the critical buckling load for the considered material properties occurs for $a/b = 1.78$ and is given by

$$(N_0)_{\text{crit}} = \frac{8.3245\pi^2 D_{22}}{b^2} \tag{7.47}$$

For an isotropic square plate under uniaxial compression, the buckling load can be calculated by substituting the isotropic properties in Eq. (7.43) and using $m = 1$:

$$(N_0)_{\text{crit}} = \frac{4\pi^2 D}{a^2} \tag{7.48}$$

Recall that N_0 has the units of force per unit length.

Case II: Buckling of a Square Plate under Biaxial Compression
Assume that a square plate is subjected to equal compressive loads on its edges such that $N_{x0} = N_{y0} = N_0$. Substitution of these values and $a = b$ in Eq. (7.41) and rearranging terms give

$$N_0(m, n) = \frac{\pi^2}{a^2} \frac{D_{11}m^2 + 2(D_{12} + 2D_{66})n^2 + D_{22}(n^4/m^2)}{1 + (n/m)^2} \tag{7.49}$$

It can be shown that the critical buckling load occurs when $m = 1$, provided that $D_{11} \geq D_{22}$. Under these conditions, the buckling load is given by

$$N_0(1, n) = \frac{\pi^2}{a^2} \frac{D_{11} + 2(D_{12} + 2D_{66})n^2 + D_{22}n^4}{1 + n^2} \tag{7.50}$$

For $D_{11}/D_{22} = 10$ and $(D_{12} + 2D_{66})/D_{22} = 1$, the critical buckling load occurs when $n = 1$ and is given by

$$(N_0)_{crit} = 6.5 \frac{\pi^2}{a^2} D_{22} \tag{7.51}$$

For an isotropic square plate under biaxial compression, the buckling load can be calculated using the isotropic material properties in Eq. (7.49) and is given by

$$N_0(m, n) = (m^2 + n^2) \frac{\pi^2 D}{a^2} \tag{7.52}$$

The critical buckling load occurs when $m = n = 1$ and is given by

$$(N_0)_{crit} = \frac{2\pi^2 D}{a^2} \tag{7.53}$$

Example 7-2: A rectangular plate of width 1 m and thickness 5 mm is simply supported at its edges. Determine and plot the buckling loads for different plate lengths when the plate is subjected to uniaxial load and equal loads in two directions. Assume that the laminate has the same lay-up and properties as given for the plate in Example 7-1.

Uniaxial buckling load is obtained from Eq. (7.43). By substitution of the stiffnesses obtained in Example 7-1 and $b = 1$ in Eq. (7.43), the uniaxial buckling load can be written as

$$N_0(m, 1) = 2{,}864.7\left[4.72m^2\left(\frac{b}{a}\right)^2 + 1.03 + \frac{1}{m^2}\left(\frac{a}{b}\right)^2\right]$$

N_0 is calculated for different values of m and the plate aspect ratio a/b and is plotted in Fig. 7-8.

Biaxial buckling load is obtained from Eq. (7.41). Substitution of $b = 1$, $N_{0x} = N_{0y} = N_0$, and the stiffness terms in Eq. (7.41) gives

$$N_0 = \frac{2{,}864.7\left[4.72m^4\left(\dfrac{b}{a}\right)^4 + 1.03m^2n^2\left(\dfrac{b}{a}\right)^2 + n^4\right]}{m^2\left(\dfrac{b}{a}\right)^2 + n^2}$$

Biaxial buckling load is calculated from the preceding equation for different values of m, n, and a/b. The biaxial buckling load is also shown in Fig. 7-8.

7.3.3 Free Vibrations

Vibration characteristics of structures are of considerable interest for their design and performance. For example, it is necessary to keep natural frequencies of structures well separated from their excitation frequencies in order

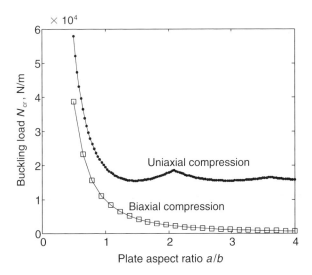

Figure 7-8. Buckling loads for rectangular plates subjected to uniaxial and biaxial compressive loads.

to avoid resonance. Free vibration of laminated plates is discussed in this section.

When a structure vibrates, the displacements change with time. Thus the structure is subjected to inertial forces produced as a result of acceleration of the structure's mass. Governing equations to study vibration characteristics take inertia forces into account. Such a governing equation is time-dependent and is called the *equation of motion*. The equation of motion for a specially orthotropic plate can be obtained from Eq. (7.15) by replacing the transverse applied load p with the inertia force given below:

$$p = -\rho_0 \frac{\partial^2 w_0}{\partial t^2} \tag{7.54}$$

where t is the time, and ρ_0 is the mass per unit area of the plate given by

$$\rho_0 = \sum_{k=1}^{n} \int_{h_{k-1}}^{h_k} \rho^k \, dz \tag{7.55}$$

where ρ^k the mass per unit volume of the kth ply of the laminate.

Substitution of Eq. (7.54) into Eq. (7.15) gives the following equation of motion for a specially orthotropic laminate:

$$D_{11} \frac{\partial^4 w_0}{\partial x^4} + 2(D_{12} + 2D_{66}) \frac{\partial^4 w_0}{\partial x^2 \partial y^2} + D_{22} \frac{\partial^4 w_0}{\partial y^4} + \rho_0 \frac{\partial^2 w_0}{\partial t^2} = 0 \tag{7.56}$$

It may be noted that w_0 in Eq. (7.56) will be a function of x, y, and t. The solution to Eq. (7.56) is obtained by assuming that the time dependence and spatial dependence of w_0 can be represented by separate functions. Further, since free vibration of an elastic continuum causes displacements that are harmonic in time, w_0 can be assumed to be periodic of the form

$$w_0(x, y, t) = W(x, y)e^{i\omega t} \tag{7.57}$$

where $W(x, y)$ is the mode shape function, ω is the frequency of vibration, and $i = \sqrt{-1}$. Substitution of Eq. (7.57) into Eq. (7.56) gives

$$D_{11} \frac{\partial^4 W}{\partial x^4} + 2(D_{12} + 2D_{66}) \frac{\partial^4 W}{\partial x^2 \partial y^2} + D_{22} \frac{\partial^4 W}{\partial y^4} - \rho_0 \omega^2 W = 0 \tag{7.58}$$

For the simply supported plate shown in Fig. 7-3, the boundary conditions can be written in terms of $W(x, y)$ as follows:

On edge AD ($x = 0$)

$$W(0, y) = M_x(0, y) = 0 \tag{7.59}$$

On edge BC ($x = a$)

$$W(a, y) = M_x(a, y) = 0 \tag{7.60}$$

On edge AB ($y = 0$)

$$W(x, 0) = M_y(x, 0) = 0 \tag{7.61}$$

On edge CD ($y = b$)

$$W(x, b) = M_y(x, b) = 0 \tag{7.62}$$

A solution of the equation [Eq. (7.58)] that also satisfies the boundary conditions [Eqs. (7.59)–(7.62)] is given by

$$W(x, y) = \sum_{m=1}^{\infty} \sum_{n=1}^{\infty} W_{mn} \sin\left(\frac{m\pi x}{a}\right) \sin\left(\frac{n\pi y}{b}\right) \tag{7.63}$$

Substitution of Eq. (7.63) into Eq. (7.58) gives an equation for natural frequencies ω_{mn} as

$$\omega_{mn}^2 = \frac{\pi^4}{\rho_0}\left[D_{11}\left(\frac{m}{a}\right)^4 + 2(D_{12} + 2D_{66})\left(\frac{m}{a}\right)^2\left(\frac{n}{b}\right)^2 + D_{22}\left(\frac{n}{b}\right)^4\right] \tag{7.64}$$

A natural frequency can be determined for any combination of m and n. The lowest natural frequency, called the *fundamental frequency*, is usually of greatest interest from the design point of view. It can be seen from Eq. (7.64) that the fundamental frequency occurs when $m = n = 1$ and is given by

$$\omega_{11} = \frac{\pi^2}{b^2\sqrt{\rho_0}}\sqrt{D_{11}\left(\frac{b}{a}\right)^4 + 2(D_{12} + 2D_{66})\left(\frac{b}{a}\right)^2 + D_{22}} \tag{7.65}$$

The mode shape corresponding to the fundamental frequency is given by

$$W(x, y) = \sin\left(\frac{\pi x}{a}\right) \sin\left(\frac{\pi y}{b}\right) \tag{7.66}$$

For an isotropic plate, the fundamental frequency is given by

$$\omega_{11} = \frac{\pi^2}{b^2} \sqrt{\frac{D}{\rho_0}} \left[\left(\frac{b}{a}\right)^2 + 1 \right] \qquad (7.67)$$

Frequencies of vibration for square plates are given in Table 7-3 for four combinations of m and n. The plates are made of a specially orthotropic $[D_{11}/D_{22} = 10, (D_{12} + 2D_{66})/D_{22} = 1]$ and an isotropic $[D_{11}/D_{22} = (D_{12} + 2D_{66})/D_{22} = 1]$ materials. It may be noted that for the isotropic plate, $\omega_{12} = \omega_{21}$, but for the specially orthotropic plate considered here, ω_{21} and ω_{12} are different. The specially orthotropic plate has a different set of the four lowest natural frequencies than does the isotropic plate. The corresponding mode shapes are shown in Fig. 7-9. The dashed lines denote the lines of zero transverse displacements and are called the *nodal lines*. It may be noted from Fig. 7-9 that the mode shapes for specially orthotropic plate and isotropic plate are identical to each other for the first and second natural frequencies, but they are different for the third and fourth natural frequencies.

Example 7-3: Determine the first five frequencies for the simply supported rectangular plate considered in Example 7-1. Assume that the material density $\rho = 1800$ kg/m³.

The natural frequencies are calculated from Eq. (7.64). The following are the plate dimensions and stiffnesses (see Example 7-1):

$$a = 0.5 \text{ m}$$
$$b = 0.25 \text{ m}$$
$$h = 0.005 \text{ m}$$

Table 7-3 Vibration frequencies of a simply supported square plate

	Specially Orthotropic $\omega = k\frac{\pi^2}{a^2}\sqrt{\frac{D_{22}}{\rho_0}}$			Isotropic $\omega = k\frac{\pi^2}{a^2}\sqrt{\frac{D}{\rho_0}}$		
Mode	m	n	k	m	n	k
First	1	1	3.62	1	1	2.00
Second	1	2	5.86	1	2	5.00
Third	1	3	10.45	2	1	5.00
Fourth	2	1	13.00	2	2	8.00

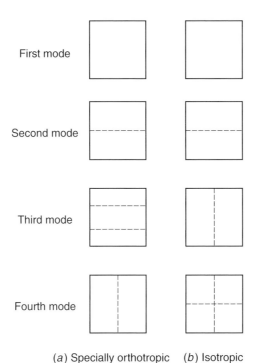

(a) Specially orthotropic (b) Isotropic

Figure 7-9. Vibration modes for simply supported square plates. Dashed lines represent nodal lines. (From Whitney [4].)

$$D_{11} = 1{,}371.4 \text{ N} \cdot \text{m}$$

$$D_{22} = 290.26 \text{ N} \cdot \text{m}$$

$$D_{12} = 33.02 \text{ N} \cdot \text{m}$$

$$D_{66} = 58.44 \text{ N} \cdot \text{m}$$

The area density ρ_0 is obtained as follows:

$$\rho_0 = \rho \times h = 1800 \times 0.005 = 9.0 \text{ kg/m}^2$$

Substitution of numerical values into Eq. (7.64) gives

$$\omega_{mn} = 455.71\sqrt{1.14m^4 + m^2n^2 + 3.87n^4}$$

The first five frequencies are calculated from the preceding equation by substituting combinations of m and n that give the lowest frequencies. These combinations and frequencies are given in Table 7-4. The mode shape corresponding to a natural frequency is obtained from Eq. (7.63). The mode shape corresponding to the first natural frequency (or fundamental frequency) is shown in Fig. 7-10.

7.4 DEFORMATIONS DUE TO TRANSVERSE SHEAR

Experimental evidence is available in the literature [11–13] that supports the classical approach to laminated-plate analysis presented in Secs. 7.2 and 7.3. Numerical results from exact elasticity solutions also agree with the laminated-plate theory for thin plates, that is, the plates with high width-to-thickness ratios. However, the results for thick plates (e.g., width-to-thickness ratios < 10) significantly differ from those obtained from the classical theory, particularly when the in-plane elastic modulus is much higher than the interlaminar shear modulus, as is commonly the case with composite materials. In particular, maximum plate deflections have been shown to be considerably larger than those predicted by classical laminated-plate theory. The discrepancies are attributed largely to the plate deformations caused by transverse shear [14]. Effects of transverse shear are discussed in this section through theories that can be applied to moderately thick plates.

7.4.1 First-Order Shear Deformation Theory

In the development of the classical lamination theory in Chap. 6 and the classical laminated-plate theory in Secs. 7.2 and 7.3, it was assumed that the out-of-plane shear strains γ_{xz} and γ_{yz} are zero. This assumption follows from the assumption that plane sections originally perpendicular to the plate midplane remain perpendicular to the plate midplane after deformation. Equations (6.3) and (6.4) for the displacements and Eq. (6.5) for the strains were derived from this assumption. These equations obviously show that the shear strains

Table 7-4 First five natural frequencies of a simply supported plate

Mode	m	n	ω_{mn} (rad/s)
First	1	1	1117.8
Second	2	1	2331.2
Third	1	2	3733.1
Fourth	2	2	4471.1
Fifth	3	1	4680.8

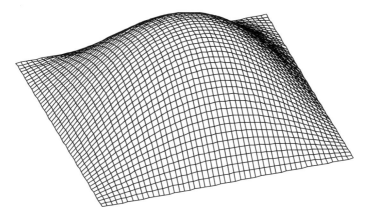

Figure 7-10. First mode shape of a simply supported plate.

γ_{xz} and γ_{yz} are zero. Therefore, the assumption concerning plate deformation needs to be modified to consider nonzero shear strains γ_{xz} and γ_{yz} in the laminate.

A theory called the *first-order shear deformation theory* (FSDT) is based on the assumption that plane sections originally perpendicular to the plate midplane remain plane after deformation, but they do not remain perpendicular to the plate midplane. The displacement field based on this assumption is obtained by assuming functions representing rotations of the plate sections. Following the procedure of Sec. 6.2, the displacement field for the FSDT can be obtained as

$$u(x, y, z) = u_0(x, y) + z\phi_x(x, y)$$
$$v(x, y, z) = v_0(x, y) + z\phi_y(x, y) \qquad (7.68)$$
$$w(x, y, z) = w_0(x, y)$$

where ϕ_x and ϕ_y are the rotations about the y and x axes, respectively. The displacement field gives the following shear strains:

$$\gamma_{xz} = \phi_x + \frac{\partial w_0}{\partial x}$$
$$\gamma_{yz} = \phi_y + \frac{\partial w_0}{\partial y} \qquad (7.69)$$

It may be noted that the shear strains γ_{xz} and γ_{yz} given by Eq. (7.69) are constant over the plate thickness. Consequently, this theory predicts uniform out-of-plane shear stresses through the thickness of each lamina. This violates the condition that the transverse shear stresses τ_{xz} and τ_{xz} must be zero on the top and bottom surfaces of the plate. This indicates an inaccuracy in the first-order shear deformation theory. Equations relating the shear-force resultants R_{xz} and R_{yz} to the shear strains γ_{xz} and γ_{yz} can be written as

$$\begin{Bmatrix} R_{yz} \\ R_{xz} \end{Bmatrix} = \begin{bmatrix} A_{44} & A_{45} \\ A_{45} & A_{55} \end{bmatrix} \begin{Bmatrix} \gamma_{yz} \\ \gamma_{xz} \end{Bmatrix} \qquad (7.70)$$

where

$$A_{ij} = K \int_{-h/2}^{h/2} \overline{Q}_{ij} \, dz$$

in which $\overline{Q}_{ij}$ are the transformed stiffness coefficients, and K is the shear correction coefficient to account for the variation in transverse shear stresses through the thickness. A value of 5/6 is used commonly for K.

It can be shown that for the displacement field given by Eq. (7.68), the plate equilibrium equations in terms of force and moment resultants do not change and are given by Eqs. (7.3), (7.4) and (7.8). However, the equilibrium equations in terms of displacements will be different because the plate constitutive equations [Eqs. (6.18) and (6.19)] are no longer valid. Constitutive equations relating force and moment resultants and the midplane strains and plate curvatures for this displacement field can be derived in a manner similar to the derivation of Eqs. (6.18) and (6.19). Development of this theory for a general laminate is beyond the scope of this book. Development is carried out for a specially orthotropic plate in the following paragraphs.

Constitutive equations for a specially orthotropic plate with a new displacement field [Eq. (7.68)] still satisfy the conditions stated in Eq. (7.12). In addition, $A_{45} = 0$. It can be shown that in view of these conditions, equilibrium Eqs. (7.5), (7.6), and (7.7) can be written in terms of the displacement field as follows:

$$D_{11} \frac{\partial^2 \phi_x}{\partial x^2} + (D_{12} + D_{66}) \frac{\partial^2 \phi_y}{\partial x \partial y} + D_{66} \frac{\partial^2 \phi_x}{\partial y^2} - A_{55} \left(\phi_x + \frac{\partial w_0}{\partial x} \right) = 0 \qquad (7.71)$$

$$D_{22} \frac{\partial^2 \phi_y}{\partial y^2} + (D_{12} + D_{66}) \frac{\partial^2 \phi_x}{\partial x \partial y} + D_{66} \frac{\partial^2 \phi_y}{\partial x^2} - A_{44} \left(\phi_y + \frac{\partial w_0}{\partial y} \right) = 0 \qquad (7.72)$$

$$A_{55} \left(\frac{\partial \phi_x}{\partial x} + \frac{\partial^2 w_0}{\partial x^2} \right) + A_{44} \left(\frac{\partial \phi_y}{\partial y} + \frac{\partial^2 w_0}{\partial y^2} \right) + p(x, y) = 0 \qquad (7.73)$$

Eqs. (7.71)–(7.73) are three coupled second-order differential equations with w_0, ϕ_x, and ϕ_y as the three unknowns. The reader may verify that substitution of $\phi_x = -(\partial w_0/\partial x)$ and $\phi_y = -(\partial w_0/\partial y)$ in these equations will reduce them to Eq. (7.15) obtained in CLPT. An exact solution of Eqs. (7.71)–(7.73) cannot be obtained for arbitrary boundary conditions. However, a closed-form series solution can be obtained for a simply supported rectangular plate, as was done for the classical laminated-plate theory. The solution is discussed below.

7.4.1.1 Transverse Shear Deformation Effects in Bending of a Simply Supported Rectangular Specially Orthotropic Plate
Consider a rectangular specially orthotropic plate simply supported on all four edges and subjected to a distributed load $p(x, y)$, as shown in Fig. 7-3. Boundary conditions for this plate are the same as those for thin plates [Eqs. (7.20)–(7.23)]. The solution technique for this problem is the same as that used for the bending of thin plates in Sec. 7.3.1. The following double Fourier series are assumed to represent w_0 and ϕ_x and ϕ_y:

$$w_0(x, y) = \sum_{m=1}^{\infty} \sum_{n=1}^{\infty} w_{mn} \sin\left(\frac{m\pi x}{a}\right) \sin\left(\frac{n\pi y}{b}\right) \tag{7.74}$$

$$\phi_x(x, y) = \sum_{m=1}^{\infty} \sum_{n=1}^{\infty} x_{mn} \cos\left(\frac{m\pi x}{a}\right) \sin\left(\frac{n\pi y}{b}\right) \tag{7.75}$$

$$\phi_y(x, y) = \sum_{m=1}^{\infty} \sum_{n=1}^{\infty} y_{mn} \sin\left(\frac{m\pi x}{a}\right) \cos\left(\frac{n\pi y}{b}\right) \tag{7.76}$$

where w_{mn}, x_{mn}, and y_{mn} are the series coefficients to be determined. It may be noted that these functions satisfy the boundary conditions of zero transverse displacement and zero moment resultant on each edge of the plate.

The applied transverse load $p(x, y)$ also can be expressed as a double Fourier sine wave:

$$p(x, y) = \sum_{m=1}^{\infty} \sum_{n=1}^{\infty} p_{mn} \sin\left(\frac{m\pi x}{a}\right) \sin\left(\frac{n\pi y}{b}\right) \tag{7.77}$$

where p_{mn} is given by Eq. (7.28).

Substitution of Eqs. (7.74)–(7.77) into Eqs. (7.71)–(7.73) gives the following matrix equation:

$$\begin{bmatrix} L_{11} & L_{12} & L_{13} \\ L_{12} & L_{22} & L_{23} \\ L_{13} & L_{23} & L_{33} \end{bmatrix} \begin{Bmatrix} x_{mn} \\ y_{mn} \\ w_{mn} \end{Bmatrix} = \begin{Bmatrix} 0 \\ 0 \\ -p_{mn} \end{Bmatrix} \tag{7.78}$$

where the coefficients of the [L] matrix are defined as

$$L_{11} = D_{11}\lambda_m^2 + D_{66}\lambda_n^2 + A_{55}$$

$$L_{12} = (D_{12} + D_{66})\lambda_m\lambda_n$$

$$L_{13} = A_{55}\lambda_m \quad (7.79)$$

$$L_{22} = D_{22}\lambda_n^2 + D_{66}\lambda_m^2 + A_{44}$$

$$L_{23} = A_{44}\lambda_n$$

$$L_{33} = A_{44}\lambda_n^2 + A_{55}\lambda_m^2$$

and

$$\lambda_m = \frac{m\pi}{a} \quad \lambda_n = \frac{n\pi}{b}$$

Solution of Eq. (7.78) gives the following coefficients:

$$x_{mn} = p_{mn}\frac{L_{12}L_{23} - L_{22}L_{13}}{\det L} \quad (7.80)$$

$$y_{mn} = p_{mn}\frac{L_{12}L_{13} - L_{11}L_{23}}{\det L} \quad (7.81)$$

$$w_{mn} = p_{mn}\frac{L_{11}L_{22} - L_{12}^2}{\det L} \quad (7.82)$$

where $\det L$ is the determinant of the coefficient matrix $[L]$. Substitution of these coefficients in Eqs. (7.74)–(7.76) gives series representing w_0, ϕ_x, and ϕ_y. Thus the plate curvatures, strains, and plate deflections can be calculated numerically.

Numerical results for a symmetric cross-ply $[0/90]_s$ simply supported square plate subjected to a sinusoidally distributed load [$p = p_0 \sin(\pi x/a) \sin(\pi y/a)$] have been obtained. The following laminate properties are used:

$$E_L = 175 \text{ GPa} \quad E_T = E_{T'} = 7 \text{ GPa}$$

$$G_{LT} = G_{LT'} = 3.5 \text{ GPa} \quad G_{TT'} = 1.4 \text{ GPa}$$

$$\nu_{LT} = \nu_{LT'} = 0.25$$

The nondimensionalized maximum deflection [$\bar{w} = w_0(E_2h^3/p_0a^410^3)$] is plotted against the side-to-thickness ratio in Fig. 7-11. For comparison, the maximum deflection predicted by the classical laminated-plate theory is also

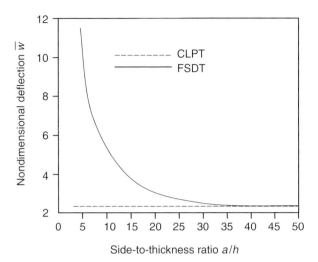

Figure 7-11. Effect of transverse shear on center deflection of a symmetric cross-ply $[0/90]_s$ laminate subjected to sinusoidally distributed transverse load. (From Whitney [14].)

given. The first-order shear deformation theory predicts significantly higher deflections for low side-to-thickness ratios.

7.4.2 Higher-Order Shear Deformation Theory

It was pointed out in the preceding section that the first-order shear deformation theory predicts uniform shear stresses τ_{xz} and τ_{yz} through the thickness of each lamina. This violates the condition that the transverse shear stresses τ_{xz} and τ_{yz} must be zero on the top and bottom surfaces of the plate. A higher-order shear deformation theory has been developed that satisfies this boundary condition. The displacement field based on the higher-order shear deformation theory is given by [15]

$$u(x, y, z) = u_0(x, y) + z\phi_x(x, y) - \frac{4}{3h^2} z^3 \left(\phi_x + \frac{\partial w}{\partial x} \right)$$

$$v(x, y, z) = v_0(x, y) + z\phi_y(x, y) - \frac{4}{3h^2} z^3 \left(\phi_y + \frac{\partial w}{\partial y} \right) \qquad (7.83)$$

$$w(x, y, z) = w_0(x, y)$$

The cubic variation of the displacements u and v predicts a quadratic variation of the transverse shear strains and transverse shear stresses through each layer. This satisfies the transverse shear-stress boundary condition at the top and bottom surfaces of the plate (shear stresses are zero on these surfaces). This also eliminates the shear correction coefficient K used in the first-order shear deformation theory. Derivation of the governing equations based on this higher-order shear deformation theory and the solution of plate problems is beyond the scope of this book. Discussion on the theory can be found in refs 9 and 15 through 17.

It is of interest to compare shear deformations and shear stresses predicted by the first-order and higher-order shear deformation theories. Consider the plate whose results were discussed in Sec. 7.4.1 and plotted in Fig. 7-11. Numerical results based on the higher-order shear deformation theory for the nondimensional center deflection $\bar{w} = w_0(E_2h^3/p_0a^4)$ of the plate are shown in Fig. 7-12. For comparison, the center deflection $\bar{w}$ predicted by the exact three-dimensional elasticity solution, first-order shear deformation theory, and classical laminated-plate theory are also shown. It may be observed that compared with the exact elasticity solution, smaller deflections are predicted by the classical, first-order, and higher-order theories. However, errors in the predictions of the first-order and higher-order theories are much smaller than those of the classical theory, particularly when the width-to-thickness ratios are small. For very thin plates, errors in the predictions of all the theories are negligible.

Results also have been obtained for transverse shear stresses τ_{xz} and τ_{yz} for this plate. Variations of the nondimensional transverse shear stresses $\bar{\tau}_{xz} = \tau_{xz}(h/p_0a)$ and $\bar{\tau}_{yz} = \tau_{yz}(h/p_0a)$ as predicted by the first-order and higher-order theories are compared in Figs. 7-13 and 7-14. It may be noted that the higher-order theory predicts a parabolic transverse shear stress variation through the

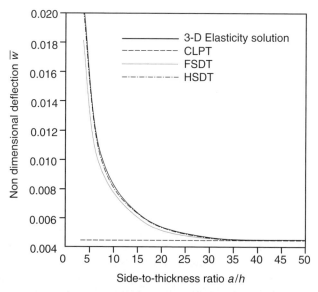

Figure 7-12. Comparison of center deflection of a symmetric cross-ply [0/90]$_s$ laminate predicted by different theories. (From Reddy [9].)

7.4 DEFORMATIONS DUE TO TRANSVERSE SHEAR **313**

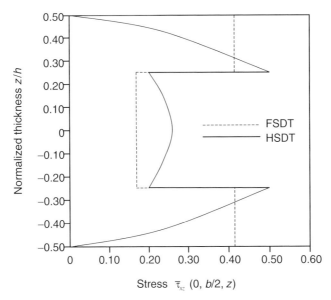

Figure 7-13. Comparison of transverse shear stress, $\bar{\tau}_{xz}$ predicted by the first-order and higher-order theories. (From Reddy [9].)

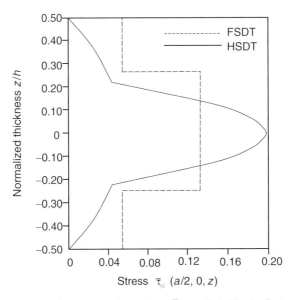

Figure 7-14. Comparison of transverse shear stress, $\bar{\tau}_{yz}$ predicted by the first-order and higher-order theories. (From Reddy [9].)

thickness of each lamina, whereas the first-order theory predicts a constant transverse shear over a lamina thickness. The higher-order theory satisfies the boundary condition that the transverse shear stresses vanish at the top and bottom surfaces of the laminate. The results of the first-order theory do not satisfy this condition.

7.5 ANALYSIS OF LAMINATED BEAMS

7.5.1 Governing Equations for Laminated Beams

A beam is a basic structural element. For composite materials, beam-type laboratory specimens are used frequently for determining their properties. Thus it is of interest to develop theories to study the behavior of laminated beams. Such beam theories are developed in this section.

Consider a laminated beam of length L, width b, and thickness h, as shown in Fig. 7-15. The laminated-beam governing equations can be obtained directly from the plate governing equations developed in Sec. 7.2 through appropriate simplifications based on the beam geometry [5,18]. Developments in this section are limited to the symmetric laminates in which $B_{ij} = 0$.

Since the beam has a high length-to-width ratio ($L/b \gg 1$), displacements and stresses are considered uniform across the width, that is, independent of the y coordinate. Therefore, some variables in the governing equations may be redefined to include the width b for simplification. The resultant force N_b, resultant moment M_b, resultant shear force R_b, and applied transverse load $p(x)$ for the beam are defined as follows:

$$N_b = bN_x$$
$$M_b = bM_x \qquad (7.84)$$
$$R_b = bR_{xz}$$
$$p(x) = bp(x, y)$$

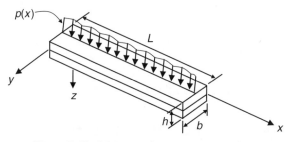

Figure 7-15. A beam under transverse loading.

In the beam analysis, moment resultants M_y and M_{xy} are considered zero. In view of all the assumptions stated here, the following beam equilibrium equations in terms of the force and moment resultants are obtained from Eqs. (7.5) and (7.7):

$$\frac{dR_b}{dx} + p(x) = 0 \tag{7.85}$$

$$\frac{dM_b}{dx} - R_b = 0 \tag{7.86}$$

Substitution of Eq. (7.86) into Eq. (7.85) gives

$$\frac{d^2M_b}{dx^2} + p(x) = 0 \tag{7.87}$$

The beam equilibrium equation in terms of transverse displacement w_0 is obtained from Eq. (7.11):

$$bD_{11}\frac{d^4w_0}{dx^4} = p(x) \tag{7.88}$$

Equations (7.85)–(7.88) are the governing equations used to predict beam behavior. Application of this beam theory is illustrated through the problems of bending, buckling, and vibration in the following subsection.

7.5.2 Application of Beam Theory

7.5.2.1 Bending Consider the laminated beam with symmetric lay-up, as shown in Fig. 7-15. The beam carries a uniformly distributed load $p(x) = p_0$. The transverse beam deflection can be obtained by integrating Eq. (7.88) as follows:

$$w_0(x) = \frac{p_0}{24bD_{11}}x^4 + C_1\frac{x^3}{6} + C_2\frac{x^2}{2} + C_3x + C_4 \tag{7.89}$$

The four constants of integration C_1, C_2, C_3, and C_4 are determined to satisfy boundary conditions at the two ends of the beam. Beam curvature (k_x), strain, and stresses can be obtained by following the procedure outlined in Chap. 6.

It is important to point out the limitation of the assumption that the transverse displacement w_0, is a function of x only and the inaccuracy produced in the solution. The constitutive equation for bending of the symmetric beam (with $M_y = M_{xy} = 0$) can be obtained from Eq. (6.21) as

$$\begin{Bmatrix} M_x \\ 0 \\ 0 \end{Bmatrix} = \begin{bmatrix} D_{11} & D_{12} & D_{16} \\ D_{12} & D_{22} & D_{26} \\ D_{16} & D_{26} & D_{66} \end{bmatrix} \begin{Bmatrix} k_x \\ k_y \\ k_{xy} \end{Bmatrix} \quad (7.90)$$

It may be written in the inverted form as

$$\begin{Bmatrix} k_x \\ k_y \\ k_{xy} \end{Bmatrix} = \begin{bmatrix} D^*_{11} & D^*_{12} & D^*_{16} \\ D^*_{12} & D^*_{22} & D^*_{26} \\ D^*_{16} & D^*_{26} & D^*_{66} \end{bmatrix} \begin{Bmatrix} M_x \\ 0 \\ 0 \end{Bmatrix} \quad (7.91)$$

where the $[D^*]$ matrix is the inverse of the $[D]$ matrix in Eq. (7.90). The $[D^*]$ matrix in Eq. (7.91) is different from that used in Eq. (6.29).

Equation (7.91) relates the plate curvatures k_y and k_{xy} to the resultant moment M_x as follows:

$$k_y = -\frac{\partial^2 w}{\partial y^2} = D^*_{12} M_x \quad (7.92)$$

$$k_{xy} = -2 \frac{\partial^2 w}{\partial x \partial y} = D^*_{16} M_x \quad (7.93)$$

Thus the deflection w cannot be independent of y. Even in homogeneous isotropic beam theory, the one-dimensional assumption is not strictly correct owing to the effect of Poisson's ratio D^*_{12} in Eq. (7.92). The effect is negligible if the length-to-width ratio is moderately large. This effect can be significant in laboratory flexural specimens made of symmetric angle-ply laminates in which the length-to-width ratio is not large. The twisting curvature induced by the D^*_{16} term in Eq. (7.93) can cause the specimen to lift off its supports at the corners. This effect should be considered for the accurate characterization of composites in the laboratory.

Example 7-4: Obtain the deflections of a beam simply supported at both ends as shown in Fig. 7-16a. Assume that the beam has a symmetric layup and carries a uniformly distributed load p_0.

Deflections of a beam in bending are given by Eq. (7.89). The constants of integration are determined from the end conditions. At the simply supported edges, deflections and bending moments are zero. Therefore, at $x = 0, L$:

7.5 ANALYSIS OF LAMINATED BEAMS 317

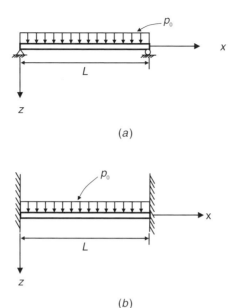

Figure 7-16. A uniformly loaded beam: (a) simply supported and (b) clamped at both ends.

$$w_0 = 0$$

$$M_b = 0 \quad \text{or} \quad \frac{d^2 w_0}{dx^2} = 0$$

Substitution of the preceding end conditions into Eq. (7.89) gives the constants of integration as

$$C_1 = -\frac{p_0 L}{2bD_{11}} \qquad C_2 = 0$$

$$C_3 = \frac{p_0 L^2}{24bD_{11}} \qquad C_4 = 0 \tag{7.94}$$

Substitution of these constants into Eq. (7.89) gives the following equation for the transverse deflection of the beam:

$$w_0(x) = \frac{p_0}{24bD_{11}} (x^4 - 2Lx^3 + L^3 x) \tag{7.95}$$

It can be shown easily that the maximum deflection occurs at $x = L/2$ and is given by

$$(w_0)_{max} = \frac{5p_0L^4}{384bD_{11}} \tag{7.96}$$

Example 7-5: Obtain the deflections of a beam clamped at both ends as shown in Fig. 7-16b. Assume that the beam has a symmetric lay-up and carries a uniformly distributed load p_0.

At a clamped edge of the beam, deflection (w_0) and slope (dw_0/dx) are zero. Therefore, the end conditions in this case at $x = 0, L$ can be written as

$$w_0 = 0 \tag{7.97}$$

$$\frac{dw_0}{dx} = 0$$

Substitution of these end conditions into Eq. (7.89) gives the following constants of integration:

$$C_1 = -\frac{p_0 L}{2bD_{11}}, \quad C_2 = \frac{p_0 L}{12bD_{11}}, \quad C_3 = 0, \quad C_4 = 0 \tag{7.98}$$

Substitution of these constants into Eq. (7.89) gives the following equation for the transverse deflection of a beam clamped at both ends:

$$w_0(x) = \frac{p_0}{24bD_{11}}(x^4 - 2Lx^3 + L^2x^2) \tag{7.99}$$

It can be shown easily that the maximum deflection occurs at $x = L/2$ and is given by

$$(w_0)_{max} = \frac{p_0 L^4}{384bD_{11}} \tag{7.100}$$

Comparison of Eq. (7.100) and Eq. (7.96) shows that the maximum beam deflection of an end-clamped beam is smaller than that of a simply supported beam.

7.5.2.2 Buckling The buckling of plates was discussed in Sec. 7.3.2. Beams also will buckle when a large compressive load is applied in their length direction. Like plates, buckling of beams cannot be predicted from the equilibrium equation [Eq. (7.88)], which does not account for the z-direction contribution of the in-plane force. An appropriate equation to predict buckling of beams may be derived in a manner similar to the derivation of Eqs. (7.34)–(7.36) for plates.

7.5 ANALYSIS OF LAMINATED BEAMS

It can be shown that the governing equation for the buckling of a simply supported beam with symmetric lay-up (Fig. 7-17), is given by

$$bD_{11}\frac{d^4w_0}{dx^4} = N_b\frac{d^2w_0}{dx^2} \tag{7.101}$$

where $N_b = bN_x$ as defined in Eq. (7.84). For the beam considered here, $N_b = -N_0$. The governing equation [Eq. (7.101)] becomes

$$bD_{11}\frac{d^4w_0}{dx^4} = -N_0\frac{d^2w_0}{dx^2} \tag{7.102}$$

For a simply supported beam, the following series will satisfy the governing equation [Eq. (7.102)] as well as the boundary conditions:

$$w_0(x) = \sum_{m=1}^{\infty} w_m \sin\left(\frac{m\pi x}{L}\right) \tag{7.103}$$

Substituion of Eq. (7.103) into Eq. (7.102) gives

$$N_0 = \frac{\pi^2 bD_{11}m^2}{L^2} \tag{7.104}$$

The critical buckling load, the smallest value of N_0, occurs when $m = 1$ and is given by

$$(N_0)_{crit} = \frac{\pi^2 bD_{11}}{L^2} \tag{7.105}$$

The corresponding buckling mode is

$$w_0(x) = \sin\left(\frac{\pi x}{L}\right) \tag{7.106}$$

7.5.2.3 Free Vibrations The equation of motion for laminated beams can be derived in a manner similar to the derivation of the equation of motion for plates [Eq. (7.56)]. It is

Figure 7-17. A simply supported beam under axial compressive load.

$$bD_{11}\frac{d^4w_0}{dx^4} + \rho A \frac{d^2w_0}{dt^2} = 0 \tag{7.107}$$

where ρ is the mass per unit volume, and A is the cross-sectional area of the beam.

Solution of Eq. (7.107) may be obtained by representing transverse displacement w_0 by the following periodic function:

$$w_0(x, t) = W(x)e^{i\omega t} \tag{7.108}$$

where ω is the frequency of vibration. Substitution of Eq. (7.108) into Eq. (7.107) gives

$$bD_{11}\frac{d^4W}{dx^4} - \rho A \omega^2 W = 0 \tag{7.109}$$

For a beam simply supported at both ends, the function $W(x)$ is given by

$$W(x) = \sum_{m=1}^{\infty} W_m \sin\left(\frac{m\pi x}{L}\right) \tag{7.110}$$

Substitution of Eq. (7.110) into Eq. (7.109) gives

$$\omega_m = \frac{m^2\pi^2}{L^2}\sqrt{\frac{bD_{11}}{\rho A}} \tag{7.111}$$

A natural frequency can be determined for any value of m from Eq. (7.111) and the corresponding mode shape from Eq. (7.110). The fundamental frequency, that is, the lowest natural frequency, occurs when $m = 1$ and is given by

$$\omega_1 = \frac{\pi^2}{L^2}\sqrt{\frac{bD_{11}}{\rho A}} \tag{7.112}$$

The mode shape corresponding to the fundamental frequency is

$$W(x) = \sin\left(\frac{\pi x}{L}\right) \tag{7.113}$$

EXERCISE PROBLEMS

7.1. A 5-mm-thick laminated plate is made of E-glass–epoxy with the properties given in Table A4-1. Calculate

(a) Stiffness matrix [Q] for the unidirectional laminae.
(b) Bending stiffness matrix [D] for the plate when lay-up is (i) [0]$_4$ and (ii) [0/90]$_s$.
(c) Center deflection and stresses in the center of the bottom layer of the plates with the preceding lay-ups. Assume that each plate is square with sides of 0.5 m, simply supported on all four edges, and carrying a uniformly distributed load of 25 Pa.

7.2. Repeat Exercise Problem 7.1 if the plates are 10 mm thick and have 1-m-long sides.

7.3. Repeat Exercise Problems 7.1 and 7.2 if the plates are made of Kevlar 49–epoxy with the properties given in Table A4-1.

7.4. Repeat Exercise Problems 7.1 and 7.2 if the plates are made of carbon–epoxy (T300/N5208) with the properties given in Table A4-1.

7.5. Repeat Exercise Problems 7.1 to 7.4 if the plates are rectangular with dimensions 1 m × 0.5 m.

7.6. Obtain two uniaxial buckling loads for each plate considered in Exercise Problems 7.1 to 7.4. Note that even for a specially orthotropic symmetric square plate, buckling load will depend on the direction of load application owing to stacking sequence.

7.7. Obtain uniaxial buckling loads for the plates considered in Exercise Problem 7.5 when the load is applied parallel to the long edge of the plate.

7.8. Obtain biaxial buckling loads for the plates considered in Exercise Problems 7.1 to 7.5 when equal (per unit length) loads are applied on all edges.

7.9. Determine the first five frequencies and corresponding mode shapes for the plates considered in Exercise Problems 7.1 to 7.5.

7.10. A rectangular specially orthotropic laminated plate of length a and width b is simply supported on all edges and is subjected to a loading $p(x, y)$ given by

$$p(x, y) = 7 \sin\left(\frac{\pi x}{a}\right) \sin\left(\frac{3\pi y}{b}\right) - 4 \sin\left(\frac{5\pi x}{a}\right) \sin\left(\frac{2\pi y}{b}\right)$$

Find the displacement function $w(x, y)$ in the Navier form [Eq. (7.30)]. (*Hint:* Note that the load is nonzero only for two combinations of m and n. Therefore, the displacement function will have two terms only.)

7.11. A specially orthotropic beam of length L, width b, and thickness h is clamped at one end ($x = 0$) and simply supported at the other ($x = L$). If it carries a uniformly distributed load p_0, obtain an expression for the deflection curve $w(x)$ for the beam in terms of bending stiffness D_{11}.

7.12. Repeat Exercise Problem 7.11 if the beam is free at the other end ($x = L$).

7.13. Determine the first three frequencies of vibration and the corresponding mode shapes for 1-m-long, 50-mm-wide, and 10-mm-thick beams of materials considered in Exercise Problems 7.1, 7.3, and 7.4. Assume that the beams are simply supported at both ends.

7.14. Determine the critical buckling loads for the beams considered in Exercise Problem 7.13. Assume that the beams are clamped at both ends.

REFERENCES

1. S. Timoshenko and S. Woinowsky-Krieger, *Theory of Plates and Shells*, 2d ed., McGraw-Hill, New York, 1959.
2. A. C. Ugural, *Stresses in Plates and Shells*, McGraw-Hill, New York, 1981.
3. J. E. Ashton and J. M. Whitney, *Theory of Laminated Plates*, Technomic, Westport, CT, 1970.
4. J. M. Whitney, *Structural Analysis of Laminated Anisotropic Plates*, Technomic, Lancaster, PA, 1987.
5. S. R. Swanson, *Introduction to Design and Analysis with Advanced Composite Materials*, Prentice-Hall, Upper Saddle River, NJ, 1997.
6. J. R. Vinson and R. L. Sierakowski, *The Behavior of Structures Composed of Composite Materials*, 2d ed., Kluwer Academic, Dordrecht, The Netherlands, 2002.
7. L. P. Kollar and G. S. Springer, *Mechanics of Composite Structures*, Cambridge University Press, New York, 2003.
8. M. E. Tuttle, *Structural Analysis of Polymeric Composite Materials*, Marcel Dekker, New York, 2004.
9. J. N. Reddy, *Mechanics of Laminated Composite Plates and Shells*, 2d ed., CRC Press, Boca Raton, FL, 2004.
10. R. M. Christensen, *Mechanics of Composite Materials*, Wiley, New York, 1979.
11. E. Reissner, "The Effect of Transverse Shear Deformation on the Bending of Elastic Plates," *J. Appl. Mech.*, **12**, 60–77 (1945).
12. R. D. Mindlin, "Influence of Rotatory Inertia and Shear on Flexural Motions of Isotropic, Elastic Plates," *J. Appl. Mech.*, **18**, 336–343 (1951).
13. P. C. Yang, C. H. Norris, and Y. Stavsky, "Elastic Wave Propagation in Heterogeneous Plate," *Int. J. Solids Struct.*, **2**, 665–684 (1966).
14. J. M. Whitney, "The Effect of Transverse Shear Deformation on the Bending of Laminated Plates," *J. Compos. Mater.*, **3**, 534–547 (1969).
15. J. N. Reddy, "A Simple Higher–Order Theory for Laminated Composite Plates," *J. Appl. Mech.*, **51**, 745–752 (1984).

16. K. Bhatia and K. Chandrashekhara, "Analytical Solutions for the Bending of Laminated Clamped Plates Using a Higher Order Theory," *J. Reinforced Plast. Compos.*, **14,** 1259–1277 (1995).
17. R. Tenneti and K. Chandrashekhara, "Large Amplitude Flexural Vibration of Laminated Plates Using a Higher Order Shear Deformation Theory," *J. Sound Vibrat.*, **176,** 279–285 (1994).
18. K. Chandrashekhara and K. M. Bangera, "Vibration of Symmetrically Laminated Clamped—Free Beam with a Mass at the Free End," *J. Sound Vibrat.*, **160,** 85–99 (1992).

8

ADVANCED TOPICS IN FIBER COMPOSITES

8.1 INTERLAMINAR STRESSES AND FREE-EDGE EFFECTS

8.1.1 Concepts of Interlaminar Stresses

The classical lamination theory (CLT) described in Chap. 6 provides a simple and direct procedure for calculating stresses and strains in a laminate. The analysis is quite adequate to predict overall behavior of the laminate. However, this analysis has limitations owing to underlying assumptions. In some cases, the assumptions may lead to inaccurate results that may be critical to predicting failure initiation. This section deals with an important area of stresses at free edges and their influence on failure initiation.

The CLT is essentially a generalized plane stress analysis and predicts in-plane stresses for given forces and moments. However, in some areas of laminates, three-dimensional stresses must be present for boundary equilibrium. This fact can be demonstrated easily through a simple example. Consider a multiply laminate (Fig. 8-1) subjected to a uniaxial load in the x direction ($N_x = N_0$). Assume the remaining in-plane loads to be zero ($N_y = N_{xy} = 0$). The midplane strains and plate curvatures can be obtained from Eq. (6.33), and then the lamina stresses can be calculated according to Eq. (6.9). Since the laminate is subjected to a uniform loading, stresses will be independent of x and y coordinates. In general, each lamina will experience nonzero in-plane stresses σ_x, σ_y, and τ_{xy} even though only a uniaxial load is applied in the x direction. Thus σ_y and τ_{xy} will be nonzero on the free boundary $y = \pm b$ (see Fig. 8-1). However, nonzero σ_y or τ_{xy} produce an unbalanced force on the boundaries $y = \pm b$ because applied forces (N_y and N_{xy}) on these boundaries are zero. It can be shown that for boundary equilibrium, out-of-plane stress components σ_z, τ_{xz}, and τ_{yz} also must be present, although CLT does not predict them.

8.1 INTERLAMINAR STRESSES AND FREE-EDGE EFFECTS

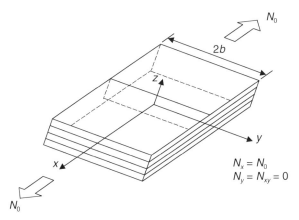

Figure 8-1. A four-ply laminate subjected to a uniaxial force in the x direction.

The stresses σ_z, τ_{xz}, and τ_{yz} are called *interlaminar stresses* and are produced primarily as a result of a mismatch of Poisson's ratios and cross-coupling coefficients of adjacent plies. They exist on the lamina surface in contact with the other lamina and within the lamina. However, their magnitude is largest at the lamina interface. It is also large at or near the free edges. The interlaminar stresses are negligible in the regions away from the free boundary. In many cases the interlaminar stresses, especially the shear stress, may be quite large near the edge and may influence failure initiation of the laminate. A large interlaminar shear strain at the laminae interface may produce matrix cracks on the free edge. These cracks start free-edge delamination and subsequent delamination growth, as shown in Fig. 8-2. This delamination failure mechanism initiated at a free edge is unique to composite laminates and is particularly significant for fatigue loading, in which damage initiation greatly influences fatigue life and fatigue strength. The methods of evaluation

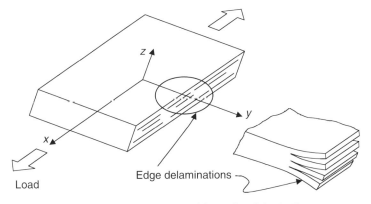

Figure 8-2. Representation of free-edge delamination.

of the interlaminar stresses and their influence on the ultimate laminate failure are discussed in this section.

8.1.2 Determination of Interlaminar Stresses

The subject of interlaminar stresses at the free edge of a finite-width angle-ply laminate has been developed by many investigators in recent years. Although some of the related studies, which later lead to important developments on the subject, were reported by Pagano [1–3] and Whitney [4,5], the first direct approach to the problem seems to have been made by Puppo and Evensen [6], who evolved an approximate formulation in which each of the anisotropic laminae of the laminate was represented by a model that contained an anisotropic plane-stress layer and an isotropic shear layer. Each anisotropic layer was assumed to carry only in-plane loads and to exist in a state of generalized plane stress, whereas the isotropic shear layers were assumed to carry the interlaminar shear stress. The interlaminar normal stress was assumed to vanish throughout the laminate.

A second solution was developed by Pipes and Pagano [7], who considered a four-ply symmetric laminate with the plies oriented only at $\pm \theta$ to the longitudinal laminate axis (Fig. 8-3). The exact equations of elasticity were derived for a uniform axial extension by assuming that the stress components are independent of x. The finite-difference method was employed to obtain numerical results for the $\pm 45°$ laminate with the following laminae properties

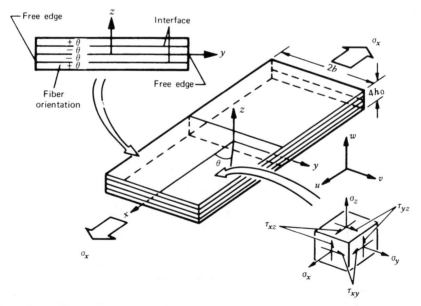

Figure 8-3. Model of a symmetric angle-ply laminate used for calculating interlaminar stresses. (From Pipes and Pagano [7].)

typical of a high-modulus graphite–epoxy system and the geometric relation $b = 8h_0$:

$$E_L = 20 \times 10^6 \text{ psi (138 GPa)} \qquad \nu_{LT} = \nu_{TL} = 0.21^*$$

$$E_T = 2.1 \times 10^6 \text{ psi (14.5 GPa)} \qquad G_{LT} = 0.85 \times 10^6 \text{ psi (5.8 GPa)}$$

The stress distributions across the width of the specimen are shown in Fig. 8-4. The stresses in the center of the cross section are the same as those predicted by the lamination theory discussed earlier. However, as the free edge is approached, σ_x decreases, τ_{xy} goes to zero, and most significantly, τ_{xz} increases from zero to infinity (a singularity exists at $y = \pm b$). The stresses σ_y, σ_z, and τ_{yz} also increase near the free edge, but their magnitudes are quite small. Numerical results on $[\pm \theta]_s$ laminates with different ply orientations (θ) also have been obtained [7]. The results show that τ_{xz} is the most significant interlaminar shear stress in these angle-ply laminates. Further, the maximum value of τ_{xz} depends on the ply orientation (θ). Maximum τ_{xz}, normalized with the highest τ_{xz} for any ply orientation (θ), is shown as a function of θ in Fig. 8-5. The stress τ_{xz} has the highest value for $\theta = 35°$ and zero for $\theta = 0°, 60°,$ and $90°$.

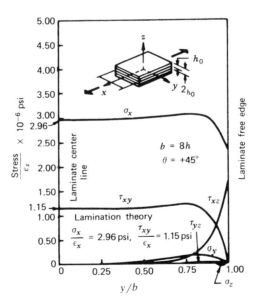

Figure 8-4. Stress variations across the width of the laminate shown in Fig. 8-3. (From Pipes and Pagano [7].)

---

*For the purpose of numerical calculations, ν_{LT} and ν_{TL} have been assumed equal by the investigators [7]. This is, however, practically inadmissible (see Chapter 5).

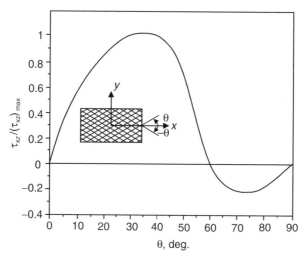

Figure 8-5. Dependence of interlaminar shear stress on ply orientation (θ) in angle-ply laminates. (From Pipes and Pagano [7].)

Through the analyses of laminates with different geometries it was established that the width of the region in which the stresses differ from those predicted by lamination theory is approximately equal to the thickness of the laminate. Thus the interlaminar stresses or the deviations from lamination theory can be regarded as an edge effect only. They are sometimes also referred to as a *boundary-layer phenomenon*. It also must be expected that the edge effects will be observed at cutouts or holes in laminates that provide internal free edges.

The theoretical results of Pipes and Pagano [7] were confirmed experimentally by Pipes and Daniels [8]. The surface displacements of the symmetric angle-ply laminate subjected to axial extension were examined by employing the Moire techniques. The experimental observations are compared with the theoretical results in Fig. 8-6. The experimental study also confirms that the interlaminar stresses can be regarded as an edge effect only because their effect is confined to a region whose width is approximately equal to the laminate thickness.

8.1.3 Effect of Stacking Sequence on Interlaminar Stresses

Pagano and Pipes demonstrated in a later study [9] that the interlaminar stresses can be influenced significantly by the laminate stacking sequence, and thus the stacking sequence may be important to a designer. Their work was motivated by observations of Foye and Baker [10] on the tensile fatigue strength of combined angle-ply, $(\pm 15/\pm 45)_s$ and $(\pm 45/\pm 15)_s$, boron–epoxy laminates. Foye and Baker reported that fatigue strength of laminates with the former stacking sequence is about 25,000 psi (175 MPa) lower than that

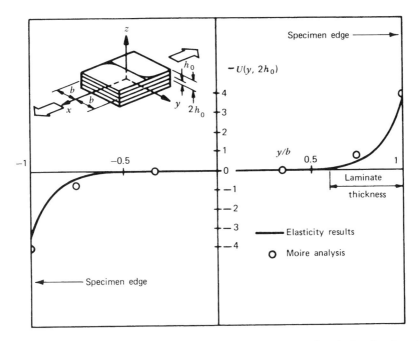

Figure 8-6. Comparison of experimentally measured and theoretically calculated surface displacements in the laminate shown in Fig. 8-3. (From Pipes and Daniel [8].)

for the latter stacking sequence. Pagano and Pipes contended that the interlaminar normal stress σ_z changes from tension to compression by changing the stacking sequence and thus accounts for the difference in strengths. The explanation seems quite reasonable in view of the fact that Foye and Baker observed delamination and stated that progressive delamination was the failure mode in fatigue. Whitney [11,12] also has suggested that the interlaminar normal stress σ_z strongly influences the delamination process during failure. During fatigue tests on graphite–epoxy laminates, he observed that a specimen that developed a tensile value of interlaminar stress showed delamination much prior to the fracture of the specimen, whereas another specimen that developed a compressive value of interlaminar stress at the free edge from change in the stacking sequence showed very little evidence of delamination even when fracture occurred. A similar dependence of both static and fatigue strengths on laminae stacking sequence also has been reported by other experimental investigators. A more elaborate discussion on the subject can be found in Whitney [12] and Pipes et al. [13].

Interlaminar stresses for two cross-ply laminates and for two $[\pm 45/0]_s$ laminates with glass and carbon fibers are shown in Figs. 8-7 and 8-8, respectively (adapted from ref. 14). In cross-ply laminates, the normal stress σ_z and shear stress τ_{yz} are more significant than the shear stress τ_{xz}, which is the more significant interlaminar stress in the angle-ply laminates (see Figs. 8-4 and 8-5). Further, in cross-ply laminates, when stacking sequence is changed

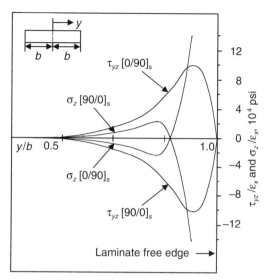

Figure 8-7. Influence of changing lay-up in cross-ply laminates on interlaminar shear stresses. (Adapted from ref. 14.)

from $[90/0]_s$ to $[0/90]_s$, interlaminar stresses change significantly. Of particular importance is the fact that σ_z is tensile on the $[0/90]_s$ laminate free edge, whereas it is compressive for $[90/0]_s$ laminate. A tensile interlaminar normal stress σ_z may promote edge delamination, as discussed in the preceding paragraphs. For $[\pm 45/0]_s$ laminates, fiber type (glass or carbon) strongly influences interlaminar stresses (see Fig. 8-8). Since interlaminar stresses are affected by various parameters, actual stress distribution should be used to assess their influence on the laminate performance.

8.1.4 Approximate Solutions for Interlaminar Stresses

Because of the importance of interlaminar stresses in the failure of laminates having free edges, many numerical techniques and approximate solutions have been developed. Isakson and Levy [15] developed a finite-element model similar to that of Pipes and Pagano [7] that incorporated nonlinear interlaminar shear response. Rybicki [16] employed a three-dimensional finite-element technique based on a complementary energy formulation in the analysis of a finite-width laminate. Pagano [17] developed an approximate method for defining the distribution of the interlaminar normal stress σ_z along the central plane of a symmetric finite-width laminate. The method takes into consideration the influence of the pertinent material and geometric parameters on the shape of the stress distribution. Pipes and Pagano [18] developed an approximate elasticity solution for the response of a multilayered symmetric angle-plied laminate under uniform axial strain. The results of the approximate

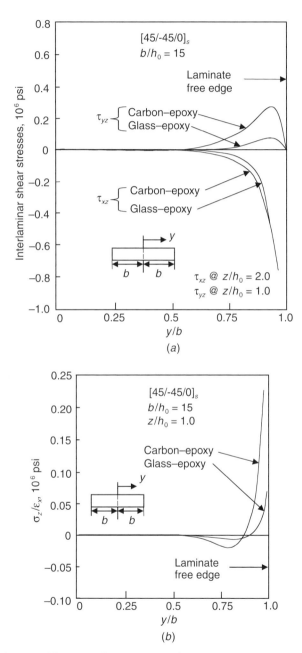

Figure 8-8. Influence of fiber type (carbon or glass) on interlaminar stresses in $[45/-45/0]_s$ laminates: (a) shear stresses τ_{xz} and τ_{yz} and (b) normal stress σ_z. (Adapted from ref. 14.)

332   ADVANCED TOPICS IN FIBER COMPOSITES

solution exhibit excellent agreement with their earlier numerical results of the exact elasticity equations [7]. Tang [19] has obtained an analytical solution for bending of a rectangular composite plate subjected to uniform transverse loading.

Whitney [12] has developed an approximate solution based on the numerical results of Pipes and Pagano [7] for an exact elasticity formulation. Whitney's approximate solution is quite simple to apply and compares reasonably well with an exact elasticity solution [7]. It is therefore discussed in the following paragraphs.

Consider a tensile specimen of length a, thickness h, and width $2b$, where $b = 4h$. A standard x, y, z coordinate system is located at the midplane of the free edge (Fig. 8-9). Equilibrium equations for this problem are (see Appendix 2)

$$\frac{\partial \sigma_x}{\partial x} + \frac{\partial \tau_{xy}}{\partial y} + \frac{\partial \tau_{xz}}{\partial z} = 0$$

$$\frac{\partial \tau_{xy}}{\partial x} + \frac{\partial \sigma_y}{\partial y} + \frac{\partial \tau_{yz}}{\partial z} = 0 \qquad (8.1)$$

$$\frac{\partial \tau_{xz}}{\partial x} + \frac{\partial \tau_{yz}}{\partial y} + \frac{\partial \sigma_z}{\partial z} = 0$$

If the origin is in the gauge section (i.e., away from the ends where the load is applied), the stresses can be assumed to be independent of x. Then the equilibrium equations take the form

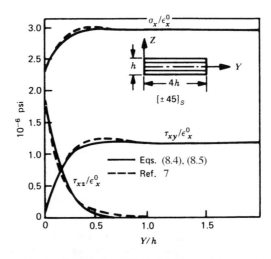

Figure 8-9. Whitney's approximate solution for interlaminar stresses compared with an elasticity solution. (From Whitney [12].)

8.1 INTERLAMINAR STRESSES AND FREE-EDGE EFFECTS

$$\frac{\partial \tau_{xy}}{\partial y} + \frac{\partial \tau_{xz}}{\partial z} = 0$$

$$\frac{\partial \sigma_y}{\partial y} + \frac{\partial \tau_{yz}}{\partial z} = 0 \qquad (8.2)$$

$$\frac{\partial \tau_{yz}}{\partial y} + \frac{\partial \sigma_z}{\partial z} = 0$$

The interlaminar stresses τ_{xz}, τ_{yz}, and σ_z can be obtained by integrating Eq. (8.2) as follows:

$$\tau_{xz} = -\int_{-h/2}^{z} \frac{\partial \tau_{xy}}{\partial y} dz$$

$$\tau_{yz} = -\int_{-h/2}^{z} \frac{\partial \sigma_y}{\partial y} dz \qquad (8.3)$$

$$\sigma_z = -\int_{-h/2}^{z} \frac{\partial \tau_{yz}}{\partial y} dz$$

Integrations of Eq. (8.3) can be carried out when variations of τ_{xy} and σ_y with y and z are known. Whitney's approach to an approximate solution involves obtaining separate functions representing variations with respect to y and z. Functions representing variations with z are obtained from classical lamination theory [Eq. (6.9)] and called $\sigma_y(z)$ and $\tau_{xy}(z)$. Functions representing variations with y are obtained by fitting curves to the numerical results of Pipes and Pagano [7]. Whitney suggested the following form of σ_y and τ_{xy} in the free-edge interval $0 \le y \le h$:

$$\sigma_y = \frac{\sigma_y(z)}{c}\left[1 - e^{-k\pi\bar{y}}\frac{k}{n}(\sin n\pi\bar{y} + \cos n\pi\bar{y})\right]$$

$$\tau_{xy} = \frac{\tau_{xy}(z)}{c}(1 - e^{-k\pi\bar{y}}\cos n\pi\bar{y}) \qquad (8.4)$$

where

$$c = [1 - (-1)^n e^{-k\pi}]$$

$$\bar{y} = \frac{y}{h}, \quad k > 0$$

Where n is a positive integer, $\sigma_y(z)$ and $\tau_{xy}(z)$ are determined from lamination theory [Eq. (6.9)]. Substituting Eq. (8.4) into Eq. (8.3) and then integrating with respect to z yields the remaining stresses:

$$\tau_{xz} = \frac{-\pi \tau_{xz}(z) e^{-ky}}{nc}(n \sin n\bar{y} + k \cos n\bar{y})$$

$$\tau_{yz} = \frac{-\pi \tau_{yz}(z)(n^2 + k^2)}{nc} e^{-k\pi\bar{y}} \sin n\pi\bar{y} \qquad (8.5)$$

$$\sigma_z = \frac{\pi^2 \sigma_z(z)(n^2 + k^2) e^{-k\bar{y}}}{nc}[n \cos n\bar{y} - k \sin n\bar{y}]$$

where $\tau_{yz}(z)$, $\sigma_y(z)$, and $\tau_{xy}(z)$ are obtained from integration of Eq. (8.3) as follows:

$$[\tau_{yz}(z), \sigma_z(z), \tau_{xz}(z)] = -\int_{-h/2}^{z}[\sigma_y(z), \tau_{yz}(z), \tau_{xy}(z)]_{,y}\, dz$$

Thus Eqs. (8.4) and (8.5) exactly satisfy the equilibrium equations as well as the free-edge boundary conditions. In addition, lamination theory is exactly recovered at $(y/h) = 1.0$. Compatibility, however, is violated. The accuracy of these approximate functions is illustrated in Fig. 8-9, where they are compared with the numerical results obtained by Pipes and Pagano [7] (also shown in Fig. 8-4). The approximate results were obtained with $n = 1$ and $k = 2$. Whitney suggested that since the character of the solution is reasonably approximated with these values of n and k, in the absence of other information, they should be used for general application.

8.1.5 Summary

In view of the preceding discussion, the following general conclusions can be drawn regarding the interlaminar stresses:

1. The interlaminar stresses are very high (sometimes singular) at the free edge of a laminate (as the edges on sides of a laminate, cutouts, holes, etc.).
2. The interlaminar normal stress σ_z has a very steep gradient near the free edge. A tensile value of σ_z at the free edge may initiate delamination and thus accelerate the failure process.
3. The stacking sequence in a laminate affects the magnitude as well as the nature of the interlaminar stresses. Thus a difference in tensile static and fatigue strengths may be observed when the stacking sequence is altered, even though the orientations of each layer do not change.

4. The interlaminar stresses can be regarded as an edge effect only because their effect is confined to a narrow region close to the edges. Predictions of the lamination theory are quite accurate in the regions away (e.g., a distance equal to the laminate thickness) from the edges.

Delamination initiation at an edge may strongly influence the performance of a laminate. It is therefore important to consider edge-delamination-suppression concepts. Since the interlaminar stresses may cause edge delamination, efforts should be made to reduce them at the design stage. Since stacking sequence affects the interlaminar stresses, the stacking sequence should be selected to minimize the interlaminar stresses. As a general rule, a stacking sequence that produces lower values of D_{16} and D_{26} without affecting other stiffness matrix elements should be selected (see Example 6-4). Free-edge delamination may be actively suppressed by strengthening edges or by modifying edges to reduce the severity of interlaminar stresses. Examples of delamination-suppression concepts are shown in Fig. 8-10. An overview of the effectiveness of these delamination-suppression concepts is given by Jones [20].

8.2 FRACTURE MECHANICS OF FIBER COMPOSITES

8.2.1 Introduction

8.2.1.1 Microscopic Failure Initiation In Chap. 5, composite materials were treated as homogeneous anisotropic bodies. A continuum analysis of anisotropic materials has led to the formulation of stress–strain relations for composite materials. In a continuum approach, the influence of a local het-

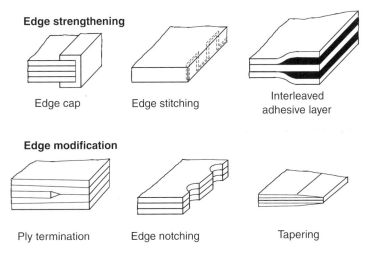

Figure 8-10. Free-edge delamination-suppression concepts.

erogeneity and microscopic flaws is neglected. Quantitatively, microscopic flaws are imperfections whose sizes are small compared with the characteristic dimensions of the body. The imperfections in composites may include voids, fiber ends, delaminations, and irregularities in fiber packing. The predictions of continuum analysis regarding deformational response of composites have proved to be accurate because the deformation is an averaged property and is not influenced by local heterogeneity. However, the applicability of a continuum approach cannot be taken for granted in a process such as failure that is initiated by localized conditions.

The discussion in Chap. 5 on strength theories or failure criteria did not consider the mechanisms responsible for the failure process. It has been assumed implicitly that the material is free from flaws. Microscopic flaw nucleation is assumed to take place over the major portion of the life of the sample, and on nucleation into a macroscopic crack, fracture is assumed to occur instantaneously. The strengths are evaluated by conducting tests on geometrically smooth specimens in which no sharp stress gradients are present. Thus the failure criteria, such as Eqs. (5.102)–(5.109), provide a rational estimate of the general structural integrity of a composite. However, experience with fracture of metals suggests that the occurrence of failure in the presence of sharp stress gradients (or flaws) is different from that in a relatively slow-varying stress field. Moreover, in practical structures, macroscopic cracks, which produce sharp stress gradients, can accrue during the various manufacturing processes as well as in service. The study of quasistatic crack growth therefore can provide useful information on the flaw sensitivity of the material and for establishing inspection requirements to determine the criticality of cracks. Fracture mechanics is the discipline concerned with failure by crack initiation and propagation. Its relevance and applicability to composite materials are discussed in this section.

8.2.1.2 Fracture Process in Composites The fracture mechanics discipline essentially has been developed to predict fracture of homogeneous materials by crack initiation and propagation. However, the fracture process in composites is significantly different from and more complex than that in homogenous materials such as metals and polymers. Therefore, to better understand and appreciate the applicability of fracture mechanics concepts to composite materials, it is necessary to understand the fracture process and crack propagation in composites.

It can be assumed that failure in a fiber composite, just as in metals, emanates from small, inherent defects in the material. These defects may be broken fibers, flaws in the matrix, or debonded interface. After initiation, the failure propagation or fracture process can be described using the simple model shown in Fig. 8-11. The model shows several possible local failure events occurring during the fracture of a fiber composite. At some distance ahead of the crack, the fibers are intact. In the high-stress region near the tip, they are broken, although not necessarily along the crack plane. Immediately

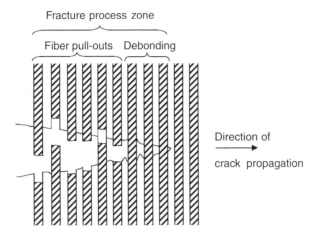

Figure 8-11. A model of fracture-process zone shows local failure events near the crack tip.

behind the crack tip, fibers pull out of the matrix. In some composites, the stress near the crack tip could cause the fibers to debond from the matrix before they break. It is also possible for a fiber to be left intact as the crack propagates. When brittle fibers are well bonded to a ductile matrix, the fibers tend to snap ahead of the crack tip, leaving bridges of matrix material that neck down and fracture in a completely ductile manner. In addition to these local failure mechanisms, on reaching the interface of two laminae in a laminated composite, a crack can split and propagate along the interface, thus producing a delamination crack.

Different failure mechanisms involved in fracture propagation as just discussed account for the total energy absorbed in the fracture. However, some mechanisms play a dominant role in one system of matrix material and fibers, whereas a different set of mechanisms may be dominant in another system of fibers and matrix material. No single mechanism can account for the observed toughness of composites. A more detailed discussion of each of the failure events and the factors influencing the energy associated with them is given in the section on impact in Chap. 9.

The most important difference in the mechanism of crack propagation in isotropic materials and composites arises from the fact that isotropic materials such as metals or polymers often exhibit self-similar crack growth; that is, the crack growth occurs by a simple enlargement of the initial crack without branching or directional changes. In composites, self-similar crack growth is not likely to occur even for unidirectional or symmetric angle-ply laminates. Self-similar crack growth may be expected in unidirectional composites only when the initial crack is parallel to a principal material direction. In laminates, the direction of crack growth generally is expected to be different in plies with different fiber orientations. Randomly oriented fiber composites are con-

338 ADVANCED TOPICS IN FIBER COMPOSITES

sidered to be macroscopically isotropic, but even in these types of composites self-similar crack growth does not take place because of local heterogeneity ahead of the crack. Therefore, one really cannot define a unique crack length for composite materials, and it may be more meaningful to consider a damage zone ahead of the crack. A typical damage zone formed ahead of the crack tip in a short-fiber composite is shown in Fig. 8-12. It is possible to control the general direction of growth of damage zones in various tests, but it is not a planar fracture of the type observed in isotropic materials. Within the damage zone, there are such energy-absorbing processes as debonding of fibers from the matrix and fiber pullout, in addition to the fracture of fibers and the matrix. It should be noted that this damage zone is comparable to the zone of plastic deformation accompanying crack growth in metals or polymers.

The preceding discussion on crack initiation, damage propagation, and the general fracture process in composites always should be considered while considering applicability of fracture mechanics concepts to composite materials.

8.2.2 Fracture Mechanics Concepts and Measures of Fracture Toughness

Fracture mechanics of materials has been studied using different concepts that hypothesize crack initiation and propagation in materials. Mathematical tech-

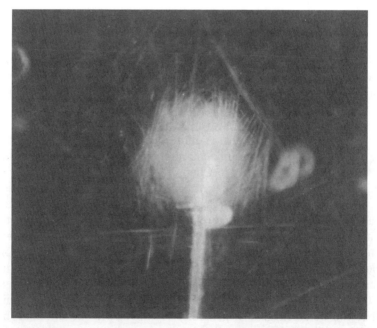

Figure 8-12. Photograph of a damage zone formed ahead of the crack tip in a short-fiber composite.

niques have been developed to predict fracture of materials according to these concepts. Three techniques that are widely used to predict fracture of metals and other homogeneous materials are discussed in this subsection, including their applicability to composites. These techniques are the strain-energy release rate, stress-intensity factor, and J-integral. Each technique leads to a measurable material property that controls crack initiation and propagation. These properties generally are accepted as the measures of fracture toughness of a material. Experimental determination of fracture toughness is discussed in Chap. 10.

8.2.2.1 Strain-Energy Release Rate (G)

The stability of a flaw or crack in a continuum can be examined by considering the energy balance. Griffith [19] was the first to provide an expression for crack instability, that is, failure by uncontrolled crack growth. His reasoning was based on the hypothesis that the free energy of a cracked body and applied forces should not increase during crack extension. This criterion can be used to derive conditions for formation of new crack surfaces, as is done in the following paragraphs.

Consider two states of a solid body subjected to external forces, as shown in Fig. 8-13. In state A, there is no macrocrack in the body. Let the total energy of the body in state A be U_A. State B is obtained by introducing a crack C prior to loading. External forces are now applied such that the deformations under the forces are the same as in state A. One may imagine that state B is obtained from state A by introducing a crack and simultaneously removing some of the load to keep the deformations under the loads unchanged so that no work is done on the body by the external forces as it goes from state A to state B. The difference in energies of states A and B appears only in surface energy and strain energy. Because of the formation of the crack, the surface energy increases by an amount S, and because of the presence of the crack, the strain energy stored in the body decreases by an amount U. This is so because smaller loads are required to cause the same deformations in the body. Thus U is the strain energy released owing to the intro-

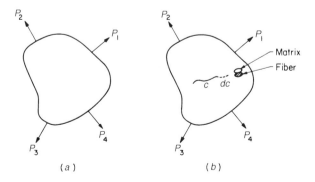

Figure 8-13. A solid body (a) without a crack and (b) with a crack subjected to external forces.

duction of a crack in the body. Summing these, the total energy of the body in state B is

$$U_B = U_A + S - U \tag{8.6}$$

When a body with a crack is loaded, crack propagation (growth, enlargement, or extension) begins at a specified load. Depending on the initial crack length, the crack propagation may be stable or unstable. When crack propagation occurs under increasing load, it is stable, and it causes the total energy U_B to increase. Thus, for stable crack propagation,

$$\frac{\partial U_B}{\partial c} > 0 \tag{8.7}$$

Substitution of Eq. (8.6) into Eq. (8.7) gives

$$\frac{\partial U}{\partial c} < \frac{\partial S}{\partial c} \tag{8.8}$$

During stable crack growth, the strain-energy release rate is less than the rate of increase in surface energy. Therefore, additional energy must be provided to the body for crack propagation by increasing external loads. Stable crack growth can be arrested by keeping the loads constant or decreasing them.

Crack propagation is unstable when it causes the total energy U_B to decrease or remain constant. That is

$$\frac{\partial U_B}{\partial c} \leq 0 \tag{8.9}$$

Substitution of Eq. (8.6) into Eq. (8.9) gives

$$\frac{\partial U}{\partial c} \geq \frac{\partial S}{\partial c} \tag{8.10}$$

Thus, for crack instability, the strain-energy release rate should be greater than the rate of increase in surface energy.

When a body with a small crack is loaded, crack propagation may occur in two stages. Initial crack propagation may be stable; that is, it may occur at increasing load. However, when the crack becomes sufficiently long, unstable crack growth occurs to cause fracture. This instability commences when the strain-energy release rate is equal to the rate of increase in surface energy, as indicated in Eq. (8.10). For ideal homogeneous isotropic materials, the surface energy is a physical property. Consequently, unstable crack propagation also will occur at a constant value of strain-energy release rate independent of position, direction, and initial length of crack. This value, called the *critical strain-energy release rate*, is often used interchangeably to denote fracture toughness of the material. The strain-energy release rate may be evaluated theoretically or experimentally using a mechanics approach.

For real solids, *surface energy* is not a good choice of words because it has been found that the total energy required to create fracture surfaces is much greater than the theoretical surface energy. The additional work required to create fracture surfaces in homogeneous isotropic materials results from plastic deformation or other irreversible deformation of the crack tip. Thus *fracture-surface work* is considered a more appropriate term. Further, in composite materials, crack extension is complex and may involve all or any of the matrix, fibers, or fiber–matrix interface. Therefore, the surface-energy term in Eq. (8.10) should be a multiparameter function consisting of the properties of fibers, matrix, and interface.

8.2.2.2 Stress-Intensity Factor (K)
Irwin [22] and others have shown that the strain-energy release rate can be correlated with the stress distribution in the neighborhood of the crack tip. Consider a plate containing a crack of length $2c$ and subjected to an arbitrary plane loading that is resolvable in σ and τ components of stress away from the crack, as shown in Fig. 8-14. The classical theory of elasticity can be employed to obtain the following stress distribution near the crack tip ($r \ll c$) in a homogeneous isotropic plate.

$$\sigma_x = \sigma \frac{\sqrt{c}}{\sqrt{2r}} \cos \frac{\theta}{2} \left(1 - \sin \frac{\theta}{2} \sin \frac{3\theta}{2}\right) - \tau \frac{\sqrt{c}}{\sqrt{2r}} \sin \frac{\theta}{2} \left(2 + \cos \frac{\theta}{2} \cos \frac{3\theta}{2}\right)$$

$$\sigma_y = \sigma \frac{\sqrt{c}}{\sqrt{2r}} \cos \frac{\theta}{2} \left(1 + \sin \frac{\theta}{2} \sin \frac{3\theta}{2}\right) + \tau \frac{\sqrt{c}}{\sqrt{2r}} \sin \frac{\theta}{2} \cos \frac{\theta}{2} \cos \frac{3\theta}{2} \quad (8.11)$$

$$\tau_{xy} = \sigma \frac{\sqrt{c}}{\sqrt{2r}} \sin \frac{\theta}{2} \cos \frac{\theta}{2} \cos \frac{3\theta}{2} + \tau \frac{\sqrt{c}}{\sqrt{2r}} \cos \frac{\theta}{2} \left(1 - \sin \frac{\theta}{2} \sin \frac{3\theta}{2}\right)$$

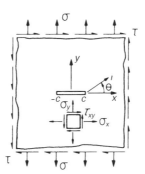

Figure 8-14. Definition of loading and coordinate system for stress analysis of a plate with a crack.

The preceding stress distributions indicate that (1) the stress singularity for all components is of the order $1/\sqrt{r}$ and (2) the stress distribution depends only on r and θ. The influence of the type of specimen and loading, as well as the magnitude of loading, can be considered through what are defined as the stress-intensity factors (not stress-concentration factor). The two stress-intensity factors k_1 and k_2 are

$$k_1 = \sigma \sqrt{c}$$
$$k_2 = \tau \sqrt{c}$$
(8.12)

The stress-intensity factor k_1 is symmetric and is associated with the opening mode of crack extension (Fig. 8-15a), whereas the skew-symmetric k_2 is associated with the shear mode (see Fig. 8-15b). These stress-intensity factors in general depend on the applied loads, the geometry of the body, and the crack length. It has been established [23] that regardless of the nature of the plane loads, the stress distribution around the crack tip always can be separated into the symmetric and skew-symmetric components even for anisotropic materials, and they differ only in magnitude according to the stress-intensity factors. Irwin has shown that if the direction of crack extension is colinear with the plane of the crack, the strain-energy release rate G is related to the stress-intensity factor k_1 by the following expression:

$$G = \frac{\pi k_1^2}{E}$$
(8.13)

Equation (8.13) is valid for the opening mode of crack extension under a plane-stress state. Similar relations for other crack-extension modes and plane strain also can be obtained. Thus Eq. (8.13) provides a means for a theoretical prediction of the strain-energy release rate in an isotropic material.

The stress analysis ahead of a crack tip in a composite material is extremely difficult. The local heterogeneity of the material prohibits obtaining a close-

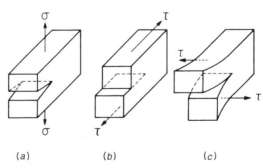

Figure 8-15. Modes of crack extension: (a) opening mode (mode I), (b) shear mode (mode II), and (c) antiplane strain or tearing mode (mode III).

bound solution. However, a close-bound solution is possible to obtain by assuming composites to be homogeneous anisotropic materials. With this assumption of homogeneity, the following stress distribution near the crack tip can be obtained [23] by following Lekhnitskii's formulation [24]:

$$\sigma_x = \sigma \frac{\sqrt{c}}{\sqrt{2r}} Re\left\{ \frac{S_1 S_2}{S_1 - S_2} \left(\frac{S_2}{\psi_2^{1/2}} - \frac{S_1}{\psi_1^{1/2}} \right) \right\}$$
$$+ \tau \frac{\sqrt{c}}{\sqrt{2r}} Re\left\{ \frac{1}{S_1 - S_2} \left(\frac{S_2^2}{\psi_2^{1/2}} - \frac{S_1^2}{\psi_1^{1/2}} \right) \right\}$$
$$\sigma_y = \sigma \frac{\sqrt{c}}{\sqrt{2r}} Re\left\{ \frac{1}{S_1 - S_2} \left(\frac{S_1}{\psi_2^{1/2}} - \frac{S_2}{\psi_1^{1/2}} \right) \right\} \quad (8.14)$$
$$+ \tau \frac{\sqrt{c}}{\sqrt{2r}} Re\left\{ \frac{1}{S_1 - S_2} \left(\frac{1}{\psi_2^{1/2}} - \frac{1}{\psi_1^{1/2}} \right) \right\}$$
$$\tau_{xy} = \sigma \frac{\sqrt{c}}{\sqrt{2r}} Re\left\{ \frac{S_1 S_2}{S_1 - S_2} \left(\frac{1}{\psi_1^{1/2}} - \frac{1}{\psi_2^{1/2}} \right) \right\}$$
$$+ \tau \frac{\sqrt{c}}{\sqrt{2r}} Re\left\{ \frac{1}{S_1 - S_2} \left(\frac{S_1}{\psi_1^{1/2}} - \frac{S_2}{\psi_2^{1/2}} \right) \right\}$$

where

$$\psi_1 = \cos\theta + S_1 \sin\theta$$

$$\psi_2 = \cos\theta + S_2 \sin\theta$$

and S_1 and S_2 are the roots of the characteristic equation that is obtained if the equations of stress equilibrium and strain compatibility are represented in terms of Airy's stress function. In general, the roots are complex and can be written in the form

$$S_1 = \alpha_1 + i\beta_1 \quad (8.15)$$
$$S_2 = \alpha_2 + i\beta_2$$

It should be noted from Eq. (8.14) that the stress distribution is controlled not only by the parameters $\sigma\sqrt{c}$ and $\tau\sqrt{c}$ but also by the roots S_1 and S_2. The roots S_1 and S_2 are functions of the elastic constants of the material and orientation of the crack with respect to the principal planes of elastic symmetry. It may be pointed out that the stress components resulting from normal and shear loads can be separated just as in the case of isotropic materials. But, for an arbitrary orientation of the crack, there exists a coupling between normal stress and shear strain and vice versa, and thus normal load produces displacements associated with the crack-opening mode, as well as the forward shear mode of crack extension. Similarly, the shear load also produces both types of displacement. Therefore, the stress-intensity factors defined by Eq. (8.12) cannot be related to the strain-energy release rate through Eq. (8.13). Thus it is questionable whether the stress-intensity factors defined in this manner can be used as material property parameters. The question can only be resolved by experimental investigations. However, for specific crack orientations, the crack-extension kinematics may decouple so that the stress-intensity factors [Eq. (8.12)] for anisotropic materials also may become as meaningful as those in isotropic materials. Such is indeed the case when the direction of the crack coincides with one of the material symmetry axes. Under these conditions $Q_{16} = Q_{26} = 0$, and the roots S_1 and S_2 fall into one of the following categories:

$$\begin{aligned} \alpha_1 = \alpha_2 = 0 & \quad \beta_1 \neq \beta_2 \\ \alpha_1 = \alpha_2 = 0 & \quad \beta_1 = \beta_2 \\ \alpha_1 = -\alpha_2 & \quad \beta_1 = \beta_2 \end{aligned} \qquad (8.16)$$

In each of these cases, the normal load produces displacements associated only with the crack-opening mode, and the shear force produces displacements associated only with the forward shear mode of crack extension. Thus in these cases the crack-tip stress distribution may be examined without further concern for the crack-tip displacement. Using the crack-tip stress analysis, Wu [25] has formulated a phenomenological failure criterion and established that crack propagation can be characterized by failure within a critical volume. He showed that the general multidimensional crack problem can be predicted by incorporation of the failure criterion into the crack-tip stress analysis. However, the formulation is quite involved and, as such, beyond the scope of this text. Besides Wu's approach, the problem of predicting the fracture stress of a composite material containing macroscopic defects has been studied by many investigators using different approaches. Some of them will be discussed in a later section.

As indicated earlier, the strain-energy release rate G and stress-intensity factor K can be calculated using linear elastic stress analysis. Application of G or K as a failure criterion therefore is limited to a class of problems, those of cracked bodies with small-scale yielding where the crack-tip plastic region is at least an order of magnitude smaller than the physical dimensions of the component [26]. It is also desirable to have a failure criterion that could predict fracture in structures in cases of both small- and large-scale plasticity. One such parameter, namely, the J-integral, is defined and discussed in the next section.

8.2.2.3 J-Integral

The J-integral is an energy-line integral defined for two-dimensional problems and is given by [27]

$$J = \int_\Gamma W\, dy - T\left(\frac{\partial u}{\partial x}\right) ds \qquad (8.17)$$

where Γ is any contour surrounding the crack tip, as shown in Fig. 8-16, W is the strain-energy density given by

$$W = W(\epsilon_{mn}) = \int_0^{\epsilon_{mn}} \sigma_{ij}\, d\epsilon_{ij} \qquad (8.18)$$

T is the traction vector defined by the outward normal n along Γ, $T_i = \sigma_{ij} n_j$, u is the displacement vector, and s is the arc length along Γ.

It has been shown [28–30] that the plastic stress and strain singularities at the crack tip are related to the J-integral. Thus the J-integral is a parameter that can be used as a failure criterion for the case of large-scale plasticity at the crack tip.

Rice [27] has proved the path independence of the J-integral. Therefore, the J-integral can be calculated accurately by using an integration path somewhat removed from the crack tip. Also, an experimental evaluation of the J-integral can be accomplished quite easily by considering the load-deflection curves of identical specimens with varying crack lengths.

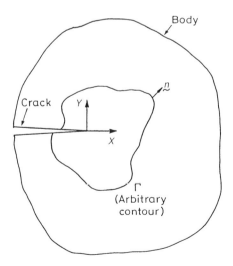

Figure 8-16. An arbitrary line contour for J-integral.

Begley and Landes [30–32] demonstrated its applicability as a fracture criterion to metals for the case of large-scale plasticity at the crack tip. Through experiments on an intermediate-strength rotor steel, they showed that the J-integral at failure for fully plastic behavior is equal to the linear elastic value of strain-energy release rate at failure for extremely large specimens. Thus the J-integral approach eliminates the necessity of testing very large specimens. For composites, it has an additional advantage in that it does not require measurement of instantaneous crack length, which is difficult to define or measure owing to the complexities of crack propagation.

8.2.3 Fracture Toughness of Composite Laminates

The fracture behavior of composite laminates has been investigated by a very large number of researchers over the years [33–55]. These studies focus on subjects such as initiation and growth of crack-tip damage, critical-damage-zone size, notch sensitivity, fracture toughness, failure modes on the micro and macro scales, and crack-arrest mechanisms using various theoretical and experimental techniques. There are several reasons for employing so many different techniques to the study of fracture mechanics in composites. First, different composite systems exhibit different failure modes and damage mechanisms and may require correspondingly different analytical tools and experimental techniques. Second, there is no consensus yet regarding the proper set of failure criteria. A number of analytical techniques, ranging from comprehensive numerical methods to simplified semiempirical fracture models, have been developed to suit different failure models and the corresponding complexity of failure processes in composite laminates. This situation is undesirable from the viewpoint of developing uniform design procedures against catastrophic failure of composite laminates, which is essential for application of composites to load-carrying structures. However, the situation seems inevitable owing to a large number of variables involved that may influence fracture behavior (toughness and notch sensitivity) of composites. These variables include intrinsic variables such as constituent properties, fiber volume fraction, fiber–matrix interface, laminate configuration, stacking sequence, and fabrication procedures and extrinsic variables such as specimen geometry, test temperature, moisture content, loading function, rate, and history.

A large amount of experimental data is available on the fracture toughness and notch sensitivity of composite laminates. Experimental fracture-toughness values of some composites are given in Table 8-1 along with the values for structural metals. However, because of the large number of variables involved, fracture-toughness test results for one type of laminate may not be useful in the design of other types of laminate. Moreover, despite the recognized importance of these variables, in most studies key information such as fiber volume fraction, constituent properties, fabrication procedures, and environmental test conditions are not reported. Consequently, unnotched and notched strength data may vary significantly among different publications even for

Table 8-1 Experimental fracture toughness values of some composites and structural metals

Material and Construction of Composite	K_{1c} (MPa·m$^{1/2}$)	Tensile Strength (MPa)
Graphite–Epoxy		
$[0/\pm 45]_S$	32.1–36.9	540–660
Quasi-isotropic	22.4–55.7	454–609
Cross-ply	42.8–54.1	637–763
Graphite–Polyimide		
$[0/45/90/-45]_{2S}$	37.0–40.5	423
Boron–Aluminum		
Unidirectional	55.7–107.0	1014–2012
Quasi-isotropic	28.2–34.9	348–409
Cross-ply	34.4–41.5	678–755
Glass–Epoxy		
Random short fibers	18.9–28.5	
2024-T6 aluminum	80.2	441
7075-T6 aluminum	58	496
Ti-6Al-4V, condition A	215	896
HP9Ni-4Co-0.30C (steel)	235	1516

seemingly identical laminate configurations and material systems. For the purpose of analyzing the complex fracture behavior of notched composite laminates and for generalizing the results obtained for specific composite laminates to other types of laminate, several simplified fracture models have been proposed in recent years [46–55]. The model parameters are expected to be related to and used as measures of the notch sensitivity of composite laminates. A good understanding of fracture models and their correlation with the experimental results will be of great help to a designer in assessing notch sensitivity of composite laminates.

A highly detailed review of several of the fracture models has been published by Awerbuch and Madhukar [56]. They have critically reviewed commonly used semiempirical fracture models that are easy to operate as predictive tools by the designers. Experimental results also have been reviewed and compared with all fracture models. The various parameters associated with the fracture models were determined for all the experimental data sets reviewed. The fracture-model parameters were correlated with the notch sensitivity of composite laminates, and their applicability as a measure of notch sensitivity has been evaluated. Even though the review was published in 1985, the major conclusions are still valid. These are

1. Most of the experimental studies on notched strength of composite laminates are limited to two types of notch geometry: circular holes and straight cracks and subjected to uniaxial quasi-static loading.
2. Practically all laminates are highly notch-sensitive. The strength drops sharply with the introduction of the smallest discontinuity. For many cases, the notch reduces strength by as much as 50% for notch length-width ratios of 0.2–0.3.
3. Notch sensitivity of composite laminates may be affected by a variety of intrinsic and extrinsic variables. However, a comprehensive evaluation of the effects of these variables on the notch sensitivity of composite laminates is still lacking.
4. Also, any comparison among the notch sensitivities of different laminates obtained from different sources is of questionable value because very few publications report fiber volume fraction, constituent (fiber) properties, fabrication conditions, environmental test conditions, and similar, all of which may affect the results.
5. Very good agreement between all fracture models reviewed and all experimental notch-strength data can be established provided that the parameters are determined properly.
6. These parameters strongly depend on laminate configuration and material system, as well as on a variety of intrinsic and extrinsic variables.
7. Among the various parameters associated with the fracture models, only the characteristic dimensions of the Waddoups–Eisenmann–Kaminski model [46] and the Whitney–Nuismer model [47,48] can be related to notch sensitivity such that the larger they are, the less notch sensitive the subject laminate is.
8. It seems, therefore, that the characteristic dimensions can serve as a relative measure of notch sensitivity of composite laminates. This relationship remains qualitative, however, until there is a more precise identification of the effects of all the variables affecting notch sensitivity.
9. Since strength does drop sharply with the introduction of a small notch in almost all laminates, the "average stress" criterion of the Whitney–Nuismer fracture model does fit the data better in the small-notch-size range.
10. None of the parameters associated with the other fracture models, including the critical stress-intensity factors, could be related to the notch sensitivity.
11. Although a large body of notched strength data is available in the literature, only a few definite conclusions can be made regarding the effect of the many variables affecting notch sensitivity because of varying objectives in different investigations.

From the preceding discussion and conclusions it is clear that a comprehensive understanding of fracture toughness and notch sensitivity of composite laminates, including the effect of different variables, is still lacking. Further, the large body of notch-strength data available in the literature from different investigations can be most appreciated and best utilized in design processes through a fracture model. Very good agreement between all the fracture models and experimental notch-strength data establishes that their predictions for notched strength are, for all practical purposes, identical. However, the fracture models proposed by Whitney and Nuismer are probably the simplest to operate, and the parameters associated with these fracture models can be related to the notch sensitivity and fracture toughness of composite laminates. The Whitney–Nuismer fracture models therefore are discussed next.

8.2.4 Whitney–Nuismer Failure Criteria for Notched Composites

The investigations of Whitney and Nuismer [47,48] were motivated by what is known as the *hole-size effect* [46]; that is, for tension specimens containing various-sized holes, larger holes cause greater strength reductions than do smaller holes. One of the explanations of the hole-size effect is based on the normal stress distribution ahead of a hole. Although the stress-concentration factor is independent of hole size, the normal stress perturbation from a uniform stress state is considerably more concentrated near the hole boundary in the case of a smaller hole. Intuitively, therefore, one might expect the plate containing a smaller hole to be stronger because greater opportunity exists in this case to redistribute high stress. By considering the stress distribution ahead of a circular hole, Whitney and Nuismer developed two criteria for the strength of notched composite materials. The first criterion is based on the stress at a point a fixed distance away from the notch and may be referred to as the *point-stress criterion*. The second criterion is based on the average stress over some fixed distance ahead of the hole and is referred to as the *average-stress criterion*. The Whitney–Nuismer failure criteria are conceptually similar to Wu's criterion, in which failure of the composite is assumed to be governed by the failure within a critical volume ahead of the notch or a macroscopic defect. Wu's criterion does predict a multidirectional crack growth, which is not the case with the Whitney–Nuismer criteria. However, the latter are much simpler and thus readily adaptable in design procedures. The two criteria can be extended to the case of a sharp crack in place of smooth holes. The criteria are developed in the following paragraphs.

Consider an infinite orthotropic plate with a hole of radius R and subjected to a uniform stress σ parallel to the y axis at infinity (Fig. 8-17). If the axes x and y are assumed normal to the planes of elastic symmetry, the normal stress σ_y along the x axis in front of the hole can be approximated by

$$\sigma_y(x,0) = \frac{\sigma}{2}\left\{2 + \left(\frac{R}{x}\right)^2 + 3\left(\frac{R}{x}\right)^4 - (k_T - 3)\left[5\left(\frac{R}{x}\right)^6 - 7\left(\frac{R}{x}\right)^8\right]\right\} \quad (8.19)$$

where k_T is the orthotropic stress-concentration factor for an infinite-width plate and can be determined from the following relationship [57]:

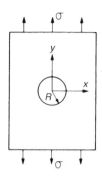

Figure 8-17. A plate with a circular hole of radius R.

$$k_T = 1 + \sqrt{\frac{2}{A_{22}}\left(\sqrt{A_{11}A_{22}} - A_{12} + \frac{A_{11}A_{22} - A_{12}^2}{2A_{66}}\right)} \tag{8.20}$$

where A_{ij} are the in-plane laminate stiffnesses as determined from laminated-plate theory discussed in Chap. 6. For a unidirectional composite or lamina, A_{ij} in Eq. (8.20) may be replaced by Q_{ij} introduced in Chap. 5. The subscript 1 denotes the direction parallel to the applied stress at infinity. In terms of effective elastic moduli, Eq. (8.20) becomes

$$k_T = 1 + \sqrt{2\left(\sqrt{\frac{E_{11}}{E_{22}}} + \nu_{12}\right) + \frac{E_{11}}{G_{12}}} \tag{8.21}$$

For an orthotropic composite loaded in the longitudinal direction, Eq. (8.21) becomes

$$k_T = 1 + \sqrt{2\left(\sqrt{\frac{E_L}{E_T}} + \nu_{LT}\right) + \frac{E_L}{G_{LT}}} \tag{8.22}$$

The point-stress criterion assumes failure to occur when σ_y at some fixed distance d_0 ahead of the hole first reaches the unnotched tensile strength of the material σ_0, that is, when

$$\sigma_y(R + d_0, 0) = \sigma_0 \tag{8.23}$$

Substitution of Eq. (8.19) into Eq. (8.23) yields the ratio of notched to unnotched strength:

8.2 FRACTURE MECHANICS OF FIBER COMPOSITES

$$\frac{\sigma_N}{\sigma_0} = \frac{2}{2 + p_1^2 + 3p_1^4 - (k_T - 3)(5p_1^6 - 7p_1^8)} \quad (8.24)$$

where

$$p_1 = \frac{R}{R + d_0} \quad (8.25)$$

and σ_N equals the applied stress σ at failure or σ_N is the notched strength of the infinite-width laminate. It may be noted that for very large holes, $p_1 \to 1$, and the classical stress-concentration result, $(\sigma_N/\sigma_0) = (1/k_T)$, is recovered. On the other hand, for vanishingly small hole sizes, $p_1 \to 0$, and the ratio $(\sigma_N/\sigma_0) \to 1$, as would be expected.

The average-stress criterion assumes failure to occur when the average value of σ_y over some fixed distance a_0 ahead of the hole first reaches the unnotched tensile strength of the material, that is, when

$$\frac{1}{a_0} \int_R^{R+a_0} \sigma_y(x, 0) \, dx = \sigma_0 \quad (8.26)$$

Substitution of Eq. (8.19) into Eq. (8.26) yields the ratio of notched to unnotched strength:

$$\frac{\sigma_N}{\sigma_0} = \frac{2(1 - p_2)}{2 - p_2^2 - p_2^4 + (k_T - 3)(p_2^6 - p_2^8)} \quad (8.27)$$

where

$$p_2 = \frac{R}{R + a_0} \quad (8.28)$$

and σ_N is again the notched strength of the infinite-width laminate. It can be seen easily that the expected limits of σ_N/σ_0 are again recovered for very small and very large holes.

The two failure criteria [Eqs. (8.23) and (8.26)] can be applied to crack problems in a similar manner. Consider an infinite plate with a crack of length $2c$ and subjected to a uniform stress σ parallel to the y axis at infinity (Fig. 8-18). The exact elasticity solution for the normal stress ahead of crack when the axes x and y are normal to the planes of elastic symmetry is given by Lekhnitskii [57]:

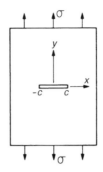

Figure 8-18. A plate with a crack of length 2c.

$$\sigma_y(x, 0) = \frac{\sigma x}{\sqrt{x^2 - c^2}} = \frac{k_1 x}{\sqrt{\pi c(x^2 - c^2)}} \qquad x > c \qquad (8.29)$$

where k_1 is the stress-intensity factor given by $k_1 = \sigma\sqrt{\pi c}$ and not by Eq. (8.12). Substitution of Eq. (8.29) into each of the failure criteria [Eqs. (8.23) and (8.26)] yields, respectively,

$$\frac{\sigma_N}{\sigma_0} = \sqrt{1 - p_3^2} \qquad (8.30)$$

where

$$p_3 = \frac{c}{c + d_0} \qquad (8.31)$$

and

$$\frac{\sigma_N}{\sigma_0} = \sqrt{\frac{1 - p_4}{1 + p_4}} \qquad (8.32)$$

where

$$p_4 = \frac{c}{c + a_0} \qquad (8.33)$$

The effect of crack size can be better visualized by writing Eqs. (8.30) and (8.32) in terms of the measured value of the fracture toughness k_Q:

8.2 FRACTURE MECHANICS OF FIBER COMPOSITES

$$k_Q = \sigma_N \sqrt{\pi c} = \sigma_0 \sqrt{\pi c (1 - p_3^2)} \tag{8.34}$$

$$k_Q = \sigma_N \sqrt{\pi c} = \sigma_0 \sqrt{\pi c \frac{1 - p_4}{1 + p_4}} \tag{8.35}$$

respectively. In both Eqs. (8.34) and (8.35), the expected limiting value of $k_Q = 0$ for vanishingly small values of crack lengths is reached, whereas for large crack lengths, k_Q asymptotically approaches a constant value. For the point- and average-stress criteria, these asymptotic values are, respectively,

$$k_Q = \sigma_0 \sqrt{2\pi d_0} \tag{8.36}$$

$$k_Q = \sigma_0 \sqrt{0.5\pi a_0} \tag{8.37}$$

At this point it should be recalled that the entire basis (or usefulness) of the models lies in the assumption that the characteristic distance d_0 or a_0 remains constant for all hole or crack sizes in at least a particular laminate of a particular material system. In such a case, the characteristic distance can be determined through one test on one hole or crack size. It is also clear that the utility of the model would be increased greatly if d_0 or a_0 can be shown to be constant for all laminates of a particular material system, and an even greater utility would be achieved if they were shown to remain constant for all laminates of all fiber-reinforced–resin matrix composites. There is some evidence [47] that such may be the case, at least for what may be called "fiber- or filament-dominated" laminates in glass–epoxy, boron–epoxy, and graphite–epoxy systems.

Nuismer and Whitney [48] have carried out experiments to examine the effect of changes in the material system, the laminate fiber orientations, and the notch shape and size on the model predictions. Experimental data have been obtained on two material systems, glass–epoxy and graphite–epoxy, in conjunction with orientations of fiber-dominated laminates, $(0/\pm 45/90)_{2S}$ and $(0/90)_{4S}$ containing through the thickness circular holes and sharp-tipped cracks of several sizes. Experimentally measured notched strengths of glass–epoxy laminates with the orientation $(0/\pm 45/90)_{2S}$ are compared with the predictions of the two failure criteria in Figs. 8-19 and 8-20. The measured and predicted values of fracture toughness for the same laminate are shown in Fig. 8-21. The predictions were made by assuming characteristic distances as $d_0 = 0.04$ in. and $a_0 = 0.15$ in. It may be pointed out that the values of characteristic distances were not picked for the best fit of the experimental results but were adopted from an earlier study [47]. It may be noted from the figures that the agreement between the experimental results and predictions of both the models is quite good. Predictions of the two models are quite close to each other, and neither model shows consistently better agreement

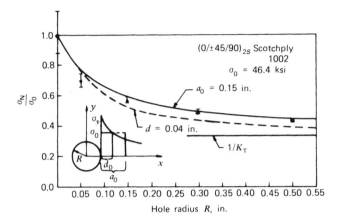

Figure 8-19. Comparison of experimentally measured and theoretically predicted strengths of [0/±45/90]$_{2S}$ glass–epoxy laminates containing circular holes. (From Nuismer and Whitney [48].)

with the experimental results. The results are typical of the various laminates used in the investigation. However, the experimental data do show a large scatter. Moreover, the agreement between experimental results and theoretical predictions for a graphite–epoxy system is not as good. In view of this, Nuismer and Whitney [48] have pointed out that a definite conclusion regarding the constancy of the characteristic lengths must await more data resulting from carefully accomplished studies. Experimental results of Brinson and Yeow [58,59] compare favorably with the predictions of the Whitney–

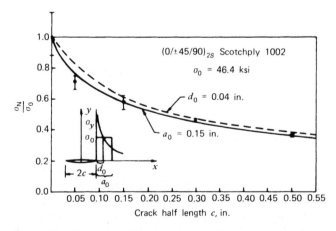

Figure 8-20. Comparison of experimentally measured and theoretically predicted strengths of [0/±45/90]$_{2S}$ glass–epoxy laminates containing sharp cracks. (From Nuismer and Whitney [48].)

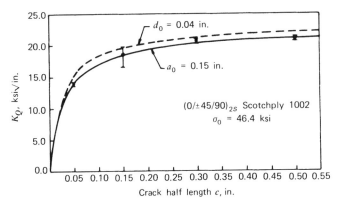

Figure 8-21. Comparison of experimentally measured and theoretically predicted fracture toughness of $[0/\pm45/90]_{2S}$ glass–epoxy laminates. (From Nuismer and Whitney [48].)

Nuismer models. Although no conclusive statement about the accuracy of the models can yet be made, there are sufficient indications that the models may become useful design tools.

8.3 JOINTS FOR COMPOSITE STRUCTURES

In the design of structures using composite materials, the stiffness and strength (particularly with respect to weight) of these materials are an important consideration. An equally important consideration for the complete design of practical structures is the development of attachment methods, joint designs, and the problems of load introduction in composite structures. Without proper joints, it is not possible to take full advantage of the high stiffness and strength of composites. This section describes various fastening methods commonly employed with composite materials, the type of joint failure, and the kind of problems that arise in the joint design because of the heterogeneous and anisotropic nature of composite materials.

Basically, there are two types of joints commonly employed with composite materials: adhesively bonded joints and mechanically fastened joints. These two types of joints are discussed in the following paragraphs.

8.3.1 Adhesively Bonded Joints

8.3.1.1 Bonding Mechanisms Adhesive bonding occurs as a result of three types of interactions between the adhesive and adherend at their interface. These are *chemical bonding, mechanical interlocking,* and *secondary bonding,* or *electrostatic bonding.* These bonding mechanisms are shown schematically in Fig. 8-22. In chemical bonding, adhesive and adherend mol-

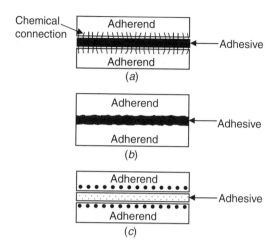

Figure 8-22. Adhesive bonding mechanisms: (a) chemical bonding, (b) mechanical interlocking, and (c) electrostatic bonding.

ecules are chemically connected to each other. It is frequently enhanced by the use of coupling agents, as discussed in Chap. 2. Chemical bonding provides cohesive strength to the bond and is primarily responsible for bond durability. Mechanical interlocking occurs when adhesive fills the micropores on the adherend surface. It is enhanced by surface roughness and provides a micromechanism to mechanically inhibit separation of adhesive and adherend. Secondary bonding forces occur owing to electrostatic forces between the adhesive and adherend molecules. These forces resist shear deformation at the interface and thus provide some shear strength to the bond line. In view of the adhesive bonding mechanisms discussed here, it is obvious that adherend surface conditioning is very important for obtaining an effective adhesive bond.

8.3.1.2 Joint Configurations Several simple bonded joint constructions are shown in Fig. 8-23. In the development of bonded joints for structures, a simple joint can be fabricated first and tested for its suitability in structures. The size of the joint can be estimated from a knowledge of the part sizes to be joined, the allotted space for the joint, and a general idea of how much overlap is required to carry the load. With such knowledge, preliminary joint designs can be made that can be refined using an iterative analysis procedure.

Adhesive joints are natural to consider for polymeric matrix composite materials because many matrix resins are also good adhesives. For example, epoxies are used as adhesives for fiber-reinforced epoxy laminates as well as for many other materials. When the matrix material of the laminates is also used as the adhesive in the joint, excellent adhesion can result.

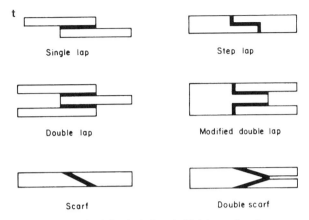

Figure 8-23. Adhesively bonded joint constructions.

8.3.1.3 Joint Failure Modes

The allowable loads on a joint are the loads at which micromechanical damage first occurs that eventually will lead to macromechanical damage. Thus the micromechanical damage can be the basis for the selection of ultimate-load-prediction techniques and the prediction of failure modes of the joints. The micromechanical damage may initiate in the adhesive layer, at the interface, or even in the adherends forming the joint.

Modes of micromechanical damage at the joint are shown in Fig. 8-24 and can be summarized as follows:

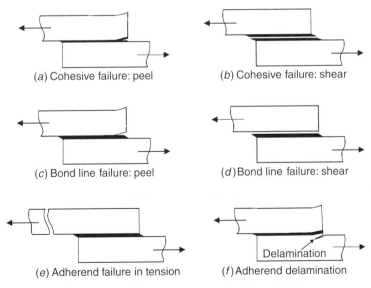

Figure 8-24. Adhesively bonded joint failure modes.

1. Cohesive failure of adhesive layer by peeling (Fig. 8-24a)
2. Cohesive failure of adhesive layer by shear (Fig. 8-24b)
3. Interface (bond line) failure by peeling (Fig. 8-24c)
4. Interface (bond line) failure by shear (Fig. 8-24d)
5. Adherend (laminate) failure in tension (Fig. 8-24e)
6. Adherend (laminate) delamination (Fig. 8-24f)

In addition to these failure modes, the matrix material in the laminae, adjacent to the adhesive, may fail, resulting in cracks in the transverse or longitudinal direction by the mechanisms discussed in Chap. 3.

The cohesive failure within the adhesive layer or in the surface layer of the adherend matrix may occur by brittle fracture or by a rubbery tearing depending on the type of adhesive used. This results in cracks perpendicular to the load and causes a reduction in the load-transferring capability of the joint. This situation is analogous to the cracks in the 90° plies of a cross-ply laminate. The adhesive–adherend interface failure occurs on a macroscale when processing or material quality are poor. This mode of failure generally is not considered in the analysis of joints because it is expected that the necessary quality-control procedures will be used to prevent its occurrence. Interlaminar failure in the laminate (not related to edge effects) may be caused by poor processing, voids, delaminations, or thermal stresses. The last three types of failures are lamina failures, which were discussed in Chap. 3.

8.3.1.4 Stresses in Joints
A joint, even when made properly, represents a discontinuity in the material, and resulting high stresses often initiate joint failure. Therefore, the joint must be analyzed carefully. There are many useful studies on the analysis of bonded joints [60–72]. Analyses have been carried out for various joint configurations and for different properties of the adherends and adhesives. Results have been obtained in closed form or as numerical values. It is beyond the scope of this introductory text to discuss the details of the analyses or their specific results. Important results are discussed here qualitatively, and the conclusions affecting the joint design are also discussed.

The primary function of a joint is to transfer load from one structural member to another. In most bonded joints, the load transfer takes place through interfacial shear. The interfacial shear gives rise to high interlaminar stresses in the adhesive layer. A qualitative variation of interlaminar normal and shear stresses for a single-lap joint is shown in Fig. 8-25. The actual magnitude of the stresses depends on many geometric and material property parameters, such as the thickness and length of the adhesive layer compared with the corresponding values for the adherend material, flexibility of adhesive, and type of load to be transferred. It can be observed in Fig. 8-25 that both interlaminar normal and shear stresses have a large stress concentration near the end of the joint. In the remainder of the joint they are distributed

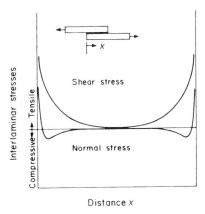

Figure 8-25. Interlaminar normal and shear stresses in a single-lap joint.

uniformly. Because of this high stress concentration in the adhesive layer, high stresses are produced in the adjacent plies of the adherend laminates. Therefore, failure may initiate in these plies. Berg [61] suggested that an effective way to reduce the local high stresses in the plies adjacent to the adhesive layer is to interleave the plies of the adherend laminates so that adhesion takes place in many layers, and consequently, stresses are distributed in many plies. Interleaving is particularly desirable when the number of plies in the laminates is large.

Based on the different types of joint failures discussed in the preceding paragraphs, joint-analysis procedures and joint-design allowables may be developed. Some of the details can be found in Grimes and Greimann [65].

8.3.1.5 Advantages and Disadvantages of Adhesively Bonded Joints

Advantages and disadvantages of adhesively bonded joints are given in Table 8-2.

Adhesively bonded joints are used routinely in most advanced structural applications in aerospace, marine, transportation, and infrastructure industries.

Table 8-2 Advantages and disadvantages of adhesively bonded joints

Advantages	Disadvantages
Relatively lightweight joint	Requires a cure cycle for joint fabrication
Negligible stress concentration in adherends	Limited adherend thickness
	Inspection is difficult
Smooth external surface	Cannot be disassembled
Superior damping characteristics	
Excellent fatigue properties	

8.3.2 Mechanically Fastened Joints

Attachments between two composite laminates or between a laminate and a metal part also can be made by means of bolts, rivets, and pins. The mechanically fastened joints are a logical carryover from the existing practice of joints in metal structures using bolts, screws, pins, and rivets. Composite materials do have some capability to withstand loads introduced by this type of joint; however, unlike isotropic materials, the design of the material itself strongly influences the allowable load transfer of the joint.

8.3.2.1 Failure Modes of Mechanically Fastened Joints

The principal failure modes of mechanically fastened joints are shown in Fig. 8-26. They are

1. Bearing failure of the material. In this type of failure, the bolt hole elongates, as shown in Fig. 8-26a.
2. Tension failure of the material in the reduced cross section through the hole (Fig. 8-26b).
3. Shear-out or cleavage failure of the material (Fig. 8-26c,d). This type of failure actually is induced by transverse tension failure of the material.
4. Shear failure of the bolt.

Composite materials have low bearing strength and low in-plane shear strength. The bearing failure in the joints may be avoided by the use of thin metal shims evenly located in the laminate in the hole area. The shims are quite effective in enhancing the bearing strength in the joint area. The tension strength of the material in the reduced cross section can be improved by

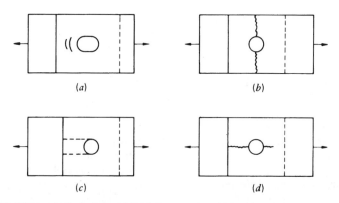

Figure 8-26. Mechanically fastened joint failure modes: (a) bearing, (b) tension, (c) shear-out, and (d) cleavage.

increasing the spacing between bolts and transferring the load through several rows of bolts so that the net shear area of the bolts is sufficient.

The low in-plane shear strength of the composite presents quite a few problems. Unidirectional composites have low shear strength in the longitudinal direction and result in the shear-out mode of joint failure. An improvement for this mode of failure can be made by the use of ±45° fiber orientation, but this results in typically low net tension capability. The use of a so-called isotropic fiber orientation, 0°, 60°, 120°, combines improved shear strength and tension strength but at the same time reduces considerably the efficiency of the material.

Besides the preceding problems related to conventional strength criteria, there are problems peculiar to composite materials. The holes in the laminates cause stress concentrations that vary with the fiber orientation relative to the load direction. The stress-concentration factors sometimes may be well above those occurring in a similar metal structure. Composite materials do not plastically deform, so stresses are not redistributed easily around the stress concentration and are thus a cause of concern. The holes also give rise to the edge effects discussed in a previous section of this chapter. The edge effects promote the tendencies of local interlaminar failures that may become critical in the presence of a corrosive environment.

8.3.2.2 Advantages and Disadvantages of Mechanically Fastened Joints
Advantages and disadvantages of mechanically fastened joints are given in Table 8-3.

Looking at the problems that they pose, mechanically fastened joints should be avoided in critical or primary structural applications. However, for secondary and noncritical applications, this type of joint functions satisfactorily and usually is less expensive. The functioning of the joint is further improved by following proper fabrication and installation procedures and by providing close tolerances.

8.3.3 Bonded-Fastened Joints

For higher reliability and better performance, joints sometimes are bonded as well as mechanically fastened. The bonding results in a reduction of the

Table 8-3 Advantages and disadvantages of mechanically fastened joints

Advantages	Disadvantages
No limit on adherend thickness	Fastener holes may become failure-initiation sites owing to large stress concentration
Easy to inspect joint	Poor fatigue properties
Can be disassembled easily	Prone to corrosion
	Fastener holes can damage the adherend

tendency of mechanical joints to shear out. The mechanical fastening decreases the tendency of debonding because of interfacial shear in bonded joints. Thus the combination joints may show better performance than either the adhesive-bonded joints or the mechanically fastened joints.

A detailed discussion on joints design practices for composite structures can be found in the literature [73,74].

EXERCISE PROBLEMS

8.1. Two laminates are constructed with stacking sequences $(0/\pm 45/90)_{2S}$ and $(0/90)_{4S}$. The lamina properties are

$$E_L = 147.5 \text{ GPa} \qquad G_{LT} = 5.3 \text{ GPa}$$

$$E_T = 11.0 \text{ GPa} \qquad \nu_{LT} = 0.29$$

Calculate the orthotropic stress-concentration factors for infinite-width plates using Eq. (8.20).

8.2. Repeat Exercise Problem 8.1 for the following lamina properties:

$$E_L = 38.6 \text{ GPa} \qquad G_{LT} = 4.1 \text{ GPa}$$

$$E_T = 8.3 \text{ GPa} \qquad \nu_{LT} = 0.26$$

8.3. Using the Whitney–Nuismer failure criteria for notched composites, construct plots of the notched-strength–unnotched-strength ratio as a function of the hole radius for the laminates considered in Exercise Problems 8.1 and 8.2. Assume that the characteristic distances are $d_0 = 1$ mm and $a_0 = 4$ mm for the point-stress and average-stress criteria, respectively. The hole radius ranges from 0–15 mm. Note that for one laminate, predictions of the two criteria should be plotted on the same graph paper.

8.4. Repeat Exercise Problem 8.3 for the sharp cracks with the half-crack length ranging from 0 to 15 mm.

8.5. Using the Whitney–Nuismer failure criteria for notched composites, construct plots of the fracture-toughness–unnotched-strength ratio, k_Q/σ_0 as a function of half-crack length. Characteristic distances are the same as those given in Exercise Problem 8.3, and the half-crack length ranges from 0–15 mm.

REFERENCES

1. N. J. Pagano and J. C. Halpin, "Influence of End Constraints in the Testing of Anisotropic Bodies," *J. Compos. Mater.*, **2**(4), 18 (1968).
2. N. J. Pagano, "Exact Solutions for Composite Laminates in Cylindrical Bending," *J. Compos. Mater.*, **3**(3), 398 (1969).
3. N. J. Pagano, "Exact Solutions for Rectangular Bidirectional Composites and Sandwich Plates," *J. Compos. Mater.*, **4**(1), 20 (1970).
4. J. M. Whitney and A. W. Leissa, "Analysis of Heterogeneous Anisotropic Plates," *J. Appl. Mech.*, **28,** 261 (1969).
5. J. M. Whitney, "The Effects of Transverse Shear Deformation on the Bending of Laminate Plates," *J. Compos. Mater.*, **3**(4), 534 (1969).
6. A. H. Puppo and H. A. Evensen, "Interlaminar Shear in Laminated Composites under Generalized Plane Stress," *J. Compos. Mater.*, **4**(2), 204 (1970).
7. R. B. Pipes and N. J. Pagano, "Interlaminar Stresses in Composite Laminates under Uniform Axial Extension," *J. Compos. Mater.*, **4**(4), 538 (1970).
8. R. B. Pipes and I. M. Daniel, "Moiré Analysis of the Interlaminar Shear Edge Effect in Laminated Composites," *J. Compos. Mater.*, **5**(2), 255 (1971).
9. N. J. Pagano and R. B. Pipes, "The Influence of Stacking Sequence of Laminate Strength," *J. Compos. Mater.*, **5**(1), 50 (1971).
10. R. L. Foye and D. J. Baker, "Design of Orthotropic Laminates," presented at the 11th Annual AIAA Structures, Structural Dynamics, and Materials Conference, Denver, CO, April 1970.
11. J. M. Whitney and E. E. Browning, "Free-Edge Delamination of Tensile Coupons," *J. Compos. Mater.*, **6**(2), 300 (1972).
12. J. M. Whitney, "Free-Edge Effects in the Characterization of Composite Materials," in *Analysis of the Test Methods for High Modulus Fibers and Composites,* ASTM STP 521, American Society for Testing and Materials, Philadelphia, PA, 1973, p. 167.
13. R. B. Pipes, B. E. Kaminski, and N. J. Pagano, "Influence of the Free Edge upon the Strength of Angle-Ply Laminates," in *Analysis of the Test Methods for High Modulus Fibers and Composites,* ASTM STP 521, American Society for Testing and Materials, Philadelphia, PA, 1973, p. 218.
14. R. B. Pipes, "Boundary Layer Effects in Composite Laminates," *Fiber Sci. Technol.*, **13,** 49 (1980).
15. G. Isakson and A. Levy, "Finite-Element Analysis of Interlaminar Shear in Fibrous Composites," *J. Compos. Mater.*, **5**(2), 273 (1971).
16. E. F. Rybicki, "Approximate Three-Dimensional Solutions for Symmetric Laminates under Inplane Loading," *J. Compos. Mater.*, **5**(3), 354 (1971).
17. N. J. Pagano, "On the Calculation of Interlaminar Normal Stress in Composite Laminate," *J. Compos. Mater.*, **8**(1), 65 (1974).
18. R. B. Pipes and N. J. Pagano, "Interlaminar Stresses in Composite Laminates: An Approximate Elasticity Solution," *J. Appl. Mech.*, **41,** Series E (3), 668 (1974).
19. S. Tang, "Interlaminar Stresses of Uniformly Loaded Rectangular Composite Plates," *J. Compos. Mater.*, **10**(1), 69 (1976).

20. R. M. Jones, *Mechanics of Composite Materials,* 2nd ed., Taylor and Francis, Philadelphia, 1999.
21. A. A. Griffith, "The Phenomena of Rupture and Flow in Solids," *Philos. Trans. R. Soc., Series A,* **221,** 163 (1920).
22. G. R. Irwin, "Fracture," in *Handbuch der Physik,* Vol. VI, Springer, Berlin, 1958.
23. F. M. Wu, "Fracture Mechanics of Anisotropic Plates," in S. W. Tsai, J. C. Halpin, and N. J. Pagano, eds., *Composite Materials Workshop,* Technomic, Stamford, CT, 1968.
24. S. G. Lekhnitskii, *Theory of Elasticity of Anisotropic Elastic Body,* English trans. by Brandstatton, Holden Day, San Francisco, 1963.
25. E. M. Wu, "Strength and Fracture of Composites," in L. J. Broutman, ed., *Fracture and Fatigue,* Academic, New York, 1974.
26. W. F. Brown and J. E. Srawley, "Plane-Strain Crack Toughness Testing of High Strength Metallic Materials," in *Fracture Toughness Testing,* ASTM STP 410, American Society for Testing and Materials, Philadelphia, PA, 1966.
27. J. R. Rice, "A Path Independent Integral and Approximate Analysis of Strain Concentration by Notches and Cracks," *J. Appl. Mech.,* **35,** 379 (1968).
28. J. W. Hutchinson, "Singular Behavior at the End of a Tensile Crack in a Hardening Material," *J. Mech. Phys. Solids,* **16,** 13 (1968).
29. J. R. Rice and G. F. Rosengren, "Plane-Strain Deformation Near a Crack Tip in a Power Law Hardening Material," *J. Mech. Phys. Solids,* **16,** 1, (1968).
30. J. A. Begley and J. D. Landes, "The J-Integral as a Fracture Criterion," in *Fracture Toughness,* ASTM STP 514, American Society for Testing and Materials, Philadelphia, PA, 1972, p. 1.
31. J. D. Landes and J. A. Begley, "The Effect of Specimen Geometry on J_{1c}," in *Fracture Toughness,* ASTM STP 514, American Society for Testing and Materials, Philadelphia, PA, 1972, p. 24.
32. J. D. Landes and J. A. Begley, "Recent Developments in J_{1C} Testing," in *Developments in Fracture Mechanics Test Methods Standardization,* ASTM STP 632, American Society for Testing and Materials, Philadelphia, PA, 1977, p. 57.
33. S. K. Gaggar and L. J. Broutman, "Crack Propagation resistance of Random Fiber Composites," *J. Compos. Mater.,* **9,** 216 (1975).
34. J. Awerbuch and H. T. Hahn, "K-Calibration of Unidirectional Metal Matrix Composite," *J. Compos. Mater.,* **12,** 222 (1978).
35. Y. J. Yeow, D. H. Morris, and H. F. Brinson, "The Fracture Behavior of Graphite/Epoxy Laminates," *Exp. Mech.,* **19,** 1, (1979).
36. S. K. Gaggar and L. J. Broutman, "Fracture Toughness of Random Glass Fiber Epoxy Composites: An Experimental Investigation," in *Flaw Growth and Fracture,* ASTM STP 631, American Society for Testing and Materials, Philadelphia, PA, 1977, pp. 310–330.
37. C. Bathias, R. Esnault, and J. Pellas, "Application of Fracture Mechanics to Graphite Fiber-Reinforced Composites," *Composites,* **12,** 195 (1981).
38. D. H. Morris and H. T. Hahn, "Fracture Resistance characterization of Graphite/Epoxy Composites," in *Composite Materials: Testing and Design,* ASTM STP 617, American Society for Testing and Materials, Philadelphia, PA, 1977, p. 5.

39. J. M. Mahishi and D. F. Adams, "Micromechanical Predictions of Crack Initiation, Propagation and Crack Growth Resistance in Boron/Aluminum Composites," *J. Compos. Mater.*, **16,** 457 (1982).
40. S. Ochiai and P. W. M. Peters, "Tensile Fracture of Center-notched Angleply $(0/\pm 45)_S$ and $(0/90)_{2S}$ Graphite–Epoxy Composites," *J. Mater. Sci.*, **17,** 417 (1982).
41. H. Yanada and H. Homma, "Study of Fracture Toughness Evaluation of FRP," *J. Mater. Sci.*, **18,** 133 (1983).
42. B. D. Agarwal and G. S. Giare, "Crack Growth Resistance of Short Fiber Composites: I. Influence of Fiber Concentration, Specimen Thickness and Width," *Fibre Sci. Technol.*, **15,** 283 (1981).
43. B. D. Agarwal and G. S. Giare, "Effect of Matrix Properties on Fracture Toughness of Short Fiber Composites," *Mater. Sci. Eng.*, **52,** 139 (1982).
44. B. D. Agarwal, B. S. Patro, and P. Kumar, "Crack Length Estimation Procedure for Short Fiber Composite: an Experimental Evaluation," *Polym. Compos.* **6,** 185 (1985).
45. B. D. Agarwal, B. S. Patro, and P. Kumar, "Prediction of Instability Point During Fracture of Composite Materials," *Compos. Technol. Rev.*, **6,** 173 (1984).
46. M. E. Waddoups, J. R. Eisenmann, and B. E. Kaminski, "Macroscopic Fracture Mechanics of Advanced Composite Materials," *J. Compos. Mater.*, **5**(4), 446 (1971).
47. J. M. Whitney and R. J. Nuismer, "Stress Fracture Criteria for Laminated Composites Containing Stress Concentrations," *J. Compos. Mater.*, **8**(2), 253 (1974).
48. R. J. Nuismer and J. M. Whitney, "Uniaxial Failure of Composite Laminates Containing Stress Concentrations," in *Fracture Mechanics of Composites,* ASTM STP 953, American Society for Testing and Materials, Philadelphia, PA, 1975, pp. 117–142.
49. R. F. Karlak, "Hole Effects in a Related Series of Symmetrical Laminates," *Proceedings of Failure Modes in Composites,* Vol. IV, The Metallurgical Society of AIME, Chicago, 1977, p. 105.
50. R. B. Pipes, R. C. Wetherhold, and J. W. Gillespie, Jr., "Notched Strength of Composite Materials," *J. Compos. Mater.*, **12,** 148 (1979).
51. R. B. Pipes, J. W. Gillespie, Jr., and R. C. Wetherhold, "Superposition of the Notched Strength of Composite Laminates," *Polym. Eng. Sci.*, **19**(16), 1151 (1979).
52. R. B. Pipes, R. C. Wetherhold, and J. W. Gillespie, Jr., "Macroscopic Fracture of Fibrous Composite," *Mater. Sci. Eng.*, **45,** 247 (1980).
53. K. Y. Lin, "Fracture of Filamentary Composite Materials," Ph.D. dissertation, Department of Aeronautics and Astronautics, Massachusetts Institute of Technology, Cambridge, MA, January 1976.
54. J. W. Mar and K. Y. Lin, "Fracture Mechanics Correlation for Tensile Failure of Filamentary Composites with Holes," *J. Aircraft,* **14**(7), 703 (1977).
55. C. C. Poe, Jr., and J. A. Sova, "Fracture Toughness of Boron/Aluminum Laminates with Various Proportions of $0°$ and $\pm 45°$ Plies," NASA technical paper 1707 (November 1980).

56. J. Awerbuch and M. S. Madhukar, "Notched Strength of Composite Laminates: Predictions and Experiments: A Review," *J. Reinf. Plast. Compos.*, **4**, 1 (1985).
57. S. G. Lekhnitskii, *Anisotropic Plates,* trans. from the second Russian edition by S. W. Tsai and T. Cheron, Gordon and Breach, New York, 1968.
58. H. F. Brinson and Y. T. Yeow, "An Experimental Study of the Fracture Behavior of Laminated Graphite-Epoxy Composites," in *Composite Materials: Testing and Design (Fourth Conference)*, ASTM STP 617, American Society for Testing and Materials, Philadelphia, PA, 1977, pp. 18–38.
59. Y. T. Yeow and H. F. Brinson, "A Study of Damage Zones or Characteristic Lengths as Related to the Fracture Behavior of Graphite-Epoxy Laminates," Technical report VPI-E-77-15, Virginia Polytechnic Institute and State University, May 1977.
60. M. Goland and F. Reissner, "The Stresses in Cemented Joints," *J. Appl. Mech.*, PA. 17-A-27 (March 1944).
61. K. R. Berg, "Problems in the Design of Joints and Attachments," in F. W. Wendt, H. Liebowitz, and N. Perrone, eds., *Mechanics of Composite Materials,* Pergamon, New York, 1970.
62. R. N. Haddock, "Joints in Composite Structures," *Proceedings of Conference on Fibrous Composites Vehicle Design,* AFFDL-TR-72-130 (AD907042), 1972, pp. 791–811.
63. L. J. Hart-Smith, "Design and Analysis of Adhesive Bonded Joints," *Proceedings of Conference on Fibrous Composites Vehicle Design,* AFFDL-TR-72-130 (AD907042), 1972, pp. 813–856.
64. A. C. Fehrle, "Fatigue Phenomena of Joints in Advanced Composites," *Proceedings of Conference on Fibrous Composites Vehicle Design,* AFFDL-TR-72-130 (AD907042), 1972, pp. 857–890.
65. G. G. Grimes and L. F. Greimann, "Analysis of Discontinuities, Edge Effects, and Joints," in C. C. Chamis, ed., *Structural Design and Analysis—Part II,* Academic, New York, 1975.
66. O. Ishai and S. Gali, "Two-Dimensional Interlaminar Stress Distribution within the Adhesive Layer of a Symmetrical Doubler Model," *J. Adhes.*, **8**, 301–312 (1977).
67. O. Ishai, D. Peretz, and S. Gali, "Direct Determination of Interlaminar Stresses in Polymeric Adhesive Layer," *Exp. Mech.*, **17**(7), 265–270 (1977).
68. S. Amijima, T. Fujii, and A. Yoshida, "Two-Dimensional Stress Analysis on Adhesive Bonded Joints," *Proceedings of Twentieth Japan Congress on Materials Research,* The Society of Materials Science, Japan, 1977.
69. S. Amijima, T. Fujii, A. Yoshida, and H. Amino, "Dynamic Response of Adhesive Bonded Joints," *Proceedings of Twentieth Japan Congress on Materials Research,* The Society of Materials Science, Japan, 1977.
70. L. J. Hart-Smith, "Adhesive Bonded Double-Lap Joints," NASA CR-112235, 1973.
71. M. Niu, *Composite Airframe Structures,* Technical Book Company, Los Angeles, CA, 1992.

72. R. B. Heslehurst and L. J. Hart-Smith, "The Science and Art of Structural Adhesive Bonding," *SAMPE Journal,* **8**(2), 60–71, 2002.
73. Structural Design Guide for Advanced Composite Applications, Air Force Materials Laboratory, Advanced Composites Division, January 1971.
74. Plastics for Aerospace Vehicles, Part I: Reinforced Plastics, MIL-HDBK-17A, Department of Defense, Washington, 1971.

9

PERFORMANCE OF FIBER COMPOSITES: FATIGUE, IMPACT, AND ENVIRONMENTAL EFFECTS

The superior strength and stiffness of composite materials can be used to full advantage in structural applications only when the behavior of these materials under different loading and environmental conditions is properly understood. Any uncertainty in this regard results in the underutilization of the material properties by the use of unusually large margins of safety in actual design. Keeping this in mind, the behavior of fiber composite materials subjected to cyclic and impact loading is discussed in the first two sections of this chapter. The third section is devoted to the understanding of material behavior under various environmental conditions, such as exposure to water, water vapor, or other corrosive environments; temperature extremes; and long term physical and chemical stability.

9.1 FATIGUE

9.1.1 Introduction

It is well known that when materials are subjected to repeated fluctuating or alternating loads, they may fail even though the maximum stress may never exceed the ultimate static strength of the material. In other words, load cycling reduces the strength of a material, or the fatigue strength of a material is lower than its static strength. This is true of almost all existing materials, including metals, plastics, and composite materials. In service, fatigue loads usually are unavoidable. For this reason, recent designs do not specify static strength alone as a primary design criterion but also include fatigue analysis. The demand for improved performance of structural materials in transportation industries, particularly in aircraft, makes fatigue analysis an important consideration. With this view, the fatigue of composite materials has been studied by a large number of investigators. Clear design criteria, similar to the ones that exist for fatigue of metals, have not yet been established, but many important aspects of the fatigue of composites are now well understood.

Unidirectional continuous-fiber-reinforced composites are known to possess excellent fatigue resistance in the fiber direction. This is so because the load in a unidirectional composite is carried primarily by the fibers, which generally exhibit excellent resistance to fatigue. In real structures, however, composites are used mostly in the form of laminates. Because of the differences in orientation of each ply, some plies are weaker than the others in the loading direction and show physical evidence of damage much before the final fracture. The evidence of damage may be in one or more forms, such as the failure of the fiber–matrix interface, matrix cracking or crazing, fiber breaking, and void growth (i.e., separation of plies or delamination). In metals, the appearance of detectable damage (e.g., a crack) generally is considered unsafe because it grows rapidly to final fracture. In composite materials, however, this is not necessarily so because although initial damage may appear very early in the fatigue life, its propagation may be arrested by the internal structure of the composite. It should be noted that in critical applications, design loads should be less than those required to cause any damage within the composite. The damage in individual plies generally causes a lowering of elastic properties of the laminate and eventually could lead to its structural failure (e.g., excessive deformation). However, this may happen long before the laminate is in danger of fracturing. Thus the definition of failure in composite materials may change from one application to another. In an application where deformation or a change in stiffness has to be limited, loss of stiffness by a fixed percent of the original stiffness may be the failure criterion rather than complete rupture. In the case of metals, the two criteria practically coincide because they exhibit little change in stiffness unless cracking is extensive. For these obvious reasons, a successful design procedure with composite materials for fatigue applications cannot be a simple extrapolation of the procedures used with metals. In the absence of well-developed design procedures, a designer has to use his or her judgment with a proper degree of caution. However, a good understanding of various aspects of the fatigue behavior of composites definitely will aid the design engineer. The presentation in this section has been made with this view in mind. Initiation and propagation of fatigue damage and its influence on composite properties are discussed first. Then the influence of material variables such as matrix material, ply orientation, fiber contents, and fiber finish and testing variables such as mean stress and frequency are discussed. A brief discussion on the trend in developing empirical relations for predicting fatigue damage and fatigue life is also presented. The last two sections are devoted to the fatigue behavior of high-modulus fiber-reinforced composites and fatigue of short-fiber composites.

9.1.2 Fatigue Damage

9.1.2.1 Damage/Crack Initiation There have been several studies [1–7] on the mechanism of damage initiation and propagation during fatigue of composite laminates. It has been established that the damage first initiates by the separation of fibers from the matrix (called *debonding*) in the fiber-rich regions of the plies in which the fibers lie perpendicular or at a large angle to the loading direction. Large stress and strain concentrations at the fiber–matrix interface are responsible for the initiation of these cracks. After initiation, the crack usually propagates between fibers primarily along the fiber–matrix interface. A typical cross-ply crack is shown in Fig. 9-1a. The crack is generally perpendicular to the direction of load and extends over the entire width of the ply. The cross-ply cracks can appear during the first cycle of loading, provided that the applied stress exceeds the local ply strength, which might happen at applied stresses as low as 20% of the ultimate stress depending on the laminate construction [4]. The number of cross-ply cracks increases with either the number of cycles or an increase in the stress level. Multiple crack formations in the cross-plies are shown in Fig. 9-1b.

The initial damage in randomly oriented fibrous composites commences in a similar manner. In a tensile test on a thin laminate made from chopped-strand mat, the first signs of damage have been observed [2] at about 30% of the expected ultimate tensile strength. In this case also the initial damage is associated with the strands lying perpendicular to the line of load. The initiation of damage can occur at any point along the length of the strand and is not particularly associated with the ends. Damage is seen to be in the form of debonding within a strand. Thus the first stage of damage in a composite laminate is formation of debonding cracks along the fibers lying perpendicular or at the largest angle to the direction of load.

9.1.2.2 Crack Arrest and Crack Branching The cross-ply cracks propagate through the entire width of the ply but are unable to propagate into the adjacent ply, particularly if it is a ply having fibers aligned in the direction of load. Thus the cross-ply cracks terminate at the interface of two plies. This feature of crack termination is well illustrated in Fig. 9-1b. However, the crack tip produces a stress concentration ahead of itself. The resulting high interlaminar stresses produce favorable conditions for starting a delamination crack along the ply interface. Figure 9-2 shows such a delamination crack being started at the tip of a cross-ply crack. More delamination cracks start and propagate as the number of cycles increases. At the time when delamination cracks appear, another type of damage is also observed. The fibers in the longitudinal plies also may start fracturing, and debonding and cracks in the longitudinal plies begin to appear. A longitudinal-ply crack as seen in the cross section of the laminate is shown in Fig. 9-3. The longitudinal-ply cracks do not follow any set path, unlike the cross-ply cracks, which generally are perpendicular to the line of load.

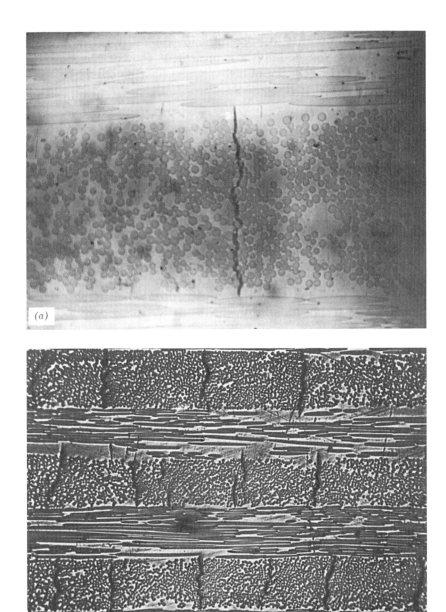

Figure 9-1. Fatigue failure initiation: (a) single cross-ply crack and (b) multiple crack formation in cross-plies.

372 PERFORMANCE OF FIBER COMPOSITES

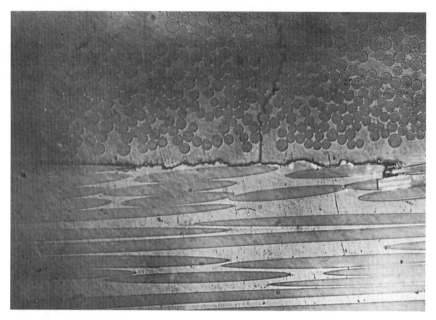

Figure 9-2. A delamination crack initiated at the tip of a cross-ply crack.

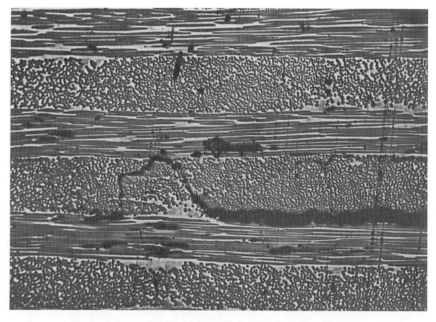

Figure 9-3. Longitudinal-ply crack appears in a cross section.

9.1.2.3 Final Fracture
The composite undergoes final fracture when it is sufficiently weakened by longitudinal-ply cracks and delamination cracks. The longitudinal-ply cracks weaken the longitudinal plies that are responsible for carrying a larger part of the load. The presence of delamination cracks prevents load distribution between plies, and the composite is essentially reduced to a number of independent longitudinal plies acting in parallel to support the applied load. The weakest of these longitudinal plies fails and triggers failure of the remaining longitudinal plies. Evidence of extensive delamination in the region adjacent to the failure zone is shown in Fig. 9-4. However, the delamination cracks, which are responsible for final fracture of the material, are clearly marked only at a late stage of the fatigue test, for example, after about 90% of the fatigue life. This observation was first made by Broutman and Sahu [4] and later confirmed by Dally and Agarwal [7].

9.1.2.4 Schematic Representation
The initiation of cracks resulting from fiber fracture and the propagation of cracks through the composite in fiber-reinforced materials are shown schematically in Fig. 9-5. A discontinuity produced by a fiber fracture causes high shear stress at the fiber–matrix interface and produces favorable conditions for a shear crack to grow, as shown in Fig. 9.5a. Depending on the relative values of bond strength and matrix strength, the shear crack may grow in the interface region or in the adjacent matrix material. In a composite with a weak interface, tensile splitting at interfaces may take place ahead of a fatigue crack in the matrix (see Fig.

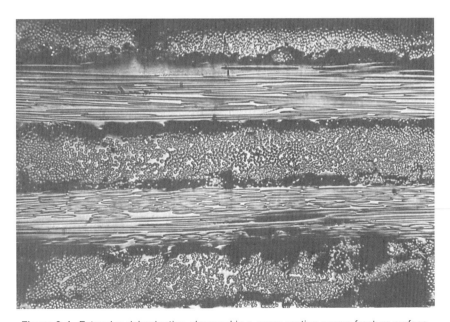

Figure 9-4. Extensive delamination observed in a cross section near a fracture surface.

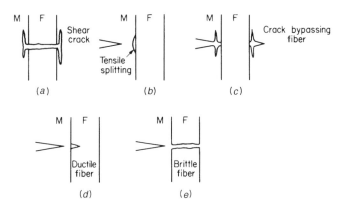

Figure 9-5. Modes of fatigue-crack growth in fiber-reinforced materials: (a) shear-crack initiation at fiber break, (b) tensile splitting of interface ahead of matrix crack, (c) matrix crack bypassing strong fiber, (d) crack initiation in ductile fiber ahead of matrix crack, and (e) fracture of brittle fiber ahead of matrix crack.

9-5b). Crack branching (see Fig. 9-5a) and tensile splitting (see Fig. 9-5b) relieve some of the stress concentration in the vicinity of the crack and enhance the fatigue life of the material as a consequence of a weak fiber–matrix interfacial bond. Plastic flow in a low-yielding stress matrix also blunts the crack tip and thus impedes crack growth.

When a fatigue crack in the matrix approaches a fiber, it may grow essentially in three ways, as shown in Fig. 9-5c,d,e. With a weak interface and strong fibers, the crack can bypass the fiber by an antiplane-strain mode of crack growth (see Fig. 9-5c). When the interface is strong, high stresses ahead of the crack tip affect the fibers. Ductile fibers are particularly sensitive to the high crack-tip stresses, and fatigue-crack growth is rapid (see Fig. 9-5d). Brittle fibers ahead of the crack fail abruptly because of the large crack-tip stresses (see Fig. 9-5e). The modes of fatigue-crack growth illustrated by Fig. 9-5d,e generally result in poor fatigue resistance of composites.

9.1.2.5 Damage Characterization The extent and character of internal cracking damage have been studied by investigators using different methods. Optical microscopy on an internal section polished by metallographic techniques is a direct method and probably the most popular technique. This method is very effective in visualization and presentation of damage characteristics. The photomicrographs shown in Figs. 9-1 to 9-4 have been obtained using this technique. The extent of damage may be represented quantitatively using this method by obtaining the average number of cracks in a specified area. Weight gain from water immersion was employed as a measure of internal damage by early investigators, such as McGarry [1], but is not always found suitable. Nondestructive inspection techniques are now being developed for detection of fatigue damage [8]. These techniques include ultrasonics, holographic interferometry, and x-ray radiography. Changes in structural properties such as static or dynamic modulus and temperature rise during fatigue loading are also considered indicative of internal damage. However, no clear quantitative correlation between structural properties and internal damage measurements has as yet been established.

9.1.2.6 Influence of Damage on Properties

Internal cracking results in lowering of the stiffness and strength of composite materials. Broutman and Sahu [4] have related the changes in residual strength and modulus to the development of cracks in a glass–epoxy cross-ply material (Fig. 9-6). The residual strength and stiffness decrease with the increasing crack density. It also has been pointed out that the stress–strain curve of a virgin cross-ply material can be approximated by two straight lines giving two elastic moduli for the material. The two moduli are referred to as *primary* and *secondary moduli* (see Chap. 6). The material exhibits a higher (primary) modulus at the beginning of the test because there are no cracks present, and both the longitudinal plies and the cross-plies contribute fully to the stiffness of the composite. As the load increases, the cross-ply cracks appear, and thus the contribution of the cross-plies to the composite stiffness decreases, causing a reduction in modulus. In fatigue tests, the modulus decreases first in the presence of cross-ply cracks and then longitudinal-ply cracks and delamination cracks. Therefore, with fatigue exposure, the stress–strain curve of the material becomes linear with a modulus close to the secondary modulus of the virgin material. The modulus may become less than the secondary modulus of the material when the longitudinal-ply cracks and delamination cracks develop owing to fatigue loading. Dally and Agarwal [7] developed a quantitative relationship between modulus change and crack density for an E-glass–epoxy cross-ply laminate. This relationship is shown in Fig. 9-7, in which the *crack pitch* is defined as the average distance between two consecutive cracks in the cross-plies.

There is a gradual decrease in the static strength of the material as it is subjected to an increasing number of cycles at a given stress level. It is obvious from Fig. 9-6 that much of the strength reduction occurs in the first 25% of the fatigue life, beyond which the rate of decrease in static strength is reduced until the fatigue life is reached and failure occurs. Once again, the reason for the initial loss of strength is the failure of the cross-plies. Development of longitudinal-ply cracks and delamination cracks is slow, and hence the loss of strength in the later part of the fatigue life is slow. Rapid loss of strength occurs in the last few cycles of fatigue life when the stronger plies fail. Prior to this rapid loss of strength, individual plies do become weakened, but the overall strength reduction is slow. The curves shown in Fig. 9-6 are typical of cross-plied materials. Tanimoto and Amijima [9,10] have studied fatigue of glass-cloth-reinforced polyester resins. Their results are very similar

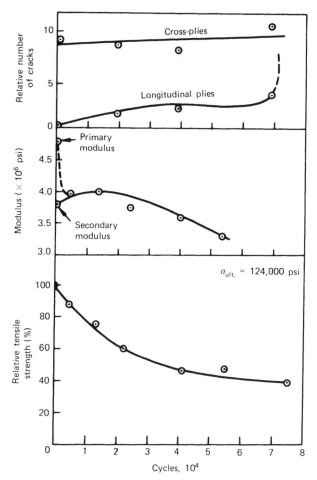

Figure 9-6. Increase in the number of cracks and loss of strength and modulus of a cross-ply laminate during fatigue. (From Broutman and Sahu [4].)

to those of Broutman and Sahu [4], as shown in Fig. 9-6. In addition, they have reported that the residual strength in interlaminar shear follows the same trend as the residual tensile strength. Hahn and Kim [11], while studying fatigue of glass–epoxy angle-ply laminates, observed that the secant modulus of the material decreases with exposure to fatigue loading and indicated that the decrease in secant modulus is related to internal damage.

Besides internal cracking damage, a rise in temperature also causes a decrease in properties of the material. Dally and Broutman [12] observed that a significant rise in temperature takes place during the fatigue of a cross-ply material, particularly when the frequency is high. Cessna et al. [13] performed constant-deflection flexural tests on glass-reinforced polypropylene and mon-

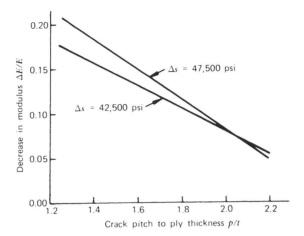

Figure 9-7. Loss of modulus as a function of crack pitch. (From Dally and Agarwal [7].)

itored the load decay (proportional to modulus decay) with cycles (Fig. 9-8). They also monitored the temperature rise caused by viscoelastic energy dissipation, which is common for polymer–matrix composites. In addition to indicating progressive fatigue damage, the temperature rise also helps to weaken the material and shorten its fatigue life. By cooling their specimens to maintain isothermal conditions, Cessna et al. were able to extend both the cycles to onset of stiffness change and the fracture life by one order of magnitude.

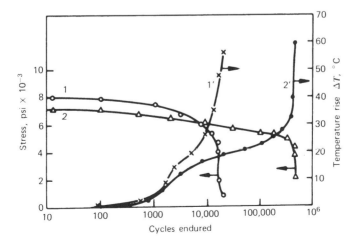

Figure 9-8. Load decay and temperature rise during constant-deflection flexural fatigue. (From Cessna et al. [13].)

9.1.3 Factors Influencing Fatigue Behavior of Composites

Results of fatigue tests typically are presented as a plot of applied stress (S) against number (N) of cycles to failure. This graph is called an *S–N curve*. The ordinate generally is the stress or strain amplitude or the maximum stress or strain in a cycle and is plotted on a linear scale. The abscissa is the number of cycles to failure for a fixed stress cycle and is plotted on a logarithmic scale. Complete separation of the specimen has been taken as the criterion for failure by most investigators. However, another approach is to record fatigue data as loss of stiffness against number of cycles and to present curves of stress versus number of cycles for fixed percent changes in stiffness [14]. The *S–N* curves for all materials including metals, polymers, and composites have a negative slope. That is, the number of cycles to failure (or the fatigue life) increases as the stress decreases. The exact shape of the curve differs from material to material. For composites, the curve is influenced by various material and testing variables as follows: (1) matrix material (type of resin), (2) ply orientation, (3) volume fraction of reinforcement, (4) interface properties, (5) type of loading, (6) mean stress, (7) frequency, and (8) environment. The first four factors are material variables, whereas the remaining are test variables.

Boller [15] has investigated the effect of matrix materials on the fatigue strengths of glass-reinforced plastic laminates. The *S–N* curves are shown in Fig. 9-9, and all the resins were reinforced with style 181 E-glass fabric. This fabric produced a balanced lamina such that E_L is approximately equal to E_T.

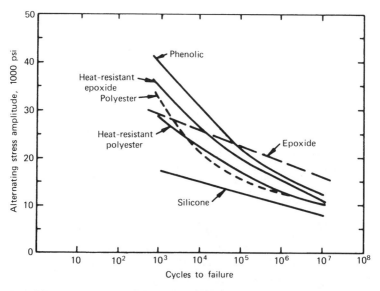

Figure 9-9. Effect of matrix material on fatigue strength of glass-reinforced plastic laminates. (From Boller [15].)

These measurements were made before 1955, and since then, improved glass-coupling agents have been developed for some of the specific resins shown in Fig. 9-9. These composites may have improved strength values compared with those shown in Fig. 9-9, but the trends shown remain the same. Strikingly similar results have been shown by Davis et al. [16]. Of the various thermosetting resins used commonly in glass-fiber laminates, the best fatigue properties are obtained with the epoxy resins. Superiority of epoxy resins is attributed to their inherent toughness and durability. In addition, they have high mechanical strength and low shrinkage during cure and form an excellent adhesive bond to glass fibers.

The effect of fiber orientation is complex. Although the tensile strength of unidirectional composites is maximum in the fiber direction, in fatigue the unidirectional construction is not optimal, as seen in Fig. 9-10. The poor performance of 0° unidirectional laminates occurs because of splitting in the fiber direction resulting from their relatively low transverse strength and imperfect testing or gripping conditions. It has been shown [15,16] that the splitting problem may be overcome and the fatigue strength improved by providing some of the plies in the 90° direction. Typical S–N curves for various types of construction are shown in Fig. 9-11. It may be noted that the cross-ply nonwoven laminates with 50% 0° and 50% 90° plies give much better fatigue strength than does the glass-fiber fabric even though both laminates have the same reinforcement pattern. In general, nonwoven materials are superior to woven materials in fatigue because fibers in nonwoven materials are straight and parallel and do not get crimped as in the fabric

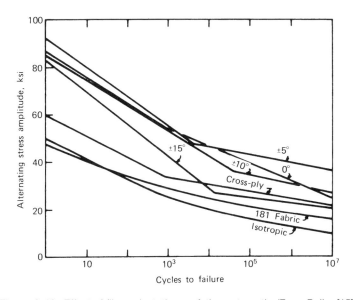

Figure 9-10. Effect of fiber orientation on fatigue strength. (From Boller [15].)

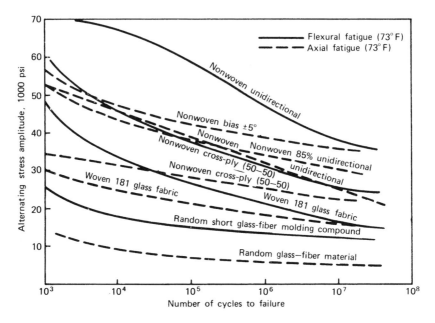

Figure 9-11. S–N curves for different laminate constructions. (From Davis et al. [16].)

construction. Thus nonwoven materials possess optimal static and fatigue properties.

Besides the orientation of plies, the stacking sequence also influences the fatigue life. Foye and Baker [17] observed that when positions of the plies in a $[\pm 15/\pm 45]_s$ laminate were changed, a difference in fatigue strength of about 25,000 psi occurred. This was explained by Pagano and Pipes [18] through analysis of interlaminar stresses. They showed that the interlaminar stress normal to the laminate changes from tension to compression by changing the stacking sequence, and this accounts for the difference in fatigue load capability. Delamination was observed to occur in the specimens that developed tensile interlaminar stress. Whitney [19,20] has made similar observations on the influence of stacking sequence on the fatigue strength and failure mode of composite laminates.

Amijima and Tanimoto [9,10,21] have studied the influence of glass content on fatigue properties of laminated glass-fiber composite materials. Their results (Figs. 9-12 and 9-13) clearly show that the fatigue strength of glass-cloth-reinforced polyester resin increases with increasing glass content in both axial fatigue (V_f range 29.3–54.2%) and rotating bending fatigue (V_f range 11.8–30.4%). This increase in fatigue strength occurs with the increase in static strength of the composite as a result of increased fiber volume fraction. Earlier studies by Boller [15] and Davis et al. [16] indicated that the fatigue strength is not related to fiber content because it varies from 63–80% in a

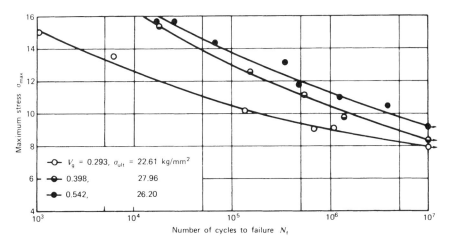

Figure 9-12. Influence of glass content on axial fatigue strength. (From Tanimoto and Amijima [10].)

glass-cloth-reinforced epoxy. It appears that an optimal fatigue strength may be achieved with 70% by weight of glass fibers in the case of fabric laminates.

The effect of interfacial bond strength between the matrix and the reinforcement on the fatigue strength of composites has been studied by Hofer et al. [22]. They studied the fatigue behavior of glass-fabric composites having four different finishes, including an untreated surface and surfaces treated with Volan A, A-1100, and S-550 finishes (organosilane coupling agents). The untreated glass exhibited the highest fatigue strength in a dry environment, but it also was the most severely affected in a humid environment. As a result,

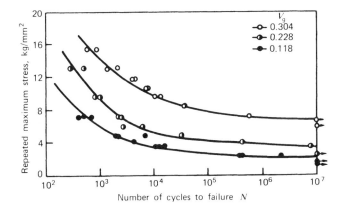

Figure 9-13. Influence of glass content on rotating bending fatigue strength. (From Amijima and Tanimoto [21].)

all fabrics tested showed a similar resistance to fatigue when tested in a humid environment. Thus, when laminates are fatigued in real environments, it is difficult to demonstrate the effectiveness of various surface treatments. This is partially a result of the stress system, which is usually such that the composite properties are fiber-dominant and not greatly dependent on the interface strength.

Like static strengths, fatigue strengths of composites in longitudinal tension and shear are quite independent. Shear fatigue has been studied recently by many investigators [23–26]. Pipes's [23] results on a unidirectional glass–epoxy composite are shown in Fig. 9-14. These results indicate that the fatigue strength of glass–epoxy in interlaminar shear is superior to that in longitudinal tension when compared with the ultimate static strength. This trend of the fatigue strengths of glass–epoxy composites is different from that of high-modulus boron- or graphite-fiber-reinforced composites, in which the longitudinal fatigue strength has been observed to be superior to the shear fatigue [23] (Fig. 9-15). This observation is highly dependent on the specific composites tested and their interfacial properties, and thus the results should not be generalized and must be used with caution. With a view of developing a failure theory to predict fatigue strength under multiaxial stress systems, the fatigue behavior of unidirectional composites in a transverse direction also has been studied [26,27]. No theory has yet been developed that can be applied with confidence. The flexural-fatigue results of Agarwal and Joneja [27] on a unidirectional glass–epoxy composite indicate that when the composite is subjected to cyclic loading that produces stresses in the transverse direction,

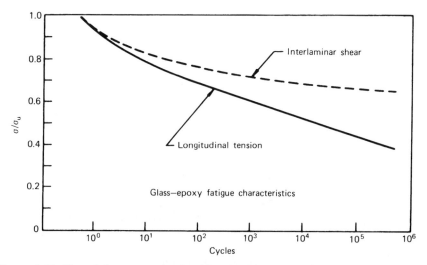

Figure 9-14. Shear fatigue strength of unidirectional glass–epoxy composite. (From Pipes [23].)

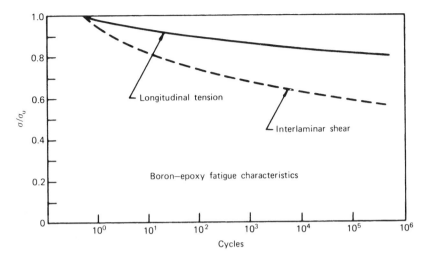

Figure 9-15. Shear fatigue strength of unidirectional boron–epoxy composite. (From Pipes [23].)

there is a sudden and drastic loss of transverse stiffness (to about 30% of its value in the first load cycle) in the initial stage of fatigue cycling (before 5% of the expected fatigue life is spent). This sudden loss of stiffness occurs when the bond between the fibers and matrix in the transverse surface layers of the specimen fails.

The influence of mean stress on fatigue strength has been studied by many investigators [2,3,9,21,28]. The influence of mean stress is usually presented through a plot of permissible stress amplitude as a function of mean stress for a fixed cyclic life. A master diagram consisting of several such plots for different fatigue lives is constructed from the S–N curves obtained for different mean stresses. A typical master diagram for a glass–polyester composite is shown in Fig. 9-16 for four different cyclic lives (namely, 10^3, 10^4, 10^5, and 10^6 cycles). The results indicate that for a fixed cyclic life, the permissible stress amplitude decreases as the mean stress increases. For a negative mean stress (i.e., a compressive mean stress), the stress amplitude is larger than that for a zero mean stress. For a given mean stress, cyclic life decreases as the stress amplitude increases. The influence of mean stress on the fatigue behavior of composites is similar to that of metallic materials. It is observed that like metals, the stress amplitude in composites is related to the mean stress through a linear relationship. The Goodman–Boller relationship, which is usually found to be in good agreement with the experimental results, assumes that for a fixed fatigue life, the decrease in stress amplitude normalized by the fatigue strength (or stress amplitude) at zero mean stress is equal to the mean stress normalized by the stress–rupture strength, defined as the

384 PERFORMANCE OF FIBER COMPOSITES

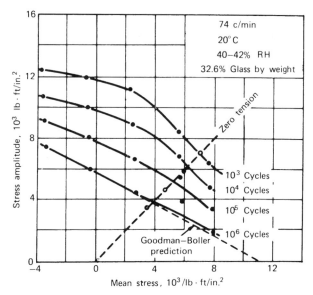

Figure 9-16. Master diagram showing influence of mean stress on fatigue strength. (From Owen et al. [3].)

constant stress that will produce fracture of the composite in a time corresponding to the fatigue life (i.e., the duration of fatigue cycling). The Goodman–Boller relationship can be written as

$$\frac{S_A}{S_E} = 1 - \frac{S_M}{S_c} \tag{9.1}$$

where S_A and S_M are stress amplitude and mean stress, respectively, S_E is the fatigue strength at zero mean stress for equal cyclic life, and S_c is the stress–rupture strength for the time corresponding to the cyclic life. Limited experimental results [21] indicate that at elevated temperatures, the influence of mean stress may be different from that predicted by the Goodman–Boller relationship.

Dally and Broutman [12] have shown that the frequency of stress cycling significantly influences the temperature rise of specimens during fatigue testing. However, the fatigue life of both cross-ply and isotropic materials is only modestly influenced by frequency effects. As the frequency and temperature rise are increased, the fatigue life decreases. Insensitivity of the fatigue life to the frequency and resulting temperature rise is probably due to the fact that the properties of glass fibers, which for the composites tested are the primary load-bearing member in the composites, remain unaffected at these temperatures.

9.1.4 Empirical Relations for Fatigue Damage and Fatigue Life

An almost infinite variety of laminates can be used for structural applications. Once it has been decided to use a specific laminate, its fatigue characteristics may be obtained through experiments. It is, of course, not practical to approach the design problem in the inverse way—that is, to characterize all possible laminates to select the proper one. It is desirable, however, to evolve some simple equations that can be used in design analysis.

Broutman and Sahu [29] have proposed a theory to predict loss of strength during fatigue. According to their theory, loss of strength is a product of the difference between static strength and fatigue strength and the applied-load-cycles–expected-fatigue-life ratio. This theory is useful in predicting cumulative fatigue damage when the load cycle changes from time to time, as actually happens in the real structures. However, this theory cannot be applied for design analysis because it requires determination of separate S–N curves for different materials. Mandell [30,31] has related the rate of crack propagation to the maximum stress-intensity factor. In this case also the fatigue strength has not been related to the fatigue life in a way that can be used for design purposes.

Hashin and Rotem [32] have suggested the following correlation between the fatigue strength and the static strength of composite materials:

$$\sigma_f = \sigma_s f(R, N, n, \theta) \tag{9.2}$$

where σ_f and σ_s are fatigue and static strengths, respectively, $f(R, N, n, \theta)$ is a function of R, stress ratio in fatigue cycling, N is the fatigue life, n is the frequency of load cycling, and θ is the fiber orientation for unidirectional composites. However, the function f has to be evaluated experimentally. This is therefore a limitation for its applicability to design analysis. Unless a simple way can be found to evaluate this function, Eq. (9.2) cannot be used very effectively.

It has been observed that the S–N curves of composite materials often can be represented by straight lines with the equation [33]

$$\frac{\Delta S}{\sigma_u} = m \log N + b \tag{9.3}$$

where ΔS is the stress range, σ_u is the ultimate tensile strength, and m and b are material constants. Some investigations [11,33] show that the values of m and b may be close to 0.1 and 1.0, respectively. However, there are not sufficient experimental data to suggest that these values may be used with confidence for design applications.

Another useful relationship to represent fatigue data is a power law:

$$N^k \Delta\epsilon = c \qquad (9.4)$$

where $\Delta\epsilon$ is the strain range, and k and c are material constants. This equation has been found to be very useful for predicting fatigue life of metallic materials [34,35]. For most metals, k is known to vary from 0.5 to 0.6, and the value of c is related to the ductility of the material. In this manner it is possible to predict the fatigue behavior of metals from their static properties. Results of Agarwal and Dally [33] and Hahn and Kim [11] do follow Eq. (9.4), but the constants k and c are not universal constants for composite materials. Further, it has not yet been possible to relate these constants to the static properties of materials. Such a correlation may be evolved in the future so as to make Eq. (9.4) as well as Eq. (9.3) very useful in design procedures.

9.1.5 Fatigue of High-Modulus Fiber-Reinforced Composites

High-modulus fiber-reinforced composites are of special interest in applications where weight saving is at a premium. For example, graphite fibers offer very attractive properties of high modulus, high strength, and low density. Graphite-fiber-reinforced polymer composites have a specific gravity of about 1.5 while achieving a tensile modulus close to that of steel. Thus the specific modulus of the composite is four to five times greater than that of steel.

High-modulus fiber-reinforced composites such as Kevlar-, boron-, and graphite-reinforced polymers display excellent fatigue resistance when tested in directions such that the properties are fiber-controlled. In other words, although the transverse tensile fatigue resistance of a unidirectional graphite composite will not differ from that of a glass-fiber composite, the longitudinal fatigue resistance will be much better. In general, it can be stated that the excellent fatigue resistance of these materials results from the environmental stability of the high-modulus fibers and their low strains to failure, which as a result produce low strains in the matrix, such as during fatigue of a unidirectional composite in the fiber direction.

Graphite fibers can be obtained from the pyrolysis of continuous polyacrylonitrile (PAN) fibers or rayon fibers. Although a continuous spectrum of fiber strength and modulus values can be obtained by varying the process details, especially the maximum temperature and stretching force during pyrolyzing, graphite fibers usually are available either as high-modulus or high-strength fibers. Tensile strength and modulus of some carbon fibers already have been given in Chap. 2. Graphite fibers are highly anisotropic. The anisotropy in strength, modulus, and thermal expansion coefficients resulting from the graphite structure is particularly important and is reflected in the properties of the composites.

Graphite-fiber-reinforced composites are much newer materials compared with glass-fiber-reinforced composites. Their commerical use is still very limited. Consequently, although their fatigue properties have been studied in laboratory tests, extensive field experience is not yet available. Early investigations [36] showed that the axial fatigue properties of high-modulus graphite-fiber composites in both unidirectional and cross-ply forms are excellent and relatively insensitive to attack by moisture or oil at ambient temperature. Typical $S-N$ curves for a unidirectional composite and a $[0_4/\pm45_2/90]$ laminate of graphite-epoxy are shown in Figs. 9-17 and 9-18, respectively. The curves show several unusual features. The curves are far more flat than the $S-N$ curves for other common structural materials, including glass-fiber-reinforced polymers. Fatigue failure appears to have occurred in only those specimens that were loaded at stresses nearly equal to the static strength, probably within the range of the static-strength scatter band. The specimens loaded at stress levels only just below the scatter band survived 10^6 cycles and in some cases even 10^7 cycles. This fatigue behavior of graphite-fiber-reinforced composites is well documented and has been observed by many investigators [37–39]. The fatigue strength of these unidirectional composites remains unaffected when the tests are carried out at 350°F instead of room temperature (see Fig. 9-17). The fatigue strength of the $[0_4/\pm45_2/90]$ laminates is only modestly lowered when the tests are carried out at 260°F instead of room temperature. The excellent fatigue resistance of graphite-fiber-reinforced epoxy composites results largely from their environmental stability. For example, longitudinal stress–rupture curves for the unidirectional composite are shown in Fig. 9-19. The rupture strengths at elevated temperatures are shown for an "as received" composite, for a composite after exposure to 98% relative humidity at 120°F for 100 h, and for a composite after a 500-h exposure to 350°F. These results indicate that the elevated temperature rupture strength of the graphite composite, even at long exposures to stress, is nearly as high as the corresponding static strength. Whereas a preexposure of the composites to 350°F for 500 h has negligible influence on the rupture strength, preexposure to 98% relative humidity seemed to improve the rupture strength.

Owen and Morris [37] have indicated that the graphite-fiber-reinforced composites do not show any significant temperature rise even at the high test frequency of 7000 cycles per minute, at which structural steels become very

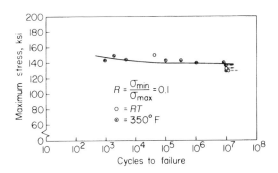

Figure 9-17. $S-N$ curve for a unidirectional graphite–epoxy laminate.

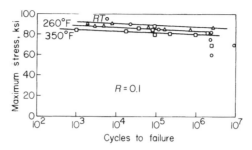

Figure 9-18. S–N curve for a [0_4/ ±45_2/90] graphite–epoxy laminate.

hot. Liber and Daniel [38] observed that these materials do not exhibit any significant deterioration of mechanical properties such as loss of stiffness and strength during fatigue loading. This observation is quite different from that for glass-fiber-reinforced polymers, in which the fatigue damage is known to be progressive. Compressive strengths of these materials have been found to be much lower than the tensile strengths, and hence their flexural fatigue is controlled by the compressive strength. The fatigue effect has been found to be much more significant under conditions of shear loading, both interlaminar and torsional [25,40,41].

Fatigue properties of a unidirectional boron–epoxy composite are shown in Fig. 9-20. The S–N curves have been given for different loading conditions and temperatures. It can be observed that for testing in alternating tension and compression ($R = -1$), at room temperature, the S–N curve (10^3–10^7 cycles) is a horizontal straight line, indicating that the fatigue strength does not decrease with increasing cyclic life. In other words, all fatigue failures occur at the same stress level, and if the tests are carried out below this stress level, the specimens will survive 10^7 cycles. This fatigue behavior is similar to that of graphite–epoxy composites discussed earlier. When testing is carried out in fluctuating tension ($R = 0.1$), the room-temperature fatigue strength decreases with increased cyclic life. The fatigue strength at elevated

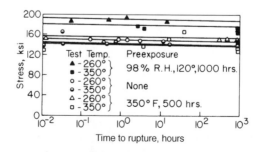

Figure 9-19. Longitudinal stress–rupture curves for unidirectional graphite–epoxy composite under different environmental conditions.

9.1 FATIGUE 389

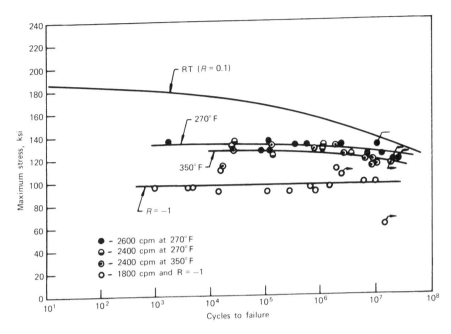

Figure 9-20. S–N curves for unidirectional boron–epoxy composite.

temperatures is lower, but the elevated-temperature S–N curves (for $R = 0.1$) are flatter than the corresponding room-temperature S–N curve. The S–N curve for a [0/±45/90] boron–epoxy laminate is shown in Fig. 9-21. It is observed that the fatigue strength of this laminate is not as good as that of the unidirectional composite.

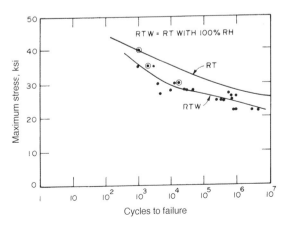

Figure 9-21. S–N curves for [0/±45/90] boron–epoxy laminate.

390 PERFORMANCE OF FIBER COMPOSITES

Comparison of the fatigue behavior for some unidirectional composites with that for aluminum is shown in Fig. 9-22. It can be observed that Kevlar–epoxy and boron–epoxy composites display far superior fatigue properties than the glass–epoxy composites and aluminum. Of course, the graphite–epoxy composites would be comparable with the Kevlar or boron fiber composites and are not shown in Fig. 9-22. An approximate comparison of the fatigue behavior of several reinforced plastics is shown in Fig. 9-23 [42]. This diagram also shows the superiority in fatigue behavior of the high-modulus fiber-reinforced polymers over the glass-fiber-reinforced plastics.

9.1.6 Fatigue of Short-Fiber Composites

Compared with continuous-fiber-reinforced composites, short-fiber composites of all kinds are much less resistant to fatigue damage because the weaker matrix is required to sustain a much greater proportion of the cyclic load. Local failures in the matrix are initiated easily. Fatigue damage in randomly oriented short-fiber composites is initiated by debonding of those fibers that lie perpendicular to the loading direction. In aligned short-fiber composites, the fiber ends and weak interfaces can become sites for fatigue crack initiation. Another important source of damage in reinforced plastics is thermal degradation from the heat dissipated in the material as a result of the large hysteresis loss per cycle and low thermal conductivity of most plastic materials. An important aspect of fatigue of short-fiber composites is that local failures in the matrix and at the interface can destroy the integrity of the composite even though the fibers remain undamaged.

A number of research studies have been performed on the fatigue of short-fiber composites by Owen and associates on glass chopped-strand mat (CSM) and polyester resin (PR) combinations, and the results have been reported in

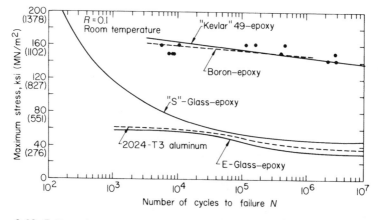

Figure 9-22. Fatigue characteristics of unidirectional composites compared with aluminum.

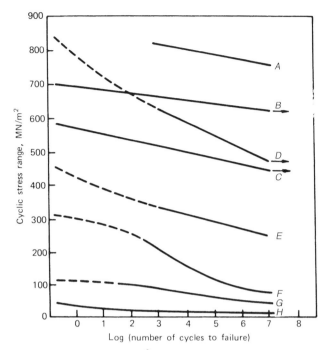

Figure 9-23. Approximate fatigue behavior of several reinforced plastics. Note that CSM refers to chopped-strand mat (curve G), which generally has 5-cm-long randomly oriented fibers, and DMC refers to dough-molding compound, which has less than 1-cm-long randomly orientated fibers. (From Harris [42].)

A —Boron–epoxy laminate; 10 ply 0 ± 45°; axial tension cycling
B —Carbon (type 1)–polyester; unidirectional, V_f = 0.40; axial tension cycling
C —as B; repeated flexure cycling
D —Carbon–epoxy laminate; 18 ply 0 ± 30°; axial tension cycling
E —as D; axial compression cycling
F —Glass–polyester; high-strength laminate from warp cloth; axial tension/compression cycling
G —Glass–polyester composite; CSM laminate, tension/compression cycling
H —Polyester DMC; $V_f \approx 0.12$; tension/compression cycling

a number of publications [2,3,5,43–55]. Important results and conclusions of these studies were discussed by Owen in a publication [56] in 1982. Some of the results are presented here.

The static and fatigue strengths of CSM–PR composites under both uniaxial and biaxial loading were studied. Failure processes, cumulative damage, macroscopic crack propagation, and fracture mechanics were considered. Fatigue test results on CSM–PR composites are presented as conventional S–N curves in Fig. 9-24. The effect of mean stress is presented in Fig. 9-16 as a conventional constant-life diagram showing the relationship between mean stress and stress amplitude. In every case the criterion of failure was separation of the specimens. The most important conclusion from the master di-

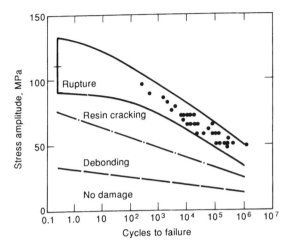

Figure 9-24. Fatigue data for CSM–PR specimens at zero mean stress. (From Owen and Smith [2].)

agram is that the Goodman law, commonly used for structural materials, is unconservative for CSM–PR. By carrying out fatigue loading for various fractions of the expected mean fatigue life and then carrying out stress–rupture tests, they showed that the prior fatiguing produced a marked reduction in the subsequent stress–rupture life. It is this effect, Owen suggested, that undoubtedly makes the Goodman law unconservative in expressing the mean-stress–stress-amplitude relationship for CSM–PR.

Owen et al. [49] produced S–N curves that showed the onset of transverse fiber debonding and the onset of resin cracking (Fig. 9-25). They also showed that the onset of debonding under a single application of tensile load occurred at approximately 0.3% strain for a wide variety of random glass-reinforced plastics (GRPs) and, furthermore, that at the onset of debonding, S–N curves for widely differing GRPs were almost superimposed on a strain basis.

Effect of a stress concentrator on static and fatigue properties of CSM–PR composites is shown in Fig. 9-26 through static-strength scatter bands and S–N curves for the three states of failure for specimens with and without holes. Owen and Bishop [48] concluded that a small hole in a CSM–PR composite was not a fully effective stress concentrator. They also suggest that, for most purposes, the onset of debonding at a hole was uneconomical as a failure criterion. They further pointed out that the S–N curves between 10^3 and 10^6 cycles usually can only be represented as a straight line that extrapolates to zero stress at a finite life. Thus it is very difficult to use the S–N curves for predicting safe loads for long lifetimes.

Fatigue data on short-fiber-reinforced thermoplastic resins have been reported only rarely in the published literature. Typical results for some glass-reinforced thermoplastics are shown in Figs. 9-27 to 9-29 [57]. It has been

9.1 FATIGUE 393

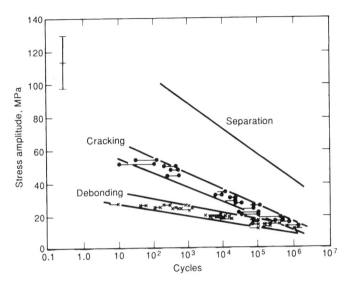

Figure 9-25. The onset of transverse fiber debonding and resin cracking in fatigue under zero-mean-stress conditions. (From Owen, Smith and Duke [44].)

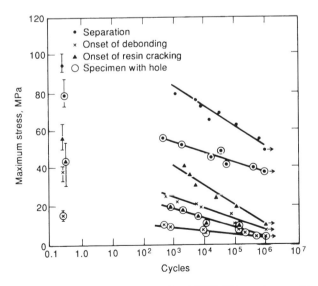

Figure 9-26. Effect of holes on fatigue strength of CSM–PR specimens. (From Owen and Bishop [48].)

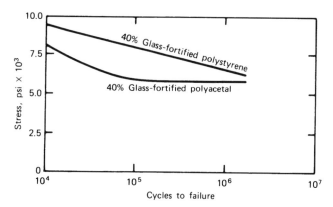

Figure 9-27. Fatigue endurance of glass-fortified polyacetal and polystyrene. (From Theberge [57].)

pointed out by Theberge [57] that fatigue endurance for many low-cost glass-fortified resin systems exceeds values obtained with "fatigue resistant" unreinforced thermoplastics. Dally and Carrillo [58] studied the fatigue behavior of glass-fortified thermoplastics with matrix materials having different ductilities. It was observed that fatigue damage was initiated by debonding of fibers in all cases, but the propagation of fatigue cracks was controlled by the toughness of the matrix materials. In the case of a brittle matrix (polystyrene), cracks could propagate easily into the resin-rich areas. In the case of a highly ductile matrix (polyethylene), no cracks were found in the matrix, and failure occurred by massive debonding. In the case of a matrix material with high strength and intermediate ductility (nylon), a limited crack propagation into the matrix was observed.

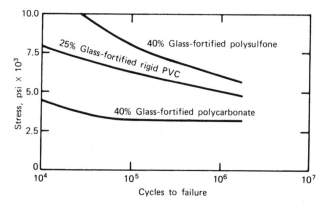

Figure 9-28. Fatigue endurance of glass-fortified rigid polyvinylchloride, polypropylene, and polysulfone. (From Theberge [57].)

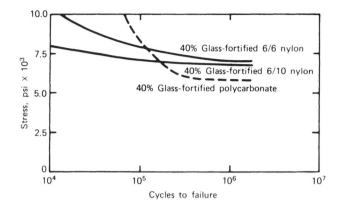

Figure 9-29. Fatigue endurance of glass-fortified polycarbonate, 6/10 nylon, and 6/6 nylon. (From Theberge [57].)

9.2 IMPACT

9.2.1 Introduction and Fracture Process

High-strain-rate or impact loads are expected in many engineering applications of composite materials. The suitability of a composite for such applications therefore is determined not only by its static strength considerations but also by its impact behavior or energy-absorbing properties. Frequently, an attempt to improve the tensile properties results in a deterioration of impact properties. For example, high-modulus fiber composites are more brittle and absorb less energy during fracture than the lower-modulus glass-fiber composites. Thus it is important to have a good understanding of impact behavior of composites for both safe and efficient design of structures and to develop new composites having good impact properties as well as good tensile properties. Discussion in this section is aimed at developing such an understanding.

The impact behavior of a material is greatly influenced by the fracture process induced by the impact loading. Thus it is important to understand the fracture process in composites to better understand their impact behavior. The fracture process in composites is much more complex than that in homogeneous materials such as metals or plastics because a number of microfailure events can occur during fracture propagation in a composite. A typical composite fracture process and various microfailure events were discussed in Chap. 8. However, because of its importance to impact properties, the discussion is being repeated in the following paragraph.

It can be assumed that failure in a fiber composite, just as in metals, emanates from small, inherent defects in the material. These defects may be broken fibers, flaws in the matrix, or debonded interfaces. After initiation, the

396 PERFORMANCE OF FIBER COMPOSITES

failure propagation or fracture process can be described using the simple model shown in Fig. 9-30. The model shows several possible local failure events occurring during the fracture of a fiber composite. At some distance ahead of the crack, the fibers are intact. In the high-stress region near the tip, they are broken, although not necessarily along the crack plane. Immediately behind the crack tip, *fibers pull out* of the matrix. In some composites, the stress near the crack tip could cause the fibers to debond from the matrix before they break. It is also possible for a fiber to be left intact as the crack propagates. When brittle fibers are well bonded to a ductile matrix, the fibers tend to snap ahead of the crack tip, leaving bridges of matrix material that neck down and fracture in a completely ductile manner. In addition to these local failure mechanisms, a crack, on reaching the interface of two laminae in a laminated composite, can split and propagate along the interface, thus producing a *delamination crack*.

Different failure mechanisms involved in the fracture process, as just discussed, account for the total energy absorbed in the fracture. However, the mechanisms playing a dominant role in one system of matrix material and fibers may be different in another system of fibers and matrix material. No single mechanism can account for the observed toughness of composites. Individual energy-absorbing mechanisms during failure of composites are discussed in the next subsection.

9.2.2 Energy-Absorbing Mechanisms and Failure Models

When a solid is subjected to any kind of loading, static or dynamic, it can absorb energy by two basic mechanisms: (1) material deformation and (2) creation of new surfaces. The material deformation occurs first. If the loading supplies a sufficient amount of energy, a crack may initiate and propagate,

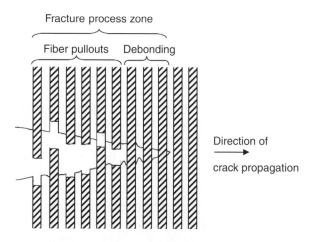

Figure 9-30. A model of fracture-process zone shows local failure events near crack tip.

creating new surfaces, which is the second source of energy absorption. The material deformation continues ahead of the crack tip during crack propagation. Brittle materials such as glass and ceramics undergo only a small amount of deformation, and the energy absorbed is small. Additional energy is absorbed as new surfaces are created during fracture. The energy associated with fracture surface creation is also relatively small. Thus brittle materials exhibit low energy-absorbing capability or low toughness. In ductile metals, plastic deformations cause large energy absorption. This results in the high fracture toughness of ductile materials.

It can be deduced easily from the preceding discussion that energy-absorbing capability or toughness of a material can be enhanced by increasing its plastic deformation capability or by increasing creation of new surfaces during fracture. In metals, the former mechanism is used frequently, and metallurgical processes are developed to enhance toughness. The toughness of a composite can be enhanced by replacing a low energy-absorbing constituent with a greater energy-absorbing constituent. For example, glass fibers are introduced into carbon-fiber composites to produce tougher *hybrid composites.* However, for a composite with a fixed system of matrix material and reinforcing fibers, the failure events taking place in the *fracture process zone,* and the fracture energies associated with them, should be studied carefully and analyzed for modifying material-impact behavior or toughness. Different failure events and associated energies are discussed in the following paragraphs.

9.2.2.1 Fiber Breakage Whenever a crack has to propagate in the direction normal to the fibers, fiber breakage eventually will occur for complete separation of the laminate. Fibers will fracture when their fracture strain is reached. Brittle fibers such as graphite have a low fracture strain and hence have a low energy-absorbing capability. The energy required per unit area of the composite for fracture of fibers in tension is given by the following expression [59]:

$$u = \frac{V_f \sigma_{fu}^2 l}{6 E_f} \tag{9.5}$$

where V_f is the fiber volume fraction, σ_{fu} is the ultimate strength of fibers, E_f is the fiber modulus, and l is the fiber length. Beaumont [60] has proposed a similar expression (with fiber length replaced by fiber critical length) for the energy-release rate caused by fiber breakage during the fracture process. Such similarity should be expected because the factors that cause an increase in the impact energy also result in the increase of fracture toughness of composites.

It may be mentioned here that although the fibers are responsible for imparting high strength to the composites, the fracture of fibers accounts for only a very small fraction of the total energy absorbed. It has been observed experimentally [60] that the number of fibers fractured has little influence on the total impact energy. It should be remembered, however, that the presence of fibers very significantly influences the failure modes and thus the total impact energy.

9.2.2.2 Matrix Deformation and Cracking The matrix material surrounding the fibers has to fracture to complete the fracture of the composite. Thermosetting resins, such as epoxies and polyesters, are brittle materials and can undergo only a limited deformation prior to fracture. However, metal matrices may undergo extensive plastic deformation. Although cracking and deformation of the matrix material both absorb energy, the energy required for plastic deformation is considerably higher than the surface-energy contribution. Thus the contribution of metal matrices to the total impact energy of composites may be significant, whereas the contribution of polymer matrices may be relatively insignificant.

The work done in deforming the matrix is proportional to the work done in deforming the matrix to rupture per unit volume U_m times the volume of the matrix deformed per unit area of the crack surface [62]. Based on the equation derived by Cooper and Kelly [63] for the volume of matrix affected by fracture, the energy required for matrix fracture per unit area of composite is given by

$$u = \frac{(1 - V_f)^2}{V_f} \frac{\sigma_{mu} d}{4\tau} U_m \qquad (9.6)$$

where σ_{mu} is the tensile strength of the matrix, d is the fiber diameter, and τ is the interfacial shear stress.

In the metal matrix composites, the contribution indicated by Eq. (9.6) to the total impact energy may be quite significant. However, in the case of brittle polymer matrices, U_m is small, and thus the energy required for matrix deformation may be only a negligible fraction of the total energy.

The total energy absorbed by matrix cracking (i.e., separation alone) is equal to the product of the surface energy and the new area produced by the crack. A typical matrix crack is shown in Fig. 9-31. When a crack propagates in one direction only, the new area produced is small, producing small fracture energy. Large crack areas may be produced by crack branching, in which case the cracks run in the direction normal to the general direction of fracture. For example, when a matrix crack encounters a strong fiber placed perpendicular (or at a large angle) to the direction of crack propagation, the crack may branch to run parallel to the fiber. In many cases the surface area produced by the secondary cracks is much larger than the area parallel to the primary cracks. This may increase the fracture energy many times and may be an effective way of increasing the toughness of composites or the total energy absorbed during fracture.

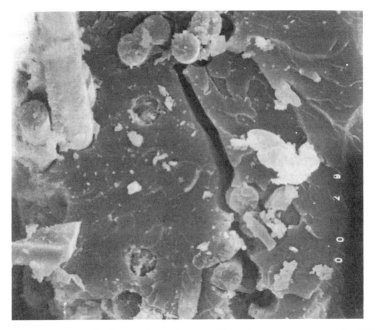

Figure 9-31. Photograph of a matrix crack formed in a glass–epoxy composite (900×).

9.2.2.3 Fiber Debonding During the fracture process, the fibers may become separated from the matrix material by cracks running parallel to the fibers (debonding cracks). In this process, the chemical or secondary bonds between the fibers and the matrix material are broken. This type of cracking occurs when fibers are strong and the interface is weak. A debonding crack may run at the fiber–matrix interface or in the adjacent matrix depending on their relative strengths. In either case, a new surface is produced. If debonding is extensive, a significant increase in the fracture energy may be obtained. An increase in impact energy may be observed with a decrease in interface strength because it promotes extensive debonding or delamination [64]. Debonding cracks also may be regarded as secondary matrix cracks branching off from the primary cracks, as already discussed.

Values of the work of debonding for a number of materials have been given by Kelly [65]; they are usually ≤ 500 J/m^2 and on the order of the interface shear strength times the failure strain of the resin. No other way of theoretically estimating the debonding energy from the properties of the components and of the interface has been given. Kelly [62] also has shown that the energy of debonding cannot be equated to the elastic energy stored in the fibers after debonding, as had been done by Outwater and Murphy [66].

9.2.2.4 Fiber Pullout Fiber pullouts occur when brittle or discontinuous fibers are embedded in a tough matrix. The fibers fracture at their weak cross sections that do not necessarily lie in the plane of composite fracture. The stress concentration in the matrix produced by the fiber breaks is relieved by matrix yielding, thus preventing a matrix crack that may join the fiber fractures at different points. In such cases, the fracture may proceed by the broken fibers being pulled out of the matrix rather than fibers fracturing again at the plane of composite fracture. This is particularly true of those fibers whose ends are within a small distance (less than half the critical fiber length) of the particular cross section at which failure occurs. A typical example of pulled-out fibers and holes remaining on a fracture surface from pulled-out fibers is shown in Fig. 9-32.

An analysis originally derived to account for the energy dissipated during extraction of discontinuous fibers of length l_c may be used to approximate the pullout energy of a continuous-fiber composite with a distribution of fiber strengths by assuming that the mean fiber pullout length is equal to $l_c/4$. Therefore, the pullout energy per unit area is given by [60]

$$u = \frac{V_f \sigma_{fu} l_c}{12} \qquad (9.7)$$

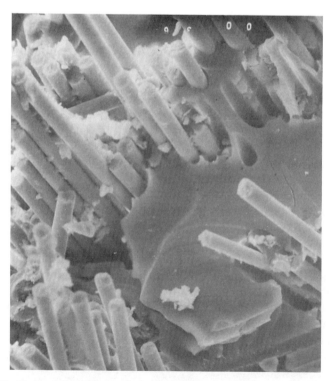

Figure 9-32. Photograph of fibers pulled out of matrix in a glass–epoxy composite (600×).

The difference between fiber debonding and fiber pullout may be clarified at this point. The fiber debonding takes place when a matrix crack is unable to propagate across a fiber, whereas fiber pullouts are a result of the inability of a crack initiated at a fiber break to propagate into the tough matrix. The fiber pullouts usually are accompanied by extensive matrix deformation, which is absent in fiber debonding. Thus fiber debonding and fiber pullout may appear to be similar phenomena because of failure taking place at the fiber–matrix interface in both cases, but they are caused by mutually exclusive conditions. However, both phenomena do significantly enhance fracture energy.

9.2.2.5 **Delamination Cracks** A crack propagating through a ply in a laminate may get arrested as the crack tip reaches the fibers in the adjacent ply. This process of crack arrestment is similar to the arrestment of a matrix crack at the fiber–matrix interface. Because of high shear stress in the matrix adjacent to the crack tip, the crack may branch off and start running at the interface parallel to the plane of the plies. These cracks are called *delamination cracks,* and whenever present, they are responsible for absorbing a significant amount of fracture energy. Delamination cracks frequently occur when laminates are tested in flexure, as is the case in Charpy and Izod impact tests. Delaminations occurring during the impact of glass-reinforced epoxy beams are shown in Fig. 9-33.

Based on the energy-absorbing mechanisms discussed in the preceding paragraphs, the energy-absorbing capabilities of composite materials can be explained. It is clear from the preceding discussion that the influence of material variables on tensile and impact properties may be mutually opposing. For example, a decrease in interface strength that adversely affects tensile and shear strengths can result in an increase in impact strength. The high-modulus fiber composites (because of lower fracture strain of fibers) absorb less energy and are more brittle than the lower-modulus glass-fiber composites. It is thus obvious that a compromise in properties can be achieved by combining both glass fibers and higher-modulus fibers in the same composite. These hybrid composites are discussed in a later section.

9.2.3 Effect of Materials and Testing Variables on Impact Properties

The study of impact behavior of fibrous composite materials did not receive much attention until the mid 1960s. The early published results on the subject were obtained with a standard Charpy impact machine without any attempt to study the phenomenon of impact [67,68]. For a fiber-reinforced polymer, it is expected that the impact behavior will be time-dependent, that is, dependent on the velocity of the hammer when striking the specimen. In conventional impact-testing machines there is no provision for studying this important aspect of impact behavior. Rotem and Lifshitz [69] showed that the

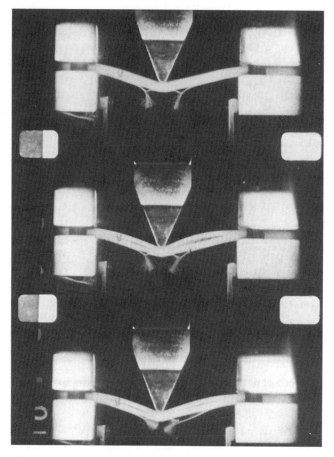

Figure 9-33. Delaminations in a composite during an impact test. Three successive frames (at a speed of 6000 frames per second) are shown. (From Broutman and Rotem [61].)

ultimate strength of a glass-fiber-reinforced plastic increases as a function of the rate of loading in the fiber direction. Later Broutman and associates [61,64,70–72] performed extensive impact studies on a specially built drop-weight impact-testing apparatus in which several experimental parameters could be varied easily.

Current practices of impact testing of composite materials are described in Chap. 10. The discussion explains various parameters mentioned in this section and how they can be varied and measured.

Broutman and associates [61,64,70–72] performed their studies on glass-fiber-reinforced epoxy and polyester resins and on hybrid graphite–Kevlar–glass composites. In either system, unidirectional as well as cross-ply lay-ups were used. Experimental parameters, which were varied, included fiber orientation, interface strength, velocity of impact, drop weight, and specimen dimensions. Through high-speed photography, important observations on the phenomenon of fracture and associated energy-absorbing mechanisms were made. Important aspects of the results on glass-fiber-reinforced polymers are discussed in this subsection along with relevant results of other investigators. The results on hybrid composites are discussed in a later section.

Fiber orientation is an important parameter influencing the impact behavior of composites. The effect of fiber orientation angle on the impact properties of off-axis composites was investigated by Mallick and Broutman [71] on E-glass–epoxy laminates. The exact configuration of what has been designated as unidirectional laminate is $[0/90/0_4/\overline{0}]_S$; that is, the unidirectional laminate actually has two layers at 90° to the others and placed beneath each surface layer. Configuration of the cross-plied laminate is $[(0/90)_3/\overline{0}]_S$. Both systems contain 13 plies, each 0.01 in. thick. Rectangular specimens were cut from the laminates so that fibers in the outer layers made angles of 0°, 15°, 45°, 75°, and 90° with the longitudinal-beam axis. This angle is referred to as the *fiber orientation angle* (θ). In all cases, the load was applied normal to the lamination plane, as shown in Fig. 9-34. The impact energy absorbed per unit width in breaking the unidirectional specimens is shown in Fig. 9-35. The lowest value is observed at $\theta = 60°$. At $\theta = 60°$ the impact energy increases in the presence of the 90° layers (beneath surface layers), in which fibers are oriented at small angles of $(90° - \theta)$ to the longitudinal-beam axis and thus are capable of absorbing higher energy. However, Agarwal and Narang's [73] Charpy impact test results on composites with all unidirectional fibers indicate that the impact energy decreases continuously with increasing fiber orientation. Minimum impact energy is observed at $\theta = 90°$. The results of Mallick and Broutman [71] for cross-ply specimens are shown in Fig. 9-36. In this case, the curve for the energy absorbed is symmetric about $\theta = 45°$, at which the lowest impact energy is observed. Also shown in Fig. 9-36 is the impact energy of unidirectional specimens for the same drop height and specimen dimensions. The impact energy absorbed by the cross-ply specimens is consistently higher than that for the unidirectional specimens except at $\theta = 0°$, at which the unidirectional specimens absorb higher energy. Lifshitz [74] also has recently determined the impact strength of angle-ply composites.

Another important material parameter is interface strength, which strongly influences the failure mode in composites. Yeung and Broutman [64] varied the interface conditions by changing the surface treatment of the glass fabric

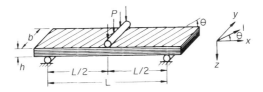

Figure 9-34. Impact test arrangement. (From Mallick and Broutman [71].)

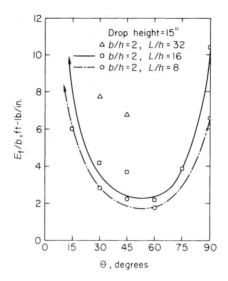

Figure 9-35. Influence of fiber orientation on energy absorbed by a unidirectional glass–epoxy composite. (From Mallick and Broutman [71].)

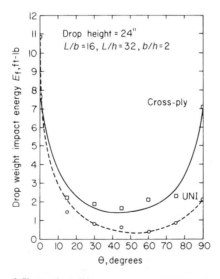

Figure 9-36. Influence of fiber orientation on energy absorbed by cross-ply laminate of a glass–epoxy composite. (From Mallick and Broutman [71].)

(style 181). Polyester and epoxy resins were used as the matrix material. The apparent shear strength as determined from a short-beam shear-strength test was used as a measure of the interface strength. In the case of polymer laminates it was observed that interface strength could be varied over a large range by changing the coupling agent on the fiber surface. However, interface strength of epoxy laminates could not be varied over the same range because the epoxy resin is capable of establishing a strong bond with the glass surface even in the absence of a coupling agent. Polyester and epoxy laminates were tested in an instrumented Charpy impact-testing machine so that initiation energy (E_i), propagation energy (E_p), and total impact energy (E_t) could be obtained. The results are shown in Figs. 9-37 and 9-38, in which the values of energies per unit area [$u_i = (E_i/bh)$, $u_p = (E_p/bh)$, and $u_t = (E_t/bh)$], where b and h are specimen width and thickness, respectively] are plotted as a function of laminate shear strength. In these experiments, the laminate is impacted on its surface (not on the edge), as shown in Fig. 9-34.

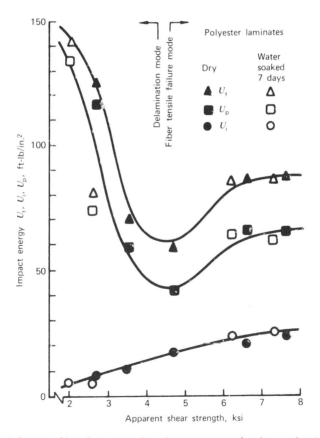

Figure 9-37. Influence of interface strength on impact energy of a glass–polyester composite. (From Yeung and Broutman [64].)

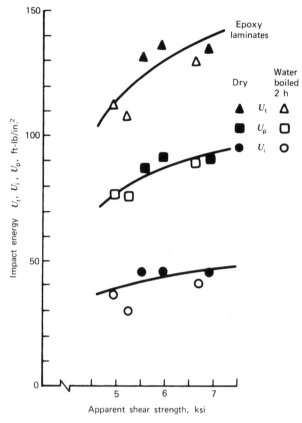

Figure 9-38. Influence of interface strength on impact energy of a glass–epoxy composite. (From Yeung and Broutman [64].)

It can be seen that u_i, the initiation energy, increases with increasing shear strength for both polyester and epoxy laminates. As the shear strength increases, the flexural strength of these fabric laminates also increases, reflecting better interfacial bonding and greater values of intralaminar strengths, particularly interlaminar tensile strengths. The initiation impact energies are much greater for the epoxy laminates, again reflecting their greater flexural strengths.

In the case of polyester laminates, the curves for propagation and total impact energy appear to have a minimum. Above a critical value of interlaminar shear strength, the total impact energy increases with increasing shear strength. Below the critical value of shear strength, the impact energy decreases with increasing shear strength. As indicated in Fig. 9-37, delamination appeared to be the dominant failure mode below the critical value, whereas fiber failure was the dominant mode above the critical value of shear strength. Thus, in the case of polyester laminates, the total impact resistance can be maximized by reducing the interfacial bonding. It is thus observed that the greatest value of impact strength is achieved when the shear strength is lowest. It should be noted that the initiation of failure requires less energy when the interfacial bond is poor and the large value of total impact energy is achieved during the delamination phase occurring after failure initiation. The specimen supports less load during propagation but absorbs more energy because of the large deflections that the specimen can sustain. In the case of epoxy laminates, the interfacial bonding is not reduced to a sufficiently low value to induce severe delamination. Thus, as in the case of polyester laminates above the critical shear-strength value, the impact energy increases with increasing shear strength. It is interesting to note that the maximum impact energy observed for a high-shear-strength epoxy laminate is nearly identical to the impact energy for a low-shear-strength polyester laminate. The failure mode for the epoxy laminate is predominantly fiber failure, whereas it is delamination for the low-shear-strength polyester laminate. It has been pointed out by Yeung and Broutman [64] that the surface treatment does not have a significant effect on the interlaminar fracture surface work. Thus the increase in impact energy caused by the reduction of interface strength is a result of more extensive delamination as opposed to changing the work of delamination.

The high energy-absorbing capability of laminates with poor interfacial bonding is of particular interest in the applications of fiber-reinforced plastics as armor materials. In such applications, high impact resistance (total energy absorbed) is most essential. Therefore, in these applications, a low interface strength may be very effective because of the increase in total energy absorbed.

9.2.4 Hybrid Composites and Their Impact Strength

In recent years, high-modulus fibers such as boron and graphite have been used widely in many aerospace applications because of their exceptionally high modulus-weight ratios. However, the impact strength of high-modulus fiber composites generally has been found to be relatively low compared with conventional steel and aluminum alloys, as well as compared with glass-fiber-reinforced composites. Comparative results obtained from a standard notched Charpy impact tests are given in Table 9-1 [75]. An effective method of enhancing the impact properties of graphite-fiber-reinforced composites is to add to them a small percentage of a low-modulus high-strength fiber, which results in higher impact performance. Glass fibers arc used frequently for this purpose. Besides improving impact performance, the incorporation of glass fibers, which is relatively inexpensive compared with graphite fibers, reduces cost, which is a limitation for the application of graphite-fiber composites.

The incorporation of two or more fibers within a single matrix is known as *hybridization,* and the resulting material generally is referred to as a *hybrid composite* or simply *hybrid.* There are basically two ways in which different types of fiber can be combined to formulate a hybrid composite. In the first method, different types of fiber are intimately mixed throughout the resin,

Table 9-1 Typical impact-energy values for various materials—standard Charpy V notched impact tests

Material Description	Impact Energy	
	kJ/m^2	ft-lb/in^2
Modmor II graphite–epoxy ($V_f = 55\%$)	114	54
Kevlar–epoxy ($V_f = 65\%$)	694	330
S-Glass–epoxy ($V_f = 72\%$)	694	330
Nomex nylon–epoxy ($V_f = 70\%$)	116	55
Boron–epoxy ($V_f = 60\%$)	78	37
4130 Steel alloy (UTS 100–160 ksi)	593	282
4340 Steel alloy [Rockwell (43–46)]	214	102
431 Stainless steel (annealed)	509	242
2024-T3 Aluminum alloy	84	40
6061-T6 Aluminum alloy (solution treated and precipitation hardened)	153	73
7075-T6 Aluminum alloy (solution treated and precipitation hardened)	67	32

with no intentional concentration of either type. However, obtaining a uniform blend is a fabrication problem. In the second method, only one type of fiber is placed in a single layer, and then the different fiber plies are dispersed through the laminate. Proper selection of the orientations of these plies offers an additional means of controlling the anisotropy of the laminate. The lay-up of hybrid laminates is almost exclusively symmetric with respect to the neutral axis; otherwise, temperature changes would lead to bending–stretching coupling and cause undesirable warping. Analysis of hybrid laminates can be carried out using the methods discussed in Chap. 6. Hybrid composites are relatively new, and only limited experimental data are available on them. A review of the literature on carbon-fiber- and glass-fiber-hybrid-reinforced plastics is presented by Summerscales and Short [76]. Important aspects of impact performance of hybrid composites are discussed in the following paragraphs.

As indicated earlier, the total impact energy E_t provides only limited information about the fracture behavior of a material. Instrumented impact tests are used to obtain the complete load history during impact so that the energy absorbed during different phases of fracture can be obtained. Through the analysis of the load history, it is observed that the total impact energy is the sum of the initiation energy E_i and the propagation energy E_p. Two materials having equal total impact energies may have completely different proportions of initiation and propagation energies. For example, a brittle, high-strength material will have a large initiation energy and a small propagation energy. Conversely, a low-strength (but high-energy-absorbing) material will have a small initiation energy and a large propagation energy. For specimens having similar geometries, the relative percentage of energy absorbed in fracture initiation and propagation provides an indication of the ductility or energy-absorbing capability of a material. Beaumont et al. [77] defined a dimensionless parameter called the *ductility index* (DI), which is found useful for ranking the impact performance of different materials with similar geometries. The DI is defined as the propagation-energy–initiation-energy ratio:

$$\mathrm{DI} = \frac{E_p}{E_i} \tag{9.8}$$

Low values of DI would mean a low value of propagation energy, and hence this is an indication of the brittleness of the material. Along with the total impact energy and maximum stress, the DI is a convenient parameter for ranking materials tested in a Charpy test when specimens have similar geometries. All these parameters are conveniently measured in an instrumented Charpy test.

Notched Charpy test results for several composites as reported by Toland [78] are shown in Fig. 9-39. The total energy absorbed during impact has been separated into initiation and propagation energies. This chart shows that the major distinction between different composites is in the propagation energies. Impact properties of several unidirectional laminates based on the results of Broutman and Mallick [72] are given in Table 9-2. E-Glass–epoxy laminates exhibit the highest energy absorbed per unit area, whereas graphite (GY-70) fiber–epoxy exhibits the lowest energy-absorption capability. Graphite GY-70 fiber–epoxy laminates exhibit a very brittle behavior, as is reflected in extremely low impact energy and a zero DI value for the material. Thornel 300–epoxy laminates exhibit a higher impact energy and a higher DI value than the GY-70 laminates because in this case the fracture was accompanied by some fiber pullout and delamination from layer to layer. The low DI value for the E-glass–epoxy laminates results because of the very high initiation energy caused by a high strain-energy-absorbing capability of the E-glass fibers. Kevlar 49–epoxy laminates also exhibit some fiber pullout and yielding

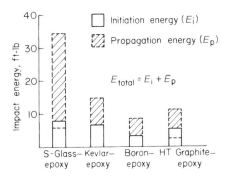

Figure 9-39. Initiation and propagation energies for several composites. (From Toland [78].)

Table 9-2 Impact properties of unidirectional fiber–epoxy composites

Fiber	L/h^a	Dynamic Flexural Strength (10^3 psi)	E_t (ft-lb)	U_t (ft-lb/in.^{2})	U_i (ft-lb/in.^{2})	DI[b]
E-Glass	16.1	281	14	296	222	.37
Graphite (Thornel 300)	14.6	229	4.8	89	40.8	1.2
Graphite (GY-70)	12.6	70	0.35	5.85	5.85	0
Kevlar 49	10.5	98	8.9	113.6	36.2	2.2

[a] Span–depth ratio of beam.
[b] Ductility index = $(E_p/E_i) = (U_p/U_i) = (U_t - U_i/U_i)$.
Source: Broutman and Mallick [72].

on the compression side, and the specimens do not fracture completely. Beaumont et al. [77] also have reported very high DI values for Kevlar 49–epoxy laminates.

A number of investigators have reported that the incorporation of fibers with high strains to failure (e.g., glass or Kevlar 49) in graphite–epoxy composites has produced materials having significantly higher Charpy impact energies. In hybrid composites, there are many variables that influence the material properties. The following variables influence the behavior of hybrid composites: (1) volume and weight fraction of each component fiber relative to both the entire hybrid laminate and the individual ply, (2) lay-up sequence and orientation, (3) relevant properties of resin and fibers and interlaminar shear strength between plies, and (4) extent and nature of any voids or other quality defects.

Glass fiber is the most common choice for incorporation in graphite-fiber-reinforced plastics, but Kevlar 49 fiber also has been found suitable for the purpose. The variation in impact energy of a unidirectional composite is shown in Fig. 9-40 as a function of S-glass stratified in a carbon-fiber-reinforced plastic composite [79]. The carbon fibers involved are high tensile strength and exhibit high fiber–matrix bond strength. Figure 9-40 shows a significant (500%) increase in impact energy for inclusion of approximately 40% S-glass. The 10–20% range of glass inclusion produces approximately a 100% increase in impact energy. This latter range is probably more representative of most structural applications. However, in designing for improved impact resistance, due consideration has to be given to the influence of hybridization on other composite properties. The improved impact resistance of carbon-fiber-reinforced plastics from the addition of glass fibers is attributed to the increase in the area of fracture surfaces created by the increased number of fracture modes available and the increased capability to store strain energy in a glass-fiber core in a sandwich composite. Beaumont et al. [77] found that unidirectional Kevlar 49–epoxy hybrid composites had significantly higher impact energies and ductility indices than had graphite–epoxy speci-

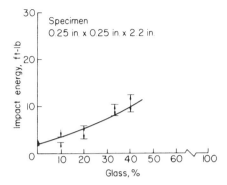

Figure 9-40. Influence of adding glass fibers in a graphite–epoxy laminate on its impact energy. (From Toland [79].)

mens with no loss in maximum stress. Broutman and Mallick [72] also have reported an increase in impact energies and ductility index by the incorporation of glass and Kevlar 49 fibers into graphite-fiber composites. They also reported that sandwich-type construction, in which all-glass or Kevlar 49 plies are placed in the core and only the skin or shell plies are of graphite fibers, are more effective in improving the impact resistance than are the stratified laminates in which alternate unidirectional layers are of graphite and E-glass or Kevlar 49 fibers.

Adams and Miller [75] have presented a procedure for analysis of the impact behavior of hybrid composite laminates. The analysis is based on fracture mechanics concepts. Calculated elastic strain energies at maximum impact load are compared with measured fracture-initiation energies. Correlation (and probably a better correlation) of such calculations for more experimental data and better understanding of linear elastic fracture mechanics (LEFM) concepts with respect to their applications to composites should be awaited before such involved procedures can be recommended.

9.2.5 Damage Due to Low-Velocity Impact

The discussion of impact behavior of composite materials in the preceding sections has considered situations where impact by a foreign object causes penetration of the impactor through the material. The penetration or puncture occurs when the impact velocity is sufficiently high to cause through-the-thickness fracture of the material, at least locally. There are many practical situations where an impact does not result in puncture of the material but causes damage to the material and results in a loss of the laminate strength, although the damage may not be clearly visible. The impacts that do not cause through-the-thickness fracture are usually termed *low-velocity impact*. When a composite structure undergoes such impacts, it is important to know the level of damage by taking into consideration the residual strength after impact. Metals, because of their ductility, are more able to absorb these impacts effectively, and the impacts do not significantly influence residual properties.

The behavior of composite materials subjected to low-velocity repeated impact has been investigated by many researchers [80–91]. The studies have considered various aspects such as damage initiation and propagation, type of damage, effect of damage on subsequent material behavior, and theoretical predictions about damage propagation. Experimental studies are carried out on rectangular or square specimens of different sizes and with different end conditions (e.g., clamped or simply supported on two or four sides). Impactors of different weights and shapes have been used. The impactor shapes include spherical, hemispherical, flat cylindrical, blunt-edged, and cantilever ball. The impactor is accelerated by a pressurized air gun or a free fall (drop-weight test). The damage is examined visually or using other techniques such as optical and electron microscopy, ultrasonic C-scanning, and acoustic imaging.

The damage modes of an impacted composite plate can be classified as fiber breakage, delamination, and matrix cracking. It is well known that penetration-induced fiber breakage is one of the major damage modes in high-velocity impact. In low-velocity impact, however, delamination accompanied by matrix cracking has been found to be the major damage mode. The delamination at the interface of two adjacent laminae is caused by mismatch in their stiffnesses as a result of their different orientations. Delamination in four three-lamina cross-ply plates subjected to impacts at different velocities is shown in Fig. 9-41 [83]. It is observed that the delamination areas depend on the impacting energy. As the impacting energy increases, the interface delamination area increases.

Matrix cracks in an impacted composite plate are a result of stress concentrations at the fiber–matrix interface and are produced by tensile stress. The higher the tensile stress, the longer and the denser are the cracks. The matrix cracking patterns in impacted plates have been observed to be completely different from quasi-statically loaded plates as a result of flexural stress waves created during impact. The degree of matrix cracking can be used as an indicator of the degree of impact and the degree of damage. It also has been shown that matrix cracking and delamination are associated phenomena in a low-velocity impact. Therefore, it may be possible to estimate the internal delamination by investigating the external matrix cracking patterns. External matrix cracking patterns on impacted and nonimpacted surfaces of a plate are shown in Fig. 9-42 [83].

The damage produced by low-velocity impact reduces residual strength of a composite plate. Results of Wyrick and Adams [81] showing reduction in tensile and compressive strengths of a carbon–epoxy composite subjected to repeated impacts are given in Figs. 9-43 and 9-44. The extent of strength reduction depends on impact energy and the number of impacts. The follow-

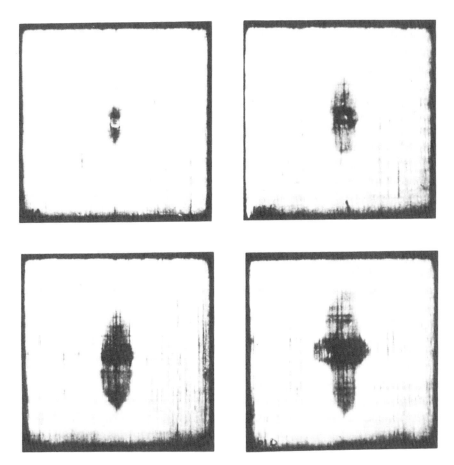

Figure 9-41. Delamination in cross-ply plates impacted at different velocities. (From Liu and Malvern [83].)

ing general comments can be made on the strength reduction owing to repeated impact:

1. Repeated impact progressively increases the damage to the composite. Impacts at higher energy levels produce more degradation than a number of lighter impacts.
2. A plate punctured at a higher energy level sustains less damage in the surrounding area than a plate punctured in a larger number of impacts with less energy per impact.
3. There appears to be a threshold impact energy required to cause a reduction in tensile strength of a laminate. This threshold will be different

414 PERFORMANCE OF FIBER COMPOSITES

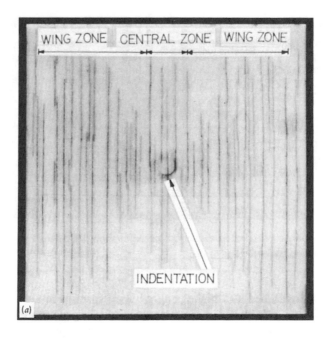

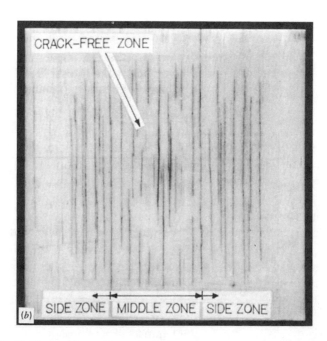

Figure 9-42. Matrix cracking patterns on (a) impacted surface and (b) surface opposite to impact. (From Liu and Malvern [83].)

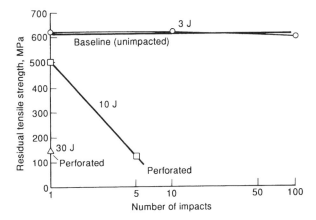

Figure 9-43. Influence of repeated impacts (with different impact energies) on residual tensile strength of a graphite–epoxy laminate. (From Wyrick and Adams [81].)

for different laminates and will depend on such parameters as the laminate thickness, type of fibers, and matrix material.
4. The residual tensile strength decreases as the impact energy or the number of impacts increases. Even below the threshold impact energy, degradation in residual strength may begin after a number of impacts.
5. The residual compressive strength also decreases with an increasing impact energy, and this drop appears significant for even very light impacts.
6. For the material investigated by Wyrick and Adams [81], the threshold impact energy for compressive-strength reduction appears to be much smaller than that for tensile-strength reduction.

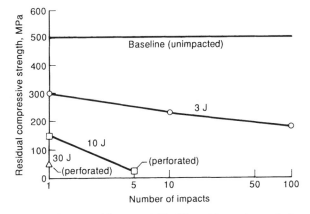

Figure 9-44. Influence of repeated impacts (with different impact energies) on residual compressive strength of a graphite-epoxy laminate. (From Wyrick and Adams [81].)

7. The residual compressive strength decreases with an increase in the number of impacts, the largest degradation occurring as a result of the first impact.

9.3 ENVIRONMENTAL-INTERACTION EFFECTS

The material behavior under cyclic and impact loading has been discussed in the preceding two sections. It is equally important to understand the material behavior under various environmental conditions, such as exposure to water, water vapor, or other corrosive environments; low and high temperatures; and long-term physical and chemical stability. This section is devoted to a discussion on the degradation of composite materials by various environmental factors.

The degradation of composite materials may result from several factors: (1) loss of strength of the reinforcing fibers by stress corrosion, (2) loss of adhesion and interfacial bond strength from degradation of the fiber–matrix interface, (3) chemical degradation of the matrix material, (4) dependence of the matrix modulus and strength on time and temperature, and (5) accelerated degradation caused by combined action of temperature and chemical environment.

As a result of these factors, the utility of composite materials is terminated when the stiffness is reduced sufficiently to cause structural instability, and/ or failure or rupture is induced. Environmental factors influence the fibers, matrix material, and interface simultaneously. Thus the degradation of composites occurs not only with the degradation of the individual constituents but also with the loss of interaction between them. The influence of environmental factors on the fibers, matrix material, and interface is discussed in the following subsections.

9.3.1 Fiber Strength

9.3.1.1 Features of Stress Corrosion In principle, the strength of materials is limited by the magnitude of forces that bind atoms together. In practice, however, the strength of most solids is found to be much smaller than the theoretical strength. The discrepancy in theoretical and experimental strengths is attributed to the presence of imperfections or flaws in the material. Moreover, it is observed experimentally that in the presence of appropriate environments, many hard amorphous or crystalline solids exhibit delayed failure, in which the strength is markedly influenced by the time during which the load is applied. For example, it is a fact that humid atmospheres reduce the breaking strength of silicate glasses, and the strength of many plastics is impaired when immersed in detergent solutions and solvents. Most studies have concluded that the delayed failure under constant load is caused by the growth of flaws under the influence of a reactive environment to a critical size at which the state of stress at the most critical flaw is sufficient to cause a spontaneous failure. In glasses and other amorphous solids, the important flaws are surface cracks or other flaws that can grow under the influence of stress and chemical attack. However, the criticality of a flaw depends on the state of stress and the bulk properties of the material. A noncritical flaw may become critical through flaw growth, stress increase, or time-dependent changes in bulk properties.

In general, details of the stress–corrosion mechanism for different materials and environments are not well understood. However, all systems do exhibit some common features:

1. In inert environments or at low temperatures, where the reaction rate of the corrosion processes should be negligibly slow, the breaking strengths of materials become independent of the duration of load application and always maintain a relatively high value. Equivalently high strengths are observed at high loading rates that do not provide sufficient time for the corrosion process to take place.
2. Exposure of these materials to reactive environments before, but not during, a test has less effect on test results, suggesting that the corrosion rate is accelerated by the stress.
3. Exposure of these materials to reactive environments during a test leads to delayed rupture at loads substantially lower than features 1 and 2 (above).
4. There is a continuous influence of temperature on the relationship between the time of loading and the failure stress. In general, a continuous loss of strength occurs with increasing temperatures.

9.3.1.2 Static Fatigue and Stress–Rupture of Fibers

The most common fibrous reinforcement material is glass. Glass is known to exhibit delayed failure under static loads. At room temperature there are no indications, prior to failure, such as creep, but most investigators agree that moisture promotes the growth of preexisting flaws under a constant load. When the flaw becomes critical in size, the failure occurs instantly. Extensive data are available on glass and glass fibers. A review article on glass fibers is given by Lowrie [92].

Static fatigue tests on E-glass fibers at room temperature and elevated temperatures were conducted by Otto [93]. His results are shown in Fig. 9-45. There is considerable scatter in the results, particularly at the lower temperatures, but reasonable approximations of the fatigue rate can be obtained. The time range covered in the experiments is from 1 min to 20 h, and the decrease in strength over this period is 40,000–65,000 psi (280–435 MPa). The glass fibers also lose their strength as the temperature increases. A large fraction

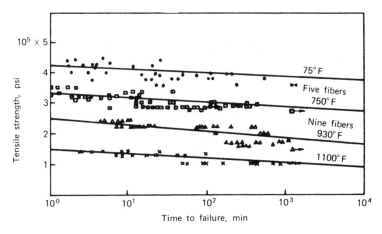

Figure 9-45. Static fatigue test results for E-glass fibers. (From Otto [93].)

of their short-term strength is lost at higher temperatures. At room temperature, the fibers lose about 3% of their short-term strength for every tenfold increase in duration of load application.

Boron and graphite filaments are being used in reinforced composite material applications where high modulus and high strength are required even at elevated temperatures. An early review of boron filaments was written by Wawner [94]. Comparatively little work has been done on the stress–rupture properties of boron filaments. Cook and Sakurai [95] reported some results of stress–rupture tests on single boron filaments at 900°F (Fig. 9-46). The tests were conducted in air and in an argon atmosphere, and the boron filaments showed a sharp decrease in stress level when the load duration increases from 3–5 h. The boron filaments showed an improved characteristic when tested at the same stress level and temperature but in an inert atmo-

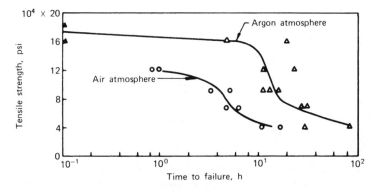

Figure 9-46. Stress–rupture properties of boron fibers. (From Cook and Sakurai [95].)

sphere. The sharp decrease in stress level occurred at a load duration of 10–20 h, as shown in Fig. 9-46. The results of a tensile test (short-term strength) showed almost no difference between the air and argon atmospheres up to 900°F (Table 2 in Wawner [94]), whereas the results in Fig. 9-46 seem to indicate that an appreciable decrease in strength takes place even at short load duration in air. There is no information for stress–rupture tests in air at a load duration greater than 20 h. The results shown in Fig. 9-46 indicate that strength degradation under a constant load at 900°F is very severe, and the filaments lose about 75% of their original strength in less than 100 h in an argon atmosphere and in 10 h in an air atmosphere.

The use of brittle high-strength fibrous materials is limited by a time-dependent distribution of flaw sizes. If, in a fibrous composite, only strength degradation of the fibers occurs, the longitudinal strength of a unidirectional composite will be seriously affected, whereas the transverse and shear strengths will be only marginally affected. The coatings placed on fibers or their incorporation into a matrix material provides an effective means of exploiting the properties of these high-strength reinforcement materials. The coating or "coupling agent" protects the fibers from abrasion or other sources of surface flaws during fabrication. The coupling agent, along with the matrix material, provides a barrier between the aggressive environment and the reinforcement. High-strength reinforcements also are developed with specific efforts devoted to achieving a chemical composition that is inert with respect to the anticipated service environment.

9.3.1.3 Stress Corrosion of Glass Fibers and GRP

The environmental stress-corrosion cracking of glass-fiber-reinforced plastics (GRP) has become an important consideration for applications involving exposure to corrosive environments. Many GRPs fail catastrophically after a critical time when exposed to acids. The catastrophic failure of GRPs occurs as a result of the time-dependent degradation of glass fibers that are highly susceptible to attack by acidic environments. Degradation of glass and GRP owing to corrosive environments is discussed in the following paragraphs.

High-strength glass fibers (10 μm in diameter) containing alkali metals decrease in strength with time when immersed in acidic solutions. It is now widely accepted that the mechanism responsible for strength reduction is an ion exchange between hydrogen ions in the surrounding medium and alkali ions, in particular sodium, in the glass [96–98]. The ion-exchange mechanism can be explained as follows.

Sodium ions (Na^+) in glass are held by only a single electric charge to the structure, whereas ions such as calcium (Ca^{2+}), which are doubly charged, are bonded more firmly. Exchange between singly charged sodium ions and singly charged hydrogen ions can occur most readily. For this to occur, the extremely small hydrogen ions penetrate the glass fibers from the surrounding environment until they find sodium ions. The sodium ions migrate back to the surface and leave the glass. A chemical equation may be written as

Glass structure $\equiv$ Si—O—Na + H$^+$ $\rightleftharpoons$ glass structure $\equiv$ Si—O—H + Na$^+$

The arrows indicate that this reaction can go either way. When large numbers of hydrogen ions are available, as in the case of an acidic environment, the reaction takes place rapidly.

The ion exchange occurs within the glass, only a few atom layers below the surface. The volume of each hydrogen ion is much smaller than that of the sodium it replaces. This leads to a contraction of molar volume to an extent depending on the availability of hydrogen ions in the surrounding medium and the initial alkali-metal content of the glass. As a result of molar contraction, the affected surface layer wants to shrink but is prevented from doing so by the bulk of the glass fiber, whose core is unaffected by the ion exchange. This produces tensile stresses on the surface layer, which can cause failure, particularly if the fiber is already under stress. The photograph in Fig. 9-47 shows a fracture surface of a GRP exposed to sulfuric acid under stress. The spiral cracks that develop on the surface of a glass fiber exposed to an acidic environment can be seen easily. Ends of fibers shown in the figure also demonstrate the delamination of the surface layers owing to the volume changes that occur.

The strength of E-glass fibers in various pH solutions compared with their strength in air is shown in Fig. 9-48 [97]. As can be seen, even short time exposures in strong acids can lower the fiber strength. Since the ion-exchange process is time-dependent, glass fibers subjected to decreasing values of strain or stress still will fail eventually, although the time will be increased, as shown in Fig. 9-49.

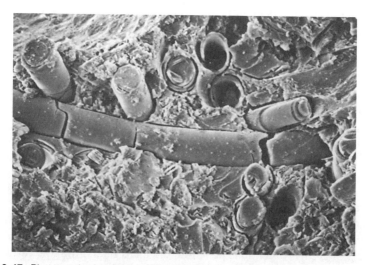

Figure 9-47. Photograph of fracture surface of a GRP exposed to sulfuric acid under stress.

9.3 ENVIRONMENTAL-INTERACTION EFFECTS

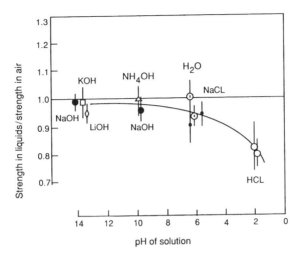

Figure 9-48. Strength of E-glass fibers in various pH solutions. (From Metcalfe and Schmitz [97].)

The degradation of glass fibers owing to environmental attack can severely affect the performance of GRP laminates. The fibers in GRP laminates are protected from the environment by the resin matrix. The degree of protection depends on the permeability of the resin to the diffusion of active species from the environment and then the ability of the resin and interface to resist

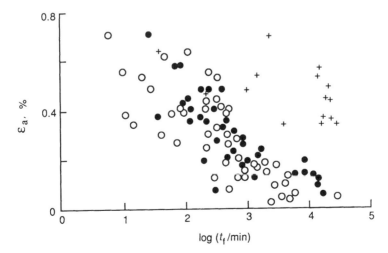

Figure 9-49. The stress-corrosion failure time (t_f) of single E-glass fibers under varying applied strains (ε_a) in 0.5 M H_2SO_4 (○), 0.5 M $NaHSO_4$ (●) and 0.01 M H_2SO_4 (+).

premature cracking, which would allow the environment to come into direct contact with the fibers. There are significant data in the literature [99–107] on the stress corrosion of GRP. Typical results of stress corrosion, shown in Fig. 9-50, illustrate the dependence of time to failure on stress parallel to the fibers for a pipe with a range of resin types tested in 0.65 M HCl at 20°C [100]. The effects of corrosion on a GRP laminate can be minimized by the use of an appropriate chemically resistant resin. The resin should be post-cured, and a nonstructural barrier should be provided. The typical barrier layers consist of a gel-coat layer (resin reinforced with a C-glass or organic fiber veil) backed by a number of low-fiber-volume-fraction chopped-strand mat plies. The size and binder on the glass fibers also should be appropriate to minimize debonding and blister formation. Laminates that fracture as a result of stress corrosion can be distinguished by the nature of their fracture surfaces. For example, Fig. 9-51 shows a comparison of samples fractured in air and in acid. Typically, the laminates exposed to a severe stress-corrosion environment produce relatively plane or flat fracture surfaces as compared with the more normal "broomlike" fractures when failed or tested in air.

9.3.2 Matrix Effects

9.3.2.1 Effect of Temperature and Moisture

Changes in the properties of the matrix owing to environmental exposure are important considerations for polymer composites. Variations in temperature and moisture content are the most frequently encountered conditions that influence the properties of

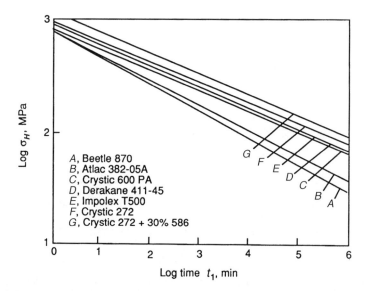

Figure 9-50. Dependence of time to pipe failure in 0.65 M HCL at 20°C on stress parallel to fibers and type of resin.

Figure 9-51. Photographs of fracture surfaces of GRP samples fractured in (a) air and (b) acid.

polymer matrices. Exposure to ambient temperature and moisture influences the distributions of temperature and moisture concentration inside the material as a function of both position and time. These distributions, in turn, influence the performance of the material. Analytical means are often used to obtain temperature and moisture concentration distributions in the material. Changes in performance resulting from these distributions can be determined experimentally. Environmental effects on composite materials have been investigated by a large number of researchers. Results from a large number of studies have been compiled by Springer [108–110]. Detailed analytical developments of the problem are beyond the scope of this book. Some important experimental results will be discussed in this subsection.

The temperature distribution in a material depends on its thermal conductivity, besides the environmental conditions, and the moisture concentration distribution on mass diffusivity. Prediction methods for thermal conductivity and mass diffusivity were discussed in Chap. 3. For many composite materials, the thermal conductivity is 10^4–10^6 times larger than mass diffusivity. Thus the temperature equilibrates much faster than the mass concentration. A typical variation in moisture content of a four-ply carbon–epoxy composite (Fig. 9-52) shows that an equilibrium (saturation) level is attained after several days of exposure [111].

The effect of moisture and temperature has been explored for many performance parameters, including tensile and shear strengths, elastic moduli, fatigue behavior, creep, stress rupture, response to dynamic impact, electrical resistance, and swelling (dimensional changes). Of these parameters, tensile and shear strengths and elastic moduli have been studied in detail. Data on other parameters are more limited and, in most cases, are inadequate to either demonstrate quantitatively the changes in these parameters or provide a database sufficient for design purposes.

A summary of the effects of moisture and temperature on the ultimate tensile strength of composites is given in Table 9-3 and on the elastic modulus

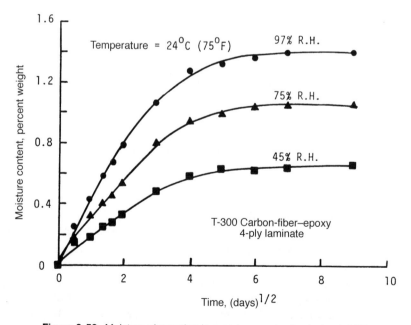

Figure 9-52. Moisture absorption in a carbon–epoxy laminate at 24°C.

Table 9-3 Summary of experimental data on the effect of moisture and temperature on the tensile strength of composites

Composite Material System and Reference	0° Moist.	0° Temp.	QIL Moist.	QIL Temp.	90° Moist.	90° Temp.	Remarks
Graphite–Epoxy							
112[b]	L	N	L	N	S	S	
113	N	N	N	N	S	S	
114	N	N	N	—	S	S	
115	—	—	—	N	S	S	Limited data (2–3 points)
122	—	—	N	N	—	—	Two data points for 90° laminates
116	L	L	N	L	S	S	
117	—	—	—	—	S	L	
118	N	L	N	L	S	S	
118	N	N	N	N	S	S	Very scattered data for 90° laminates
119	—	—	L	S	—	—	Only two data points for temperature
119	—	—	N	N	—	—	
119	—	—	L	N	—	—	
121	—	—	—	—	S	S	
119	—	—	N	S	—	—	Only two data points for temperature
115	—	N	—	N	—	N	Only two data points for 90° laminates
119	—	—	—	L	—	—	
Boron–Epoxy							
118	L	N	L	L	S	S	
123	—	L	—	—	—	S	
119	—	—	N	N	—	—	
Kevlar 49–Epoxy							
120	—	L	—	—	—	—	

[a] N = negligible effect (<30%); S = strong effect (<30%); L = little effect (<30%); QIL = quasi-isotropic lamina.
[b] Reference number at the end of the chapter.

Source: G. S. Springer [108].

in Table 9-4. On the basis of these data the following general conclusions can be drawn:

1. Temperature effects
   a. Temperature in the range −40 to 190°C has a negligible effect on the strengths of the 0° laminates and quasi-isotropic laminates (QILs) regardless of the moisture content of the material. There may be a slight decrease in strength (<20%) as the temperature increases from 190 to 230°C.
   b. Temperature in the range −40 to 230°C has a negligible effect on the elastic modulus of 0° laminates and QIL regardless of the moisture content.
   c. The strength and modulus of the 90° laminates decrease significantly as the temperature increases from −40 to 230°C.
2. Moisture effects
   a. Moisture content below 1% has negligible effect on the strength of 0° laminates and QILs. When moisture content is above 1%, the strength of these laminates decreases with increasing moisture content. This reduction in strength seems to be insensitive to the temperature of the material.
   b. Moisture content, regardless of temperature in the range −40 to 230°C, has very little effect on the modulus of 0° laminates and QILs.
   c. Strength and modulus of the 90° laminates decrease significantly with increasing moisture content.

From the preceding discussion a general conclusion may be stated that the moisture content and elevated temperature (up to about the glass-transition temperature of the polymer matrix) do not have a significant effect on fiber-dominated properties but may reduce matrix-dominated properties. Other studies have shown that the exposure history (cycling or spiking) also does not have a major influence on fiber-dominated properties.

9.3.2.2 Degradation at Elevated Temperatures

In general, organic materials are unstable at elevated temperatures and undergo a chemical breakdown resulting from degradation. If the degradation reactions persist for a sufficiently long time or are sufficiently rapid, substantial degradation will occur such that the matrix material is removed by volatilization in a gaseous form. Such drastic degradation severely affects the mechanical integrity of composite systems and limits the temperatures at which they can be used. The degradation can be characterized by measuring the amount of volatiles given off by the material as a function of time and temperature or by recording the weight loss. Such data [124] are shown in Fig. 9-53 for a phenolic–glass system. At temperatures of 300°F and above, decomposition of the matrix is

Table 9-4 Summary of experimental data on the effects of moisture and temperature on the elastic modulus of composite materials

Composite Material System and Reference	Laminate Lay-up Orientation[a]					
	0°		QIL		90°	
	Moist.	Temp.	Moist.	Temp.	Moist.	Temp.
Tension Test						
Graphite–Epoxy						
113[b]	L	N	L	N	S	S
114	N	N	N	—	S	S
115	—	N	—	N	—	—
116	N	N	N	N	N	N
117	—	—	—	—	S	S
118	N	N	N	N	S	S
118	N	N	N	N	N	S
119	—	—	N	S	—	—
119	—	—	N	N	—	—
119	—	—	N	L	—	—
121	—	—	—	—	N	S
119	—	—	N	S	—	—
115	—	N	—	N	—	—
119	—	—	—	L	—	—
Boron–Epoxy						
118	N	N	N	N	S	S
119	—	—	N	N	—	—
Kevlar 49–Epoxy						
120	—	S	—	—	—	—
Compression Test						
Graphite–Epoxy						
114	N	N	—	—	L	S
116	L	N	N	N	L	N
118	N	N	N	N	S	S
118	N	N	N	N	S	S
Boron–Epoxy						
118	N	N	N	N	S	S

[a] N = negligible effect; S = strong effect (>30%); L = little effect (<30%); QIL = quasi-isotropic laminate.
[b] Reference number at the end of the chapter.
Source: Springer [108].

428 PERFORMANCE OF FIBER COMPOSITES

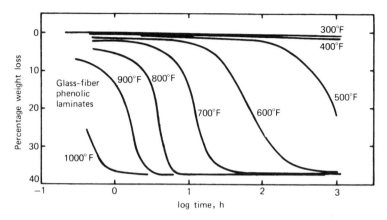

Figure 9-53. Thermal degradation of glass–phenolic system. (From Boller [124].)

occurring at a rate that is uniformly accelerated with increasing temperature. It is observed that the weight-loss curves at different temperatures have the same shape. Therefore, it may be concluded that the temperature influences only the rate constant and that a time–temperature superposition applies to these data. A master curve obtained [125] by shifting all the data to the 300°F curve is shown in Fig. 9-54, in which log a_D represents the distance by which the data obtained at different temperatures are shifted on the log t scale, and t is the time of the process.

The matrix decomposition leads to a loss in stiffness and strength with the loss of matrix integrity. Boller [124] has obtained data in tension, compression, and flexure for the phenolic–glass composite samples that were exposed at the same temperatures presented in Fig. 9-53 for different periods of time.

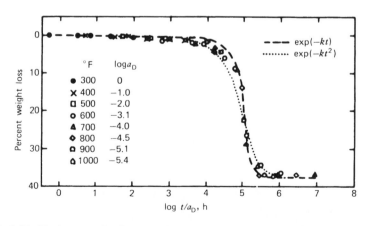

Figure 9-54. Master curve for decomposition of a phenolic–glass system. (From Tsai [125].)

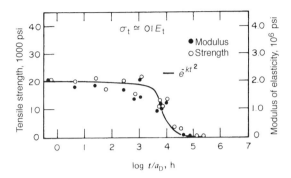

Figure 9-55. Master curve for stiffness and strength of a phenolic–glass system. (From Tsai [125].)

The data generally show the same trends as presented in Fig. 9-53. The results from Tsai [125] are shown in Fig. 9-55, where it also can be seen that the mechanical property loss follows the same rate law as the decomposition reaction. Using the experimental data of Stevens [126] on fatigue of phenolic–glass composite at elevated temperatures, Tsai [125] has shown that the time–temperature superposition is applicable for fatigue strength also.

From the preceding discussion it can be seen that the maximum design temperature of a high-strength fiber composite generally will be limited by the properties of the matrix. Thus composites for applications at elevated temperatures (e.g., >200–250°C) are at present made of metal and ceramic matrices or carbon–carbon composites.

9.3.2.3 Stress–Rupture Characteristics at Modest Temperatures

Polymer matrix degradation at elevated temperatures results in severe deterioration of all composite properties, whether they are fiber-dominated or matrix-dominated. However, polymer–matrix composites display good stress–rupture characteristics at modest temperatures.

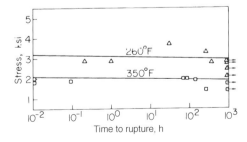

Figure 9-56. Stress–rupture behavior of unidirectional graphite–epoxy composite subjected to transverse load.

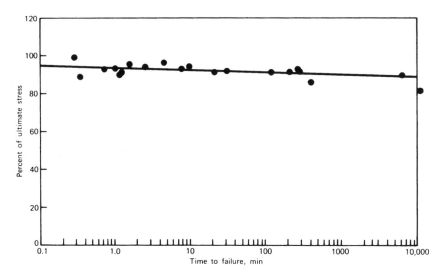

Figure 9-57. Stress–rupture behavior of unidirectional boron–epoxy composite subjected to transverse load.

Stress–rupture curves for unidirectional graphite–epoxy composites are given in Fig. 9-19. It was noted that when the composites are loaded in the fiber direction, they exhibit excellent stress–rupture characteristics, which is evident from a negligible drop in strength even for 1000 h of loading time (see Fig. 9-19). A transverse rupture diagram of the same unidirectional graphite–epoxy composite is shown in Fig. 9-56. The transverse strength of the composite is much smaller than the longitudinal strength. However, at the temperatures considered, the stress–rupture characteristics in the transverse

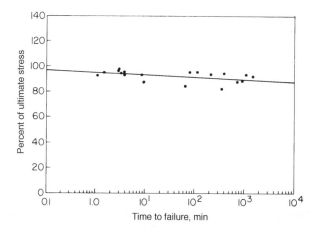

Figure 9-58. Stress–rupture behavior of [±45] boron–epoxy laminate.

direction are also good. The drop in strength for long exposures to stress is only modest, but there is a larger scatter. Stress–rupture diagrams for a unidirectional boron–epoxy composite in the transverse direction and for [±45] laminates are shown in Figs. 9-57 and 9-58, respectively. As is clear from these figures, boron–epoxy composites also display good stress–rupture characteristics at room temperature.

EXERCISE PROBLEMS

9.1. Describe the progression of fatigue damage in the following laminates: $[0]_8$ and $[0/\pm 45/90]_s$. Assume cyclic stress in tension ($R = 0.1$) equivalent to 75% of the laminate ultimate strength.

9.2. Fatigue strengths with zero mean stress of a glass–polyester composite at 10^3, 10^4, 10^5, and 10^6 cycles have been found to be 84, 70, 60, and 52 MPa, respectively. Stress–rupture strength of the same composite follows the relationship

$$S_c = 21.8 - 4 \log t$$

where t is the time in minutes, and S_c is the stress rupture strength in megapascals. Using these data and the Goodman–Boller relationship [Eq. (9.1)], construct a master diagram showing the influence of mean stress on permissible stress amplitude at different cyclic lives. Assume that the fatigue tests were performed at 2000 cycles per minute.

9.3. If the laminate whose fatigue characteristics are shown by Fig. 9-6 is interrupted after 20,000 cycles and subjected to a tensile strain of 2%, calculate whether it will fracture at this strain level.

9.4. Explain why a high-modulus unidirectional graphite-fiber-reinforced epoxy beam fractured in impact breaks cleanly into two halves without delamination and with little fiber pullout, whereas an equivalent glass-fiber composite exhibits considerable delamination on failure.

REFERENCES

1. F. J. McGarry, "Crack Propagation in Fiber Reinforced Plastic Composites," in R. T. Schwartz and H. S. Schwartz, eds., *Fundamental Aspects of Fiber Reinforced Plastic Composites,* Interscience, New York, 1968.
2. M. J. Owen and T. R. Smith, "Some Fatigue Properties of Chopped-Strand-Mat Polyester-Resin Laminates," *Plast. Polym.,* **36**, 33–44 (February 1968).
3. M. J. Owen, R. Dukes, and T. R. Smith, "Fatigue and Failure Mechanisms in GRP with Special Reference to Random Reinforcements," SPI, 23rd Annual Technical Conference, Washington, February 1968, Sec. 14-A.

4. L. J. Broutman and S. Sahu, "Progressive Damage of a Glass Reinforced Plastic during Fatigue," SPI, 24th Annual Technical Conference, Washington, February 1969, Sec. 11-D.
5. T. R. Smith and M. J. Owen, "Progressive Nature of Fatigue Damage in RP," *Mod. Plast.,* **46**(5), 1281–1331 (1969).
6. J. W. Dally and D. H. Carrillo, "Fatigue Behavior of Glass Fiber Fortified Thermoplastics," *Polym. Eng. Sci.,* **9**(6), 434–444 (1969).
7. J. W. Dally and B. D. Agarwal, "Low Cycle Fatigue Behavior of Glass Fiber Reinforced Plastics," Proceedings of the Army Symposium on Solid Mechanics, AMMRC MS 70-5, 1970.
8. J. J. Nevadunsky, J. J. Lucas, and M. J. Salkind, "Early Fatigue Damage Detection in Composite Materials," *J. Compos. Mater.,* **9**(4), 394–408 (1975).
9. T. Tanimoto and S. Amijima, "Fatigue Properties of Laminated Glass Fiber Composite Materials," SPI, 29th Annual Technical Conference, Washington, February 1974, Sec. 17-B.
10. T. Tanimoto and S. Amijima, "Progressive Nature of Fatigue Damage of Glass Fiber Reinforced Plastics," *J. Compos. Mater.,* **9**(4), 380–390 (1975).
11. H. T. Hahn and R. Y. Kim, "Fatigue Behavior of Composite Laminate," *J. Compos. Mater.,* **10**(2), 156–180 (1976).
12. J. W. Dally and L. J. Broutman, "Frequency Effects on the Fatigue of Glass Reinforced Plastics," *J. Compos. Mater.,* **1**, 424–442 (1967).
13. L. Cessna, J. Levens, and J. Thompson, "Flexural Fatigue of Glass Reinforced Thermoplastics," SPI, 24th Annual Technical Conference, Washington, February 1969, Sec. 1-C.
14. M. J. Salkind, "Fatigue of Composites," in *Composite Materials: Testing and Design* (2nd Conference), ASTM STP 497, American Society for Testing and Materials, Philadelphia, PA, 1972, pp. 143–169.
15. K. H. Boller, "Fatigue Characteristics of RP Laminates Subjected to Axial Loading," *Mod. Plast.,* **41**, 145 (1964).
16. J. W. Davis, J. A. McCarthy, and J. N. Schurb, "The Fatigue Resistance of Reinforced Plastics," *Mater. Des. Eng.,* **60**, 87–91 (1964).
17. R. L. Foye and D. J. Baker, "Design of Orthotropic laminates," paper presented at the 11th Annual AIAA Structures, Structural Dynamics, and Materials Conference, Denver, CO, April 1970.
18. N. J. Pagano and R. B. Pipes, "The Influence of Stacking Sequence on Laminate Strength," *J. Compos. Mater.,* **5**(1), 50 (1971).
19. J. M. Whitney and E. E. Browning, "Free-Edge Delamination of Tensile Coupons," *J. Compos. Mater.,* **6**(3), 300 (1971).
20. J. M. Whitney, "Free-Edge Effects in the Characterization of Composite Materials," in *Analysis of the Test Methods for High Modulus Fibers and Composites,* ASTM STP 521, American Society for Testing and Materials, Philadelphia, PA, 1973, p. 167.
21. S. Amijima and T. Tanimoto, "The Effect of Glass Content and Environmental Temperature on the Fatigue Properties of Laminated Glass Fiber Composite Materials," *Mechanical Behavior of Materials, Proceedings of the 1971 Interna-*

tional Conference on Mechanical Behavior of Materials, Vol. V, The Society of Materials Science, Japan, 1972, pp. 269–278.
22. K. E. Hofer, Jr., L. C. Benett, and M. Stander, "Effect of Various Fiber Surface Treatments on the Fatigue Behavior of Glass Fabric Composites in High Humidity Environments," SPI, 31st Annual Technical Conference, Washington, February 1976, Sec. 6-A.
23. R. B. Pipes, "Interlaminar Fatigue Characteristics of Fiber-Reinforced Composite Materials," *Composite Materials: Testing and Design* (3rd Conference), ASTM STP 546, American Society for Testing and Materials, Philadelphia, PA, 1974, pp. 419–432.
24. L. G. Bevan, "Axial and Short Beam Shear Fatigue Properties of CFRP Laminates," *Composites,* **8**(4), 227–232 (1977).
25. D. C. Phillips and J. M. Scott, "The Shear Fatigue of Unidirectional Fibre Composites," *Composites,* **8**(4), 233–236 (1977).
26. D. F. Sims and V. H. Brogdon, "Fatigue Behavior of Composites Under Different Loading Modes," *Fatigue of Filamentary Composite Materials,* ASTM STP 636, K. L. Reifsnider and K. N. Lauraitis, eds., American Society for Testing and Materials, Philadelphia, PA, 1977, pp. 185–205.
27. B. D. Agarwal and S. K. Joneja, "Flexural Fatigue Properties of Unidirectional Composites in Transverse Direction," *Composites,* **10**(1), 28–30 (1979).
28. K. H. Boller, "Fatigue Properties of Fibrous Glass Reinforced Plastics Laminates Subjected to Various Conditions," *Mod. Plast.,* **34,** 163 (1957).
29. L. J. Broutman and S. Sahu, "A New Theory to Predict Cumulative Fatigue Damage in Fiberglass Reinforced Plastics," *Composite Materials: Testing and Design* (Second Conference), ASTM STP 497, American Society for Testing and Materials, Philadelphia, PA, 1972, pp. 170–188.
30. J. F. Mandell and U. Meier, "Fatigue Crack Propagation in 0°/90° E-Glass/Epoxy Composites," *Fatigue of Composite Materials,* ASTM STP 569, American Society for Testing and Materials, Philadelphia, PA, 1975, pp. 28–44.
31. J. F. Mandell, "Fatigue Crack Propagation in Woven and Non-Woven Fiberglass Laminates," *Composite Reliability,* ASTM STP 580, American Society for Testing and Materials, Philadelphia, PA, 1975, p. 515.
32. Z. Hashin and A. Rotem, "A Fatigue Failure Criterion for Fiber Reinforced Materials," *J. Compos. Mater.,* **7**(4), 448–464 (1973).
33. B. D. Agarwal and J. W. Dally, "Prediction of Low-Cycle Fatigue Behaviour of GFRP: An Experimental Approach," *J. Mater. Sci.,* **10**(1), 193–199 (1975).
34. S. S. Manson, "Fatigue: A Complex Subject—Some Simple Approximations," *Exp. Mech.,* **5**(7), 193–226 (1965).
35. J. F. Tavernelli and L. F. Coffin, Jr., "Experimental Support for Generalized Equation Predicting Low Cycle Fatigue," *J. Basic Eng.,* **84**(4), 533–541 (1962).
36. M. J. Owen, "Fatigue of Carbon-Fiber-Reinforced Composites," in Lawrence J. Broutman, ed., *Fracture and Fatigue,* Academic, New York, 1974.
37. M. J. Owen and S. Morris, SPI, 25th Annual Technical Conference, Washington, February 1970, Sec. 8-E.
38. T. Liber and I. M. Daniel, "Effects of Tensile Load Cycling on Advanced Composite Angle-Ply Laminates," SPI, 31st Annual Technical Conference, Washington, February 1976, Sec. 21-E.

39. T. Tanimoto and S. Amijima, "Static and Fatigue Properties of Carbon Fiber Reinforced Plastics," *Proceedings of the Second International Conference on Mechancial Behavior of Materials,* Boston, MA, August 1976, pp. 946–950.
40. M. J. Owen and S. Morris, "Some Interlaminar Shear Fatigue Properties of Carbon-Fibre-Reinforced Plastics," *Plast. Polym.,* 209–216 (August 1972).
41. D. C. Phillips and J. M. Scott, "Shear Fatigue of CFRP Under Low Frequency Torsion," *Fiber Sci. Technol.,* **11**(1), 23–54 (1979).
42. B. Harris, "Fatigue and Accumulation of Damage in Reinforced Plastics," *Composites,* **8**(4), 214–220 (1977).
43. M. J. Owen and R. Dukes, *J. Strain Anal.,* **2**(4), 272–279 (1967).
44. M. J. Owen, T. R. Smith, and R. Dukes, *Plast. Polym.* **37,** 227–233 (1969).
45. M. J. Owen and R. G. Rose, "The Effect of Resin Flexibility on the Fatigue Behavior of GRP," Paper 8, British Plastics Federation 7th International Reinforced Plastics Conference, Brighton, October 1970; also published in *Mod. Plast.,* **47**(11), 130 (1970).
46. M. J. Owen and R. J. Howe, *J. Phys. (D): Appl. Phys.,* **5,** 1637–1649 (1972).
47. R. J. Howe and M. J. Owen, "Cumulative Damage in Chopped Strand Mat/Polyester Resin Laminates," 8th International R.P. Conference, British Plastics Federation, Brighton, October 1972, Paper 21.
48. M. J. Owen and P. T. Bishop. *J. Phys. (D): Appl. Phys.,* **5,** 1621–1636 (1972).
49. M. J. Owen and M. S. Found, "Static and Fatigue Failure of Glass Fibre Reinforced Polyester Resins under Complex Stress Conditions," in *Faraday Special Discussions of the Chemical Society,* No. 2, 1972, pp. 77–89.
50. M. J. Owen and R. G. Rose, *J. Phys. (D): Appl. Phys.,* **6,** 42–53 (1973).
51. M. J. Owen and P. T. Bishop, *J. Phys. (D): Appl. Phys.,* **6,** 2057–2069 (1973).
52. M. J. Owen and P. T. Bishop, *J. Phys. (D): Appl. Phys.,* **7,** 1214–1224 (1974).
53. M. J. Owen, J. R. Griffiths, and M. S. Found, "Biaxial Stress Fatigue Testing of Thin-Walled GRP Cylinders," *Proceedings, 1975 International Conference on Composite Materials,* Vol. 2, American Institute of Mining, Metallurgical, and Petroleum Engineers, New York, 1976, pp. 917–941.
54. M. J. Owen and J. R. Griffiths, *Composites,* 89–94 (April 1979).
55. M. J. Owen and R. J. Cann, *J. Mater. Sci.,* **14,** 1982–1996 (1979).
56. M. J. Owen, "Static and Fatigue Strength of Glass Chopped Strand Mat/Polyester Resin Laminates," in *Short Fiber Reinforced Composite Materials,* ASTM STP 772, American Society for Testing and Materials, Philadelphia, PA, 1982, pp. 64–84.
57. J. E. Theberge, "Fatigue Endurance and Creep Characteristics of Fiber Glass Fortified Thermoplastics," LNP Engineering Plastics, Malvern, PA.
58. J. W. Dally and D. H. Carrillo, "Fatigue Behavior of Glass Fiber Fortified Thermoplastics," *Polym. Eng. Sci.,* **9**(6), (1969).
59. R. C. Novak and M. A. DeCrescente, "Impact Behavior of Unidirectional Resin Matrix Composites Tested in Fiber Direction," in *Composite Materials: Testing and Design* (Second Conference), ASTM STP 497, American Society for Testing and Materials, Philadelphia, PA, 1972, pp. 311–323.
60. P. W. R. Beaumont, "A Fracture Mechanics Approach to Failure in Fibrous Composites," *J. Adhes.,* **6,** 107–137 (1974).

61. L. J. Broutman and A. Rotem, "Impact Strength and Toughness of Fiber Composite Material," in *Foreign Object Impact Damage to Composites,* ASTM STP 568, American Society for Testing and Materials, Philadelphia, PA, 1975, pp. 114–133.
62. A. Kelly, *Strong Solids,* Clarendon, Oxford, 1973.
63. G. A. Cooper and A. Kelly, *J. Mech. Phys. Solids,* **15,** 279 (1967).
64. P. Yeung and L. J. Broutman, "The Effect of Glass-Resin Interface Strength on the Impact Strength of Fiber Reinforced Plastics," *Polym. Eng. Sci.,* **18**(2), 62–72 (1978).
65. A. Kelly, *Proc. R. Soc.,* **A319,** 95 (1970).
66. J. O. Outwater and M. C. Murphy, 24th Annual Technical Conference, SPI, Washington, 1969.
67. A. J. Barker, "Charpy Notched Impact Strength of Carbon-Fiber/Epoxy-Resin Composites," First International Conference on Carbon Fibers, London, 1971, Paper 20.
68. G. R. Sidney and F. J. Bradshaw, "Some Investigations on Carbon-Fibre Reinforced Plastics under Impact Loading and Measurement of Fracture Energies," First International Conference on Carbon Fibers, London, 1971.
69. A. Rotem and J. M. Lifshitz, "Longitudinal Strength of Unidirectional Fibrous Composite under High Rate of Loading," SPI, 26th Annual Technical Conference, Washington, 1971, Sec. 10-G.
70. L. J. Broutman and A. Rotem, "Impact Strength and Fracture of Carbon Fiber Composite Beams," SIP, 28th Annual Technical Conference, Washington, 1973, Sec. 17-B.
71. P. K. Mallick and L. J. Broutman, "Impact Properties of Laminated Angle Ply Composites," SPI, 30th Annual Technical Conference, Washington, 1975, Sec. 9-C.
72. L. J. Broutman and P. K. Mallick, "Impact Behavior of Hybrid Composites," AFOSR TR-75-0472, November 1974.
73. B. D. Agarwal and J. N. Narang, "Strength and Failure Mechanism of Anisotropic Composites," *Fiber Sci. Technol.,* **10**(1), 37–52 (1977).
74. J. M. Lifshitz, "Impact Strength of Angle Ply Fiber Reinforced Materials," *J. Compos. Mater.,* **10**(1), 92–101 (1976).
75. D. F. Adams and A. K. Miller, "An Analysis of the Impact Behavior of Hybrid Composite Materials," *Mater. Sd. Eng.,* **19,** 245–260 (1975).
76. J. Summerscales and D. Short, "Carbon Fibre and Glass Fibre Hybrid Reinforced Plastics," *Composites,* **9**(3), 157–166 (1978)
77. P. W. R. Beaumont, P. G. Reiwald, and C. Zweben, "Methods for Improving the Impact Resistance of Composite Materials," in *Foreign Object Impact Behavior of Composites,* ASTM STP 568, American Society for Testing and Materials, Philadelphia, PA, 1974, pp. 134–158.
78. R. H. Toland, "Instrumented Impact Testing of Composite Materials," ASTM STP 563, American Society for Testing and Materials, Philadelphia, PA, 1974, pp. 133–145.
79. R. H. Toland, "Failure Modes in Impact Loaded Composite Materials," paper presented at Failure Modes in Composites Symposium AIME Spring Meeting, May 1972.

80. J. C. Aleszka, "Low Energy Impact Behavior of Composite Panels," *J. Test. Eval.,* **6**(3), 202–210 (1978).

81. D. A. Wyrick and D. F. Adams, "Residual Strength of a Carbon/Epoxy Composite Material Subjected to Repeated Impact," *J. Compos. Mater.,* **22**(8), 749–765 (1988).

82. *Foreign Object Impact Damage to Composites,* ASTM STP 568, American Society for Testing and Materials, Philadelphia, PA, 1974.

83. D. Liu and L. E. Malvern, "Matrix Cracking in Impacted Glass/Epoxy Plates," *J. Compos. Mater.,* **21**(7), 594–609 (1987).

84. S. P. Joshi and C. T. Sun, "Impact-Induced Fracture in a Quasi-Isotropic Laminate," *J. Compos. Technol. Res.,* **19**(2), 40–46 (1987).

85. S. P. Joshi and C. T. Sun, "Impact Induced Fracture in a Laminated Composite," *J. Compos. Mater.,* **19,** 51–66 (1985).

86. K. M. Lal, "Residual Strength Assessment of Low Velocity Impact Damage of Graphite–Epoxy Laminates," *J. Reinf. Plast. Compos.,* **2,** 226–238 (1986).

87. W. J. Cantwell, P. T. Curtis, and J. Morton, "Impact and Subsequent Fatigue Damage Growth in Carbon Fiber Laminates," *Int. J. Fatigue,* **6**(2), 301–305 (1984).

88. S. K. Chaturvedi and R. L. Sierakowski, "Effect of Impactor Size on Impact Damage—Growth and Residual Properties in an SMC-50 Composites," *J. Compos. Mater.,* **19**(1), 100–113 (1985).

89. G. I. Caprino, V. Crivelli, and A. D'Ilio, "Elastic Behavior of Composite Structures under Low Velocity Impact," *Composites,* **15,** 231–234 (1984).

90. M. W. Wardle and E. W. Tokarsky, "Drop Weight Impact Testing of Laminates Reinforced with Kevlar Aramid Fiber, E-Glass, and Graphite," *Compos. Technol. Rev.,* **5**(1), 4–10 (1983).

91. V. S. Avva, J. Rao Vala, and M. Jeyasulam, "Effect of Impact and Fatigue Loads on the Strength of Graphite/Epoxy Composites," in *Composite Materials: Testing and Design* (Seventh Conference), ASTM STP 893, American Society for Testing and Materials, Philadelphia, PA, 1986, pp. 196–206.

92. R. E. Lowrie, "Glass Fibers for High Strength Composites," in L. J. Broutman and R. H. Krock, eds., *Modern Composite Materials,* Addison-Wesley, Reading, MA, 1967.

93. W. H. Otto, "Properties of Glass Fibers at Elevated Temperatures," *Proceedings of 6th Sagamore Ord. Materials Research Conference,* 1959, p. 277.

94. F. W. Wawner, Jr., "Boron Filaments," in L. J. Broutman and R. H. Krock, eds., *Modern Composite Materials,* Addison-Wesley, Reading, MA, 1967.

95. J. L. Cook and T. T. Sakurai, SAMPE National Symposium, 10th, 1966, 10, Sec. H-1.

96. A. G. Metcalfe, M. E. Gulden, and G. K. Schmitz, "Spontaneous Cracking of Glass Filaments," *Glass Technol.,* **12**(1), 15–23 (1971).

97. A. G. Metcalfe and G. K. Schmitz, "Mechanism of Stress Corrosion in E-Glass Filaments," *Glass Technol.,* **12**(1), 5–16 (1972).

98. F. R. Jones and J. W. Rock, "On the Mechanism of Stress Corrosion of E-Glass Fibers," *J. Mater. Sci. Lett.,* **2,** 415–418 (1983).

99. P. J. Hogg and D. Hull, "Micromechanisms of Crack Growth in Composite Materials under Corrosive Environments," *Metal. Sci.*, **14**, 441–449 (1980).
100. P. J. Hogg, "Factors Affecting the Stress Corrosion of GRP in Acid Environments," *Composites*, **14**(3), 254–261 (1983).
101. J. N. Price and D. Hull, "Propagation of Stress Corrosion Cracks in Aligned Glass Fiber Composite Materials," *J. Mater. Sci.*, **18**, 2798–2810 (1983).
102. H. H. Collins, "Strain–Corrosion Cracking of GRP Laminates," *Plast. Rubbers: Mater. Appl.*, **3**, 6–10 (February 1978).
103. S. Torp and R. Arvesen, "Influence of Glass Fiber Quality on Mechanical and Strain Corrosion Properties of FRP-Proposed Method for Quality Control," *Proceedings of the 34th SPI Conference*, 1979, Sec. 13-D, pp. 1–9.
104. H. Hojo and K. Tsuda, "Effects of Chemical Environments and Stress on Corrosion Behaviors of Glass Fiber-Reinforced Plastics and Vinyl Ester Resin," *Proceedings of the 34th SPI Conference*, 1979, Sec. 13-B, pp. 1–6.
105. S. Torp, O. Stromsodd, and M. Onarheim, "Influence of Glass Fiber Type on Laminate Corrosion Resistance," *Proceedings of the 37th SPI Conference*, 1982, Sec. 9-E, pp. 1–5.
106. B. Noble, S. J. Harris, and M. J. Owen, "Stress Corrosion Cracking of GRP Pultruded Rods in Acid Environments," *J. Mater. Sci.*, **18**, 1244–1254 (1983).
107. F. R. Jones, J. W. Rock, and J. E. Bailey, "The Environmental Stress Corrosion Cracking of Glass Fiber-Reinforced Laminates and Single E-Glass Filaments," *J. Mater. Sci.*, **18**, 1059–1071 (1983).
108. G. S. Springer, ed., *Environmental Effects on Composite Materials*, Technomic, Westport, CT, 1981.
109. G. S. Springer, ed., *Environmental Effects on Composite Materials*, Vol. 2, Technomic, Lancaster, PA, 1984.
110. G. S. Springer, ed., *Environmental Effects on Composite Materials*, Vol. 3, Technomic, Lancaster, PA, 1988.
111. C. D. Shirrell, J. C. Halpin, and C. E. Browning, "Moisture: An Assessment of Its Impact on Design of Resin Based Advanced Composites," NASA technical report, NASA-44-TM-X-3377, April 1976.
112. C. H. Shen and G. S. Springer, "Moisture Absorption and Desorption of Composite Materials," *J. Compos. Mater.*, **10**, 2 (1976).
113. C. E. Browning, G. E. Husman, and J. M. Whitney, "Moisture Effects in Epoxy Matrix Composites," *Composite Materials: Testing and Design*, ASTM, STP 617 (1976).
114. R. M. Verette, "Temperature/Humidity Effects on the Strength of Graphite/Epoxy Laminates," AIAA Paper No. 75-1011, AIAA 1975 Aircraft Systems and Technology Meeting, Los Angeles, CA, August 4–7, 1975.
115. J. R. Kerr, J. F. Haskins and B. A. Stein, "Program Definition and Preliminary Results of a Long-Term Evaluation Program of Advanced Composites for Supersonic Cruise Aircraft Applications," *Environmental Effects on Advanced Composite Materials*, ASTM, STP 602, American Society for Testing and Materials, Philadelphia, PA, 1975, p. 3.
116. K. E. Hofer, Jr., D. Larsen, and V. E. Humphreys, "Development of Engineering Data on the Mechanical and Physical Properties of Advanced Composite Mate-

rials," Technical Report AFML-TR-74-266, February 1975, Air Force Materials Laboratory, Air Force Systems Command, Wright-Patterson Air Force Base, Dayton, OH.

117. G. E. Husman, "Characterization of Wet Composite Materials," presented at the Mechanics of Composites Review, Bergamo Center, Dayton, OH, January 28–29, 1976.

118. K. E. Hofer, Jr., N. Rao, and D. Larsen, "Development of Engineering Data on the Mechanical and Physical Properties of Advanced Composite Materials," Technical Report AFML-TR-72-205, Part II, February 1974, Air Force Materials Laboratory, Air Force Systems Command, Wright-Patterson Air Force Base, Dayton, OH.

119. C. E. Browning, "The Effects of Moisture on the Properties of High Performance Structural Resins and Composites," Technical Report AFML TR-72-94, September 1972, Air Force Materials Laboratory, Air Force Systems Command, Wright-Patterson Air Force Base, Dayton, OH.

120. M. P. Hanson, "Effect of Temperature on the Tensile and Creep Characteristics of PRD 49 Fiber/Epoxy Composites," in B. R. Norton, ed., *Composite Materials in Engineering Design,* Proceedings of 6th St. Louis Symposium, May 11–12, 1972, p. 717. Published by The American Society for Metals.

121. J. Hertz, "Investigation into the High-Temperature Strength Degradation of Fiber-Reinforced Resin Composite During Ambient Aging," Convair Aerospace Division, General Dynamics Corporation, Report No. GDCA-DBG73-005, Contract NAS8-27435, June 1973.

122. R. Y. Kim and J. M. Whitney, "Effect of Environment on the Notch Strength of Laminated Composites," presented at the Mechanics of Composites Review, Bergamo Center, Dayton, OH, January 28–29, 1976.

123. B. E. Kaminski, "Effects of Specimen Geometry on the Strength of Composite Materials," in *Analysis of the Test Methods for High Modulus Fibers and Composites,* ASTM STP 521, American Society for Testing and Materials, Philadelphia, PA, 1973, p. 181.

124. K. H. Boller, "Strength Properties of Reinforced Plastic Laminates at Elevated Temperatures," Wright Air Development Center Technical Report No. 59-569, 1960.

125. S. W. Tsai, "Environmental Factors in the Design of Composite Materials," in F. W. Wendt, H. Liebowitz, and N. Perrone, eds., *Mechanics of Composite Materials,* Pergamon, New York, 1970, pp. 749–767.

126. G. H. Stevens, "Fatigue Tests of Phenolic Laminate at High Stress Levels and Elevated Temperatures," Forest Product Laboratory Report No. 1884, 1961.

10

EXPERIMENTAL CHARACTERIZATION OF COMPOSITES

10.1 INTRODUCTION

Experimental characterization here refers to the determination of material behavior and properties through tests conducted on suitable material specimens. Experimental characterization serves various purposes, for example, providing data needed for analysis and design, understanding the material response under different loading and environmental conditions, ensuring quality control of fabrication procedures and incoming materials, assessing material uniformity, and allowing comparison of different candidate materials. In the case of composite materials, it also serves as a tool to verify and validate micromechanics analysis models and procedures through which composite properties are often predicted from the constituent (fibers and matrix) properties. Physical and mechanical characterization of composite materials is discussed in this chapter.

Elastic constants and strengths are basic mechanical properties of materials. For a unidirectional lamina or composite, there are four independent elastic constants—the elastic moduli (E_L and E_T) in the longitudinal and transverse directions, the shear modulus (G_{LT}), and the major Poisson ratio (ν_{LT})—and five independent strengths, namely, tensile and compressive strengths (σ_{LU}, σ_{TU}, σ'_{LU}, σ'_{TU}) in the longitudinal and transverse directions and in-plane shear strength (τ_{LTU}). In the case of composite laminates, the interlaminar shear strength determines an important mode of failure, namely, the delamination failure. It is usually considered a basic property. These basic properties are established for minimum characterization of a unidirectional lamina. They are generally established by subjecting suitable material specimens to in-plane loads. However, since composite material structures often are subjected to bending loads, it is desirable to establish flexural properties in addition to the properties just mentioned. It is desirable that all the properties be established for a single ply or lamina of the composite material that is the basic building block for composite structures. Then laminate theory, as discussed in Chaps. 5 and 6, can be used to calculate the properties of laminates. However, practical considerations often prevent the construction of single-layer test specimens. Thus it becomes necessary to conduct tests on multilayer specimens and use approximate laminate theory to reduce the results in terms of lamina properties. If the laminates are unidirectional, of course, their behavior simulates the lamina behavior. In addition to the basic properties of composites, fracture toughness and impact properties are also important to predict performance of composite structures in various applications. Measurement procedures for fracture toughness and impact properties are also discussed in this chapter.

It is well known that during load application, damage in composites can commence at very low loads and grow progressively until fracture. This damage accumulation affects the stress–strain behavior as strain increases. For an efficient use of composites through advanced analysis procedures, their actual stress–strain behavior with accruing damage may be used in analyses. It is therefore helpful to identify and characterize damage in composite materials. One section in this chapter has been devoted to the discussion of nondestructive evaluation techniques commonly employed for damage identification and characterization.

The test procedures commonly employed for evaluating various composite properties are described in this chapter. Discussions on different tests include their advantages, limitations, and considerations that go into the designing of specimens, loading mechanisms, and measuring techniques. Methods of reducing the experimental data also are discussed. However, the details of specimen preparation, instrumentation, and measuring techniques are omitted, and the discussion is limited to only static properties. A more detailed discussion can be found in other literature [1–20]. Standard procedures for the testing of composites and determination of the properties from these tests have been adopted by various standard-setting bodies such as the American Society for Testing and Materials (ASTM) in the United States. These standards should be consulted for detailed test protocols. Relevant ASTM standards are cited in the discussions of this chapter. Standards are updated periodically to incorporate new developments in materials, analyses, instrumentation, and measuring techniques.

10.2 MEASUREMENT OF PHYSICAL PROPERTIES

10.2.1 Density

The *density* of a material is defined as its mass per unit volume. The procedure for measuring the density of a composite material is same as that used for any other solid and is based on ASTM Standard D792-00. The procedure involves obtaining weights of a suitable size specimen in air (w_a) and in water (w_w). The density can be calculated from these weights and the densities of air and water. The density of air is assumed negligible, whereas that of water is 1 g/cm³. The density of composite ρ_c is then

$$\rho_c = \frac{w_a}{w_a - w_w} \tag{10.1}$$

10.2.2 Constituent Weight and Volume Fractions

The constituent weight and volume fractions are very important physical characteristics of composite materials that influence their properties. The constituent weight fractions can be obtained by following the methods of ASTM Standard D3171-99. In these methods, the matrix material is physically removed from a composite specimen while the reinforcement remains essentially unaffected. The matrix material is removed either by dissolving it in a hot-liquid medium or by its combustion in a furnace. The *matrix-dissolution method*, also called the *matrix-digestion method*, requires a careful selection of the medium that will dissolve the matrix but will not attack the fibers. When the matrix is dissolved in the medium, the residue containing the fibers is then filtered, washed, dried, cooled, and weighed. The fiber weight fraction now can be calculated from the initial weight of the composite specimen and the weight of the fibers. The matrix weight fraction is obtained by subtracting the fiber weight fraction from one.

The *combustion method*, often called the *burnoff method*, can be used only with polymer-matrix composites and with fibers that are not affected by high-temperature environments. In this method, a preweighed composite specimen is heated in a furnace to a temperature compatible with the composite system that will burn off the matrix and not affect the reinforcement. The remainder after burnoff is cooled to room temperature and weighed to obtain fiber and matrix weight fractions. This method is used widely with glass-fiber-reinforced polymer composites. The burnoff method is essentially the same as the ignition loss of cured reinforced resin described in ASTM Standard D2584-02.

The equations for converting weight fractions to volume fractions and vice versa were discussed in Chap. 3. The conversions require a knowledge of the constituents' densities. The void volume, if not negligible, also will have to be determined, as discussed in the next section, before weight fractions can be converted to volume fractions.

Fiber volume fraction of a unidirectional composite also can be measured by an optical method. In this method, a cross section of the composite perpendicular to the fibers is polished and photographed in a microscope such as those shown in Fig. 3-2. The ratio of the total fiber area to the total composite area as determined from the enlarged photograph is taken as the fiber volume fraction (V_f). The total fiber area can be determined either by using a computerized image-analysis technique or by manually counting the number of fibers and multiplying it by the average fiber cross-sectional area.

10.2.3 Void Volume Fraction

Polymeric- and ceramic-matrix composites typically have voids after fabrication. A well-fabricated composite may have a void volume of 1% or less, and a poorly made composite can have a much higher void content. The void content of a composite can affect its properties and performance. Thus it is important to know the void content for assessing fabrication quality.

Void content of a composite can be measured using the method of ASTM Standard D2734-94 (reapproved 2003). In this method, the densities of the resin, the fibers, and the composite are measured separately. The resin and fiber weight fractions are obtained as discussed in Sec. 10.2.2. The theoretical density of the composite with no voids is then calculated using Eq. (3.4). This is compared with the measured composite density. The difference in the densities indicates the void content as follows:

$$V_v = \frac{\rho_{ct} - \rho_{ce}}{\rho_{ct}} \quad (10.2)$$

where V_v = void volume fraction
ρ_{ct} = theoretical density of composite with no voids
ρ_{ce} = experimentally measured density of composite

10.2.4 Thermal Expansion Coefficients

The coefficient of linear thermal expansion of isotropic materials is measured by measuring elongation of a rod with temperature. Unidirectional composites have, as discussed in Chap. 3, two principal coefficients of thermal expansion in the longitudinal direction α_L and in the transverse direction α_T. The coefficients α_L and α_T are determined using a rectangular specimen of dimensions 50 mm × 50 mm made with eight plies. The specimen is heated in an oven at a rate of about 1°C/min. The temperature is monitored with a thermocouple or a temperature sensor. In place of measuring length changes in the specimen, it has been found more convenient and appropriate to record strains with the help of strain gauges. For this purpose, two strain gauges are mounted in the longitudinal and transverse directions. From the measured strains and temperatures, the coefficients α_L and α_T can be determined using Eq. (6.54) as follows:

$$\varepsilon_L = \alpha_L \, \Delta T \qquad (10.3)$$

$$\varepsilon_T = \alpha_T \, \Delta T$$

Since moisture induces dimensional changes in many resin systems, it is important to dry the specimens before measuring the thermal strains.

There is no ASTM standard exclusive to the measurement of coefficients of thermal expansion of composite materials. However, ASTM Standards E289-04 and E831-03 may be useful.

10.2.5 Moisture Absorption and Diffusivity

Moisture absorption and desorption characteristics of polymer-matrix composites are of considerable practical importance. Studies on the effects of moisture content on composite laminates require diffusivities of a unidirectional composite in the longitudinal and transverse directions and the equilibrium moisture content [or parameters a and b in Eq. (3.80)]. The diffusivity of the matrix material can be determined experimentally and the diffusivities of composites calculated using Eqs. (3.81) and (3.82).

In most engineering applications, moisture diffusion is through a large surface area with very few edges. As a result, diffusion through the thickness is of primary interest. For such a case, thin composite specimens are used for diffusion measurements with edges sealed with an impermeable coating or foil.

Diffusion tests consist of measuring weight gain as a function of time when the specimen is exposed to a constant temperature and humidity environment. The specimens are dried completely in a desiccator before starting the moisture absorption test. The moisture content C is plotted as a function of $\sqrt{t}$. A relationship between C and $\sqrt{t}$ is shown schematically in Fig. 10-1. The

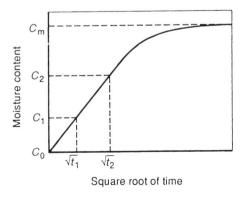

Figure 10-1. Moisture-absorption curve for diffusivity measurement.

diffusivity D is calculated from the initial linear relationship. For this purpose, Eq. (3.78) can be approximated by the following relationship for short times:

$$G = \frac{C - C_0}{C_m - C_0} = 4\left(\frac{Dt}{\pi S^2}\right)^{1/2} \quad (10.4)$$

Now two moisture contents C_1 and C_2 corresponding to times t_1 and t_2, respectively, are chosen from the linear region of the curve and substituted in Eq. (10.4). It can be shown easily that the diffusivity D is

$$D = \frac{\pi}{16}\left(\frac{C_1 - C_2}{C_m - C_0}\right)^2\left(\frac{S}{\sqrt{t_1} - \sqrt{t_2}}\right)^2 \quad (10.5)$$

The equilibrium moisture content C_m, as shown in Fig. 10-1, depends on the relative humidity of the environment. Therefore, tests may be conducted with different relative humidities. Constants a and b of Eq. (3.80) are determined by plotting C_m as a function of relative humidity.

ASTM Standard D5229/D5229M-92 (reapproved 1998) describes the procedure for evaluating moisture absorption properties and diffusivity.

10.2.6 Moisture Expansion Coefficients

A coefficient of moisture expansion β has been defined in Sec. 3.7.2 as the change in linear dimension of a body per unit initial length per unit change in moisture concentration (moisture concentration being defined as the weight of moisture present per unit weight of the body). Like thermal expansion coefficients, there are two principal coefficients of moisture expansion for unidirectional composites, β_L in the longitudinal direction and β_T in the transverse direction. However, as discussed in Sec. 3.7.2, the longitudinal moisture expansion coefficient β_L is often taken to be zero for the high-modulus inorganic-fiber-reinforced polymer-matrix composites. The transverse moisture expansion coefficient β_T then is related to the moisture expansion coefficient of the matrix β_m through Eq. (3.71). Therefore, moisture expansion coefficients for such composites can be obtained by measuring the moisture expansion coefficient of the matrix.

Moisture expansion coefficients of composites can be determined directly by measuring the strains in the principal directions as a function of moisture concentration. For this purpose, oven-dried specimens are immersed in a water bath at a moderately high temperature, for example, 50°C (120°F). Specimen expansion or swelling may be measured by a micrometer or a caliper gauge. The application of strain gauges, as used for measuring thermal expansion coefficients, is questionable for two reasons. First, the presence of a gauge on the specimen surface will locally interfere with the moisture diffusion, and second, the strain gauge adhesive may be attacked by moisture and may compromise the accuracy of the strain measurement. However, both these problems may be solved by the use of encapsulated strain gauges and embedding them in the midplane of the specimen. With this technique, the strain gauge does not interfere with the moisture diffusion because the gauge is located at a plane of symmetry. Further, the encapsulated gauge does not require additional adhesive and thus provides accurate strain values. In this technique, two specimens of identical size are prepared, one with and the other without the embedded strain gauges. The specimens are dried and then immersed in a water bath inside an oven at 50°C (120°F) so that both specimens are exposed to the same environment. The specimen without the gauge is removed periodically from the water and weighed to determine moisture concentration. The embedded strain gauges in the other immersed specimen are monitored continuously to obtain strain values. These data provide a relationship between moisture concentration and expansions of the composite owing to moisture. The moisture expansion coefficients then can be calculated.

10.3 MEASUREMENT OF MECHANICAL PROPERTIES

10.3.1 Properties in Tension

The static uniaxial tension test is probably the simplest and most widely used mechanical test. This test is conducted to determine elastic modulus, tensile strength, and Poisson's ratio of the material. In the case of composite materials, the tension test generally is performed on flat specimens. The most commonly used specimen geometries are the dog-bone specimen and the straight-sided specimen with end tabs, as shown in Fig. 10-2. A uniaxial load is applied through the ends by providing a pin-type or serrated-jaw-type end connection as shown in Fig. 10-3.

The dog-bone specimens may fail at the neck radius, particularly when testing uniaxial specimens, because of stress concentrations and poor axial shear properties of the specimen. The pin-type end connections also tend to fail at low loads by shear. A single pin-type end connection is sometimes replaced by multihole-type end connections, particularly in research work, to provide a uniform distribution of load. However, it is not normally used in full-scale experimental characterization work. The straight-sided specimen with end tabs (see Fig. 10-2b), along with the serrated-jaw-type end connections (see Fig. 10-3b), relieves these problems.

The ASTM standard test method for tensile properties of fiber–resin composites has the designation D 3039/D 3039M-00. It recommends that the specimens with fibers parallel to the loading direction be 12.7 mm wide and made with 6–8 plies and that the specimens with fibers perpendicular to the

446 EXPERIMENTAL CHARACTERIZATION OF COMPOSITES

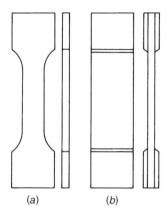

Figure 10-2. Tension-test specimens: (a) dog-bone and (b) straight-sided with end tabs.

load be 25.4 mm wide and be made with 8–16 plies. Length of the test section in both cases should be 153 mm. End tabs made from unidirectional nonwoven E-glass fibers (balanced cross-ply laminate) have proved to be satisfactory. The tabs should be at least 38 mm long and have a thickness 1.5–4 times the thickness of the test specimen.

The data recording in a tension test consists of measuring the applied load and strain both parallel and perpendicular to the load. The applied load is usually measured by means of a load cell that generally is provided with the testing machine. The strains can be measured by means of an extensometer or an electrical-resistance strain gauge. Extensometers, mechanically attached, tend to slip at times, although they are quite simple to use. Strain gauges may be used for a more accurate measurement of strains. From these data taken until, failure, a stress–strain curve can be plotted easily for the material and the required material properties determined. If the applied load is in the longitudinal direction, the initial slope of the stress–strain curve gives the longitudinal modulus (E_L). Similarly, the transverse modulus (E_T) can be determined by applying the load in transverse direction. Ultimate longitudinal and transverse tensile strengths (σ_{LU} and σ_{TU}) are obtained from the knowledge of load at fracture in the two tests. The Poisson's ratio (ν_{LT}) is obtained

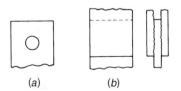

Figure 10-3. End connections for tension-test specimens: (a) pin type and (b) serrated-jaw type.

from the strains parallel and perpendicular to the load measured at the same axial load.

While testing a unidirectional composite in the longitudinal direction, it should be ensured that the load direction does coincide with the fiber direction. Misalignment by only a few degrees may result in considerably lower values of elastic modulus and ultimate tensile strength. This problem is not as critical for tests in the transverse direction.

Another problem that is sometimes encountered during the determination of longitudinal tensile strength of high-strength unidirectional composites concerns the transfer of load through the end tabs, which requires a good adhesive bond. Therefore, longer tabs and better adhesives may be necessary for longitudinal specimens. Dog-bone specimens with a large radius of curvature also may be used.

Since the elastic constants and strengths of composite materials change with direction, the elastic constants in an arbitrary direction may be calculated from the four constants in the longitudinal and transverse directions. Transformation equations for this purpose were derived in Chap. 5. A knowledge of the shear modulus is required, and the measurement techniques are described later. It may be of considerable interest to obtain off-axis elastic constants and strength directly by conducting tension tests on specimens in which fibers are oriented to the load direction at angles other than 0° or 90°. Conducting tension tests on off-axis specimens, however, require greater care because the coupling between normal stress and shear strain presents problems in obtaining uniform states of stress and strain. The problems can be best illustrated by considering states of stress and strain near the specimen ends. A clamped end, which is normally used in tension tests, produces the following state of strain in its vicinity:

$$\epsilon_x \neq 0, \qquad \epsilon_y = \gamma_{xy} = 0 \qquad (10.6)$$

where the direction parallel to the load (also perpendicular to the edge of the clamp) has been denoted as the x axis. Boundary conditions for the clamped end are

$$\sigma_x \neq 0 \quad \text{and} \quad \tau_{xy} \neq 0 \qquad (10.7)$$

Since the net force is applied only in the direction of the x axis, the shear-stress distribution is such that the net force produced by it will be zero. Now it can be seen easily from Eq. (5.20) that in the presence of τ_{xy}, the ratio σ_x/ϵ_x does not represent E_x but represents $\overline{Q}_{11}$, as given in Eq. (5.94). Further, it should be noted that in this case E_x cannot be evaluated from Eq. (5.20) because the magnitude of τ_{xy} is unknown. This difficulty in conducting off-axis tension tests is overcome either by the use of a long (compared with width) specimen when the ends are clamped or by making the ends free to deform in the manner shown in Fig. 10-4a. In a specimen with free ends and

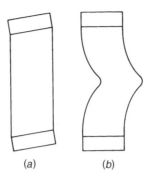

Figure 10-4. Effects of end constraints in off-axis tension test (effect shown exaggerated): (a) grips free to rotate in-plane and (b) grips restrained from rotating.

at sections away from the clamped ends in the case of a long specimen (St. Venant principle), the following state of uniaxial stress exists:

$$\sigma_x \neq 0 \qquad \sigma_y = \tau_{xy} = 0 \qquad (10.8)$$

Equation (5.20) gives strains as

$$\epsilon_x = \frac{\sigma_x}{E_x}$$

$$\epsilon_y = -\nu_{xy} \frac{\sigma_x}{E_x} \qquad (10.9)$$

$$\gamma_{xy} = -m_x \frac{\sigma_x}{E_L}$$

Therefore, by measuring ϵ_x, ϵ_y, and γ_{xy}, E_x and ν_{xy} can be evaluated, as well as m_x if E_L is already known. Richards et al. [21] showed that satisfactory results can be achieved for 45° specimens by having a value of 12 for the ratio of the specimen length between the grips to its width. A more detailed discussion on the end effects and the corrections necessary for shear coupling is presented by Pagano and Halpin [22] and Wu and Thomas [23].

It may be pointed out that besides influencing the state of stress, the clamped ends can result in unusual deformations in an off-axis specimen because of coupling between in-plane shear and tension, as shown in exaggeration in Fig. 10-4b.

The tensile properties of composites can be determined by using a sandwich-beam specimen subjected to bending in which a thin layer of composite material is bonded to the top and bottom of a thick substrate such as an aluminum honeycomb core [2,24,25]. The composite on one side is loaded in tension, and that on the other side, in compression. Thus a single specimen can be used to determine both the tensile and compressive properties. An additional advantage of a sandwich-beam specimen is that it can be used conveniently for determining off-axis properties and transverse properties, as shown by Lantz [25]. Also, sandwich beams using composites as faces are commonly employed structures in the aerospace industry and can be readily fabricated and simulate end-use application.

In the early days of the development of filament-wound composites, it was recognized that it would be more realistic to use filament-wound specimens for characterization than to use flat specimens made by hand lay-up. Therefore, filament-wound tubes and the rings cut from the tube have been used as test specimens (ASTM D 2290). However, their use has been limited because of the complex loading mechanism involved. Moreover, filament-winding facilities are not available in many laboratories.

10.3.2 Properties in Compression

Static uniaxial compression tests are similar to the tension tests but present many more problems. The biggest problem is the necessity to prevent geometric buckling of the specimens. This requirement is particularly relevant to thin, flat specimens and usually is met by providing multiple side supports that prevent the specimen from buckling out of its plane. The use of side supports can be avoided by using a block- or bar-type specimen rather than a plate. However, the block-type specimens are more difficult to prepare.

Compression tests on unidirectional composites pose one more problem. When the composite is subjected to compressive loads in the fiber direction, premature failure occurs by localized "brooming" at the ends, as shown in Fig. 10-5. This problem can occur with metal-matrix composites as well as polymer-matrix composites and is present even with the block-type specimens. One means of reducing end-brooming action is to embed the ends of the specimens into either a polymer or a low-melting-point metal alloy. A more successful method of eliminating brooming is to clamp the ends. Various designs are available for end-clamping methods [26,27].

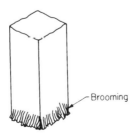

Figure 10-5. End brooming under compressive loading represents premature failure.

A number of loading fixtures and specimen configurations have been developed to measure the compressive strengths of composites [28]. Alignment of the test fixture and specimen is an essential consideration in any compression test. The most common test fixture, the IITRI test fixture, developed by the Illinois Institute of Technology Research Institute [29], is shown schematically in Fig. 10-6. A relatively short (gauge length 12.7 mm), unsupported test specimen is used with this fixture. The fixture employs linear bearings and hardened-steel shafts to ensure colinearity of the loading direction. The IITRI fixture mounted in a loading frame is shown in Fig. 10-7. ASTM Standard D 3410/D 3410M-03 describes the test method for compressive properties of polymer-matrix composite materials with an unsupported gauge section by shear loading such as the method using the IITRI test fixture. A newer ASTM standard (D 6641/D 6641M-01) describes the test method for determining the compressive properties of polymer-matrix composites using a combined loading compression (CLC) test fixture. This loading mechanism has some advantages over the shear loading.

The data recording in compression tests is also similar to that in tension tests. The strain in the direction of loading may be measured by a compres-

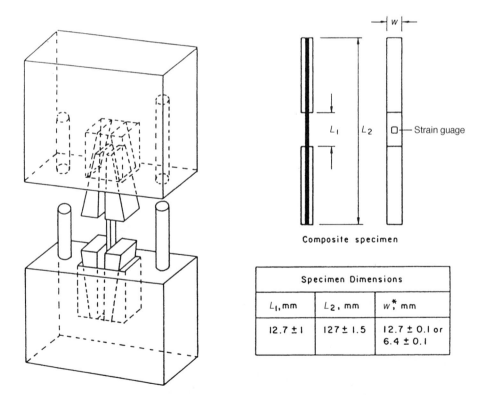

Figure 10-6. Modified IITRI compression test fixture and specimen dimensions.

Figure 10-7. Photograph of an IITRI compression test fixture. (From Hofer and Rao [29].)

someter or electrical-resistance strain gauges. The testing-machine head movement is not a reliable measure of strain because some error may result from crushing of the specimen ends. The strain in the direction perpendicular to the load is measured using strain gauges only because of space limitations. From these measurements, the elastic moduli and Poisson ratios of the material in compression can be determined and the stress–strain curve in compression plotted. A straight-sided specimen is well suited for elastic modulus and Poisson ratio determination. For compressive-strength determination, reduced-center-section specimens often are used to ensure that failure does not occur near the end of the specimen. Again, it should be mentioned that composite properties in compression also may be determined by using sandwich-beam specimens [ASTM D 5467/D 5467M-97 (reapproved 2004)].

10.3.3 In-Plane Shear Properties

The tests in which shear distortion takes place entirely in the plane of the composite material sheet are termed *in-plane shear tests.* The properties that are determined through these tests are the shear modulus and shear strength. In these tests, a material specimen is subjected to loads that produce a pure shear state of stress, and the resulting strains are measured.

10.3.3.1 Torsion Tube Test The easiest way to produce a state of pure shear is to subject a thin-walled circular tube to a torque about its axis, as shown schematically in Fig. 10-8. This produces a uniform shear on the surface of the tube. The relationship between the torque T and the shear stress τ_{xy} is given as

$$\tau_{xy} = \frac{T}{2\pi r^2 t} \tag{10.10}$$

where r is mean radius, and t is the thickness of the tube. Since the wall thickness is small compared with the mean radius, the shear-strain variation through the thickness may be neglected. Therefore, the torsion tube is a widely used specimen for in-plane shear tests. Pagano and Whitney [30] have shown that the torsion tube is the most desirable shear test specimen from an applied mechanics standpoint. However, care should be taken to ensure that only pure torque is applied to the specimen. The specimen must be mounted concentrically and free to move axially so that bending moments and axial forces are not developed if coupling exists because of laminate construction or fiber angle.

The test requires means for accurate measurement of applied torque and the resulting shear strain. The shear strain generally is measured by means of electrical-resistance strain gauges. Since the strain gauges cannot measure shear strain directly, the normal strains at 45° to the tube axis are measured, from which the shear strain can be calculated by use of the Mohr strain circle. ASTM Standard D 5448/D 5448M-93 (reapproved 2000) provides more details of this test procedure.

Two additional precautions for the test are to prevent failure of the bond between the specimen and the end attachment through which the torque is applied and to prevent buckling of the specimen. The former is achieved by

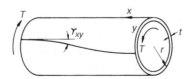

Figure 10-8. Torsion tube for shear test.

providing a long adhesive joint bonded on the inside and outside surfaces of the tube. The buckling may be avoided by either providing sufficient wall thickness or keeping the length small.

Given the difficulties associated with the fabrication and testing of tubular specimens, other test methods are employed for determination of the in-plane shear properties of unidirectional composites. Among them, the Iosipescu shear test, the $\pm 45°$ laminate test, and the off-axis tensile test are currently the most commonly employed methods. These and some other tests are described in the following subsections.

10.3.3.2 Iosipescu Shear Test The Iosipescu shear test specimen is shown schematically in Fig. 10-9 along with the associated shear-force and bending-moment diagrams. A state of pure shear is achieved at the specimen

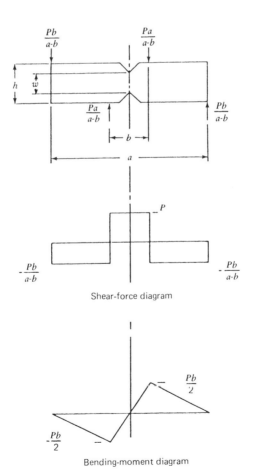

Figure 10-9. Schematic representation of the Iosipescu shear test method and associated shear-force and bending moment diagrams.

midlength by application of two counteracting moments produced by two force couples. In the middle section of the test specimen, a constant shear force of magnitude P is induced. The bending moments exactly cancel at the midlength of the specimen, producing a pure shear. An Iosipescu test fixture (Fig. 10-10) is designed to restrain each end of the test specimen from rotating. The 90° notches on each edge of the specimen produce a constant shear–stress distribution between the two notches instead of a parabolic shear–stress distribution for a constant-cross-section beam. Therefore, the value of shear stress τ for the test shown in Figs. 10-9 and 10-10 is given by the shear force divided by the net cross-sectional area:

$$\tau = \frac{P}{wt} \qquad (10.11)$$

where w is the net width between the two notches, and t is the thickness of the test specimen. An Iosipescu shear test fixture is shown in Fig. 10-11. ASTM Standard D 5379/D 5379M-98 describes the test method for measuring shear properties employing this procedure.

Details regarding the development of the Iosipescu test for composites can be seen in various publications of Adams and Walrath [31–34]. A typical specimen for the Iosipescu test is 51 mm long and 12.7 mm wide. A 90° notch is cut to a depth of 2.5 mm on each edge of the specimen at the midlength. In general, the specimen thickness should be about 2.5 mm to avoid buckling-induced failures. Very thin materials may be tested by bonding several layers together or by using reinforcing tabs in the loading regions. Shear strain in the test section is obtained by measuring normal strains at

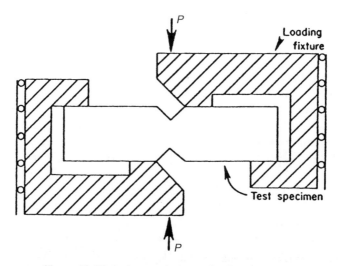

Figure 10-10. Iosipescu shear test loading mechanism.

Figure 10-11. Photograph of an Iosipescu shear test fixture. (Courtesy of Wyoming Test Fixtures, Inc.)

±45° to the longitudinal axis of the specimen. For this purpose, two strain gauges oriented at ±45° to the longitudinal axis of the specimen and centered between the notches are used. Other methods of measuring shear strains so far have not produced satisfactory results.

The Iosipescu shear test is relatively easy to use. Specimens are small, easy to fabricate, and easy to install in the test fixture. Both shear modulus and shear strength can be determined through the test. However, for highly orthotropic materials (i.e., large value of E_L/E_T), there is a nonuniformity in the stress distribution in the test section, which introduces an error in the determination of property values. Pindera et al. [35] have suggested correcting the property values by incorporating a factor obtained through finite-element analysis. Adams and Walrath [31] suggest using a cross-ply laminate, $[0/90]_S$, instead of a unidirectional composite. In theory, shear properties of the cross-ply laminate are the same as those of the unidirectional composite.

10.3.3.3 [±45]$_S$ Coupon Test In-plane shear properties of a unidirectional composite can be determined by conducting a tension test on a $[\pm 45]_S$ laminate (Fig. 10-12). The test results are interpreted using laminate stress analysis. It can be shown (see Exercise Problems 6.2 and 6.3) that for an applied stress σ_x, the laminae stresses are

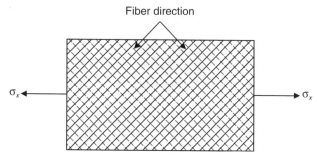

Figure 10-12. Tension test on $[\pm 45]_S$ coupon for shear modulus measurement.

$$\sigma_L = \tfrac{1}{2}(\sigma_x + 2\tau_{xy})$$
$$\sigma_T = \tfrac{1}{2}(\sigma_x - 2\tau_{xy}) \quad (10.12)$$
$$\tau_{LT} = -\tfrac{1}{2}\sigma_x$$

where τ_{xy}, the shear stress induced in the laminae, is not known. The lamina strains are related to the laminate strains as follows:

$$\varepsilon_L = \tfrac{1}{2}(\varepsilon_x^0 + \varepsilon_y^0)$$
$$\varepsilon_T = \tfrac{1}{2}(\varepsilon_x^0 + \varepsilon_y^0) \quad (10.13)$$
$$\gamma_{LT} = (\varepsilon_y^0 - \varepsilon_x^0)$$

These results show that for the $[\pm 45]_S$ laminate, the laminae shear stress τ_{LT} and the corresponding shear strain γ_{LT} can be obtained from laminate stress and strains. Thus the results of a tension test on a $[\pm 45]_S$ laminate can be employed to establish shear stress–strain response of the lamina. This test method is quite suitable for determination of shear modulus (G_{LT}) but not for shear strength (τ_{LTU}) because the lamina is in a state of combined stress rather than pure shear. This test procedure was suggested by Rosen [36], and details are presented in ASTM Standard D3518/D3518M-94 (reapproved 2001).

10.3.3.4 Off-Axis Coupon Test The off-axis tension test is used often to determine the in-plane shear response of unidirectional composites. The test is shown schematically in Fig. 10-13. For an applied stress σ_x, the stresses along the longitudinal and transverse directions can be written as

$$\sigma_L = \sigma_x \cos^2 \theta$$
$$\sigma_T = \sigma_x \sin^2 \theta \quad (10.14)$$
$$\tau_{LT} = -\sigma_x \sin \theta \cos \theta$$

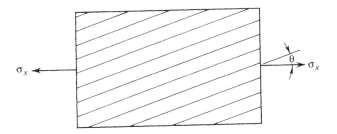

Figure 10-13. Tension test on off-axis coupon for shear-stress–strain response.

It can be shown that the strains along the longitudinal and transverse directions are related to the strains in the x and y directions and the strain in a direction oriented at 45° to the x axis:

$$\varepsilon_L = \cos\theta(\cos\theta - \sin\theta)\varepsilon_x + \sin\theta(\sin\theta - \cos\theta)\varepsilon_y + 2\sin\theta\cos\theta\,\varepsilon_{45}$$

$$\varepsilon_T = \sin\theta(\cos\theta + \sin\theta)\varepsilon_x + \cos\theta(\sin\theta + \cos\theta)\varepsilon_y - 2\sin\theta\cos\theta\,\varepsilon_{45}$$

$$\gamma_{LT} = -(\cos^2\theta + 2\cos\theta\sin\theta - \sin^2\theta)\varepsilon_x$$
$$- (\cos^2\theta - 2\sin\cos\theta - \sin^2\theta)\varepsilon_y + 2(\cos^2\theta - \sin^2\theta)\varepsilon_{45} \quad (10.15)$$

Equation (10.15) shows that the strains along the longitudinal and transverse directions can be obtained from the strains (ε_x, ε_y, ε_{45}) measured using a rectangular three-element strain-gauge configuration. Thus, the off-axis tension test can be used to obtain the shear stress–strain response of the unidirectional composite. Quite often $\theta = 45°$ is used for off-axis specimens. In this case, the shear stress and shear strain can be written as

$$\tau_{LT} = \tfrac{1}{2}\sigma_x \quad (10.16)$$

$$\gamma_{LT} = \varepsilon_y - \varepsilon_x \quad (10.17)$$

By combining Eqs. (10.16) and (10.17) and using the stress–strain relations, it can be shown that for a 45° off-axis specimen,

$$G_{LT} = \frac{E_x}{2(1 + \nu_{xy})} \quad (10.18)$$

Pindera and Herakovich [37] have suggested the use of 45° off-axis specimen for the measurement of G_{LT}. In the case of low off-axis configurations such as the 10° off-axis coupon, a specimen with very high aspect ratio must be employed. The 45° off-axis specimen is not suitable for measurement of shear strength owing to the presence of σ_L and σ_T. Chamis and Sinclair [38] rec-

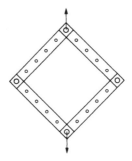

Figure 10-14. Picture frame test for in-plane shear properties.

ommended the use of the 10° off-axis test to minimize the effects of σ_L and σ_T on the shear response.

10.3.3.5 Other Tests   Besides the tests described in earlier sections, other tests are employed for determination of the in-plane shear properties. For example, a thin, square plate is loaded at its edges by means of a loading fixture called a *picture frame*, as shown in Fig. 10-14. The loading fixture is bonded or bolted to the edges of the specimen. This type of loading produces a state of pure shear near the edges, but the state of stress in the central portion deviates substantially from pure shear. Hence this method may be used for determining shear strength but not the shear modulus.

The rail shear type of test shown schematically in Fig. 10-15 has been used widely to measure the thickness properties of various sandwich-core materials. This type of test may be used for in-plane shear properties also. ASTM Standard D 4255/D 4255M-01 describes this test procedure.

Whitney et al. [39] concluded through a theoretical stress analysis that it is a valid test for determining shear modulus, provided that the length–width ratio is at least 10. However, an additional requirement should be satisfied for determining shear strength: The major Poisson ratio with respect to the specimen edges should be less than unity. If this condition is not met, as in a ±45° laminate made from unidirectional laminae, a severe stress concentration is developed, which results in an excessively low value of shear strength.

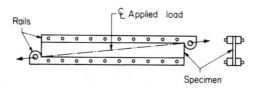

Figure 10-15. Rail shear test.

10.3 MEASUREMENT OF MECHANICAL PROPERTIES

The four-point loaded flat plate and sandwich cross-beam specimens also have been used [2,40] for determining in-plane shear properties. In the central portion of a sandwich cross-beam, a uniform pure shear is produced at 45° to the specimen edges, as shown in Fig. 10-16. However, because of stress concentration at the corners of the cross, a uniform stress state is approached in the very center of the cross, whereas failure initiates at the corners. Thus the cross-beam test is no longer regarded as an adequate test for measuring shear strength and shear stiffness. The flat-plate test also has not been used widely.

10.3.4 Flexural Properties

The two most popular flexural tests, the three-point and four-point bending tests, are shown schematically in Figs. 10-17 and 10-18, respectively. In these tests, a flat specimen is simply supported at the two ends and is loaded by either a central load (three-point bending) or by two symmetrically placed loads (four-point bending). The shear-force and bending-moment diagrams for the two tests are also shown in Figs. 10-17 and 10-18. Variations across the beam thickness of normal stress (often called *bending stress*) owing to bending moment and shear stress resulting from shear force are shown in Fig. 10-19, where the cross section of the beam has been assumed rectangular. Material properties also have been assumed uniform through the thickness, as in unidirectional composites or iostropic materials (bending stresses in laminated composites were discussed in Chap. 6). The normal stress varies linearly from maximum compression on one surface of the beam to an equal

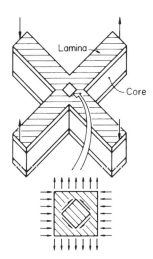

Figure 10-16. Sandwich cross-beam test.

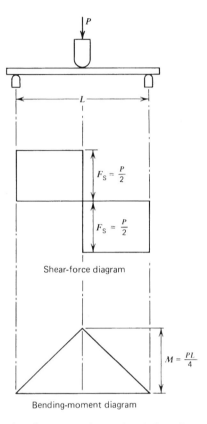

Figure 10-17. Three-point bending test and associated shear-force and bending-moment diagrams.

tensile value on the other surface, being zero at the midplane or neutral axis. The maximum normal stress is given by

$$\sigma = \frac{6M}{bh^2} \qquad (10.19)$$

where M is the bending moment on the cross section, b is the specimen width, and h is the specimen thickness. The shear stress has a parabolic distribution with maximum value at the midplane and zero on the outer surfaces of the beam. The maximum shear stress is given by

$$\tau = \frac{3F_s}{2bh} \qquad (10.20)$$

where F_s is the shear force on the section. Flexural response of the beam is obtained by measuring applied load and corresponding strain. The strain may

10.3 MEASUREMENT OF MECHANICAL PROPERTIES

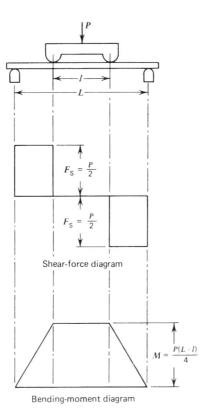

Figure 10-18. Four-point bending test and associated shear-force and bending-moment diagrams.

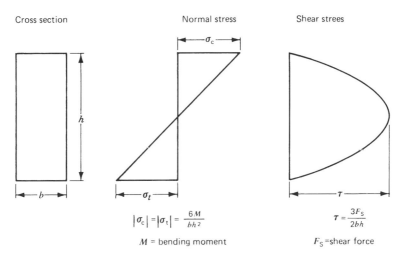

Figure 10-19. Normal-stress and shear-stress variations across thickness of a homogeneous rectangular-cross-section beam during bending.

be measured using a strain gauge bonded to the beam on the tension side, that is, the surface opposite to the applied load.

The bending moment M is determined from the measured load and specimen geometry, and the stress then is calculated from Eq. (10.19). Thus the complete stress–strain behavior in bending can be obtained.

The three-point and four-point bending tests differ from each other with respect to the state of stress and may give slightly different results. In a three-point bending test, the bending moment on the beam varies linearly from zero at the supports to a maximum value at the center (see Fig. 10-17). The shear force and therefore the interlaminar shear stress [Eq. (10.20)] at the midplane are uniform all along its length. This may promote interlaminar shear failure. In four-point bending (see Fig. 10-18), the bending moments increase linearly from zero at the supports to a maximum value under the load. The bending moment remains constant between the loads. In this case, the shear force and consequently the interlaminar shear stress between the loads are zero, and thus this portion of the beam is subjected to a pure bending moment. Thus, from the standpoint of state of stress, a four-point bending test is more desirable, whereas the three-point bending test is easier to conduct.

Flexural strength is the theoretical value of stress on the surface of the specimen at failure. It is calculated from the maximum bending moment by assuming a straight-line stress–strain relation to failure. Thus, for a beam of rectangular cross section, it is calculated using Eq. (10.19), where M is the bending moment corresponding to the failure load.

To obtain the correct value of flexural strength, it should be ensured that the failure takes place by breaking of fibers and not by interlaminar shear. This is accomplished by providing a large span–depth ratio. This can be explained by considering the influence of span on normal stress and interlaminar shear stress. The span of the beam does not influence the interlaminar shear, whereas a larger span results in higher bending moment and thus an increased tendency for longitudinal failure. However, the larger span–depth ratios result in larger deflections, which, in turn, make it necessary to account for the horizontal forces developed at the supports in the calculation of bending moment. Most standard methods for three-point bending suggest a span–thickness ratio of 16. ASTM Standards D 790-03 and D 6272-02 describe the methods for three-point and four-point bending tests, respectively.

The problems associated with flexural testing of off-axis fiber-reinforced composites have been studied by Halpin et al. [41] and Whitney and Dauksys [42]. The bending, twisting, coupling, or warping effects in such specimens (as discussed in Chap. 6) tend to partially lift the specimen off at the supports. However, if the lift-off is suppressed using double knife edges at the supports, additional twisting moments are induced in the specimen. It is possible to minimize the lift-off effects by using a sufficiently narrow specimen and a high span–thickness ratio.

10.3.5 Measures of In-plane Fracture Toughness

Fracture toughness of a material refers to the resistance offered by the material to its failure by crack initiation and propagation. Three different approaches to crack initiation and propagation were discussed in Chap. 8. Each of these approaches assumes the existence of a measure of fracture toughness that controls failure by crack initiation and propagation. These measures are the critical strain-energy release rate (G_C), critical stress-intensity factor or crack-growth resistance (K_R), and critical J-integral (J_C). Experimental measurement of these measures of fracture toughness is discussed in the following paragraphs. It may be pointed out that discussion is limited to crack-opening mode (mode I). The remaining two failure modes can be treated in a similar manner.

10.3.5.1 Critical Strain-Energy Release Rate (G_C)

Consider a plate (Fig. 10-20) that contains a centrally located crack of length $2C$ subjected to a load P. The strain energy stored in the plate is

$$U = \tfrac{1}{2} P \delta \qquad (10.21)$$

where δ is the elongation between the load points. Consider now the strain-energy release rate G per unit crack extension assuming that the position of the grips through which the load is applied is fixed; that is, δ is constant. Thus

$$G = -\frac{\partial U}{\partial C} = -\frac{1}{2} \delta \frac{\partial P}{\partial C} \qquad (10.22)$$

The spring constant k of the plate may be defined by

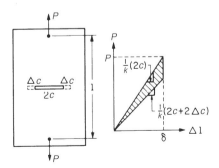

Figure 10-20. Tension test on a plate with a crack to measure G_{IC}.

$$\delta = \frac{P}{k} \qquad (10.23)$$

It follows that for the fixed-grip case,

$$\frac{\partial \delta}{\partial C} = \frac{1}{k}\frac{\partial P}{\partial C} + P\frac{\partial(1/k)}{\partial C} = 0$$

$$\frac{\partial P}{\partial C} = -P \cdot k \frac{\partial(1/k)}{\partial C} \qquad (10.24)$$

Substituting Eqs. (10.23) and (10.24) into Eq. (10.22) gives

$$G = -\frac{\partial U}{\partial C} = \frac{1}{2} P^2 \frac{\partial(1/k)}{\partial C} \qquad (10.25)$$

Equation (10.25) provides a convenient means for measuring strain-energy release rate G. The load P at crack initiation is directly measurable, and the spring constant k or the compliance $1/k$ can be obtained from a compliance curve, that is, a diagram of $1/k$ as a function of crack length C. It can be shown easily that the same relationship for G can be derived if instead of fixed grips the load is assumed to be constant.

It should be noted that this experimental technique is applicable for isotropic as well as anisotropic materials. However, for composite materials, the strain-energy release rate, as already mentioned, depends on many factors, such as the orientation of the initial crack with respect to loading and material axes, the geometry of crack extension, the type of loading, and other variables. These parameters are not significant for homogeneous isotropic materials. A more detailed discussion of the fracture process in composite materials is given in Sec. 8.2.1.2

10.3.5.2 Critical Stress-Intensity Factor or Crack Growth Resistance (K_R)

The critical value of the stress-intensity factor is determined through tension or bend tests on notched specimens. The notch is provided on one or both edges of the specimen in the case of a tension test and on the tension side in the case of a bend test, as shown in Fig. 10-21. Load-elongation data are recorded in the tension or bend test. For calculating the stress-intensity factor in a tension test, applied stress is calculated as the load divided by cross-sectional area and as the maximum flexure stress in a bend test. However, Eq. (8.12) cannot be used for calculating the stress-intensity factor because this equation is applicable when loading and the edges of the plate are far (infinitely) away from the crack. The following relationship, incorporating a correction factor to account for the finite width of the specimen, is used to calculate the stress-intensity factor:

10.3 MEASUREMENT OF MECHANICAL PROPERTIES

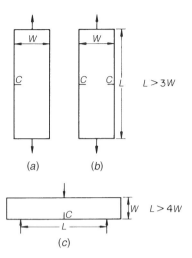

Figure 10-21. Fracture toughness testing specimens: (a) single-edge notched (SEN), (b) double-edge notched (DEN), and (c) notch bend test specimen.

$$K_1 = Y\left(\frac{P}{tw}\right)\sqrt{c} \qquad (10.26)$$

where t and w are specimen thickness and width, respectively, P is applied load, c is crack length, and Y is the width correction factor, called the K-*calibration* factor. The Y can be obtained experimentally. Quite often the values of Y are taken to be the same as the ones used for isotopic materials given below for the three types of specimens shown in Fig. 10-21:

For SEN specimens:

$$Y = 1.99 - 0.41\left(\frac{c}{w}\right) + 18.70\left(\frac{c}{w}\right)^2 - 38.48\left(\frac{c}{w}\right)^3 + 53.85\left(\frac{c}{w}\right)^4 \qquad (10.27)$$

For DEN specimens:

$$Y = 1.98 + 0.36\left(\frac{2c}{w}\right) - 2.12\left(\frac{2c}{w}\right)^2 + 3.42\left(\frac{2c}{w}\right)^3 \qquad (10.28)$$

For notched-bend tests:

$$Y = 1.93 - 3.07\left(\frac{c}{w}\right) + 14.53\left(\frac{c}{w}\right)^2 - 25.11\left(\frac{c}{w}\right)^3 + 25.80\left(\frac{c}{w}\right)^4 \qquad (10.29)$$

The applied stress-intensity factor K_1 tends to extend the crack, and therefore, K_1 is often referred to as the *crack-extension* or *crack-driving force*. It is opposed by the internal material resistance called the *crack-growth resistance* K_R. This resistance can be calculated from knowledge of the applied force in equilibrium with an instantaneous crack length C_i and is given by

$$K_R = Y\left(\frac{P}{tw}\right)\sqrt{C_i} \qquad (10.30)$$

The crack growth resistance is an experimentally determined value of the resistance offered by the material to the applied stress intensity-factor and therefore is calculated from the actual values of load and corresponding instantaneous crack length observed in a fracture test. The expression for K_R is identical to that for K_1, but K_1 may be calculated for any practical or impractical combination of load and crack length, whereas K_R is calculated for equilibrium values of load and crack length.

A plot of crack growth resistance against crack length is referred to as a *crack-growth-resistance curve* (R curve). A plot of crack extension force, for a constant load, against crack length is a crack extension-force curve (K_1 curve).

A typical R curve and several K_1 curves for different loads are shown in Fig. 10-22. One of the K_1 curves is tangent to the R curve. Beyond the point of tangency (*I*), crack extension causes a larger increase in crack extension force K_1 than the crack-growth resistance K_R, and consequently, crack growth can occur without increasing applied load. In other words, unstable crack growth starts at this point. Or the point of tangency is the point of instability on the R curve. The crack growth resistance corresponding to point *I* is critical and is related to the fracture toughness (critical strain-energy release rate) of the material. Thus experimental determination of the R curve for a notched specimen provides a means of experimental measurement of fracture toughness.

ASTM Standard E 1922-97 provides a test method for translaminar fracture toughness of laminated polymer-matrix composite materials under some restricted conditions. It may be pointed out that since the fracture process in a composite can be greatly influenced by a small change in material parameters, a single test method cannot be expected to cover measurement of fracture toughness of different types of composites. Discussion in this section may be helpful in understanding the factors influencing the measurements and in designing the tests.

The R curve for a material is obtained by testing a notched specimen and measuring instantaneous crack length with increasing load. In the case of homogeneous materials, the crack is distinct, well defined, and easily measurable at any instant. However, the fracture process in composite materials does not proceed by a simple enlargement of the original crack. It progresses

10.3 MEASUREMENT OF MECHANICAL PROPERTIES

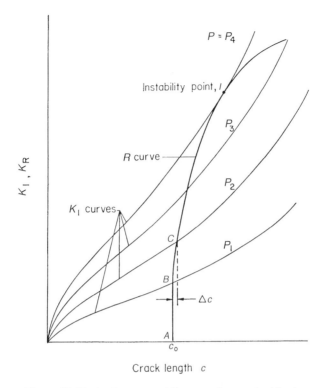

Figure 10-22. An R curve and K_I curves for constant loads.

by the formation and growth of a damage zone in front of the crack tip caused by a large number of microcracks owing to debonding, matrix cracking, and fiber breaks (see Fig. 8-12). An instantaneous crack length is difficult to define or measure. Consequently, a procedure for estimating instantaneous crack length may be adopted. A procedure involving matching of the compliance of a damaged specimen to that of a fresh specimen, generally referred to as the *compliance-matching procedure,* has been used extensively by researchers to transform the test data (e.g., the crack-opening displacement, COD) into an estimated crack length [43–53]. The procedure is explained in the following paragraphs.

Typical load–COD curves for single-edge notched (SEN) specimens of a composite material in tension are shown in Fig. 10-23 for different initial crack lengths. In the compliance-matching procedure, initial compliance (obtained from the curves in Fig. 10-23) is first plotted against crack length (Fig. 10-24). This plot is referred to as the *compliance curve* or the *crack-length-estimation curve.* To estimate instantaneous crack length at points P_1, P_2, and P_3 on a load–COD curve (Fig. 10-25), compliances C_1, C_2, and C_3 of the lines joining P_1, P_2, and P_3 to the origin are obtained, and then the crack

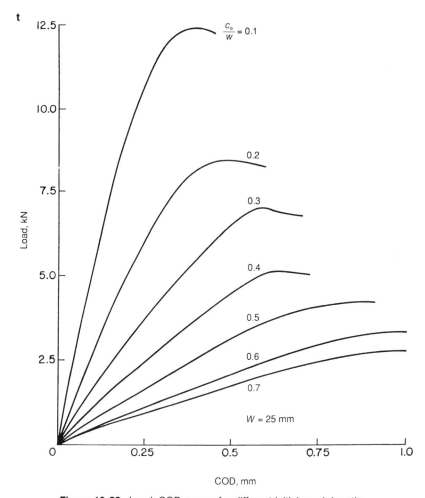

Figure 10-23. Load–COD curves for different initial crack lengths.

lengths are estimated from the crack-length-estimation curve (Fig. 10-24) corresponding to the compliances C_1, C_2, and C_3.

The compliance-matching procedure assumes that the damaged specimen behaves like an undamaged specimen having a machined crack of length equal to the estimated crack length. This underlying assumption has been examined by Agarwal et al. [54] for short-fiber composites by studying the behavior of damaged specimens. It has been established that use of the compliance-matching procedure is justified for estimating instantaneous crack length in short-fiber composites.

To obtain the fracture toughness of composite materials, the complete load–COD curve is transformed to an R curve through estimation of instantaneous crack length, as explained earlier. Location of the point of instability

10.3 MEASUREMENT OF MECHANICAL PROPERTIES

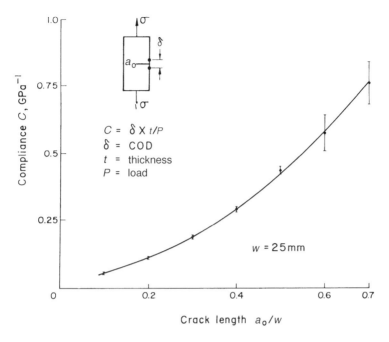

Figure 10-24. Crack-length-estimation curve for 25-mm-wide specimens.

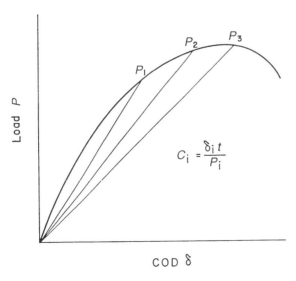

Figure 10-25. Instantaneous compliance-evaluation procedure.

on the R curve gives fracture toughness of the materials. The transformation process is tedious, time-consuming, and consequently, error-prone. It has been established mathematically by Agarwal et al. [55] that the peak load point on the load–COD curve corresponds to the instability point on the R curve. This result eliminates the necessity of transforming the entire load–COD curve to an R curve. Thus fracture toughness can be measured more easily and accurately by knowing the peak load and estimating the corresponding instantaneous crack length. It also has been pointed out that in order to make use of this result more effectively, displacement-controlled fracture tests should be preferred over load-controlled tests.

10.3.5.3 Critical J-Integral (J_c)

A method for experimental measurement of J_c has been suggested by Begley and Landes [56–58]. It is usually measured through the energy-rate interpretation of the J-integral given by Rice [59]. He showed that the J-integral may be interpreted as the potential-energy difference between two identically loaded bodies having neighboring crack sizes. This is stated mathematically as

$$J = -\frac{\partial U}{\partial c} \tag{10.31}$$

where U is the potential energy, and c is the crack length. In the linear elastic case, as well as for small-scale yielding, J therefore is equal to G, the crack-driving force. For any nonlinear elastic body, J may be interpreted as the energy available for crack extension.

The potential energy per unit thickness of a two-dimensional elastic body of area A with a boundary S is given by

$$U = \int_A W \, dx \, dy - \int_{S_T} T(u \, ds) \tag{10.32}$$

where W is the strain-energy density and S_T that portion of the boundary over which the traction T is prescribed. When the displacements are prescribed, the negative term in Eq. (10.32) drops out because S_T then is nonexistent. The potential energy then is equal to the strain energy, the area under the load-deflection curve. This energy-rate interpretation permits the experimental determination of the J-integral.

In order to determine the J-integral experimentally, several notched specimens with neighboring crack lengths are tested, and load-displacement (at the point of load application) curves are obtained. Areas under the load-displacement curves are obtained for different displacements. A typical load-displacement curve (Fig. 10-26a) shows that for an initial crack length C_1, A_1 is the strain energy at displacement δ_1, and ($A_1 + A_2$), at δ_2. These values of strain energies are plotted against crack lengths for constant displacements

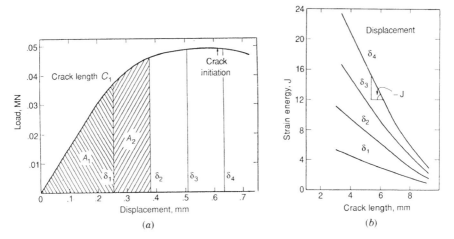

Figure 10-26. (a) Determination of strain energies at different displacements, and (b) energy curves at constant displacements.

(see Fig. 10-26b). These curves usually are referred to as the *energy curves*. Slope of an energy curve gives $\partial U/\partial c$ for a constant displacement, and thus the J-integral $(-\partial U/\partial c)$ is obtained. In many cases energy curves are approximated, for simplicity, by straight lines. A plot of the J-integral against displacement is a J curve. The critical value of the J-integral J_c is obtained corresponding to the critical displacement beyond which the load decreases monotonically.

10.3.6 Interlaminar Shear Strength and Fracture Toughness

The stresses acting on the interface of two adjacent laminae are called *interlaminar stresses*. The interlaminar stresses are illustrated in Fig. 10-27, where $\sigma_{T'}$ is the interlaminar normal stress on plane $ABCD$, and $\tau_{T'L}$ and $\tau_{T'T}$ are the interlaminar shear stresses. These stresses cause relative deformations between the laminae 1 and 2. If these stresses are sufficiently high, they may cause failure along plane $ABCD$ (see Chap. 8 for details). A tensile interlaminar normal stress may cause a tensile failure at the laminae interface. The stress required to cause this failure mode usually is taken to be the same as the transverse tensile strength of the unidirectional lamina. Large interlaminar shear stresses may cause delamination failure, which is an important failure mechanism for composite laminates. It is therefore of considerable interest to evaluate interlaminar shear strength through tests in which failure of laminates initiates in a shear mode.

It was pointed out in the preceding section that a large span–depth ratio in a flexural test increases the maximum normal stress without affecting the interlaminar shear stress and thereby increases the tendency for longitudinal

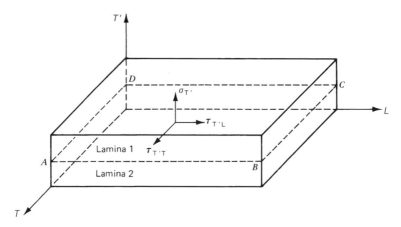

Figure 10-27. Representation of interlaminar shear stresses.

failure. Thus, if the span is short enough, failure initiates and propagates by interlaminar shear failure, and the test can be used to evaluate interlaminar shear strength. A flexural test with a short span is therefore called a *short-beam shear test* and is the most widely used test to evaluate interlaminar shear strength. ASTM Standard D 2344/D 2344M-00 provides further details of this test method. As was pointed out in an earlier section, maximum shear stress in a beam occurs at the midplane. Therefore, in a short-beam shear test, failure consists of a crack running along the midplane of the beam so that the crack plane is parallel to the lamination plane. It may be emphasized that a short-beam shear test becomes invalid if the tensile failure of fibers precedes the shear failure or if tensile failure and shear failure occur simultaneously.

Another type of test used for evaluating interlaminar shear (sometimes called *thickness shear*) strength is a notched-plate test. This test is shown schematically in Fig. 10-28. In this test, two notches are provided in the thickness direction from the opposite faces of the specimen, and the specimen is tested in tension or compression. The distance between the notches is adjusted such that the failure load corresponding to interlaminar shear failure between the notches is smaller than the failure load corresponding to tensile failure of the notched cross sections. ASTM Standard D 3846-02 provides further details of this method. The rail shear test, which was discussed earlier (see Fig. 10-15), is also used to evaluate inter-laminar shear strength. However, the short-beam shear test is much simpler to perform than the notched-plate test or the rail shear test and hence is very widely used.

Because of the importance of the delamination failure mode in laminated composite structures, the static interlaminar fracture toughness of unidirectional composites is determined. The double-cantilever-beam (DCB) test used commonly for this purpose is shown schematically in Fig. 10-29. ASTM Standard D 5528-01 provides details of this test method. The specimen is

10.3 MEASUREMENT OF MECHANICAL PROPERTIES

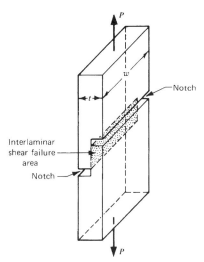

Figure 10-28. Tension test on staggered two-notch specimen for interlaminar shear-strength determination.

made of an even number of layers. A starter crack at one end of the specimen is introduced by placing a 0.025-mm-thick Teflon film at the midthickness during fabrication. The specimen is usually 3 mm thick, 38 mm wide, and 229 mm long. At the crack end of the specimen, two hinges are mounted for load application. The load-displacement curve during loading is recorded on a chart. The displacements are measured from an extensometer or linear voltage differential transformer (LVDT) attached to the specimen. The loading is stopped when the crack extends about 10 mm. The actual crack length is measured using a traveling microscope or a precision-dial caliper and is marked on the chart for identification. The specimen is unloaded and then reloaded. This procedure is repeated for 10-mm crack extensions each time until the crack is approximately 150 mm long. A typical complete load-

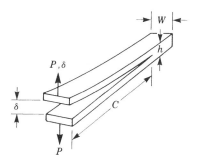

Figure 10-29. Double-cantilever-beam (DCB) specimen for fracture toughness measurement.

displacement record for a DCB specimen is shown in Fig. 10-30. The critical strain-energy release rate is obtained following the derivations in Chap. 8 and is discussed further in the following paragraphs.

The strain-energy release rate is given by Eq. (10.25). For the specimen, the compliance ($1/k$) is given by

$$\frac{1}{k} = \frac{\delta}{P} = \frac{2C^3}{3EI} \tag{10.33}$$

where E is the elastic modulus and $I = \frac{1}{12}W(h/2)^3$.

The following strain-energy release rate per unit width of the specimen now can be obtained by combining Eqs. (10.33) and (10.25):

$$G = \frac{P^2 C^2}{WEI} \tag{10.34}$$

The critical strain-energy release rate G_{IC} is determined for $P = P_c$:

$$G_{IC} = \frac{P_c^2 C^2}{WEI} \tag{10.35}$$

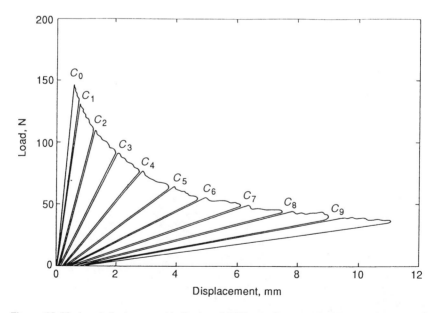

Figure 10-30. Load-displacement behavior of DCB specimens with different crack lengths.

The constant ($3EI/2$) in Eq. (10.33) is obtained by plotting ($1/k$) against C on a log–log scale, and the constant ($\sqrt{WG_{IC}EI}$) in Eq. (10.35) is obtained by plotting P_c against C on a log–log scale. Thus the critical strain-energy release rate can be obtained. A more detailed data-analysis procedure is given in ref. 60.

The critical strain-energy release rate can be determined directly through the area method from the definition of energy-release rate [61]:

$$G_{IC} = \frac{\Delta A}{W(C_2 - C_1)} \quad (10.36)$$

where ΔA is the area indicated in Fig. 10-31, and $C_2 - C_1$ is the crack growth from A to B. Values of G_{IC} are obtained from all the loading and unloading curves shown in Fig. 10-30, and an average G_{IC} is determined. The two methods may give slightly different values of G_{IC} owing to nonlinear load-displacement behavior and the inaccuracies in determining the displacements at the onset of crack growth and arrest.

10.3.7 Impact Properties

A very common way to evaluate impact properties is to determine material toughness by measuring the energy required to break a specimen of a particular geometry. The well-known Charpy and Izod impact tests developed for isotropic materials are used widely for this purpose. The test arrangements

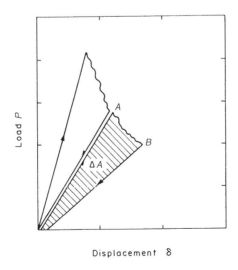

Figure 10-31. Area method for evaluation of G_{IC}.

are shown schematically in Fig. 10-32. In a Charpy test, the beam specimen is simply supported at the two ends and struck by a pendulum at the midspan. In an Izod test, the specimen is fixed at one end as a cantilever beam and struck by a pendulum at the free end. For homogeneous isotropic materials, the tests usually are performed on specimens with a notch on the tension side. The notch produces a high stress concentration and thus minimizes the energy required for initiation of fracture. The total measured energy required for fracture then is essentially the energy for propagation of the fracture.

The Charpy and Izod impact tests are very useful for a comparative study of different materials and, for this purpose, are quite adequate for studying impact behavior of isotropic metals or polymers. However, for composites, in which the fracture phenomenon is much more complex, conventional impact tests may not be sufficient for providing data of basic physical significance. The particular mode of fracture determines various energy-absorbing mechanisms operative during material failure under impact. The mode of fracture and thus the energy absorbed are influenced by various test variables such as fiber orientation, specimen geometry, velocity of impact, and other test arrangements. For a better understanding, Charpy and Izod impact tests frequently are instrumented to record the load history during the impact event. In conventional impact tests, however, such test variables as velocity of impact and available impact energy are held constant. This has resulted in the development of other types of impact-testing systems, described in the following paragraphs.

A test used increasingly for studying the impact behavior of composite materials is the drop-weight impact test, where the specimen is placed on rigid supports and a known weight is dropped on the specimen from a desired height. The drop height can be adjusted to achieve the desired impact velocity. Provisions usually are made to change the shape of the striking edge and/or to alter its weight so that the amount of available impact energy can be varied. The specimen supports sometimes are instrumented so that they also serve as

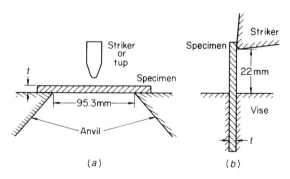

Figure 10-32. Impact test arrangements: (a) Charpy test and (b) Izod test.

load cells [62,63]. In other testing systems [64,65], the tup or striking edge itself, rather than the support, is instrumented. It is usually more effective to instrument the tup. In either case, the load history during the impact event can be recorded. In addition, the load signal can be integrated to produce a second signal that will be proportional to the area under the load–time curve. By adjustment for an appropriate velocity, the second signal can be used to calculate the energy absorbed at any instant during the impact loading of the test piece. This same information also can be obtained in the instrumented Charpy and Izod impact tests.

ASTM Standard D 3763-02 describes a test method for high-speed puncture properties of plastics using load and displacement sensors. This test method also may be used for composite materials.

A typical load history in an impact test is shown schematically in Fig. 10-33. The load–time history can be divided into two distinct regions, a region of fracture initiation and a region of fracture propagation. As the load increases during the fracture-initiation phase, elastic strain energy is accumulated in the specimen, and no gross failure takes place, but failure mechanisms on a microscale—for example, microbuckling of the fibers on the compression side or debonding at the fiber–matrix interface—are possible. When a critical load is reached at the end of the initiation phase, the composite specimen may fail either by a tensile or a shear failure depending on the relative values of the tensile and interlaminar shear strengths. At this point the fracture propagates either in a catastrophic "brittle" manner or in a progressive manner continuing to absorb energy—at smaller loads. The total impact energy E_t as recorded on the impact machine or on the energy–time curve on the oscilloscope thus is the sum of the initiation energy E_i and propagation energy E_p. Since a high-strength brittle material—which has a large initiation energy but a small propagation energy—and a low-strength ductile material—which has a small initiation energy but a large propagation energy—may have the same total impact energy, knowing the value of E_t alone is not sufficient to

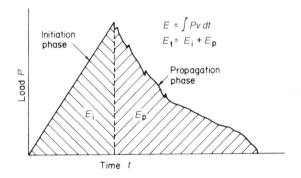

Figure 10-33. Typical load history during impact test on composite laminates.

478 EXPERIMENTAL CHARACTERIZATION OF COMPOSITES

interpret the fracture behavior of the material. An oscilloscope recording showing the load–time trace and the integrated energy–time trace is shown in Fig. 10-34 for a Charpy impact test.

The energy absorbed by the specimen (E) at any time is given by the following equation:

$$E = \int Pv \, dt \qquad (10.37)$$

where P and v, respectively, represent instantaneous load and velocity. However, the energy signal usually recorded on the energy–time curve on the oscilloscope (see Fig. 10-34) is the product of impulse (area under the load–time curve, $\int P \, dt$) and initial impact velocity (v_0). Therefore, this energy signal E_a is given by

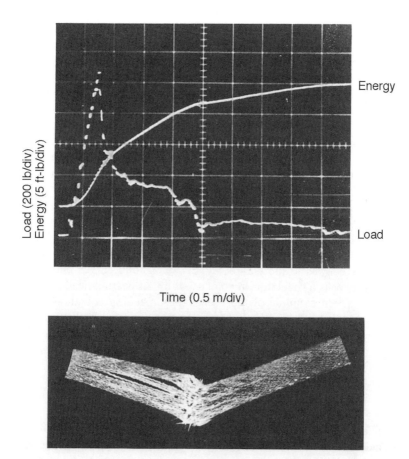

Figure 10-34. Load and energy histories during impact on a glass–epoxy laminate obtained from an instrumented Charpy test. Tested specimen is also shown.

10.3 MEASUREMENT OF MECHANICAL PROPERTIES

$$E_a = v_0 \int P \, dt \tag{10.38}$$

Thus it should be observed that to obtain the actual energy absorbed by the specimen (E), E_a must be corrected for the reduction in velocity of the tup during contact with the specimen. As an approximation, the right-hand side of Eq. (10.37) can be replaced by the product of average tup velocity ($\bar{v}$) and impulse, so that

$$E = \bar{v} \int P \, dt \tag{10.39}$$

where $\bar{v} = \frac{1}{2}(v_0 + v_f)$, and v_f is the velocity at the specific instant for which E is to be calculated. Using the principles of impulse and momentum, it can be shown easily that

$$\frac{\bar{v}}{v_0} = 1 - \frac{E_a}{4E_0} \tag{10.40}$$

where E_0 is the maximum available impact energy defined by

$$E_0 = \tfrac{1}{2} m v_0^2 \tag{10.41}$$

where m is the mass of the drop weight. Combining Eqs. (10.37), (10.39), and (10.40) yields

$$E = E_a \left(1 - \frac{E_a}{4E_0}\right) \tag{10.42}$$

Equation (10.42) is the relationship employed most commonly for reduction of instrumented impact data.

These corrections to the recorded energy-absorbed curves become significant when there is a reduction in velocity of the impacting head as it fractures the specimen. An example of the use of the preceding equations is given in studies performed by Aleszka [64] on damage done to graphite–epoxy laminates when the laminate surface is impacted (low-level impacts) with a 0.5-lb weight (hemispherical head). Load–time and energy–time traces are shown in Fig. 10-35 for an impact when the weight was dropped from 0.5 ft. The available energy E_0 was 0.22 ft-lb. At point 1 in Fig. 10-35, which corresponds to peak load, the energy absorbed by the specimen is also 0.22 ft-lb, thereby indicating that at the time of peak load the specimen has absorbed all the available energy. At point 2, the absorbed energy is zero; thus all the available energy has been returned to the drop weight, and none has been retained by the specimen. This implies that the specimen has not suffered any damage or permanent deformation from this impact.

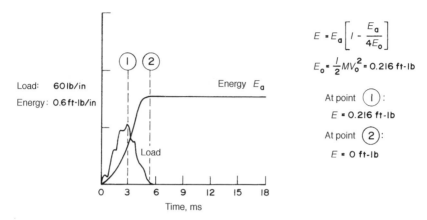

Figure 10-35. Load and energy histories during low-energy impact on a graphite–epoxy laminate. (From Aleszka [64].)

Figure 10-36 is a load–time and energy–time trace produced when the weight was dropped from 7 ft onto a panel. The available energy was 2.22 ft-lb. At point 1, the energy absorbed by the specimen is 0.71 ft-lb. The oscillations on the load–time trace at this point indicate visible incipient damage. At peak load (point 2), the absorbed energy is 1.75 ft-lb. The absorbed energy at point 3 is 2.22 ft-lb, which is equal to the initial impact energy. The indenter is stopped at this point, and the specimen has incurred major damage. At point 4, the absorbed energy calculated from Eq. (10.42) is 1.90 ft-lb. This result indicates that 0.32 ft-lb, the difference between the available

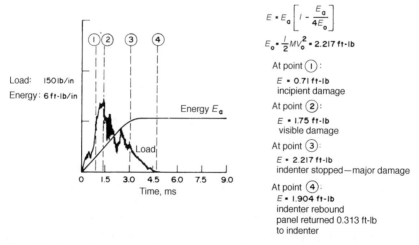

Figure 10-36. Load and energy histories during high-energy impact on a graphite–epoxy laminate. (From Aleszka [64].)

impact energy E_0 and the final absorbed energy E has been returned to the indenter.

10.4 DAMAGE IDENTIFICATION USING NONDESTRUCTIVE EVALUATION TECHNIQUES

Damage or defects in a material affect its performance. In fiber composites, defects or flaws can accrue during various manufacturing processes, as well as in service. Because of their heterogeneity, the damage in composites in the form of voids, fiber breaks, interface failure, matrix cracking, and delamination can occur more readily than in homogeneous materials. It is therefore of considerable interest to identify damage in composite materials. The damage identification can be used to determine the material quality or adequacy of the fabrication process and its control, the effects of defects on performance of the composite, establishment of inspection requirements for acceptance or rejection of material, and repair procedures for service-induced flaws. Nondestructive evaluation (NDE) techniques can be used for such damage identification in composite materials.

The NDE techniques of identifying damage involve directing energy into the structure and analyzing the response to obtain information regarding size, location, and orientation of damage or cracks. NDE techniques differ from each other in the form of energy transmitted to the structure and receiving and analyzing the energy after it interacts with the structure and any defects therein. Tapping thin composite structures or adhesive bonds with a coin and listening to the sound frequency is one of the oldest methods of nondestructive examination. Some of the NDE techniques currently employed for composite materials are discussed briefly in the following subsections.

10.4.1 Ultrasonics

In this technique, a high-frequency sound wave (1–25 MHz) is generated by a transducer in periodic short bursts of a few cycles and transmitted into the material being examined. A coupling medium such as water is often used to minimize energy losses in air between the transducer and material surface. The wave, as it travels through the specimen, is reflected by the discontinuities in the material such as voids, delaminations, and cracks, as well as the far surface of the component. The reflected wave returns to the transducer when the waves are traveling normal to the specimen surface, and the transmitting transducer acts as the receiver as well (Fig. 10-37). If the waves are directed to the surface at some other angle, the reflected waves are received by a second transducer (usually of the same type as the transmitter) called the *receiver* (Fig. 10-38). The former method is called the *pulse–echo method,* and the latter, the *pitch–catch method.* In another method, the waves transmitted through the specimen are received by a second transducer on the far

482 EXPERIMENTAL CHARACTERIZATION OF COMPOSITES

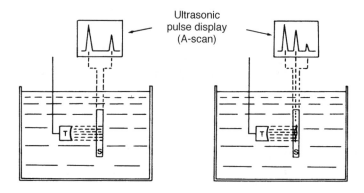

T - Transducer
S - Specimen

Figure 10-37. Ultrasonic pulse–echo method.

side of the specimen (Fig. 10-39). This method is called the *through-transmission method.*

The information contained in the attenuated pulse emerging from the specimen is processed electronically and displayed for the evaluation of presence, size, and location of damage. The amplitude of the emerging wave displayed by an oscilloscope as a function of time for one position of the transducer is called the *A-scan* in NDE terminology. A comparison between the A-scans from a test specimen and a standard (unflawed) specimen provides information regarding the defect at that location. A-scans are included in Figs. 10-37 and 10-39. A number of A-scans are obtained at points along a line, and a new scan, called a *B-scan,* is constructed that gives a cross-sectional view of the material and the defects present in the cross section. The scan constructed from A-scans made across a two-dimensional grid on the specimen

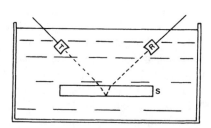

T - Transmitting transducer
R - Receiving transducer
S - Specimen

Figure 10-38. Ultrasonic pitch–catch method.

10.4 DAMAGE IDENTIFICATION USING NONDESTRUCTIVE EVALUATION TECHNIQUES

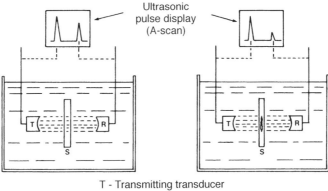

Figure 10-39. Ultrasonic through-transmission method.

is called a *C-scan*. The C-scan produces a plan view of the material and an image of the defect in it. A typical C-scan of a graphite–epoxy specimen is shown in Fig. 10-40. A flaw in the form of a square film patch was purposely embedded between two plies in the specimen to represent a delamination. The specimen had other natural defects as well.

ASTM Standards E 114-95 (reapproved 2001) and E 214-01 describe standard practice for ultrasonic examinations.

10.4.2 Acoustic Emission

An *acoustic emission* is a transient elastic wave generated by the rapid release of energy within a material. It sometimes can be audible to the unaided ear,

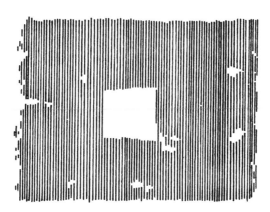

Figure 10-40. A C-scan of a graphite–epoxy laminate. The blank spaces indicate defects.

for example, the creaking of timber subjected to loads near failure or the sound produced by the failure of rocks. The latter have been used to detect the impending failure of mine shafts and the onset of land slides. In composite materials, the easily recognizable sources of acoustic emissions of practical importance are fiber fracture, matrix cracking, fiber–matrix interface failure, and delamination. The acoustic event is detected with appropriately designed equipment (transducer and associated instrumentation). However, the events are abrupt and discontinuous and are characterized by high-frequency transient acoustic signals. The information regarding the signal source may be obtained by amplitude and frequency analysis of the detected signals.

In a standard experiment, the acoustic emissions emanating from the test section are recorded in such a way that their numbers and distributions can be correlated with the test parameters, such as applied load in a tension test. A schematic diagram of a typical experimental setup is presented in Fig. 10-41. The cut-ins marked (*a*) and (*b*) indicate the appearance of a single event

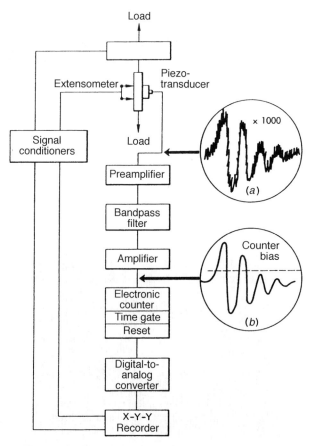

Figure 10-41. Diagram of an acoustic emission system.

as it might be viewed on an oscilloscope. In addition, (*b*) shows a typical bias level to which the electronic counter might be set (this acoustic emission would register two counts). In modern acoustic emission equipment, all the components beyond the bandpass filter are replaced by a cathode-ray oscilloscope (CRO). The bandpass filter is used to suppress low-frequency noises. The signals usually are recorded in a microcomputer interfaced with the CRO and can be analyzed using appropriate software. The parameters recorded in an acoustic emission experiment include peak amplitudes, rise time during events, ring-down counts, and duration of an event. The records of these parameters are analyzed to relate them to specific failure events, such as a fiber break or matrix cracking. A distinct correlation between the acoustic emission parameters and the failure events has not yet been established. However, it is generally accepted that the fracture of individual fibers results in high-amplitude events.

The acoustic emission technique, although still requiring further improvements, is very convenient to study damage initiation and progression during fatigue loading because it does not require interruption of the test for NDE examination, unlike ultrasonic C-scanning or x-radiography.

Acoustic emission (AE) monitoring has been adopted as an NDE technique for inspecting new and in-service fiberglass-reinforced plastic storage tanks and booms in aerial man-lift devices [66,67]. This procedure involves mounting AE sensors at key locations and proof-loading the structure by a prescribed loading sequence. The sensors detect AE signals generated by defect propagation within the structure. These signals are transmitted to an AE analysis instrument that processes the data to determine the type of defect present and its approximate location. For each type of structure and application, an acceptance criterion related to the AE parameters for the loading sequence is first established.

10.4.3 x-Radiography

Intensity of a monochromatic (one wavelength) beam of x-rays as it propagates through any matter decreases because of absorption. The relationship between the intensity I and the distance traveled x is given by

$$I = I_0 e^{-\mu x} \qquad (10.43)$$

where I_0 is the initial intensity of the beam, and μ is the linear absorption coefficient, which depends on the wavelength of the beam and on the material. Absorption coefficients of some elements for x-rays of interest in composite materials are given in Table 10-1.

In the x-radiography (NDE) technique, an x-ray beam is passed through the material, and the emerging beam exposes a photographic film. The exposure patterns on the film provide information about the material. The highly exposed areas are indicative of material with smaller absorption coefficients,

Table 10-1 Absorption coefficients for 0.098-Å wavelength x-rays

Material	μ (cm^{-1})
Boron	0.35
Carbon	0.33
Aluminum	0.42
Titanium	0.98
Tungsten	56.00

and vice versa. Thus voids and delaminations in the material can be distinguished from defect-free areas.

A parameter useful for comparing absorption by different materials is the thickness at which the incident intensity is reduced by one-half, that is, $I = \frac{1}{2}I_0$. This thickness $x_{1/2}$ can be obtained easily using Eq. (10.43) as

$$x_{1/2} = \frac{0.69}{\mu} \qquad (10.44)$$

The half-thicknesses for some materials of interest to composites, irradiated with x-rays of three different wavelengths, are given in Table 10-2. The large difference between half-thicknesses of composites such as carbon-fiber-reinforced polymers ($\approx$3 cm) and air, effectively infinity for short-wavelength x-rays, is used for the detection of voids in composite materials. To enhance the contrast at defects that emerge at the external surface, the defects may be filled with radiopaque material such as barium salt. This technique, called *penetrant-enhanced radiography*, provides contrast between the composite and a heavy metal rather than that between the composite and air.

ASTM Standard E 94-04 provides guidelines for general radiographic examination.

10.4.4 Thermography

Thermography is an inspection technique that makes use of one of the following facts: (1) The thermal conductivity of a material at a flaw location is

Table 10-2 Thickness of material at which x-ray intensities are reduced to half the incident intensity

Material	$x_{1/2}$ (cm)		
	$l = 0.1$ Å	$l = 0.7$ Å	$l = 2.0$ Å
Air	—	410	26
Polymer (cellophane)	4.3	0.4	0.05
Carbon	2.1	—	—
Aluminum	1.6	0.05	0.0025

10.4 DAMAGE IDENTIFICATION USING NONDESTRUCTIVE EVALUATION TECHNIQUES

different from that at a location without a flaw, and (2) heat is generated at flaws under cyclic loading. These two material characteristics influence thermal patterns on the inspection surface when the material is heated either by an external source (active heating) or by the internal heat generation at the flaws during cyclic loading (passive heating) [68,69].

A few different approaches to thermographic inspection using active heating are possible. In one approach, the component is suddenly heated by flash lamps that pulse for few milliseconds to provide heat to the surface. High-voltage tungsten–halide lamps are used frequently that raise the surface temperature by 10–30°F. The temperature of the heated surface is monitored using an infrared imaging system. Since a defect beneath the surface does not conduct as much heat as the bulk material does, the surface directly above it will remain warmer than the surroundings. A typical inspection arrangement is shown schematically in Fig. 10-42. This method may be called a *pulse–echo thermographic inspection*. In another approach, the temperature on the other side of the heated surface is monitored (*through-transmission thermographic inspection*). Cooler regions in an otherwise essentially homogeneous temperature field indicate defects underneath the surface. Yet another possibility is to heat the entire component to a homogeneous temperature in an oven and then monitor its heat loss on cooling; a relatively warmer region indicates a subsurface defect.

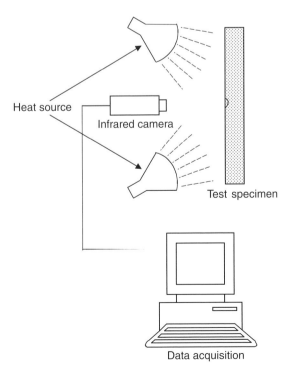

Figure 10-42. A typical thermography inspection arrangement.

With passive heating, the component is subjected to cyclic loads that make the walls of flaws within the component rub against each other to generate heat through friction. Although technically possible, passive heating is not common except for monitoring damage propagation in fatigue tests.

In comparison with ultrasonic and radiographic inspection, thermographic inspection provides poor resolution [70]. However, it is a relatively fast and noncontact inspection technique that has a wide coverage area. It can be used to detect delaminations, impact damage, water ingression into a honeycomb, inclusions, and density variations [71].

10.4.5 Laser Shearography

Shearography is essentially an optical method of surface-strain measurement. The subsurface defects are detected by the presence of concentrations of surface strain owing to an applied stress or stresses. This method may be explained as follows: When an object is illuminated by a laser beam, it produces a characteristic granular appearance on the object surface, usually referred to as the *speckle effect*. This speckle pattern changes when the object is stressed. The changes in the speckle pattern from the unstressed state to the stressed state of the object are used to determine the change in displacement gradients or the strains. In the application of a shearographic inspection technique, a reference image of the unstressed object is obtained first using a shearographic video laser interferometer and stored electronically. A typical system configuration is shown schematically in Fig. 10-43. A uniform stress then is applied to the object by means of vacuum, pressure, vibration, sound, or heat. An image of the object under stress is obtained and compared with the reference image. When the two images are overlapped, the composite image yields a fringe pattern that depicts loci of displacement gradients. A typical shearographic fringe pattern of a strain concentration appears as a double bulls-eye or butterfly pattern, as shown in Fig. 10-44. The outer periphery of the fringe pattern provides the size, shape, and location of a subsurface defect, whereas the number of fringes provides information concerning the depth and type of defect [72].

Laser shearography is a very sensitive, rapid, full-field inspection technique. A shearography camera imaging 700 square inches can detect surface deformations of 2×10^{-5} in. per fringe (half a wavelength of laser light) or 40 microstrains (i.e., 10^{-6} in./in.). It provides video images of flaws in near real time. It has been found to be particularly suited to detection of delamination flaws in composites [73–76].

10.5 GENERAL REMARKS ON CHARACTERIZATION

The present discussion of experimental characterization of composites has only a limited objective of familiarizing readers with the most common type

10.5 GENERAL REMARKS ON CHARACTERIZATION

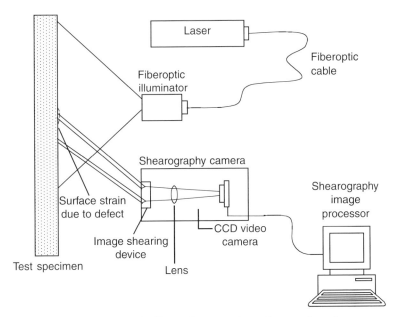

Figure 10-43. A typical laser shearography system configuration.

of tests that are performed on composite materials. The tests described herein are only static tests conducted to evaluate static properties. The discussion on static tests is also by no means complete. There are many static tests that are performed to evaluate properties needed in design under certain circumstances but have not been discussed here; for instance, tests often are performed to evaluate properties in the thickness direction. Moreover, the discussion is more qualitative in nature, and specific details of specimen size and shape, instrumentation and data recording, and data analysis have not been given.

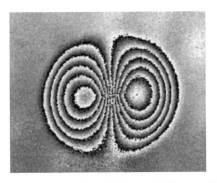 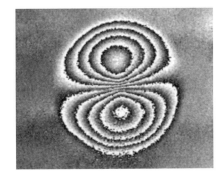

Figure 10-44. Typical "butterfly" fringe patterns of a strain concentration obtained by shearography (Courtesy of Laser Technology, Inc.)

This is partly because widely acceptable test standards have not yet been evolved. Composite materials are relatively new and more complex compared with homogeneous and isotropic metallic materials. Their characterization poses many problems that are not present with isotropic materials. Experience with these materials also is limited. The American Society for Testing and Materials (ASTM) organizes many conferences that are devoted solely to the subject of testing and design of composite materials. The special technical publications (STP) [6–20] based on the proceedings of the conferences are very helpful in understanding various tests, test variables, and their influence on the material properties. It is suggested that any particular test procedure for characterization of composites should be selected only after a careful review of the literature on the test procedure under consideration.

EXERCISE PROBLEMS

10.1. Show how you think a dog-bone, pin-ended specimen of a unidirectional composite would fail in a tension test.

10.2. Assume that straight-sided off-axis specimens (12.5 mm × 4 mm) of the composite considered in Exercise Problem 5.14 are tested in uniaxial tension with ends clamped. An axial force of 500 N produces a strain 0.0926% near the clamped end in a 30° specimen, 0.105% in a 45° specimen, and 0.150% in a 60° specimen. Calculate the apparent off-axis elastic modulus in each case. Compare your results with the corresponding moduli values obtained by transformation in Exercise Problem 5.13. Explain discrepancies, if any.

10.3. Derive Eq. (10.15).

10.4. The variables E_x and ν_{xy} are the elastic modulus and Poisson ratio of a $[\pm 45]_s$ laminate obtained in a tension test. Show that the shear modulus G_{LT} of the constituent laminae is given by Eq. (10.18).

10.5. The values of E_x and ν_{xy} of a boron–epoxy $[\pm 45]_s$ laminate, measured in a tension test, are 20.98 GPa and 0.69, respectively. The two properties of a $[45]_n$ laminate of the same material are 15.87 GPa and 0.28, respectively. Calculate the shear modulus G_{LT} of the constituent laminae in both the cases. Comment on the differences in values of E_x and ν_{xy} in the two cases.

10.6. Derive an expression for failure load in terms of in-plane shear strength and specimen dimensions for a rail shear test.

10.7. Discuss why flexural strength determined from a four-point bend test is always less than that determined from a three-point bend test.

10.8. A three-point bend test is conducted on a beam of unidirectional graphite–epoxy composite. The beam is 100 mm long, 12.5 mm wide, and 3 mm thick and deflects 5.1 mm for a load of 1 kN. Calculate the longitudinal modulus of the composite.

10.9. Derive Eq. (10.25) by assuming that the load remains constant during crack extension.

10.10. Derive an expression for span–depth ratio in terms of tensile and interlaminar shear strengths to ensure that a short-beam shear specimen will fail as a result of interlaminar shear. From properties of glass and graphite composites shown in Chap. 3, choose appropriate span–depth ratios.

10.11. Derive an expression for the maximum distance between two notches (for a notched tension test) in terms of plate thickness and its tensile and interlaminar shear strengths to ensure that the specimen will fail as a result of interlaminar shear.

10.12. Derive Eq. (10.40).

10.13. In an impact testing system, an instrumented tup weighing 2 kg freely drops through a height of 3 m before making contact with the specimen. An energy signal proportional to the product of impulse and initial impact velocity is recorded continuously on a storage oscilloscope during the impact. In a certain test, values of this signal at the peak contact load between the tup and the specimen and at fracture of specimen were recorded to be 6.38 and 7.91 N·m, respectively. Calculate initiation energy, propagation energy, and total impact energy for the specimen.

10.14. In the impact testing system described in Exercise Problem 10.13, a smaller tup weighing 0.25 kg is used without altering the drop height. If the values of the recorded energy signal at peak load and at fracture are 6.38 and 7.91 N·m, respectively, calculate initiation energy, propagation energy and total impact energy for the specimen.

10.15. In a drop-weight impact experiment on a fiber composite beam, the drop weight passes completely through the beam, and the ratio of final impacter velocity to initial velocity is 0.8. What error is involved in using the value E_a rather than E to describe the energy-absorbing capability of the beam.

REFERENCES

1. C. W. Bert, "Experimental Characterization of Composites," in C. C. Chamis, ed., *Structural Design and Analysis—Part II,* Academic, New York, 1975.
2. M. E. Waddoups, "Characterization and Design of Composite Materials," in S. W. Tsai, J. C. Halpin, and N. J. Pagano, eds., *Composite Materials Workshop,* Technomic, Stamford, CT, 1968.

3. G. Epstein, "Testing of Reinforced Plastics," in G. Lubin, ed., *Handbook of Fiberglass and Advanced Plastics Composites,* Van Nostrand, Princeton, NJ, 1969.
4. G. C. Grimes and M. E. Bronstad, "Testing Methods for Advanced Composites," in G. Lubin, ed., *Handbook of Fiberglass and Advanced Plastics Composites,* Van Nostrand, Princeton, NJ, 1969.
5. J. C. Halpin, *Primer on Composite Materials: Analysis* (revised), Technomic, Lancaster, PA, 1984.
6. *Composite Materials: Testing and Design,* ASTM STP 460, American Society for Testing and Materials, Philadelphia, PA, 1969.
7. *Composite Materials: Testing and Design (Second Conference),* ASTM STP 497, American Society for Testing and Materials, Philadelphia, PA, 1972.
8. *Composite Materials: Testing and Design (Third Conference),* ASTM STP 546, American Society for Testing and Materials, Philadelphia, PA, 1974.
9. *Composite Materials: Testing and Design (Fourth Conference),* ASTM STP 617, American Society for Testing and Materials, Philadelphia, PA, 1977.
10. *Composite Materials: Testing and Design (Fifth Conference),* ASTM STP 674, American Society for Testing and Materials, Philadelphia, PA, 1979.
11. *Composite Materials: Testing and Design (Sixth Conference),* ASTM STP 787, American Society for Testing and Materials, Philadelphia, PA, 1982.
12. *Composite Materials: Testing and Design (Seventh Conference),* ASTM STP 893, American Society for Testing and Materials, Philadelphia, PA, 1986.
13. *Composite Materials: Testing and Design (Eighth Conference),* ASTM STP 972, American Society for Testing and Materials, Philadelphia, PA, 1988.
14. *Composite Materials: Testing and Design (Ninth Volume),* ASTM STP 1059, American Society for Testing and Materials, Philadelphia, PA, 1990.
15. *Composite Materials: Testing and Design (Tenth Volume),* ASTM STP 1120, American Society for Testing and Materials, Philadelphia, PA, 1992.
16. *Composite Materials: Testing and Design (Eleventh Volume),* ASTM STP 1206, American Society for Testing and Materials, Philadelphia, PA, 1993.
17. *Composite Materials: Testing and Design (Twelfth Volume),* ASTM STP 1274, American Society for Testing and Materials, Philadelphia, PA, 1996.
18. *Composite Materials: Testing and Design (Thirteenth Volume),* ASTM STP 1242, American Society for Testing and Materials, Philadelphia, PA, 1997.
19. *Composite Materials: Testing and Design (Fourtheenth Volume),* ASTM STP 1436, American Society for Testing and Materials, Philadelphia, PA, 2003.
20. *Composite Materials: Testing, Design, and Acceptance,* ASTM STP 1416, American Society for Testing and Materials, Philadelphia, PA, 2002.
21. G. L. Richards, T. P. Airhart, and J. E. Ashton, "Off-Axis Tensile Coupon Testing," *J. Compos. Mater.,* **3**(3), 586–589 (1969).
22. N. J. Pagano and J. C. Halpin, "Influence of End Constraint in the Testing of Anisotropic Bodies," *J. Compos. Mater.,* **2**(1), 18 (1968).
23. E. M. Wu and R. L. Thomas, "Off-Axis Test of a Composite," *J. Compos. Mater.,* **2**(4), 523 (1968).
24. R. G. Hill, "Evaluation of Elastic Moduli of Bilaminate Filament Wound Composites," *Exp. Mech.,* **8**(2), 75 (1968).

25. R. B. Lantz, "Boron–Epoxy Laminate Test Methods," *J. Compos. Mater.*, **3**(4), 642 (1969).
26. N. Fried and R. R. Winans, in *Symposium on Standard Filament-Wound Reinforced Plastics*, ASTM STP 327, American Society for Testing and Materials, Philadelphia, PA, 1963, pp. 83–95.
27. R. N. Hadcock and J. B. Whiteside, in *Composite Materials: Testing and Design*, ASTM STP 460, American Society for Testing and Materials, Philadelphia, PA, 1969, p. 27.
28. J. M. Whitney, I. M. Daniel, and R. B. Pipes, *Experimental Mechanics for Fiber Reinforced Composite Materials*, SESA Monograph No. 4, Society of Experimental Stress Analysis, Brookfield Center, CT, 1981.
29. K. E. Hofer, Jr., and P. N. Rao, "A New Static Compression Fixture for Advanced Composite Materials," *J. Testing Eval.*, **5**(4), 278–283 (1977).
30. N. J. Pagano and J. M. Whitney, "Geometric Design of Composite Cylindrical Characterization Specimens," *J. Compos. Mater.*, **4**(3), 360 (1970).
31. D. F. Adams and D. E. Walrath, "Further Development of the Iosipescu Shear Test Method," *Exp. Mech.*, **27**(2), 113–119 (1987).
32. D. E. Walrath and D. F. Adams, "The Iosipescu Shear Test as Applied to Composite Materials," *Exp. Mech.*, **23**(1), 105–110 (1983).
33. D. F. Adams and D. E. Walrath, "Iosipescu Shear Properties of SMC Composite Materials," in *Composite Materials: Testing and Design* (Sixth Conference), ASTM STP 787, American Society for Testing and Materials, Philadelphia, PA, 1982, pp. 19–33.
34. D. E. Walrath and D. F. Adams, "Analysis of the Stress State in an Iosipescu Shear Test Specimen," NASA-CR-176745.
35. M. J. Pindera, G. Choksi, J. S. Hidde, and C. T. Herakovich, "A Methodology for Accurate Shear Characterization of Unidirectional Composites," *J. Compos. Mater.*, **21**(12), 1164–1184 (1987).
36. B. W. Rosen, "A Simple Procedure for Experimental Determination of the Longitudinal Shear Modulus of Unidirectional Composites," *J. Compos. Mater.*, **6**(4), 552–554 (1972).
37. M. J. Pindera and C. T. Herakovich, "Shear Characterization of Unidirectional Composites with the Off-Axis Tension Test," *Exp. Mech.*, **26**(1), 103 (1986).
38. C. C. Chamis and J. H. Sinclair, "Ten-Degree Off-Axis Test for Shear Properties in Fiber Composites," *Exp. Mech.*, **17**(9), 339–346 (1977).
39. J. M. Whitney, D. L. Stansbarger, and H. B. Howell, "Analysis of the Rail Shear Test—Applications and Limitations," *J. Compos. Mater.*, **5**(1), 24 (1971).
40. J. M. Whitney, "Experimental Determination of Shear Modulus of Laminated Fiber-Reinforced Composites," *Exp. Mech.*, **7**(10), 447 (1967).
41. J. C. Halpin, N. J. Pagano, J. M. Whitney, and E. M. Wu, in *Composite Materials: Testing and Design*, ASTM STP 460, American Society for Testing and Materials, Philadelphia, PA, 1969, pp. 37–47.
42. J. M. Whitney and R. J. Dauksys, "Flexure Experimental of Off-Axis Composites," *J. Compos.*, **4**(1), 135 (1970)
43. S. K. Gaggar and L. J. Broutman, "Crack Propagation Resistance of Random Fiber Composites," *J. Compos. Mater.*, **9**, 216 (1975).

44. J. Awerbuch and H. T. Hahn, "K-Calibration of Unidirectional Metal Matrix Composite," *J. Compos. Mater.,* **12,** 222 (1978).
45. Y. J. Yeow, D. H. Morris, and H. F. Brinson, "The Fracture Behavior of Graphite/Epoxy Laminates," *Exp. Mech.,* **19,** 1 (1979).
46. S. K.. Gaggar and L. J. Broutman, "Fracture Toughness of Random Glass Fiber Epoxy Composites: An Experimental Investigation," in *Flaw Growth and Fracture,* ASTM STP 631, American Society for Testing and Materials, Philadelphia, PA, 1977, pp. 310–330.
47. C. Bathias, R. Esnault, and J. Pellas, "Application of Fracture Mechanics to Graphite Fiber-Reinforced Composites," *Composites,* **12,** 195 (1981).
48. D. H. Morris and H. T. Hahn, "Fracture Resistance Characterization of Graphite/Epoxy Composites," in *Composite Materials: Testing and Design,* ASTM STP 617, American Society for Testing and Materials, Philadelphia, PA, 1977, p. 5.
49. J. M. Mahishi and D. F. Adams, "Micromechanical Predictions of Crack Initiation, Propagation and Crack Growth Resistance in Boron/Aluminum Composites," *J. Compos. Mater.,* **16,** 457 (1982).
50. S. Ochiai and P. W. M. Peters, "Tensile Fracture of Centre-notched Angleply (0/$\pm 450)_S$ and $(0/90)_{2S}$ Graphite–Epoxy Composites," *J. Mater. Sci.,* **17,** 417 (1982).
51. H. Yanada and H. Homma, "Study of Fracture Toughness Evaluation of FRP," *J. Mater. Sci.,* **18,** 133 (1983).
52. B. D. Agarwal and G. S. Giare, "Crack Growth Resistance of Short Fiber Composites: I. Influence of Fiber Concentration, Specimen Thickness and Width," *Fibre Sci. Technol.,* **15,** 283 (1981).
53. B. D. Agarwal and G. S. Giare, "Effect of Matrix Properties on Fracture Toughness of Short Fiber Composites," *Mater. Sci. Eng.,* **52,** 139 (1982).
54. B. D. Agarwal, B. S. Patro, and P. Kumar, "Crack Length Estimation Procedure for Short Fiber Composite: an Experimental Evaluation," *Polym. Compos.* **6,** 185 (1985).
55. B. D. Agarwal, B. S. Patro, and P. Kumar, "Prediction of Instability Point During Fracture of Composite Materials," *Compos. Technol. Rev.,* **6,** 173 (1984).
56. J. A. Begley and J. D. Landes, "The J Intergral as a Fracture Criterion," in *Fracture Toughness,* ASTM STP 514, American Society for Testing and Materials, Philadelphia, PA, 1972, p. 1.
57. J. D. Landes and J. A. Begley, "The Effect of Specimen Geometry on J_{1C} in *Fracture Toughness,* ASTM STP 514, American Society for Testing and Materials, Philadelphia, PA, 1972, p. 24.
58. J. D. Landes and J. A. Begley, "Recent Developments in J_{1C} Testing," in *Developments in Fracture Mechanics Test Methods Standardization,* ASTM STP 632, American Society for Testing and Materials, Philadelphia, PA, 1977, p. 57.
59. J. R. Rice, "A Path Independent Integral and the Approximate Analysis of Strain Concentration by Notches and Cracks," *J. Appl. Mech.,* **35,** 379 (1968).
60. D. F. Adams, L. A. Carlsson, and R. B. Pipes, *Experimental Characterization of Advanced Composite Materials,* 3d ed., CRC Press, Boca Raton, FL, 2003.
61. J. M. Whitney, C. E. Browning, and W. Hoogsteden, "A Double Cantilever Beam Test for Characterization of Composite Materials," *J. Reinf. Plast. Compos.,* **1,** 297 (1982).

62. L. J. Broutman and A. Rotem, "Impact Strength and Fracture of Carbon Fiber Composite Beams," SPI, 28th Annual Technical Conference, Washington, 1973, Sec. 17-B.
63. L. J. Broutman and A. Rotem, "Impact Strength and Toughness of Fiber Composite Material," in *Foreign Object Impact Damage to Composites,* ASTM STP 568, American Society for Testing and Materials, Philadelphia, PA, 1975, pp. 114–133.
64. J. C. Aleszka, "Low Energy Impact Behavior of Composite Panels," *J. Test. Eval.,* **6**(3), 202–210 (1978).
65. R. H. Toland, "Failure Modes in Impact Loaded Composite Materials," paper presented at Failure Modes in Composites Symposium AIME Spring Meeting, May 1972.
66. Recommended Practice for Acoustic Emission Testing of Fiberglass Reinforced Plastics Resin (RP) Tanks/Vessels, The Composites Institute of the Society of the Plastics Industry, New York, 1988.
67. J. R. Mitchell and D. G. Taggart, "Acoustic Emission Testing of Fiberglass Bucket Truck Booms," 39th Annual Conference, Reinforced Plastics/Composites Institute, Society of the Plastics Industry, 1984, Sec. 16-B, pp. 1–7.
68. P. Cawley, "The Rapid Non-Destructive Inspection of Large Composite Structures," *Composites,* **25,** p. 351–357, (1994).
69. T. S. Jones and H. Berger, "Application of Nondestructive Inspection Methods to Composites," *Mater. Eval.,* **47,** 390–400 (1989).
70. B. T. Astrom, *Manufacturing of Polymer Composites,* Nelson Thomes, Cheltenham, U.K., 2002.
71. F. C. Campbell, *Manufacturing Processes for Advanced Composites,* Elsevier, Oxford, U.K., 2004.
72. J. Tyson II, "Advances in Evaluation of Composites and Composite Repairs," 18th International SAMPE/JEC Symposium, 1997.
73. S. L. Toh, H. M. Shang, F. S. Chau, C. J. Tay, "Flaw detection in Composites Using Time-Average Shearography," *Optics Laser Technol.,* **23,** 25–31 (1991).
74. J. W. Newman, "Shearography Inspection of Aircraft Structure," *Mater. Eval.,* **49,** 1106–1109 (1991).
75. S. Waldner, "Removing the Image-Doubling in Shearography by Reconstruction of the Displacement Field," *Optics Commun.,* **127,** 117–126 (1996).
76. K. Chandrashekhara, S. E. Watkins, and F. Akhavan, "Impact Characterization of Composite Plates Using Fiber Optic Sensors and Shearography," *Proceedings of the International Conference on Fiber Optics and Photonics,* New Delhi, India, 1998, pp. 716–719.

11

EMERGING COMPOSITE MATERIALS

11.1 NANOCOMPOSITES

Nanocomposites are materials filled with nano-sized particles, that is, particles with at least one dimension in the nanometer scale (a nanometer is 10^{-9} m, or a billionth of a meter). There are two types of nanoparticles that are used most often for this purpose, *platelets* and *nanotubes,* although spherical particles are also being considered. The platelets are obtained from a layered-silicate clay, an inexpensive natural silicate mineral. The individual sheets in the silicates generally are stacked together and are not compatible with the organic-matrix polymers. Therefore, the layered silicates are modified by an ion-exchange reaction to make them compatible with the polymer matrix. When this modified clay is placed in a polymer matrix, polymer fills the spaces between the individual sheets or the platelets. This increases the distance between the platelets and causes the clay to swell. Further swelling causes the platelets to be exfoliated and dispersed throughout the polymer. The platelets thus obtained are ultrathin with a specific surface area (ratio of the surface area to volume) much larger than the conventionally sized reinforcement (a typical platelet may be 1 nm thick and 70–150 nm across). One of the most commonly used clays in the nanocomposites is *montmorillonite.*

The large surface area of the platelets and their high strength impart enhanced properties to the nanocomposites. Their dispersion in the polymer matrix and the interfacial bonding between the individual sheets and the polymer matrix facilitate stress transfer to improve mechanical properties. The platelets' dispersion hinders diffusion pathways and makes nanocomposites exhibit lower permeability to moisture, gases, and hydrocarbons. The platelets' constraining effects cause improvements in thermal stability and flame retardancy.

Carbon nanotubes, discovered in 1991, can be visualized as a sheet of graphite (i.e., a single layer of hexagonally arranged carbon atoms) rolled into a cylindrical tube and welded together. They derive stiffness and strength from the C—C bond. Nanotubes exist as either single-walled or multiwalled structures. The multiwalled carbon nanotubes (MWCNTs) are simply composed of concentric single-walled carbon nanotubes (SWCNTs).

Primary synthesis methods for single and multiwalled carbon nanotubes include arc-discharging, laser ablation, and gas-phase catalytic growth from carbon monoxide and chemical vapor deposition (CVD) from hydrocarbons. For their application in composites, large quantities of nanotubes are required. The gas-phase processes tend to produce nanotubes with fewer impurities and more amenable to large-scale processing. Thus gas-phase techniques, such as CVD, offer the greatest potential for the scaling up of nanotube production for the manufacturing of nanocomposites.

Selected properties of SWCNTs are given in Table 11-1. The modulus of nanotubes is reported to be extremely high, greater than 1000 GPa. Tensile strength is predicted to reach higher than the reported value. In addition to their exceptional modulus and strength, the carbon nanotubes possess superior thermal and electrical properties: thermally stable up to 2800°C in vacuum, thermal conductivity about twice as high as diamond, and electric-current-carrying capacity about 1000 times higher than copper wires.

Nanocomposites formed of polymer reinforced with exfoliated clays and carbon nanotubes are considered for applications such as interior and exterior automotive accessories, structural components for electronic portable devices, and film for food packaging. An attractive feature for nanocomposites is that they provide improvement in a variety of properties with only a small amount of reinforcement. Typical loadings in nanocomposites are in the range of 1–10% by weight.

Nanocomposites are still a nascent technology. Some of the critical issues hindering realization of their full potential include dispersion and alignment of nanotubes and clay platelets and bonding at the particle–matrix interface. Owing to the nano size of the reinforcement, some of the micromechanics analysis models may not be valid. Therefore, much work remains to be done in developing and validating theoretical models with experimental data. Some recent publications are listed at the end of this chapter.

Table 11-1 Properties of single-walled carbon nanotubes

Property	Value
Young's modulus	~1000 GPa
Tensile strength	30 GPa
Electrical resistivity	10^{-4} Ω-cm
Thermal conductivity	~2000 W/mK

The development of nanostructural materials creates tremendous opportunities for the design of multifunctional material systems. Composite matrix resins can be modified with nanoparticles to optimize properties of interest. The modified resin acts as the matrix material for advanced composites. Thus nanocomposites can be used in conjunction with conventional fillers, such as glass fibers and talc, to achieve customized property sets. A nanocomposite with 6% nanoclay by weight that also contains 10–12% glass fibers can have a property set similar to that of a 30% straight glass-filled composite material but at a lower specific gravity and part weight.

The nanocomposites appear positioned to compete with a wide range of materials, including filled and reinforced compounds, flame-retardant compounds, and barrier materials. Packaging, primarily for beer, carbonated beverages, food, and condiments, is expected to be a major growth market for nanoclay compounds, which can improve moisture-barrier and odor-barrier characteristics, whereas carbon nanotubes can increase strength and electrostatic dissipation. Growth is also expected from increased demand for conductive compounds and miniaturization of electronic components.

11.2 CARBON–CARBON COMPOSITES

Carbon–carbon composites are ultra-high-temperature composites that use carbon fibers in a carbon matrix. They are especially valued for high-abrasion and high-temperature environments, such as heat shields, aircraft brakes, and engine turbines. Their primary advantage lies in their ability to withstand high temperatures ($\leq 3300°C$). Actually, their strength increases with temperatures up to about 2050°C and begins to decline above 2200°C, but they still can serve as a heat shield. They have extremely low creep at high temperatures, several orders of magnitude less than ceramics. In high-temperature applications, they do not require external cooling liquids. They resist thermal shock without cracking or distorting. They can withstand temperature cycling from subzero temperatures to 1500°C.

Carbon–carbon composites are lightweight, only half the density of aluminum. They tend to have low shear strength but good tensile and compressive strengths. They are self-lubricating and resist abrasion, wear, and fatigue. They can be machined, drilled, or sawed. They are less sensitive to flaws or impact than ceramics.

Carbon–carbon composites are commonly produced by two methods: liquid impregnation and chemical-vapor deposition (CVD). In liquid impregnation, phenolic resin is infiltrated into carbon fibers and then pyrolyzed. Phenolic resin is chosen because it has a high-carbon yield. In CVD, the matrix is built up layer by layer. With both liquid impregnation and CVD, liquid or vapor must be added several times to fill in the pores, each time pyrolyzing to increase density.

CVD tends to produce a porous structure because carbon that deposits on the surface of the skeleton preform tends to seal off the interior. Liquid impregnation costs less and is better suited for thicker parts than CVD. However, in liquid impregnation, the liquid shrinks when it is carbonized, resulting in residual stresses. In CVD, these stresses are smaller, leading to higher strength and modulus; CVD, on the other hand, produces a highly anisotropic matrix. At present, manufacturing carbon–carbon composites is an extremely time-consuming process, requiring slow pyrolyzing to drive off gases without cracking the matrix.

Unprotected carbon–carbon composites will react with oxygen, burning rapidly at temperatures around 450°C. Therefore, a carbon–carbon composite is coated with ceramics to protect it from oxidizing. The coatings include SiC, silicides, boron carbide, boron nitrate, phosphates, and alumina.

Carbon–carbon composites are fabricated from both two-dimensional laminates and three-dimensional fiber preforms. The latter are important because the carbon matrix is inherently brittle, and a three-dimensional preform adds toughness. Carbon–carbon composites have been used in rocket nozzles and nose cones, brakes, wear guides, and many other industrial and high-speed applications.

11.3 BIOCOMPOSITES

A product that is derived from renewable resources, stable in its intended lifetime, but would biodegrade after disposal in composting conditions generally is considered to be a biobased product or simply a bioproduct. Thus biocomposites consist of biofibers and biomatrix and are expected to be biodegradable. There are very few composites that are truly biodegradable and can be used in engineering applications. However, composites containing biofibers are being used in many engineering applications. Important aspects of biofibers and bioresins are discussed in this section.

11.3.1 Biofibers

In recent years, there has been mounting interest in the use of natural fibers from renewable sources for the reinforcement of plastics. Such fibers have their origin in plants. The advantages of these fibers over the manmade fibers such as glass and carbon are low cost, low density, acceptable specific strength properties, high toughness, good thermal properties, reduced dermal and respiratory irritation, low energy content, and biodegradability. Plant-fiber composites are finding use in building products, infrastructure, transportation, industrial, and consumer applications.

Biofibers may be classified in two broad categories: wood fibers and non-wood fibers. Wood fibers and wood-fiber composites will be discussed in a separate section later. Important nonwood fibers are kenaf, flax, jute, hemp, coir and sisal. Straw and grass fibers are also attracting attention as reinforcing fibers.

It is well known that the tensile strength of plant fibers is lower than the tensile strength of E-glass fibers. However, the specific strength of some of the natural fibers is more comparable with that of glass fibers because the density of E-glass is much higher than the density of most of the natural fibers. Strength, modulus, and density of common plant fibers are given in Table 11-2 along with the properties of E-glass fibers for comparison.

The disadvantages of plant fibers are their water reactivity, finite length and large diameter, and variability. Plant fibers strongly attract water, and the resulting sorption adversely affects their performance in a composite. This problem can be overcome either by encapsulating water-reactive fibers in a water-resistant matrix to keep water out or by reducing water reactivity of fibers by chemical modification. However, both methods have their advantages and disadvantages with respect to cost, performance, and reliability. The plant fibers are small in length and large in diameter, which makes them inefficient as reinforcing material. Short fibers (2–5 mm long) are difficult to align, which is essential to maximize strength of a product. Plant fibers vary widely in chemical composition, structure, and dimensions, and originate from different parts of the plant. Therefore, properties of plant fibers vary considerably. This makes it difficult to produce parts with consistent strength. The traditional approach to variability in properties is to overdesign and allow for a worst-case scenario. This results in considerable inefficiency. However, this has proved viable commercially with a number of existing plant-fiber-reinforced products.

Plant fibers possess excellent sound absorption efficiency, are more shatter resistant, and have better energy-management characteristics than glass fibers in their respective composite structures. These characteristics make them suitable for automotive parts. Because biofibers are derived from renewable resources, material cost can be reduced markedly with their large scale use.

Table 11-2 Properties of various natural fiber reinforcements

Fiber	Density ρ (g/cm³)	Tensile Strength σ_u (MPa)	Specific Tensile Strength σ_u/ρ	Tensile Modulus E (GPa)	Specific Tensile Modulus E/ρ
Flax	1.5	1100	733	27	18
Kenaf	1.5	900	600	22	15
Hemp	1.5	700	467	22	15
Sisal	1.45	600	414	18	12
Jute	1.5	500	333	23	15
Coir	1.2	150	125	5	4
E-glass	2.54	3500	1378	72.4	28.5

The ecofriendly nature of these fibers is also an important driving force for their development.

11.3.2 Wood–Plastic Composites (WPCs)

The term *wood–plastic composite* refers to any composite that contains wood (in any form) and a plastic matrix. Because of the limited thermal stability of wood, only thermoplastics that melt or can be processed at temperatures below 200°C are used in WPCs. Currently, most WPCs are made with polyethylene, both recycled and virgin, for use in exterior building components. WPCs made with polypropylene typically are used in automotive applications and consumer products. Wood–polyvinyl chloride (PVC) composites typically used in window manufacturing are now being used in decking as well.

Wood flour and very short fibers, rather than long individual wood fibers, are used most often in WPCs. Products typically contain approximately 50% wood. The relatively high bulk density and free-flowing nature of wood flour compared with wood fibers or other longer natural fibers, low cost, and availability are attractive features for its use in WPCs. Common wood species used include pine, maple, and oak. Typical particle sizes are 10–80 mesh. Additives such as coupling agents, ultraviolet (UV) stabilizers, pigments, lubricants, fungicides, and foaming agents are used frequently to improve processing and performance.

The majority of WPCs are manufactured by profile extrusion, in which molten composite material is forced through a die to make a continuous profile of the desired shape. Extrusion lends itself to processing the high viscosity of the molten WPC blends and to shaping the long, continuous profiles common to building materials. Preblended, free-flowing pellets of wood and other natural fibers mixed with thermoplastics are now available commercially for manufacturing WPCs. Other processing technologies such as injection molding are also used to produce WPCs, but to a much lesser extent than extrusion.

Wood flour is frequently added to thermoplastics to reduce cost. It also reduces shrinkage and improves part stiffness but often makes it more brittle. Most commercial WPC products are considerably less stiff than solid wood. Adding wood fibers rather than flour increases mechanical properties such as strength, elongation, and impact energy. There is a significant variability in the properties of WPCs reported in the literature, probably owing to differences in the constituent properties. Therefore, properties should be determined experimentally to assess suitability of a WPC for a specific application.

The greatest growth potential for WPCs is in building products that have limited structural requirements, such as decking, fencing, industrial flooring, landscape timbers, railings, and moldings. Although WPC decking is more expensive than pressure-treated wood, WPC decking requires lower maintenance, does not crack or splinter, and has higher durability. However, creep resistance, stiffness, and strength are lower than those of solid wood. Wood fibers, wood flour, and rice hulls are the most common organic fillers used in decking.

Window and door profiles represent another important area of WPC application. PVC is used most often as the thermoplastic matrix in these applications because of its balance of thermal stability, moisture resistance, and stiffness. Sometimes wood-filled PVC is coextruded with an unfilled outside layer for increased durability. Another variation in this type of product has wood-filled PVC and a composite with a foamed interior for easy nailing and screwing.

Research and development efforts are being directed toward new applications of WPCs. Some of these include roof shingles, roofing timber, wall studs, and waterfront applications. There is a strong movement in research toward more highly engineered WPCs with greater structural performance and more efficient design.

11.3.3 Biopolymers

Biopolymers or biodegradable polymers may be suitable as a matrix material for composite applications. Biopolymers may be obtained from renewable resources, synthesized microbially, or synthesized from petroleum-based chemicals. Biopolyester (polylactides and microbial polyesters), cellulosic plastics, soy-based plastics, and starch plastics are promising biopolymers obtained from renewable resources. Chemical structures of some biopolyesters are given in Fig. 11-1. Polylactic acid (PLA) is a versatile biopolymer derived from corn. Biocomposites from natural fibers and PLA are attracting interest. The bacterial polyester PHAs (polybeta-hydroxy alkanoate) are produced from sugar by fermentation process. They are naturally produced by transgenic plants from CO_2 and sunlight. The copolymer of PHA, that is, poly-

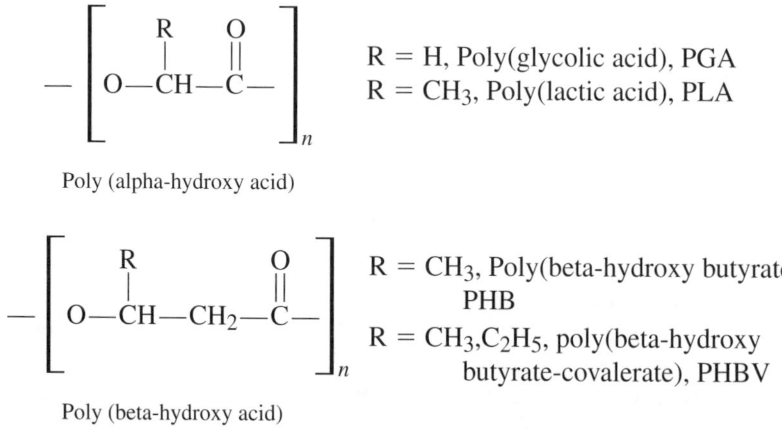

$$-\left[O-\underset{R}{CH}-\underset{\parallel}{\overset{O}{C}}-\right]_n$$

R = H, Poly(glycolic acid), PGA
R = CH_3, Poly(lactic acid), PLA

Poly (alpha-hydroxy acid)

$$-\left[O-\underset{R}{CH}-CH_2-\underset{\parallel}{\overset{O}{C}}-\right]_n$$

R = CH_3, Poly(beta-hydroxy butyrate), PHB
R = CH_3, C_2H_5, poly(beta-hydroxy butyrate-covalerate), PHBV

Poly (beta-hydroxy acid)

Figure 11-1. Structure of some important biopolyesters.

β-hydroxy butyrate-CO-valerate (PHBV) is used successfully as matrix material in jute-fiber-based biocomposites.

Nonfood applications of soybeans are being investigated at various U.S. universities. Soybeans typically contain about 20% oil and 40% protein. Both protein and oil from soybeans can be converted to biodegradable plastics usable in composites. Resin suitable for natural-fiber composites is produced by functionalization of soy oil. Epoxidized soybean oil (ESO) is used as a plasticizer or stabilizer for PVC. It also can be used as a reactive modifier, diluent, and toughener of the epoxy-resin system. Direct structural applications of ESO are limited owing to its low cross-linking density and mechanical properties.

11.4 COMPOSITES IN "SMART" STRUCTURES

A "smart" structure ideally is one that is capable of sensing changes occurring to the structure due to its environment, analyzing this information, deciding on the actions necessary to optimize its performance, and then commanding the appropriate devices to initiate those actions. Thus a smart structure involves distributed sensors and actuators and one or more microprocessors that analyze the responses from the sensors and use control theory to command the actuators to act. The smart structures, often called *intelligent structures,* are modeled on biological systems with

- Sensors acting as a nervous system
- Actuators acting like muscles
- Microprocessors acting as a brain to control the system

The sensors in a smart structure can pick up changes to the external environment (such as loads or shape change), as well as a changing internal environment (such as damage or failure). The actuators may bring about the alterations of system characteristics (such as stiffness or damping), as well as of system response (such as strain or shape), in a controlled manner. The actuating, sensing, and signal-processing elements are incorporated into a structure for the purpose of influencing its states or characteristics, be they mechanical, thermal, optical, chemical, electrical, or magnetic. Many types of actuators and sensors are being considered, such as piezoelectric materials, shape-memory alloys, electrostrictive materials, magnetostrictive materials, electrorheologic fluids, and fiberoptics. These can be integrated with the main load-carrying structure by surface bonding or embedding without causing any significant changes in mass or structural stiffness of the system.

To date, intelligent structures have not been built, but their feasibility is accepted. Laminated and composite materials provide an important advancement, helping the realization of intelligent structures. In the past, structures were manufactured from large pieces of monolithic materials, which were machined, forged, or formed to a final structural shape. It is difficult to incorporate active elements such as sensors and actuators into metallic components and structures. However, modern composite materials are built up from constitutive elements and allow for incorporation of active elements within the structural form. One can envision the incorporation of an intelligent ply in a laminate carrying actuators, sensors, processors, and interconnections within the laminated composite. The developments in microelectronics, bus architectures, switching circuitry, and fiberoptics are other important technological advancements leading to intelligent structures. Development in the fields of information processing, artificial intelligence, and control disciplines also are central to the emergence of intelligent structures.

Much of the work relating to smart structures and materials is still at the research stage, with the aim of demonstrating the concepts and providing the feasibility of workable schemes in real structures. The advanced composite materials are an obvious choice to host smart materials technologies. Polymeric-matrix-based composites, with carbon, glass, and Kevlar fibers as the reinforcement, present the most benign host materials. Reinforced metals, ceramics, and carbon are much more difficult systems to incorporate smart concepts in view of the high temperatures involved both in processing and in operation. The first use of embedded sensors may be in process monitoring and control of the fabrication stage. This is particularly important in thick laminates based on thermosetting resin matrices, where exothermic reactions can cause high temperatures in the center of the laminate, producing nonuniform cure through the thickness and high thermal stresses that can lead to cracking.

Potential applications of smart structures include actively controlling vibration, noise, damping, aeroelastic stability, shape, and stress distribution. Applications range from space systems, fixed-wing and rotary-wing aircraft, automotive, civil structures, and machine tools. However, there are major barriers to be overcome before this technology can be fully realized.

SUGGESTED READING

Nanocomposites

1. T. W. Ebbesen, ed., *Carbon Nanotubes,* CRC Press, Boca Raton, FL, 1997.
2. Y. Saito, G. Dresselhaus, M. S. Dresselhaus, *Physical Properties of Carbon Nanotubes,* Imperial College Press, London, 1999.
3. R. A. Vaia, G. Price, R. N. Ruth, H. T. Nguyen, and J. Lichtenhan, "Polymer/Layered Silicate Nanocomposites as High Performance Ablative Materials," *Appl. Clay Sci.,* **15,** 67–92 (1999).
4. T. J. Pinnavaia and G. W. Beal, eds., *Polymer Clay Composites,* Wiley, New York, 2000.

5. E. T. Thostenson, Z. Ren, and T. W. Chou, "Advances in the Science and Technology of Nanotubes and their Composites: A Review," *Compos. Sci. Technol.,* **61,** 1899–1912 (2001).
6. R. A. Vaia, T. Benson Tolle, G. F. Schmitt, et al., "Nanoscience and Nanotechnology: Materials Revolution for the 21st Century," *SAMPE J.,* **37,** 24–31 (2001).
7. B. Maruyama and K. Alam, "Carbon Nanotubes and Nanofibers in Composite Materials," *SAMPE J.,* **38,** 59–67 (2002).
8. R. Stewart, "Nanocomposites," *Plast. Eng.,* 22–30 (May 2004).
9. J. H. Koo and L. A. Pilato, "Polymer Nanostructured Materials for High Temperature Applications," *SAMPE J.,* **41**(2), 7–19 (March–April 2005).

Carbon–Carbon Composites

1. A. J. Klein, "Carbon/Carbon Composites," *Adv. Mater. Proc.,* **11,** 64–68 (1986).
2. A. J. Klein, "Carbon/Carbon: An Ultrahigh-Temperature Composite," *Adv. Compos.,* 38–44 (March–April 1989).
3. J. D. Buckley and D. D. Edie, eds., *Carbon-Carbon Materials and Composites,* Noyes Publications, Park Ridge, NJ, 1993.
4. A. Buchman and R. G. Bryant, "Molded Carbon–Carbon Composites Based on Microcomposite Technology," *Appl. Compos. Mater.,* **6,** 309–326 (1999).
5. S. R. Dhakate, R. B. Mathur, and T. L. Dhami, "Mechanical Properties of Unidirectional Carbon–Carbon Composites as a Function of Fiber Volume Content," *Carbon Sci.,* **3,** 127–132 (2002).

Biocomposites

1. A. J. Bolton, "Natural Fibers for Plastic Reinforcement," *Mater. Technol.,* **19,** 12–20 (1994).
2. J. J. Balatinecz and B. D. Park, "The Effects of Temperature and Moisture Exposure on the Properties of Wood-Fiber Thermoplastic Composites," *J. Thermoplast. Compos. Mater.,* **10,** 476–487 (1997).
3. G. I. Williams and R. P. Wool, "Composites from Natural Fibers and Soy Oil Resins," *Appl. Compos. Mater.,* **17,** 421–432 (2000).
4. A. K. Mohanty, M. Misra, and L. T. Drzal, "Sustainable Bio-composites from Renewable Resources: Opportunities and Challenges in the Green Materials World," *J. Polymers Environ.,* **10,** 19–26 (2002).
5. C. Clemens, "Wood–Plastic Composites in the United States—The Interfacing of Two Industries," *Forest Products J.,* **52,** 10–18 (2002).
6. J. Zhu, K. Chandrashekhara, V. Flanigan, and S. Kapila, "Manufacturing and Mechanical Properties of Soy-Based Composites Using Pultrusion," *Composites* Part A, **35,** 95–101 (2004).

Composites in Smart Structures

1. M. W. Hiller, M. D. Bryant, and J. Umegaki, "Attenuation and Transformation of Vibration Through Active Control of Magnetostrictive Terfenol," *J. Sound Vibrat.,* **134,** 507–519 (1989).

2. M. V. Gandhi and B. S. Thomson, *Smart Materials and Structures: The Impending Revolution,* Technomic, Lancaster, PA, 1990.
3. C. A. Rogers, C. Liang, and J. Jia, "Structural Modification of Simply-Supported Laminated Plates Using Embedded Shape Memory Alloy Fibers," *Comput. Struct.,* **38,** 569–580 (1991).
4. E. Udd, ed., *Fiber Optic Sensors: An Introduction for Engineers and Scientists,* Wiley, New York, 1991.
5. R. M. Measures, "Advances toward Fiber Optic Based Smart Structures," *Opt. Eng.,* **31,** 34–47 (1992).
6. D. Damjanovic and R. E. Newnham, "Electrostrictive and Piezoelectric Materials for Actuator Applications," *J. Intell. Syst. Struct.,* **3,** 190–208 (1992).
7. K. Chandrashekhara and A. Agarwal, "Active Vibration Control of Laminated Composite Plates Using Piezoelectric Devices—A Finite Element Approach," *J. Intell. Mater. Syst. Struct.,* **4,** 496–508 (1993).
8. K. D. Weiss, J. D. Carlson, and J. P. Coulter, "Material Aspects of Electrorheological Systems," *J. Intell. Mater. Syst. Struct.,* **4,** 13–34 (1993).
9. E. F. Crawley, "Intelligent Structures for Aerospace: A Technology Overview and Assessment," *AIAA J.,* **32,** 1689–1699 (1994).
10. F. Akhavan, S. E. Watkins, and K. Chandrashekhara, "Measurement and Analysis of Impact Induced Strain Using Extrinsic Fabry-Perot Fiber Optic Sensors," *Smart Mater. Struct.,* **7,** 745–751 (1998).

APPENDIX 1

MATRICES AND TENSORS

MATRIX DEFINITIONS

A rectangular array of numbers of the form

$$\begin{bmatrix} a_{11} & a_{12} & \cdots & a_{1n} \\ a_{21} & a_{22} & \cdots & a_{2n} \\ a_{m1} & a_{m2} & \cdots & a_{mn} \end{bmatrix}$$

is called a *matrix*. The numbers $a_{11} \cdots a_{mn}$ are called the *elements* of the matrix. The horizontal lines are called *rows* or *row vectors*, and the vertical lines are called *columns* or *column vectors* of the matrix. A matrix with m rows and n columns is called an $(m \times n)$ matrix (read "m by n matrix").

Matrices are denoted by capital letters A, B, and so on or by (a_{ij}), (b_{ij}), and so forth, that is, by writing the general element of the matrix.

In the double-subscript notation for the elements, the first subscript always denotes the row and the second subscript the column containing the given element.

Example A1-1: Let

$$A = \begin{bmatrix} 4 & 7 & 3 & 10 \\ 8 & 11 & 6 & 2 \\ 15 & 9 & 21 & 5 \end{bmatrix}$$

where A is a (3×4) matrix. The element a_{23} refers to 6 or, similarly, a_{31} refers to 15.

A matrix

$$(a_1 \quad a_2 \quad \cdots \quad a_n)$$

having only one row is called a *row matrix* or *row vector*. A matrix

$$\begin{Bmatrix} b_1 \\ b_2 \\ \vdots \\ b_n \end{Bmatrix}$$

having one column is called a *column matrix* or *column vector*.

The transpose of an $(m \times n)$ matrix $A = (a_{ij})$ is defined as the $(n \times m)$ matrix formed by interchanging rows and columns and is denoted by A^T. Thus the transpose of a row matrix is a column matrix and vice versa.

Example A1-2: If

$$A = \begin{bmatrix} 5 & -8 & 1 \\ 4 & 0 & 2 \end{bmatrix}$$

then

$$A^T = \begin{bmatrix} 5 & 4 \\ -8 & 0 \\ 1 & 2 \end{bmatrix}$$

Example A1-3: If

$$b = (2 \quad 5 \quad -7)$$

then

$$b^T = \begin{Bmatrix} 2 \\ 5 \\ -7 \end{Bmatrix}$$

A matrix having the same number of rows and columns is called a *square matrix*, and the number of rows is called its *order*. Thus

$$A = \begin{bmatrix} a_{11} & a_{12} & a_{13} \\ a_{21} & a_{22} & a_{23} \\ a_{31} & a_{32} & a_{33} \end{bmatrix}$$

is a square matrix of order 3.

MATRIX DEFINITIONS

The principal or main diagonal of a square matrix goes from the upper left- to the lower right-hand corner of the matrix. Thus the principal diagonal contains the elements $a_{11}, a_{22}, \ldots, a_{nn}$.

A square matrix $A = (a_{ij})$, whose elements other than those in the principal diagonal are all zero; that is, $a_{ij} = 0$, for all $i \neq j$, is called a *diagonal matrix*. For example,

$$\begin{bmatrix} 2 & 0 & 0 \\ 0 & 1 & 0 \\ 0 & 0 & -3 \end{bmatrix} \quad \text{and} \quad \begin{bmatrix} 4 & 0 & 0 \\ 0 & 4 & 0 \\ 0 & 0 & 4 \end{bmatrix}$$

are diagonal matrices.

A diagonal matrix whose elements in the principal diagonal are all 1 is called a *unit matrix* or *identity matrix* and is denoted by I. For example, the three-rowed unit matrix is

$$I = \begin{bmatrix} 1 & 0 & 0 \\ 0 & 1 & 0 \\ 0 & 0 & 1 \end{bmatrix}$$

A square matrix $A = (a_{ij})$ is said to be symmetric if it is equal to its transpose:

$$A^T = A$$

this is, $a_{ji} = a_{ij}$ for all values of i and j.

A square matrix $A = (a_{ij})$ is said to be antisymmetric or skew–symmetric if

$$A^T = -A$$

that is, $a_{ji} = -a_{ij}$ for all values of i and j.

Note that for $i = j$, $a_{ii} = -a_{ii}$, which implies that the elements in the principal diagonal of a skew-symmetric matrix are all zero.

Example A1-4: The matrices

$$A = \begin{bmatrix} 1 & -2 \\ -2 & 0 \end{bmatrix} \quad \text{and} \quad B = \begin{bmatrix} -3 & 1 & -2 \\ 1 & 0 & 5 \\ -2 & 5 & 4 \end{bmatrix}$$

are symmetric matrices.

Example A1-5: The matrices

$$A = \begin{bmatrix} 0 & -2 \\ 2 & 0 \end{bmatrix} \quad \text{and} \quad B = \begin{bmatrix} 0 & 1 & -2 \\ -1 & 0 & 5 \\ 2 & -5 & 0 \end{bmatrix}$$

are skew-symmetric matrices.

Two ($m \times n$) matrices $A = (a_{ij})$ and $B = (b_{ij})$ are said to be equal if, and only if, corresponding elements are equal, that is,

$$a_{ij} = b_{ij}$$

for all values of i and j. Then the following can be written:

$$A = B$$

Note that this definition of equality refers to matrices that have the same number of rows and the same number of columns.

Example A1-6: The simple algebraic equations

$$a = p + 1$$
$$b = q + 6$$
$$c = r - 3$$
$$d = s + 6$$

may be given in the matrix form as

$$\begin{bmatrix} a & b \\ c & d \end{bmatrix} = \begin{bmatrix} p+1 & q+6 \\ r-3 & s+6 \end{bmatrix}$$

A determinant of order n is a square array of n^2 quantities enclosed between two vertical bars

$$D = \begin{vmatrix} a_{11} & a_{12} & \cdots & a_{1n} \\ a_{21} & a_{22} & \cdots & a_{2n} \\ \cdots & \cdots & \cdots & \cdots \\ \cdots & \cdots & \cdots & \cdots \\ a_{n1} & a_{n2} & \cdots & a_{nn} \end{vmatrix}$$

which has a definite value, as defined in the paragraphs that follow.

By deleting the ith row and the jth column from the determinant D, an $(n - 1)$th-order determinant is obtained, which is called the *minor of the element* a_{ij} (which belongs to the deleted row and column) and is denoted by M_{ij}.

The minor M_{ij} multiplied by $(-1)^{i+j}$ is called the *cofactor of* a_{ij} and is denoted by A_{ij}; thus

$$A_{ij} = (-1)^{i+j} M_{ij}$$

A determinant D represents the sum of the products of the elements of any row or column and their respective cofactors. Thus the determinant D of order n (as defined earlier) means

$$D = a_{i1}A_{i1} + a_{i2}A_{i2} + \cdots + a_{in}A_{in} \qquad (i = 1, 2, \ldots, \text{or } n)$$

or

$$D = a_{1j}A_{1j} + a_{2j}A_{2j} + \cdots + a_{nj}A_{nj} \qquad (j = 1, 2, \ldots, \text{or } n)$$

Example A1-7: Let a second-order determinant be

$$D = \begin{vmatrix} a_{11} & a_{12} \\ a_{21} & a_{22} \end{vmatrix}$$

Cofactors and minors of the determinant are

$$A_{11} = M_{11} = a_{22}$$
$$A_{12} = -M_{12} = -a_{21}$$
$$A_{21} = -M_{21} = -a_{12}$$
$$A_{22} = M_{22} = a_{11}$$

The value of D may be calculated by its development by any row or column, for example, by the first row:

$$D = a_{11}a_{22} + a_{12}(-a_{21})$$

Or by the first column:

$$D = a_{11}a_{22} + a_{21}(-a_{12})$$

Example A1-8: In the third-order determinant

$$D = \begin{vmatrix} 1 & 2 & -1 \\ 3 & 6 & 0 \\ 0 & 4 & 2 \end{vmatrix}$$

the cofactors of the elements of the first row are

$$A_{11} = M_{11} = \begin{vmatrix} 6 & 0 \\ 4 & 2 \end{vmatrix} = 12$$

$$A_{12} = -M_{12} = -\begin{vmatrix} 3 & 0 \\ 0 & 2 \end{vmatrix} = -6$$

$$A_{13} = M_{13} = \begin{vmatrix} 3 & 6 \\ 0 & 4 \end{vmatrix} = 12$$

Thus the value of the determinant is

$$D = 1 \times 12 + 2 \times (-6) - 1 \times 12 = -12$$

MATRIX OPERATIONS

Addition and subtraction of matrices are defined only for matrices having the same number of rows and the same number of columns. The sum of two $(m \times n)$ matrices $A = (a_{ij})$ and $B = (b_{ij})$ is defined as the $(m \times n)$ matrix $C = (c_{ij})$ whose elements are

$$c_{ij} = a_{ij} + b_{ij} \qquad \begin{aligned} i &= 1, 2, \ldots, n \\ j &= 1, 2, \ldots, n \end{aligned}$$

and

$$C = A + B$$

can be written. Similarly, the matrix

$$D = A - B = (a_{ij} - b_{ij})$$

is called the *difference of A and B*.

A square matrix A may be written as the sum of a symmetric matrix R and a skew-symmetric matrix S, where

$$R = \tfrac{1}{2}(A + A^{\mathrm{T}}) \quad \text{and} \quad S = \tfrac{1}{2}(A - A^{\mathrm{T}})$$

Example A1-9: Algebraic equations

$$a = p + 1$$
$$b = q + 6$$
$$c = r + 3$$
$$d = s + 5$$

are equivalent to

$$\begin{bmatrix} a & b \\ c & d \end{bmatrix} = \begin{bmatrix} p & q \\ r & s \end{bmatrix} + \begin{bmatrix} 1 & 6 \\ 3 & 5 \end{bmatrix}$$

Example A1-10: Let a square matrix be

$$A = \begin{bmatrix} -3 & -3 & 6 \\ 5 & 0 & -7 \\ 4 & 3 & 4 \end{bmatrix}$$

The transpose of A is

$$A^{\mathrm{T}} = \begin{bmatrix} -3 & 5 & 4 \\ -3 & 0 & 3 \\ 6 & -7 & 4 \end{bmatrix}$$

The matrix A may be written in the form $A = R + S$, where

$$R = \frac{1}{2} \left(\begin{bmatrix} -3 & -3 & 6 \\ 5 & 0 & -7 \\ 4 & 3 & 4 \end{bmatrix} + \begin{bmatrix} -3 & 5 & 4 \\ -3 & 0 & 3 \\ 6 & -7 & 4 \end{bmatrix} \right) = \begin{bmatrix} -3 & 1 & 5 \\ 1 & 0 & -2 \\ 5 & -2 & 4 \end{bmatrix}$$

and $\quad S = \dfrac{1}{2} \left(\begin{bmatrix} -3 & -3 & 6 \\ 5 & 0 & 7 \\ 4 & 3 & 4 \end{bmatrix} - \begin{bmatrix} -3 & 5 & 4 \\ -3 & 0 & 3 \\ 6 & -7 & 4 \end{bmatrix} \right) = \begin{bmatrix} 0 & -4 & 1 \\ 4 & 0 & -5 \\ -1 & 5 & 0 \end{bmatrix}$

The product of a matrix $A = (a_{ij})$ by a number of c is defined as the matrix (ca_{ij}) and is denoted by cA or Ac; thus

$$cA = Ac = \begin{bmatrix} ca_{11} & ca_{12} & \cdots & ca_{1n} \\ ca_{21} & ca_{22} & \cdots & ca_{2n} \\ \vdots & \vdots & \vdots & \vdots \\ \vdots & \vdots & \vdots & \vdots \\ ca_{m1} & ca_{m2} & \cdots & ca_{mn} \end{bmatrix}$$

Let $A = (a_{ij})$ be an $(m \times n)$ matrix and let $B = (b_{ij})$ be an $(r \times p)$ matrix; then the product AB (in this order) is defined only when $r = n$ and is the $(m \times p)$ matrix $C = (c_{ij})$ whose elements are

$$c_{ij} = a_{i1}b_{1j} + a_{i2}b_{2j} + \cdots + a_{in}b_{nj} = \sum_{k=1}^{N} a_{ik}b_{kj}$$

The process of matrix multiplication is illustrated by the following example:

Example A1-11: Let

$$A = \begin{bmatrix} 1 & 5 & 6 \\ 0 & 3 & 4 \\ 2 & 1 & 2 \end{bmatrix} \quad \text{and} \quad B = \begin{bmatrix} 3 & 0 \\ 2 & 4 \\ -1 & 6 \end{bmatrix}$$

Then the product AB is

$$AB = \begin{bmatrix} 1 & 5 & 6 \\ 0 & 3 & 4 \\ 2 & 1 & 2 \end{bmatrix} \begin{bmatrix} 3 & 0 \\ 2 & 4 \\ -1 & 6 \end{bmatrix} = \begin{bmatrix} 7^* & 56 \\ 2 & 36 \\ 6 & 16 \end{bmatrix}$$

Example A1-12:

$$\begin{bmatrix} 1 & 5 & 6 \\ 0 & 3 & 4 \\ 2 & 1 & 2 \end{bmatrix} \begin{Bmatrix} 1 \\ 3 \\ 2 \end{Bmatrix} = \begin{Bmatrix} 28 \\ 17 \\ 9 \end{Bmatrix}$$

Example A1-13:

$$(2 \ 4 \ 1) \begin{Bmatrix} 3 \\ 5 \\ 9 \end{Bmatrix} = (35)$$

$$\begin{Bmatrix} 3 \\ 5 \\ 9 \end{Bmatrix} (2 \ 4 \ 1) = \begin{bmatrix} 6 & 12 & 3 \\ 10 & 20 & 5 \\ 18 & 36 & 9 \end{bmatrix}$$

Matrix multiplication is associative and distributive; that is,

*$1 \times 3 + 5 \times 2 + 6(-1) = 7$.

$$(AB)C = A(BC) = ABC$$
$$(A + B)C = AC + BC$$
$$C(A + B) = CA + CB$$

provided A, B, and C are such that the expressions on the left are defined.

Matrix multiplication is not *commutative;* that is, if A ad B are matrices such that both AB and BA are defined, then

$$AB \neq BA \text{ in general}$$

Example A1-14:

$$\begin{bmatrix} 1 & 0 \\ 0 & 0 \end{bmatrix} \begin{bmatrix} 0 & 1 \\ 2 & 0 \end{bmatrix} = \begin{bmatrix} 0 & 1 \\ 0 & 0 \end{bmatrix}$$

but

$$\begin{bmatrix} 0 & 1 \\ 2 & 0 \end{bmatrix} \begin{bmatrix} 1 & 0 \\ 0 & 0 \end{bmatrix} = \begin{bmatrix} 0 & 0 \\ 2 & 0 \end{bmatrix}$$

Thus the order of matrices in a matrix multiplication is very important. To be precise, it is stated that in the product AB the matrix B is premultiplied by the matrix A or, alternatively, that A is postmultiplied by B.

An important property of matrix multiplication relates to the transpose of the matrices. The transpose of a product equals the product of the transposed matrices taken in reverse order; thus

$$[A \quad B]^T = [B^T][A^T]$$

The inverse of a square matrix A of order n is defined as another matrix of order n that, when premultiplied or postmultiplied by the matrix A, results in an identity matrix. The inverse of the matrix A is denoted by A^{-1}. Thus

$$AA^{-1} = A^{-1}A = I$$

A squre matrix A has an inverse if and only if the determinant (det) $A \neq 0$. The inverse of such a matrix $A = (a_{ij})$ may be obtained in the following manner:

1. Replace each element a_{ij} by its cofactor A_{ij} in det A.
2. Divide each element of this cofactor matrix by det A.
3. Transpose of the result is the inverse matrix A^{-1}.

Example A1-15: If

$$A = \begin{bmatrix} a_{11} & a_{12} \\ a_{21} & a_{22} \end{bmatrix}$$

$$A^{-1} = \frac{1}{a_{11}a_{22} - a_{12}a_{21}} \begin{bmatrix} a_{22} & -a_{12} \\ -a_{21} & a_{11} \end{bmatrix}$$

Example A1-16: Inverse of a diagonal matrix

$$A = \begin{bmatrix} a_{11} & 0 & \cdots & 0 \\ 0 & a_{22} & \cdots & 0 \\ \cdot & \cdot & \cdots & \cdot \\ \cdot & \cdot & \cdots & \cdot \\ 0 & 0 & \cdots & a_{nn} \end{bmatrix}$$

is simply

$$A^{-1} = \begin{bmatrix} \dfrac{1}{a_{11}} & 0 & \cdots & 0 \\ 0 & \dfrac{1}{a_{22}} & \cdots & \cdot \\ \cdot & \cdot & \cdots & \cdot \\ \cdot & \cdot & \cdots & \cdot \\ 0 & 0 & \cdots & \dfrac{1}{a_{nn}} \end{bmatrix}$$

Example A1-17:

$$A = \begin{bmatrix} -3 & 3 & 4 \\ 6 & -4 & -8 \\ -11 & 6 & 13 \end{bmatrix}$$

$$\det A = 10$$

$$A^{-1} = \begin{bmatrix} -0.4 & -1.5 & -0.8 \\ 1.0 & 0.5 & 0 \\ -0.8 & -1.5 & -0.6 \end{bmatrix}$$

Example A1-18:

$$A = \begin{bmatrix} \cos\theta & -\sin\theta \\ \sin\theta & \cos\theta \end{bmatrix}$$

$$\det A = \cos^2\theta + \sin^2\theta = 1$$

$$A^{-1} = \begin{bmatrix} \cos\theta & \sin\theta \\ -\sin\theta & \cos\theta \end{bmatrix}$$

Example A1-19:

$$A = \begin{bmatrix} \cos\theta & -\sin\theta & 0 \\ \sin\theta & \cos\theta & 0 \\ 0 & 0 & 1 \end{bmatrix}$$

$$A^{-1} = \begin{bmatrix} \cos\theta & \sin\theta & 0 \\ -\sin\theta & \cos\theta & 0 \\ 0 & 0 & 1 \end{bmatrix}$$

Example A1-20:

$$A = \begin{bmatrix} 0 & 1 & 0 \\ 1 & 0 & 0 \\ 0 & 0 & 1 \end{bmatrix}$$

$$A^{-1} = \begin{bmatrix} 0 & 1 & 0 \\ 1 & 0 & 0 \\ 0 & 0 & 1 \end{bmatrix}$$

Note that the inverse matrices in Examples A1-18–A1-20 satisfy the relationship

$$A^{-1} = A^{\mathrm{T}}$$

A matrix that satisfies this relationship is called an *orthogonal matrix*.

Example A1-21:

$$A = \begin{bmatrix} \cos^2\theta & \sin^2\theta & 2\sin\theta\cos\theta \\ \sin^2\theta & \cos^2\theta & -2\sin\theta\cos\theta \\ -\sin\theta\cos\theta & \sin\theta\cos\theta & \cos^2\theta - \sin^2\theta \end{bmatrix}$$

$\det A = 1$

$$A^{-1} = \begin{bmatrix} \cos^2\theta & \sin^2\theta & -2\sin\theta\cos\theta \\ \sin^2\theta & \cos^2\theta & 2\sin\theta\cos\theta \\ \sin\theta\cos\theta & -\sin\theta\cos\theta & \cos^2\theta - \sin^2\theta \end{bmatrix}$$

The matrix A in this example is a very useful matrix. It is known as the *transformation matrix*. Its application is illustrated later while discussing the tensor transformations and the transformation of stresses and strains. It may be noted that the inverse of the transformation matrix is obtained by replacing θ by $-\theta$ in the transformation matrix.

Matrix operations discussed in this appendix are almost always required in the laminate analysis calculations, as illustrated in Chap. 6. Accurate and efficient calculations required for the matrix operations have a direct influence on the accuracy and efficiency of laminate analysis. The matrix operations were illustrated in this section using simple examples. The operation could be completed easily by hand calculations because the matrices involved are of lower order. When higher-order matrices are involved, the calculations become tedious and time-consuming. Such calculations should be carried out with the help of computers for accurate results. Software programs for general calculations, such as MATLAB and Mathcad, are available commercially as mathematical tools that facilitate matrix operations and many other calculations. It is recommended that the calculations such as matrix inversions and multiplications be carried out using these tools. Their use improves accuracy of results and minimizes probability of an error.

TENSORS

In engineering and physics, mathematical models are developed to describe physical conditions and phenomena. In the process, a number of useful concepts and definitions are evolved. Properties or quantities such as density, temperature, pressure, velocity, displacement, force, stresses, and strains are well known. It is interesting to examine some of these and the mathematical concepts involved. Properties such as density, temperature, and pressure are described completely when their magnitudes are given. Such properties are known as *scalars* and are described completely by a single number. Velocity, displacement, and force involve slightly advanced concepts. In addition to the magnitude, a direction has to be assigned to them to complete their description. Such quantities are called *vectors*. Even more complicated quantities such as stress and strain require for their complete description not only the magnitude and direction but also the definition of the plane on which they act. A convenient way of defining a plane is by describing the direction of a normal to the plane. Thus a complete description of stress or strain requires a magitude and two directions.

In many situations, the method of describing properties or quantities through one magnitude and one or more directions is inconvenient. Alternative methods of description are adopted quite frequently. One of the methods is to describe a quantity through its components. For example, a vector may be represented by a line segment with its length proportional to the magnitude V of the vector and the direction parallel to the vector. In the case of a two-dimensional representation, the direction is indicated by an angle between the line segment and an arbitrarily fixed direction (or reference axis), as shown in Fig. A1-1a. Alternatively, the vector may be represented by the length of its projections along two mutually perpendicular directions, such as the reference axes x_1 and x_2 shown in Fig. A1-1b. The projections V_1 and V_2 are called the *components of the vector* in the x_1–x_2 coordinate system. The vector is then indicated by V_i, where the subscript i can assume values 1 or 2 only in a two-dimensional representation. In a three-dimensional representation, a vector is represented by three components along three mutually perpendicular

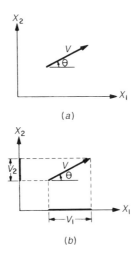

Figure A1-1. Representation of a vector by (a) a directed line segment and (b) components.

axes, and consequently, the subscript i assumes values 1, 2, or 3. In a similar manner, stress may be represented by an average force vector and unit length vector normal to the plane on which the force is acting (shown in Fig. A1-2a). Alternatively, the stress may be represented by four components in a two-dimensional problem (see Fig. A1-2b) or by nine components in a three-

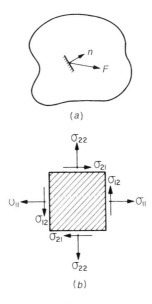

Figure A1-2. Representation of stress as (a) an average force vector and a unit normal and (b) components.

dimensional problem. Each component of stress signifies the magnitude of the average force component in one of the arbitrarily fixed directions or the reference axes acting on a plane perpendicular to one of the axes. Therefore, the stress is represented by a symbol with two subscripts such as σ_{ij}, where each subscript signifies a particular direction. The subscripts i and j can independently assume values 1 or 2 in a two-dimensional case and 1, 2, or 3 in a three-dimensional case. By convention, the first subscript always indicates the direction of the normal, and the second subscript, the direction of the force component. Arbitrarily chosen directions frequently are taken to be mutually perpendicular, as in the Cartesian coordinate system. Thus the Cartesian components of stresses in three dimensions are shown in Fig. A1-3. It may be noted that a typical component σ_{23} signifies the component along the x_3 axis of the average force acting on a plane perpendicular to the x_2 axis.

It was pointed out in the preceding discussion that the choice of coordinate system is completely arbitrary. However, it is clear that the magnitude of the components of physical entities is influenced by the orientation of the coordinate system. It also should be clear that the physical phenomenon taking place at a point is not influenced by the choice of a coordinate system. This means that all operations performed with physical entities should be independent of the orientation of the coordinate system, and the components in it must be obtainable from the components in the original coordinate system by means of suitable transformation equations. The physical entities such as density, temperature, pressure, velocity, displacement, force, stress, and strain transform according to specific transformation laws. A mathematical or physical entity that with a change in the coordinate system transforms according to a specific law of transformation is called a *tensor*. The transformation laws are the specialty of a tensor. A precise definition of a tensor is given later. It may be mentioned here that the physical entities discussed earlier are all tensors, although of different orders. It becomes clear in a later discussion

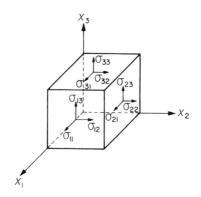

Figure A1-3. Cartesian components of stress.

that the scalar quantities such as density, temperature, and pressure are tensors of zero order; the vectors such as velocity, displacement, and force are first-order tensors; and stress and strain are second-order tensors. The transformation laws for first-order tensors (i.e., vectors) are derived first in this section and then are generalized for tensors of any order.

Consider an arbitrary vector V, as shown in Fig. A1-4. It components in the x_1, x_2 coordinate system are

$$V_1 = V \cos \alpha \quad \quad (A1.1)$$
$$V_2 = V \sin \alpha$$

Consider a second coordinate system x_1', x_2' obtained by rotating the x_1, x_2 system counterclockwise through an angle θ, as shown in Fig. A1-4. The components of the vector in the x_1', x_2' coordinate systems are

$$V_1' = V \cos (\alpha - \theta) = V \cos \alpha \cos \theta + V \sin \alpha \sin \theta \quad (A1.2)$$
$$V_2' = V \sin (\alpha - \theta) = V \sin \alpha \cos \theta - V \cos \alpha \sin \theta$$

Substituting Eq. (A1.1) in Eq. (A1.2) gives

$$V_1' = V_1 \cos \theta + V_2 \sin \theta \quad \quad (A1.3)$$
$$V_2' = V_2 \cos \theta - V_1 \sin \theta$$

Equation (A1.3) represents the law for transforming components of a vector in one coordinate system to those in another. The equations are written in terms of the single angle θ. Looking ahead to the generalization in three dimensions, it is seen that a single angle will be insufficient to describe the relative position of two sets of axes. As a concept that will be easily generalized, the four possible angles are introduced between the two coordinate systems, as indicated in Fig. A1-5. Define the cosines of the angles as

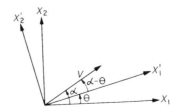

Figure A1-4. Definition of two-coordinate systems.

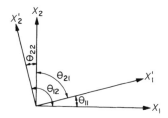

Figure A1-5. Definition of transformation angles in two dimensions.

$$a_{11} = \cos \theta_{11} = \cos \theta$$
$$a_{12} = \cos \theta_{12} = \cos (90 + \theta) = -\sin \theta$$
$$a_{21} = \cos \theta_{21} = \cos (90 - \theta) = \sin \theta \quad \text{(A1.4)}$$
$$a_{22} = \cos \theta_{22} = \cos \theta$$

Substituting Eq. (A1.4) in Eq. (A1.3) yields

$$V'_1 = a_{11} V_1 + a_{21} V_2$$
$$V'_2 = a_{12} V_1 + a_{22} V_2 \quad \text{(A1.5)}$$

Equations (A1.5) can be written in the index notations as a single equation as

$$V'_i = a_{ji} V_j \quad \text{(A1.6)}$$

It may be pointed out here that while writing expanded equations in index notation and vice versa, the following rules must be followed:

Rule 1. If a subscript occurs precisely once in one term of an expression or equation, it must occur precisely once in each term. It is to be successively assigned each value in its range. It is known as a *live* subscript or a *free index*.

Rule 2. If a subscript occurs precisely twice in one term of an expression, it may or may not occur precisely twice in any other term. It is to be summed over its range of values. It is known as a *dummy* or *summation* subscript.

With the help of these rules, the expanded form of Eq. (A1.6) as Eq. (A1.5) and vice versa can be obtained easily.

It may be observed that Eq. (A1.5) can be written in the matrix form as

$$\begin{Bmatrix} V'_1 \\ V'_2 \end{Bmatrix} = \begin{bmatrix} a_{11} & a_{21} \\ a_{12} & a_{22} \end{bmatrix} \begin{Bmatrix} V_1 \\ V_2 \end{Bmatrix} \tag{A1.7}$$

or

$$\begin{Bmatrix} V'_1 \\ V'_2 \end{Bmatrix} = (a_{ji}) \begin{Bmatrix} V_1 \\ V_2 \end{Bmatrix} \tag{A1.8}$$

where the matrix

$$(a_{ij}) = \begin{bmatrix} a_{11} & a_{12} \\ a_{21} & a_{22} \end{bmatrix} = \begin{bmatrix} \cos\theta_{11} & \cos\theta_{12} \\ \cos\theta_{21} & \cos\theta_{22} \end{bmatrix} \tag{A1.9}$$

is known as the *transformation* or *rotation matrix*. It is also called the *direction cosines matrix* because its columns represent the direction cosines of the new axes with respect to the original axes. Further, with the help of definitions [Eq. (A1.4)], it is easy to see that the matrix (a_{ij}) is an orthogonal matrix (see Example A1-18) so that its transpose is its inverse also. Therefore, Eq. (A1.7) can be written in the inverse form as

$$\begin{Bmatrix} V_1 \\ V_2 \end{Bmatrix} = \begin{bmatrix} a_{11} & a_{12} \\ a_{21} & a_{22} \end{bmatrix} \begin{Bmatrix} V'_1 \\ V'_2 \end{Bmatrix} \tag{A1.10}$$

Equation (A1.10) now can be written in the index notation as

$$V_i = a_{ij} V'_j \tag{A1.11}$$

Thus it has been established that the components V_i and V'_j of a vector V in any two right-handed coordinate systems are related by Eq. (A1.6) or, equivalently, Eq. (A1.11).

The laws of transforming components of stress in a coordinate system to the components in another coordinate system may be established by considering equilibrium of an appropriate region whose boundaries are parallel to the axes in the two coordinate systems. Details of the procedure may be obtained in Myklestad [1] and Hodge [2]. The resulting transformation laws may be written in the index notation as

$$\sigma'_{ij} = a_{mi} a_{nj} \sigma_{mn} \tag{A1.12}$$

or, alternatively,

$$\sigma_{mn} = a_{mi} a_{nj} \sigma'_{ij} \tag{A1.13}$$

where σ_{ij} are the components in the $x_1 x_2$ coordinate system and σ'_{ij} in the $x'_1 x'_2$ coordinate system. The two coordinate systems are the same as indicated in earlier discussions and shown in Fig. A1-5. The transformation matrix (a_{ij}) is the same as in the case of vector transformation [Eq. (A1.9)].

Stress is an example of a second-order tensor. The transformation laws represented by Eqs. (A1.12) and (A1.13) are applicable to all second-order tensors; that is, σ_{ij} and σ'_{ij} may represent components of any second-order tensor in the coordinate systems.

The transformation laws for the first-order and second-order tensors [given by Eqs. (A1.6) and (A1.11)–(A1.13)] have been obtained for a two-dimensional case. Equations may be derived in a similar manner for a three-dimensional case also. However, in the index notations the transformation equations for tensors in three dimensions are identical to those in two dimensions, but the subscripts now may assume values 1, 2, and 3 instead of just 1 and 2. Thus the transformation matrix now consists of nine elements because there are nine angles between the axes of the two coordinate systems. The transformation matrix is

$$(a_{ij}) = \begin{bmatrix} \cos \theta_{11} & \cos \theta_{12} & \cos \theta_{13} \\ \cos \theta_{21} & \cos \theta_{22} & \cos \theta_{23} \\ \cos \theta_{31} & \cos \theta_{32} & \cos \theta_{33} \end{bmatrix} \qquad (A1.14)$$

where the definition of angles has been illustrated in Fig. A1-6 through the angles in the first column. Thus the three columns of the transformation matrix give the direction cosines of the three axes in the x'_i coordinate system with respect to the x_i coordinate system. The nine elements of the transformation matrix are not independent, however. The relations between them exist because both coordinate systems are Cartesian coordinates with mutually perpendicular axes. The nine direction cosines may be determined in terms of the three Eulerian angles. The procedure is quite involved, however, and an

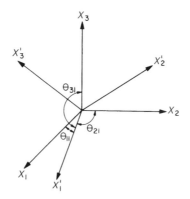

Figure A1-6. Definition of transformation angles in three dimensions.

interested reader may refer to a standard text on the subject such as that by Myklestad [1]. General rotation of the coordinate axes is not considered in this book. For the cases of rotation about one of the coordinate axes, the transformation matrix may be obtained directly by the definition [Eq. (A1.14)]. This is illustrated through some examples later in this section. It may be pointed out that the transformation matrix in three dimensions is also an orthogonal matrix like the one in two dimensions.

The transformation laws may be generalized to higher-order tensors also, and thus the following definition of an nth-order tensor may be given. A tensor of order n is a quantity **T** that satisfies the following two conditions:

1. In any right-handed coordinate system x_i, **T** is described by components $T_{i_1 i_2 \cdots i_n}$ (2^n components in two dimensions and 3^n components in three dimensions).
2. The components $T_{i_1 i_2 \cdots i_n}$ and $T'_{i_1 i_2 \cdots i_n}$ in any two right-handed coordinate systems are related by

$$T'_{j_1 j_2 \cdots j_n} = a_{i_1 j_1} a_{i_2 j_2} \cdots a_{i_n j_n} T_{i_1 i_2 \cdots i_n} \tag{A1.15}$$

or, equivalently, by

$$T_{i_1 i_2 \cdots i_n} = a_{i_1 j_1} a_{i_2 j_2} \cdots a_{i_n j_n} T'_{j_1 j_2 \cdots j_n} \tag{A1.16}$$

It now may be observed that the scalar quantities also satisfy the preceding definition of a tensor. Thus a scalar is a tensor of order zero. The tensor transformation equation for a third-order tensor is

$$T'_{ijk} = a_{li} a_{mj} a_{nk} T_{lmn} \tag{A1.17}$$

or

$$T_{lmn} = a_{li} a_{mj} a_{nk} T'_{ijk} \tag{A1.18}$$

and for a fourth-order tensor is

$$T'_{ijkl} = a_{pi} a_{qj} a_{rk} a_{sl} T_{pqrs}$$

or

$$T_{pqrs} = a_{pi} a_{qj} a_{rk} a_{sl} T'_{ijkl}$$

Now some applications of tensor transformations of interest are illustrated through examples.

Example A1-22: Write Eq. (A1.12) in its expanded form.

$$\sigma'_{ij} = a_{mi}a_{nj}\sigma_{mn}$$

First, summing over the dummy subscripts m,

$$\sigma'_{ij} = a_{1i}a_{nj}\sigma_{1n} + a_{2i}a_{nj}\sigma_{2n}$$

Summing over the dummy subscript n,

$$\sigma'_{ij} = a_{1i}a_{1j}\sigma_{11} + a_{1i}a_{2j}\sigma_{12} + a_{2i}a_{1j}\sigma_{21} + a_{2i}a_{2j}\sigma_{22}$$

Now the live subscripts i and j may be assigned values in four pairs as (1, 1), (1, 2), (2, 1), and (2, 2) to yield the desired equations:

$$\sigma'_{11} = a_{11}a_{11}\sigma_{11} + a_{11}a_{21}\sigma_{12} + a_{21}a_{11}\sigma_{21} + a_{21}a_{21}\sigma_{22}$$
$$\sigma'_{12} = a_{11}a_{12}\sigma_{11} + a_{11}a_{22}\sigma_{12} + a_{21}a_{21}\sigma_{21} + a_{21}a_{22}\sigma_{22}$$
$$\sigma'_{21} = a_{12}a_{11}\sigma_{11} + a_{12}a_{21}\sigma_{12} + a_{22}a_{11}\sigma_{21} + a_{22}a_{21}\sigma_{22}$$
$$\sigma'_{22} = a_{12}a_{12}\sigma_{11} + a_{12}a_{22}\sigma_{12} + a_{22}a_{12}\sigma_{21} + a_{22}a_{22}\sigma_{22}$$

It may be noted that if σ_{ij} is symmetric in the x_i coordinate system, σ'_{ij} will be symmetric in any x'_i coordinate system.

Example A1-23: Write Eq. (A1.12) in the matrix form, and then, taking the definition of the elements (a_{ij}) from Eq. (A1.4), carry out the matrix multiplication to obtain four equations relating σ'_{ij} and σ_{ij}.

It may be noted that the cosine factors a_{ij} and stress components σ_{ij} can be written separately in the matrix form. However, to obtain the expanded form of the right-hand side of Eq. (A1.12) through matrix multiplication, it will have to be written in a form consistent with the laws of matrix multiplication, as discussed earlier in this appendix. A careful observation will show that for this purpose, Eq. (A1.12) needs to be written as

$$\sigma'_{ij} = (a_{mi})^T(\sigma_{mn})(a_{nj})$$

Writing in full:

$$\begin{bmatrix} \sigma'_{11} & \sigma'_{12} \\ \sigma'_{21} & \sigma'_{22} \end{bmatrix} = \begin{bmatrix} a_{11} & a_{21} \\ a_{12} & a_{22} \end{bmatrix} \begin{bmatrix} \sigma_{11} & \sigma_{12} \\ \sigma_{21} & \sigma_{22} \end{bmatrix} \begin{bmatrix} a_{11} & a_{12} \\ a_{21} & a_{22} \end{bmatrix}$$

Substituting Eq. (A1.4):

$$\begin{bmatrix} \sigma'_{11} & \sigma'_{12} \\ \sigma'_{21} & \sigma'_{22} \end{bmatrix} = \begin{bmatrix} \cos\theta & \sin\theta \\ -\sin\theta & \cos\theta \end{bmatrix} \begin{bmatrix} \sigma_{11} & \sigma_{12} \\ \sigma_{21} & \sigma_{22} \end{bmatrix} \begin{bmatrix} \cos\theta & -\sin\theta \\ \sin\theta & \cos\theta \end{bmatrix}$$

Now the matrix multiplication may be carried out to obtain

$$\sigma'_{11} = \sigma_{12}\cos^2\theta + (\sigma_{12} + \sigma_{21})\sin\theta\cos\theta + \sigma_{22}\sin^2\theta$$

$$\sigma'_{12} = (\sigma_{22} - \sigma_{11})\sin\theta\cos\theta + \sigma_{12}\cos^2\theta - \sigma_{21}\sin^2\theta$$

$$\sigma'_{21} = (\sigma_{22} - \sigma_{11})\sin\theta\cos\theta - \sigma_{12}\sin^2\theta + \sigma_{21}\cos^2\theta$$

$$\sigma'_{22} = \sigma_{11}\sin^2\theta - (\sigma_{12} + \sigma_{21})\sin\theta\cos\theta + \sigma_{22}\cos^2\theta$$

In view of the symmetry of the stress tensor ($\sigma_{12} = \sigma_{21}$), the preceding four equations reduce to the following equations:

$$\sigma'_{11} = \sigma_{11}\cos^2\theta + \sigma_{22}\sin^2\theta + 2\sigma_{12}\sin\theta\cos\theta$$

$$\sigma'_{22} = \sigma_{11}\sin^2\theta + \sigma_{22}\cos^2\theta - 2\sigma_{12}\sin\theta\cos\theta$$

$$\sigma'_{12} = \sigma'_{21} = -\sigma_{11}\sin\theta\cos\theta + \sigma_{22}\sin\theta\cos\theta + \sigma_{12}(\cos^2\theta - \sin^2\theta)$$

The preceding transformation equations are frequently written in the matrix form as

$$\begin{Bmatrix} \sigma'_{11} \\ \sigma'_{22} \\ \sigma'_{12} \end{Bmatrix} = \begin{bmatrix} \cos^2\theta & \sin^2\theta & 2\sin\theta\cos\theta \\ \sin^2\theta & \cos^2\theta & -2\sin\theta\cos\theta \\ -\sin\theta\cos\theta & \sin\theta\cos\theta & \cos^2\theta - \sin^2\theta \end{bmatrix} \begin{Bmatrix} \sigma_{11} \\ \sigma_{22} \\ \sigma_{12} \end{Bmatrix}$$

The matrix in the preceding equation is frequently referred to as the *transformation matrix*. However, it has nothing to do with the tensor transformation matrix in three dimensions that is given by Eq. (A1.14). It is purely a coincidence that the transformation equations for a second-order two-dimensional *symmetric* tensor can be written in the preceding form. The array of (σ_{11}, σ_{22}, σ_{12}) cannot be regarded as a vector or a three-dimensional first-order tensor. It always should be borne in mind that the preceding matrix equation has been derived from the general transformation equation for a two-dimensional second-order tensor and has been written in the present form for convenience only.

It is interesting to note that the inverse of the preceding transformation matrix is obtained by replacing θ by $-\theta$ (see Example A1-21) so that the transformation equations may be written as

$$\begin{Bmatrix} \sigma_{11} \\ \sigma_{22} \\ \sigma_{12} \end{Bmatrix} = \begin{bmatrix} \cos^2 \theta & \sin^2 \theta & -2 \sin \theta \cos \theta \\ \sin^2 \theta & \cos^2 \theta & 2 \sin \theta \cos \theta \\ \sin \theta \cos \theta & -\sin \theta \cos \theta & \cos^2 \theta - \sin^2 \theta \end{bmatrix} \begin{Bmatrix} \sigma'_{11} \\ \sigma'_{22} \\ \sigma'_{12} \end{Bmatrix}$$

This equation can be obtained directly by expansion of Eq. (A1.13). An interested reader easily may find from arguments based on the geometry of the two coordinate systems why the inverse of the transformation matrix is obtained by replacing θ by $-\theta$.

Example A1-24: Obtain the transformation matrix when the x'_i coordinate axes are obtained by a rotation of an angle θ of the x_i coordinate axes about the x_3 axis as shown in Fig. A1-7.

The angles between the axes of the two coordinate systems may be noted from the figure as

$$\theta_{11} = \theta \qquad \theta_{12} = 90° + \theta \qquad \theta_{13} = 90°$$
$$\theta_{21} = 90° - \theta \qquad \theta_{22} = \theta \qquad \theta_{23} = 90°$$
$$\theta_{31} = 90° \qquad \theta_{32} = 90° \qquad \theta_{33} = 0°$$

Therefore, the transformation matrix becomes

$$(a_{ij}) = \begin{bmatrix} \cos \theta & -\sin \theta & 0 \\ \sin \theta & \cos \theta & 0 \\ 0 & 0 & 1 \end{bmatrix}$$

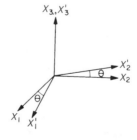

Figure A1-7. Rotation of coordinate axes about x_3 axis.

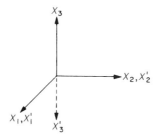

Figure A1-8. Definition of coordinate systems for Example A1-25.

Example A1-25: Obtain the transformation matrix when the x_i' coordinate axes are obtained by reversing the direction of the x_3 axis in the x_i coordinate axes, as shown in Fig. A1-8.

The required angles may be noted from Fig. A1-8 as

$$\theta_{11} = 0° \quad \theta_{12} = 90° \quad \theta_{13} = 90°$$
$$\theta_{21} = 90° \quad \theta_{22} = 0° \quad \theta_{23} = 90°$$
$$\theta_{31} = 90° \quad \theta_{32} = 90° \quad \theta_{33} = 180°$$

The transformation matrix becomes

$$(a_{ij}) = \begin{bmatrix} 1 & 0 & 0 \\ 0 & 1 & 0 \\ 0 & 0 & -1 \end{bmatrix}$$

REFERENCES

1. N. O. Myklestad, *Cartesian Tensors,* Van Nostrand, Princeton, NJ, 1967.
2. P. G. Hodge, Jr., *Continuum Mechanics,* McGraw-Hill, New York, 1970.

APPENDIX 2

EQUATIONS OF THEORY OF ELASTICITY

ANALYSIS OF STRAIN

Whenever the relative positions of points in a body change from any cause, the body is in a state of deformation. The problem of determining the relative change in the positions of the points is purely geometric. Neither the causes that give rise to the deformations nor the laws according to which the body resists it are of any importance to its study.

Discussion in this appendix is confined to only small deformations such as those that occur commonly in engineering structures and are of primary interest in this book. The small displacement of a particle, such as P in Fig. A2-1, can be resolved into components u, v, and w parallel to the coordinate axes x, y, and z, respectively. Consider a line element PA of length Δx originally lying parallel to the x axis. The displacement in the x direction of the point A, accurate to the first order in Δx, is

$$u + \frac{\partial u}{\partial x} \Delta x$$

The increase in length of the element PA owing to deformation is thus $(\partial u/\partial x) \Delta x$. The change in length–initial-length ratio of a straight-line element is defined as the longitudinal strain ϵ. Hence the strain at point P in the x direction is $\partial u/\partial x$. By considering the line elements originally lying parallel to the y and z axes, it can be shown that the strains in the y and z directions are given by $\partial v/\partial y$ and $\partial w/\partial z$.

The shearing strain γ is defined as the decrease in value of the initial right angle between two line elements. To obtain an expression for the shearing strain, consider the distortion of the angle between the line elements PA and PB, as shown in Fig. A2-2. The displacements of the point A in the y direction

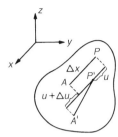

Figure A2-1. Deformation of a line segment.

and that of the point B in the x direction are $v + (\partial v/\partial x)\, \Delta x$ and $u + (\partial u/\partial y)\, \Delta y$, respectively. The deformed line element $P'A'$ is inclined to the initial direction of PA by the small angle α indicated in the figure, equal to $\partial v/\partial x$. In the same manner, the angle between $P'B'$ and PB is $\beta = (\partial u/\partial y)$. Thus the initial right angle APB is diminished by the angle $(\partial v/\partial x) + (\partial u/\partial y)$. This is the shearing strain γ_{xy} between the planes xz and yz. The shearing strains between the planes xy and xz and the planes yx and yz can be obtained in the same manner.

Thus the following six equations can be written to relate the six components of strain to the three components of displacement:

$$\epsilon_x = \frac{\partial u}{\partial x} \qquad \epsilon_y = \frac{\partial v}{\partial y} \qquad \epsilon_z = \frac{\partial w}{\partial z}$$
$$\gamma_{yz} = \frac{\partial v}{\partial z} + \frac{\partial w}{\partial y} \qquad \gamma_{zx} = \frac{\partial u}{\partial z} + \frac{\partial w}{\partial x} \qquad \gamma_{xy} = \frac{\partial u}{\partial y} + \frac{\partial v}{\partial x} \tag{A2.1}$$

These equations are often called the *strain–displacement relations*.

The longitudinal and shearing strains can be found in any direction in a manner such as that illustrated in the preceding paragraphs. However, when the six strain components [defined by Eq. (A2.1)] are known, the strain in an arbitrary direction can be obtained in terms of the strains ϵ_x, ϵ_y, ϵ_z, γ_{yz}, γ_{xz}, and γ_{xy} and the angles between the arbitrarily chosen direction and the x, y,

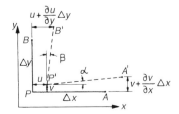

Figure A2-2. Deformation of two orthogonal line segments.

and z axes. In a two-dimensional case, the strain ϵ'_x, ϵ'_y, and γ'_{xy} along the x' and y' axes (shown in Fig. A2-3) can be obtained easily as

$$\epsilon'_x = \epsilon_x \cos^2 \theta + \epsilon_y \sin^2 \theta + \gamma_{xy} \sin \theta \cos \theta$$

$$\epsilon'_y = \epsilon_x \sin^2 \theta + \epsilon_y \cos^2 \theta - \gamma_{xy} \sin \theta \cos \theta \quad \text{(A2.2)}$$

$$\gamma'_{xy} = 2(\epsilon_y - \epsilon_x) \sin \theta \cos \theta + \gamma_{xy}(\cos^2 \theta - \sin^2 \theta)$$

Similar transformation equations can be obtained for a three-dimensional case. A derivation of the transformation equations and other details relating to this appendix can be found in a standard text on the theory of elasticity [1, 2].

It may be noted that the transformations here [Eqs. (A2.2)] are not identical to the transformation equations for a two-dimensional, second-order tensor as obtained in Appendix 1. However, if a factor of $\frac{1}{2}$ is associated with the shearing strains, Eq. (A2.2) may be rewritten as

$$\epsilon'_x = \epsilon_x \cos^2 \theta + \epsilon_y \sin^2 \theta + (\tfrac{1}{2}\gamma_{xy}) 2 \sin \theta \cos \theta$$

$$\epsilon'_y = \epsilon_x \sin^2 \theta + \epsilon_y \cos^2 \theta - (\tfrac{1}{2}\gamma_{xy}) 2 \sin \theta \cos \theta \quad \text{(A2.3)}$$

$$(\tfrac{1}{2}\gamma'_{xy}) = (\epsilon_y - \epsilon_x) \sin \theta \cos \theta + (\tfrac{1}{2}\gamma_{xy})(\cos^2 \theta - \sin^2 \theta)$$

The transformation equations [Eq. (A2.3)] are now identical to those for a two-dimensional, second-order tensor. It can be shown easily that the three-dimensional strain components also obey the transformation equations for a second-order tensor, provided that a factor of $\frac{1}{2}$ is associated with the shearing strains. Thus a second-order, symmetric strain tensor can be defined as follows:

$$\epsilon_{ij} = \begin{pmatrix} \epsilon_x & \tfrac{1}{2}\gamma_{xy} & \tfrac{1}{2}\gamma_{xz} \\ \tfrac{1}{2}\gamma_{xy} & \epsilon_y & \tfrac{1}{2}\gamma_{yz} \\ \tfrac{1}{2}\gamma_{xz} & \tfrac{1}{2}\gamma_{yz} & \epsilon_z \end{pmatrix} \quad \text{(A2.4)}$$

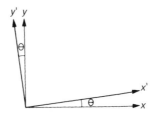

Figure A2-3. Definition of coordinate axes.

where i and j can assume values 1, 2, and 3 and the directions 1, 2, and 3 correspond to the axes x, y, and z, respectively. The strains ϵ_{ij}, defined by Eq. (A2.4), are called the *tensorial strains*, whereas the strains ϵ_x, ϵ_y, ϵ_z, γ_{xy}, γ_{xz}, and γ_{yz}, defined by Eq. (A2.1), are the *engineering strains*. It may be noted that the difference between the engineering and tensorial strains arises only in the shearing strains. A tensorial shearing strain is half the corresponding engineering shearing strain. Tensorial strains may be written in terms of the displacement gradients by substituting Eq. (A2.1) into Eq. (A2.4) so that

$$\epsilon_{ij} = \tfrac{1}{2}(u_{i,j} + u_{j,i}) \qquad (A2.5)$$

where the comma denotes partial differentiation.

Strain–displacement relations represent six strain components expressed in terms of three displacements, u, v, and w. If the strain components are given functions of x, y, and z, the relations are a system of six partial differential equations for the determination of the three displacements. Further, if the strains are prescribed arbitrarily, the six equations cannot, in general, be expected to yield single-values, continuous solutions for u, v, and w. Hence certain conditions of noncontradiction must exist among the strains. These are known as *conditions of strain compatibility*. These can be derived by eliminating u, v, and w from Eq. (A2.1) through partial differentiation of strains and algebraic manipulation and with the knowledge that the displacements u, v, and w are continuous and well-behaved functions, and hence their order of differentiation is immaterial. In the case of a three-dimensional strain field, the following six equations represent the compatibility conditions:

$$\begin{aligned}
\frac{\partial^2 \gamma_{xy}}{\partial x \, \partial y} &= \frac{\partial^2 \epsilon_x}{\partial y^2} + \frac{\partial^2 \epsilon_y}{\partial x^2} \\[4pt]
\frac{\partial^2 \gamma_{yz}}{\partial y \, \partial z} &= \frac{\partial^2 \epsilon_y}{\partial z^2} + \frac{\partial^2 \epsilon_z}{\partial y^2} \\[4pt]
\frac{\partial^2 \gamma_{zx}}{\partial z \, \partial x} &= \frac{\partial^2 \epsilon_z}{\partial x^2} + \frac{\partial^2 \epsilon_x}{\partial z^2} \\[4pt]
\frac{\partial^2 \epsilon_x}{\partial y \, \partial z} &= \frac{1}{2} \frac{\partial}{\partial x} \left(-\frac{\partial \gamma_{yz}}{\partial x} + \frac{\partial \gamma_{zx}}{\partial y} + \frac{\partial \gamma_{xy}}{\partial z} \right) \\[4pt]
\frac{\partial^2 \epsilon_y}{\partial z \, \partial x} &= \frac{1}{2} \frac{\partial}{\partial y} \left(\frac{\partial \gamma_{yz}}{\partial x} - \frac{\partial \gamma_{zx}}{\partial y} + \frac{\partial \gamma_{xy}}{\partial z} \right) \\[4pt]
\frac{\partial^2 \epsilon_z}{\partial x \, \partial y} &= \frac{1}{2} \frac{\partial}{\partial z} \left(\frac{\partial \gamma_{yz}}{\partial x} + \frac{\partial \gamma_{zx}}{\partial y} - \frac{\partial \gamma_{xy}}{\partial z} \right)
\end{aligned} \qquad (A2.6)$$

ANALYSIS OF STRESS

When external forces are applied to a solid body, their effect is transmitted throughout the body by producing internal forces. The magnitude of internal forces at a point P of a body in equilibrium under the action of external forces may be studied by considering an elemental area ΔA around P, as shown in Fig. A2-4. The forces acting across this elemental area, because of the action of the external forces on the body, can be reduced to a resultant ΔT. The ratio $\Delta T/\Delta A$ gives the average force on the area ΔA. The limiting value of the ratio $\Delta T/\Delta A$ as ΔA approaches zero gives the intensity of the internal force at P and is called the *stress*. The limiting direction of the resultant ΔT is the direction of the stress. In general, the direction of ΔT is inclined to the area ΔA, and thus the resultant ΔT can be resolved into components parallel and perpendicular to the area. The intensity of the force component perpendicular to the area is called the *normal stress* and is denoted by σ. The intensity of the parallel force component is called the *shearing stress* and is denoted by τ.

When the orientation of ΔA is changed, the resultant force ΔT acting on the area will, in general, be different. The new resultant force cannot be calculated from the knowledge of the resultant force on only one orientation of the elemental area. Thus the description of the state of stress or the intensity of internal force at P is incomplete when the intensity of force at a point is known only for one orientation of the area ΔA. It can be shown that in a three-dimensional continuum, the state of stress at a given point can be described completely when the magnitude of stresses on any three mutually perpendicular planes passing through the point is known. That is, the stress components on an arbitrarily oriented plane can be calculated from the magnitude of stresses on three mutually perpendicular planes. It is common to take the three planes parallel to the three reference planes in the Cartesian coordinates. Now the resultant force ΔT acting on an elemental area ΔA parallel to one of the reference planes can be resolved into three components in the direction of the three reference axes, giving one normal and two shearing stresses. The stress components are represented by symbols with two sub-

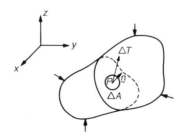

Figure A2-4. Internal forces at a point.

scripts such as σ_{ij}. By convention, the first subscript indicates the orientation of the plane (i.e., the direction of an outward normal to the plane), and the second subscript indicates the direction of the force component. The three-dimensional stress components are shown in Fig. A2-5. It may be noted that a typical component σ_{xy} signifies the component in the y direction of the average force acting on a plane perpendicular to the x axis. It also may be pointed out that the normal stresses σ_{xx}, σ_{yy}, and σ_{zz} are sometimes represented by single-subscripted symbols such as σ_x, σ_y, and σ_z. The shearing-stress components are always denoted by the double-subscripted symbols but more often are denoted by τ, as in this book.

The positive directions of stress components are shown in Fig. A2-5. The sign convention can be stated in words as "on a plane where the outward normal is in the positive direction of a coordinate axis, all the stress components acting in the positive directions of the axes are positive." According to this convention, when the outward normal is in the negative direction of a coordinate axis, the stress components are positive when they act in the negative directions of the axes. Therefore, the stress components on the hidden faces of the parallelopiped shown by dotted lines in Fig. A2-5 are also positive. This sign convention is consistent with the convention of taking normal tensile stress to be positive. This sign convention is almost universally accepted.

The solid body considered here is in static equilibrium under the action of external forces. Therefore, each elemental volume of the body considered separately also will be in equilibrium under the action of internal forces. Equilibrium conditions are obtained by making the resultant forces and moments acting on an elemental volume vanish. In the limit, when the dimensions of the elemental volume approach zero, the equilibrium conditions lead to what are known as the *equilibrium equations*. The equilibrium equations are partial differential equations relating variations of stress components in different directions. The most useful information concerning equilibrium conditions is obtained by considering equilibrium of a cubic volume element in

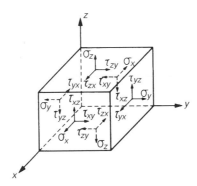

Figure A2-5. Three-dimensional stress components.

the interior of the body and a tetragonal volume element on the surface of the body. The results of an equilibrium analysis are summarized in the following paragraphs.

Consider a parallelopiped whose faces are parallel to the reference planes, as shown in Fig. A2-6. The stress components on the six faces are also shown in the figure. In the limit when the sides of the parallelopiped approach zero, the equilibrium of moments about the reference axes leads to the following symmetry conditions:

$$
\begin{aligned}
\tau_{xy} &= \tau_{yx} \\
\tau_{yz} &= \tau_{zy} \\
\tau_{zx} &= \tau_{xz}
\end{aligned}
\quad (A2.7)
$$

Thus there are only six independent stress components, namely, σ_x, σ_y, σ_z, τ_{xy}, τ_{yz}, and τ_{zx}. The state of stress at a given point is described completely by giving these six stress components.

Now the equilibrium of forces in the direction of the reference axes yields the following differential equations:

$$
\begin{aligned}
\frac{\partial \sigma_x}{\partial x} + \frac{\partial \tau_{yx}}{\partial y} + \frac{\partial \tau_{zx}}{\partial z} + F_x &= 0 \\
\frac{\partial \tau_{xy}}{\partial x} + \frac{\partial \sigma_y}{\partial y} + \frac{\partial \tau_{zy}}{\partial z} + F_y &= 0 \\
\frac{\partial \tau_{xz}}{\partial x} + \frac{\partial \tau_{yz}}{\partial y} + \frac{\partial \sigma_z}{\partial z} + F_z &= 0
\end{aligned}
\quad (A2.8)
$$

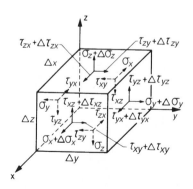

Figure A2-6. Equilibrium of a parallelopiped inside the body.

ANALYSIS OF STRESS   537

where F_x, F_y, and F_z are the x, y, and z components, respectively, of the intensity of body forces (e.g., the gravitational and magnetic forces). In many problems, the body forces are negligible compared with the externally applied forces. The term *equilibrium equations* is employed commonly to refer to Eq. (A2.8) only. Equation (A2.8) can be written in the index notation as

$$\sigma_{ji,j} + F_i = 0 \qquad (A2.9)$$

where, again, the comma refers to partial differentiation. Equilibrium of forces on a tetragonal volume element on the surface of the body, such as the one shown in Fig. A2-7, yields in the limit when the tetragon shrinks to a point P on the surface the following conditions:

$$\begin{aligned}
\sigma_x n_x + \tau_{yx} n_y + \tau_{zx} n_z &= T_x \\
\tau_{xy} n_x + \sigma_y n_y + \tau_{zy} n_z &= T_y \\
\tau_{xz} n_x + \tau_{yz} n_y + \sigma_z n_z &= T_z
\end{aligned} \qquad (A2.10)$$

where n_x, n_y, and n_z are the direction cosines of the outward normal to the surface at point P, and T_x, T_y and T_z are the components of external forces acting at point P. Equilibrium conditions [Eq. (A2.10)] are known as the *boundary conditions* because they relate internal stresses to the external surface forces. Equation (A2.10) can be written in the index notation as

$$\sigma_{ji} n_j = T_i \qquad (A2.11)$$

It was pointed out earlier that the components of stress in an arbitrary direction may be obtained from the six components of stress in a reference coordinate system. The transformation laws may be established by considering equilibrium of appropriate infinitesimal volume elements. In a two-

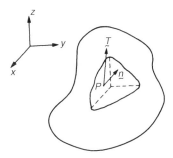

Figure A2-7. Equilibrium of a tetragonal surface volume element.

dimensional case, the following transformation equations may be obtained by considering equilibrium of triangular elements such as shown in Fig. A2-8:

$$\sigma'_x = \sigma_x \cos^2\theta + \sigma_y \sin^2\theta + 2\tau_{xy}\sin\theta\cos\theta$$
$$\sigma'_y = \sigma_x \sin^2\theta + \sigma_y \cos^2\theta - 2\tau_{xy}\sin\theta\cos\theta \quad\quad (A2.12)$$
$$\tau'_{xy} = (\sigma_y - \sigma_x)\sin\theta\cos\theta + \tau_{xy}(\cos^2\theta - \sin^2\theta)$$

Transformation equations for three-dimensional stress components also may be obtained in a similar manner. It will be observed that the transformation equations for stresses are the transformation equations for a second-order tensor as obtained in Appendix 1. The stress-transformation equations may be written in the index notation as

$$\sigma'_{ij} = a_{mi}a_{nj}\sigma_{mn} \quad\quad (A2.13)$$

where a_{ij} are the direction cosines of the x', y', and z' axes with respect to the x, y, and z axes [Eq. (A1.14)]. Thus the stress is a second-order symmetric tensor. Further, since no assumption was made concerning material properties while deriving Eq. (A2.7), the stress tensor is symmetric for all materials (e.g., isotropic, orthotropic, or anisotropic materials).

STRESS–STRAIN RELATIONS FOR ISOTROPIC MATERIALS

The generalized Hooke law was stated in Chap. 5 [Eq. (5.40)]. It relates stresses and strains in an anisotropic material through 21 independent elastic constants. Subsequently, it was shown that for an orthotropic material, the number of independent elastic constants reduces to 9 [Eq. (5.61)] because of the symmetry in material properties with respect to certain planes. In an isotropic material, all planes are planes of symmetry, and thus the properties are independent of direction. It can be shown that the number of independent elastic constants for an isotropic material further reduces to only 2. The

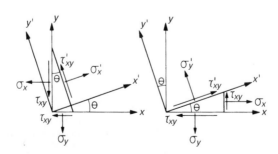

Figure A2-8. Equilibrium of two-dimensional triangular elements.

stress–strain relations for an isotropic material can be written in terms of engineering constants as

$$\epsilon_x = \frac{1}{E}[\sigma_x - \nu(\sigma_y + \sigma_z)]$$

$$\epsilon_y = \frac{1}{E}[\sigma_y - \nu(\sigma_x + \sigma_z)]$$

$$\epsilon_z = \frac{1}{E}[\sigma_z - \nu(\sigma_x + \sigma_y)] \quad \text{(A2.14)}$$

$$\gamma_{xy} = \frac{\tau_{xy}}{G}$$

$$\gamma_{yz} = \frac{\tau_{yz}}{G}$$

$$\gamma_{zx} = \frac{\tau_{zx}}{G}$$

where E is Young's modulus or elastic modulus, G is shear modulus or modulus of rigidity, and ν is the Poisson ratio. Out of the three material constants in Eq. (A2.14), only two are independent. The following equation relates the three constants with one another:

$$G = \frac{E}{2(1+\nu)} \quad \text{(A2.15)}$$

Equation (A2.14) may be solved to obtain the stress components as explicit functions of strain components as follows:

$$\sigma_x = 2G\left(\epsilon_x + \frac{\nu}{1-2\nu}e\right)$$

$$\sigma_y = 2G\left(\epsilon_y + \frac{\nu}{1-2\nu}e\right)$$

$$\sigma_z = 2G\left(\epsilon_z + \frac{\nu}{1-2\nu}e\right)$$

$$\tau_{xy} = G\gamma_{xy} \quad \text{(A2.16)}$$

$$\tau_{yz} = G\gamma_{yz}$$

$$\tau_{zx} = G\gamma_{zx}$$

where $e = \epsilon_x + \epsilon_y + \epsilon_z$.

Using the stress–strain relations [Eq. (A2.16)] and the strain–displacement relations [Eq. (A2.1)], the equilibrium equations [Eq. (A2.8)] can be derived in terms of displacements as

$$\nabla^2 \mathbf{u} + \frac{1}{1 - 2\nu} \nabla(\nabla \cdot \mathbf{u}) + \frac{1}{G} \mathbf{F} = 0 \qquad (A2.17)$$

where $\mathbf{u} = u\mathbf{i} + v\mathbf{j} + w\mathbf{k}$

$\mathbf{F} = F_x\mathbf{i} + F_y\mathbf{j} + F_z\mathbf{k}$

$\nabla = \frac{\partial}{\partial x}\mathbf{i} + \frac{\partial}{\partial y}\mathbf{j} + \frac{\partial}{\partial z}\mathbf{k}$

$\nabla^2 = \frac{\partial^2}{\partial x^2} + \frac{\partial^2}{\partial y^2} + \frac{\partial^2}{\partial z^2}$

$\mathbf{i}$, $\mathbf{j}$, and $\mathbf{k}$ are unit vectors in the direction of the reference axes x, y, and z, respectively. Equation (A2.17) may be written in index notations as

$$u_{i,jj} + \frac{1}{1 - 2\nu} u_{j,ji} + \frac{1}{G} F_i = 0 \qquad (A2.18)$$

For an imcompressible material, when $\nu = \frac{1}{2}$ and $e = 0$, Eqs. (A2.17) and (A2.18) have to be modified by writing e ($= \nabla \cdot \mathbf{u}$) in terms of the mean stress $\frac{1}{3}(\sigma_x + \sigma_y + \sigma_z)$. Equations (A2.17) and (A2.18) are quite useful forms of the equilibrium equations for many problems of interest.

The purpose of solving an elasticity problem is to obtain stresses, strains, and displacements at every point in a body subjected to known boundary loads or displacements. Therefore, in a three-dimensional problem there are a total of 15 unknowns (6 stresses, 6 strains, and 3 displacements). The 15 unknowns can be solved for by using the elasticity equations discussed earlier. Specifically, the 15 equations needed for the solution are the 3 equilibrium equations [Eq. (A2.8)], 6 stress–strain relations [Eq. (A2.14) or (A2.16)], and 6 strain–displacement relations [Eq. (A2.1)]. Note that the compatibility conditions are not considered as the equations to solve for the unknowns. They are the constraints on the strains and are used to eliminate arbitrariness in the solution for strains so that a unique solution for displacements is obtained. An elasticity problem may be viewed as having only 9 unknowns (6 stresses and 3 displacements) that may be solved for by using 3 equilibrium equations and 6 stress–displacement gradient relations. In this case, the strain–displacement relations are used to obtain the six strain components, and thus the compatibility conditions do not come into the picture because they are identically satisfied. In addition to the elasticity equations, 3 boundary conditions [Eq. (A2.10)] must be satisfied for a complete solution to the problem.

In fact, the elasticity equations are the same for all the problems, and what makes one problem different from another is the geometry of the body and the boundary conditions. Depending on the geometry and the boundary conditions, simplifying assumptions can be made and solution techniques suitable for different classes of problems developed. A large number of three-dimensional elasticity problems may be reduced to two- or one-dimensional problems and their solution obtained by solving a reduced number of equations. A discussion, however, on the solution techniques of various classes of problems is beyond the scope of this book. An interested reader may, for this purpose, refer to a standard text on the theory of elasticity [1, 2].

REFERENCES

1. S. P. Timoshenko and J. N. Goodier, *Theory of Elasticity*, 3d ed., McGraw-Hill, New York, 1970.
2. I. S. Sokolnikoff, *Mathematical Theory of Elasticity*, 2d ed., McGraw-Hill, New York, 1956.

APPENDIX 3

LAMINATE ORIENTATION CODE

Each laminate is unique in its properties and characteristics and hence must be distinctly identified whenever it is to be associated with specific quantitative or numerical data. Positive and concise identification of a laminate can be achieved through the use of a laminate orientation code. An adequate code must be able to specify as concisely as possible (1) the orientation of each lamina relative to a reference axis (the x axis in the text), (2) the number of laminae at each orientation, and (3) the exact geometric sequence of laminae. In the Standard Laminate Code, which is described in this appendix, it is assumed that all laminae are identical in thickness and properties. Special notations are used to indicate hybrid laminates.

The Standard Laminate Code is best defined by the following detailed description of its features.

STANDARD CODE ELEMENTS

1. Each lamina is denoted by a number representing the angle in degrees between its fiber direction and the x axis.
2. Individual adjacent laminae are separated in the code by a slash if their angles are different.
3. The laminae are listed in sequence from one laminate face to the other, starting with the first lamina laid up, with brackets indicating the beginning and end of the code.
4. Adjacent laminae of the same orientation are denoted by a numerical subscript.

Laminate	Code
45°	
0°	[45/0/45/90$_2$/30]
45°	
90°	
90°	
30°	

POSITIVE AND NEGATIVE ANGLES

When adjacent laminae are oriented at angles equal in magnitude but opposite in sign, the appropriate use of plus (+) and minus (−) signs is employed. Each + or − sign represents one lamina. Note that a numerical subscript is used only when the angles are of the same sign. Convention for the positive and negative angles should be consistent with the coordinate system chosen. This means that an orientation denoted positive in a right-handed coordinate system may be negative in another right-handed coordinate system. This is illustrated in Fig. A3-1, in which the signs of ±45° laminae are reversed when the directions of y and z axes are reversed, whereas the x axis remains unaltered.

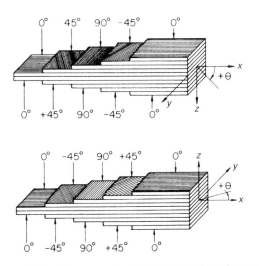

Figure A3-1. Sign convention for laminate orientation code.

Laminate	Code
45° / −45° / −30° / 30° / 0°	$[\pm 45/\mp 30/0]$
45° / 0° / −60° / −60° / 30°	$[45/0/-60_2/30]$
45° / 45° / −45° / −45° / 0°	$[45_2/-45_2/0]$
45° / −45° / −45° / 45° / 45° / −45° / 0°	$[\pm\mp\pm 45/0]$

SYMMETRIC LAMINATES

The laminates possessing symmetry of laminae orientations about the geometric midplane require specifying only half the stacking sequence. Symmetric laminates with an even number of laminae list the laminae in sequence, starting at one face, but stopping at the plane of symmetry instead of continuing to the other face. The subscript S to the bracket indicates that only one-half the laminate is shown, with the other half symmetric about the midplane.

Laminate	Code
90°	
0°	$[90/0_2/45]_S$
0°	
45°	
45°	
0°	
0°	
90°	

Symmetric laminates with an odd number of laminae are coded the same as even symmetric laminates, except that the center lamina, listed last, is overlined to indicate that half of it lies on either side of the plane of symmetry:

Laminate	Code
0°	
45°	$[0/45/\overline{90}]_S$
90°	
45°	
0°	

SETS

Repeating sequences of laminae are called *sets* and are enclosed in parentheses. A set is coded in accordance with the same rules that apply to a single lamina.

Laminate	Code
45°	
0°	
90°	$[(45/0/90)_2]_S$
45°	or
0°	$[45/0/90]_{2S}$
90°	
90°	

Laminate	Code
0°	
45°	
90°	
0°	
45°	
45°	
0°	
90°	$[(45/0/90)_4]$
45°	or
0°	$[45/0/90]_4$
90°	
45°	
0°	
90°	
45°	
0°	
90°	

HYBRID LAMINATES

When referring to hybrid laminates, the standard code is partially modified. The modification generally consists of augmented subscripts to the normal lamina angle callouts that designate not only the number of laminae at each angle but also the generic fiber material of each. (Note that it is not the function of the code to define specific material systems, either for hybrid or nonhybrid laminates.) The code for hybrids is illustrated as follows:

Laminate*		Code
0°	B/Ep	
45°	Gr/Ep	
−45°	Gr/Ep	$[0_B / \pm 45_{Gr} / 90_{Gr}]_S$
90°	Gr/Ep	
90°	Gr/Ep	
−45°	Gr/Ep	

45°	Gr/Ep	
0°	B/Ep	
0°	B/Ep	
0°	B/Ep	$[0_{2B}/45_{Gr}/\overline{90}_{Gr}]_S$
45°	Gr/Ep	
90°	Gr/Ep	
45°	Gr/Ep	
0°	B/Ep	
0°	B/Ep	

*B = boron; Gr = graphite; Ep = epoxy.

APPENDIX 4

PROPERTIES OF FIBER COMPOSITES

Physical, mechanical, and hygrothermal properties of 10 commercial composites are given in Table A4-1. Eight of these composites are unidirectional, whereas the other two are fabric-reinforced. Fiber volume fraction ranges from 45–70%. The properties given here can be used for initial design purposes. However, final properties may be different owing to manufacturing variables and should be evaluated by experimental measurements.

Stress–strain curves for carbon–epoxy (T300/N5208) laminates are shown in Figs. A4-1–A4-6. The curves for the unidirectional laminates (Figs. A4-1 and A4-2) are linear to failure because they are controlled by the fibers. The in-plane shear–stress–strain curve (Fig. A4-3) is nonlinear because it is dominated by the matrix properties. In the transverse direction, the tensile curve (Fig. A4-4) is linear, whereas the compression curve (Fig. A4-5) is nonlinear because it can reach large strain values. The longitudinal tension and compression stress-strain curves for [±45] angle-ply laminate (Fig. A4-6) are nonlinear because the matrix is again dominant when the applied stress is not in the fiber directions.

Table A4-1 Properties of some commercial fiber composites

Material Description[a]	Fiber Volume Fraction V_f	Density (g/cm³)	Elastic Constants					Strengths					Hygrothermal Expansion Coefficients			
			E_L (GPa)	E_T (GPa)	ν_{LT}	G_{LT} (GPa)		σ_{LU} (MPa)	σ'_{LU} (MPa)	σ_{TU} (MPa)	σ'_{TU} (MPa)	τ_{LTU} (MPa)	α_L ($10^{-6}/°C$)	α_T ($10^{-6}/°C$)	β_L	β_T
Carbon–epoxy T300/N5208	0.70	1.60	181.0	10.30	0.28	7.17		1500	1500	40	246	68	0.02	22.5	0	0.6
Carbon–epoxy AS/H3501	0.66	1.60	138.0	8.96	0.30	7.10		1447	1447	51.7	206	93	−0.3	28.1	0	0.4
Carbon–PEEK AS4/APC2	0.66	1.60	134.0	8.90	0.28	5.10		2130	1100	80	200	160	—	—	—	—
Carbon–epoxy IM6/epoxy	0.66	1.60	203.0	11.20	0.32	8.40		3500	1540	56	150	98	—	—	—	—
Carbon–epoxy T300/Fiberite 934	0.60	1.50	148.0	9.65	0.30	4.55		1314	1220	43	168	48	—	—	—	—
Boron–epoxy B-4/N5505	0.50	2.00	204.0	18.50	0.23	5.59		1260	2500	61	202	67	6.10	30.30	0	0.6
Glass–epoxy E-glass–epoxy	0.45	1.80	38.6	8.27	0.26	4.14		1062	610	31	118	72	8.60	22.10	0	0.6
Aramid–epoxy Kevlar 49/epoxy	0.60	1.46	76.0	5.50	0.34	2.30		1400	235	12	53	34	−4.00	79.0	0	0.6
Carbon–epoxy T300/Fiberite 934 (13-mil)	0.60	1.50	74.0	74.0	0.05	4.55		499	352	458	352	46	—	—	—	—
Carbon–epoxy T-300/Fiberite 934 (7-mil)	0.60	1.50	66.0	66.0	0.04	4.10		375	279	368	278	46	—	—	—	—

[a]The first eight materials are unidirectional, while the last two are fabric reinforced.

550 PROPERTIES OF FIBER COMPOSITES

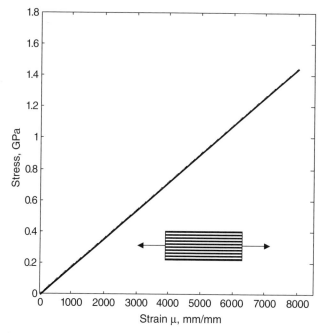

Figure A4-1. Longitudinal tensile stress–strain curve for a unidirectional graphite-fiber (T300)–epoxy laminate.

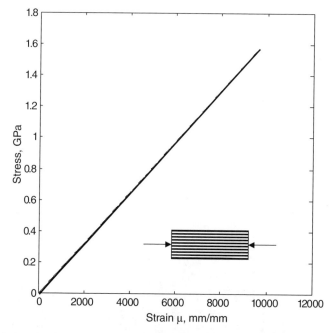

Figure A4-2. Longitudinal compression stress–strain curve for a unidirectional graphite-fiber (T300)–epoxy laminate.

PROPERTIES OF FIBER COMPOSITES **551**

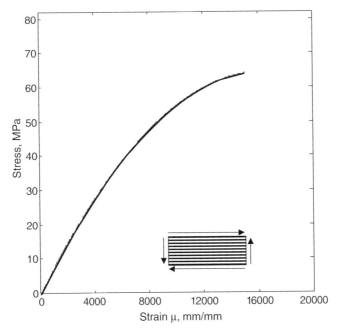

Figure A4-3. Shear stress–strain curve for a unidirectional graphite-fiber (T300)–epoxy laminate.

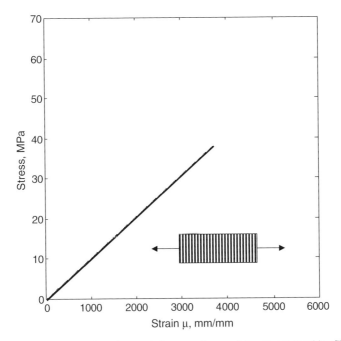

Figure A4-4. Transverse tension stress–strain curve for a unidirectional graphite-fiber (T300)–epoxy laminate.

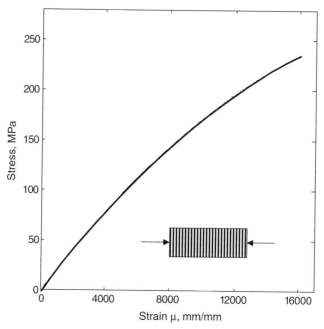

Figure A4-5. Transverse compression stress–strain curve for a unidirectional graphite-fiber (T300)–epoxy laminate.

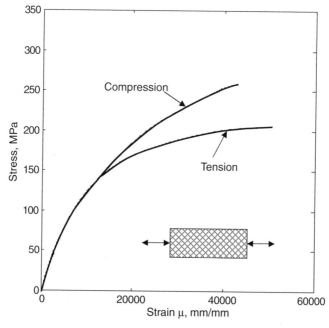

Figure A4-6. Longitudinal stress–strain curves for [±45] graphite–epoxy laminate.

APPENDIX 5

COMPUTER PROGRAMS FOR LAMINATE ANALYSIS

Laminate analysis calculations are performed with the help of computers for accurate results. There are two types of software programs available for the calculations. The first type are essentially mathematical tools for general calculations (e.g., MATLAB, Mathcad, and MAPLE). Matrix inversions and multiplications, vector and matrix manipulations, and obtaining numerical values using given formulas, which are frequently required for laminate analyses, can be carried out easily using these tools. These programs are easy to use and relatively inexpensive. Their use improves accuracy of results and minimizes probability of an error.

The second type of software includes the finite-element analysis (FEA) codes to carry out complete structural analysis, including acoustic and thermal analyses. The FEA codes can be used to solve one-, two- and three-dimensional static and dynamic, linear and nonlinear problems involving isotropic and anisotropic materials. A list of commercial software programs, along with their sources, is given in Table A5-1. These programs also have several other useful features, such as flexibility regarding the use of a failure theory, iterative procedure for design analysis and strength calculations, hygrothermal stress calculations, and excellent graphics for presentation of results. These software programs require greater training in their application.

Table A5-1 Commercially available software packages

Applications	Program Name	Source
Mathematical tools	MATLAB	The MathWorks 3 Apple Hill Drive Natick, MA 01760 *www.mathworks.com/*
	Mathcad	Mathsoft Engineering & Education, Inc. 101 Main Street Cambridge, MA 02142 *www.mathsoft.com/*
	MAPLE	Maplesoft 615 Kumpf Drive, Waterloo, Ontario, Canada N2V 1K8 *www.maplesoft.com/*
Finite-element analysis codes	ABAQUS	ABAQUS, Inc. 1080 Main Street Pawtucket, RI 02860 *www.abaqus.com/*
	ANSYS	ANSYS, Inc. 275 Technology Dr. Canonsburg, PA 15317 *www.ansys.com/*
	NASTRAN	MSC.Software Corporation 2 MacArthur Place Santa Ana, CA 92707 *www.mscsoftware.com/*
	COSMOS	SolidWorks Corporation 300 Baker Avanue Concord MA 01742 *www.solidworks.com/*
	LS-DYNA	Livermore Software Technology Corporation 7374 Las Positas Road Livermore, CA 94551 *www.lstc.com/*

INDEX

A-scan, 482
Acoustic emission, 483
Aelotropic material, 175
Anisotropic material, 175
Anisotropy, 3
   controlled, 10, 11
   of thermal expansion, 111
Aramid fibers, see Kevlar
Autoclave, 46, 47, 48
Average fiber stress, 139–140
Average stress criterion, 349
Axes of symmetry, 63, 160

B-scan, 482
Bag molding, 46–48
   pressure, 46, 47
   processes, 46–48
   vacuum, 46, 47
Bagging, 46
Balanced orthotropic lamina, 172
Bending
   of beams, 315
   of general laminates, 294
   of plates, 288
Biocomposites, 499–503
Bio fibers, 499–501
Bio product, 499
Bonded-fastened joints, 361
Bonding mechanisms, 355–356
Boron fibers, 27–28
   by CVD, 27
   properties of, 28
Boundary-layer phenomenon, 328
Buckling, 296
   load, 296
      critical, 298
   mode, 296
   of plates, 295
   of square plate, 300
Burn off method, 441

C-scan, 483
Carbon black, 6
Carbon-carbon composites, 498–499

Carbon fibers, 23–26 also see graphite fibers
   carbon yield, 24
   form of, 26
   PAN, 24
   precursor, 24
   process of making, 24, 35
   properties of, 25, 26
   roving of, 26
   yarn of, 26
Cermets, 6
   oxide-based, 6
   carbide-based, 6
Ceramic fibers, 28–29
   Alumina, 28, 29
   Fibers FP, 28, 29
   properties of, 29
   silicon carbide, 28, 29
Charpy tests, 475
Classical lamination theory, 214
CLT, see classical lamination theory
COD, see crack-opening displacement
Code, laminate orientation, 544–548
Coefficient of thermal expansion, 111
Cold solders, 6
Combustion method, 441
Compliance curve, 467
Compliance matching procedure, 467
Compliance matrix, 184
   invariant form of, 194
   transformation of, 189
Compliance tensor, 184
Composite materials:
   applications of, 10–14
   bridge, 12–14
   characteristics of, 2–3
   classification of, 3–5
   consumption of, 12
   continuous fiber, 9
   cross-ply, 3
   definition of, 1–2
   degradation of, 426–429

   density of, 65–66, 440
   discontinuous fiber, 7, 9
   fabrication of, 41–60
   fiber-reinforced, 5, 7–10
   fibrous composite, 7–11
   growth of, 11
   hybrid, 9, 407–411
   multilayered, 7, 8
   particulate, 1, 5–7
   particle-reinforced, 5
   properties of, 2, 123–124, 550–554
   quasi-isotropic, 147
   ribbon reinforced, 152–155
   short-fiber, 132–155
   single layer, 7, 8, 9
   tape reinforced, see ribbon reinforced
   unidirectional, 3, 9, 62–131
   use in U.S. industries, 12
Compounding, 57
Compression molding, 56
Computer:
   laminate analysis through, 272
   programs, 550
   software, commercial, 556
Concentration, 2
Constant-stress model, 80
Constitutive equations, 192
   for laminates, 221, 224–225
Contact lay-up, 43
Continuous fiber reinforced composite, 9
Coupling agents, 16, 19
Coupling coefficients, see Cross-coefficients
Crack-driving force, 466
Crack extension, 342
Crack-extension force, 466
Crack-extension force curve, 466
Crack-growth resistance, 466
Crack-growth resistance curve, 466, 467
Crack-length-estimation curve, 467, 469

555

Crack-opening displacement, 467
Crack opening mode, 342
Creel, 49
Critical buckling load, 298
Critical fiber length, 136
Critical volume fraction, 75, 76, 146
Cross-coefficients, 168–170
Cross-linking, 42
Cumulative weakening, 78
Curing agent, 35
Curing stresses, 268

Damage:
 due to low velocity impact, 411–416
 fatigue, 368
 identification, 481–488
 initiation, 96
DCB see double-cantilever-beam
Debonding, 96, 370
 of fibers, 399
Degradation of composites, at elevated temperatures, 426–429
Delamination, 96
Delamination crack, 370, 373, 396, 401
DEN, see double edge notched
Density, 57–58, 440
Diffusion, mass, 106
Diffusivity, 120
 longitudinal, 120, 121
 measurement of, 443
 transverse, 120, 121
Discontinuous fiber reinforced composites,
 see Short-fiber composites
Double-cantilever-beam, 472, 473
Double edge notched specimen, 465
Drop-weight impact test, 476
Ductility index, 409

Edge delamination suppression, 335
Edge effects, 324–335
Effect of temperature and moisture on composite properties, 422–426
Effective modulus, 252
Elastic constants, 175, see also Engineering constants
 number of, 182
 relations with compliance matrix, 185
 relations with stiffness matrix, 185–187
 restrictions on, 187–189
 symmetry of, 175
 variation of, 170–172
Elasticity methods, 83
Electrical conductivity, 115, 116, 118
End effects, 78
End tabs, 445, 446
Energy absorbing mechanisms, 396
 fiber breakage, 397
 fiber debonding, 399
 matrix cracking, 398
 matrix deformation, 398
Energy curves, 471
Engineering constants, 160
 determination of, see Testing of composites
 extremum value of, 170
 for orthotropic lamina, 160, 161
 restrictions on, 187–189
 relations with compliance matrix, 185
 relations with stiffness matrix, 185–187
 transformation of, 166–172
 variation of, 170–172
 variations with fiber orientation, 169–172
Environmental interaction effects, 416–431
Epoxy, 36–38
 properties of, 37
Epoxy resin, 36
Equations of motion, 302
Experimental characterization, 431–495
 See also Testing of composites

Fabrication of composites, 41–60
 by bag molding, 46–48
 by contact lay-up, 43
 by hand lay-up, 43–45
 by resin transfer molding, 49
 by stamping, 57
 by thermoforming, 57, 58
 ceramic matrix, 59–60
 filament winding, 49–51
 metal matrix, 58–59
 molding compounds, 53
 prepregs, 55
 pultrusion, 51–53
 thermoplastic resin matrix, 55–58
 thermosetting resin matrix, 42–55
Failure:
 envelop, 204–205, 206
 initiation, 88, 96
  microscopic, 335
 internal, 96
 load, 97
 models, 78–79
 modes of, 96–108
 shear, 107, 109
Failure criteria, see also Failure theories
 for biaxial stress field, 197–205
 for notched composites, 349
 maximum distortion energy, 91
 Whitney-Nuismer, 349–355
Failure theories:
 maximum strain, 200–203
 maximum stress, 197–200
 maximum work, 203–205
 Tsai-Hill, 203
Fatigue, 368–395
 characterization of, 374–375
 crack arrest in, 370
 crack branching in, 370
 cross-ply cracks in, 371, 372
 damage, 370
 damage initiation, 370
 delamination crack, 370
 empirical relations for, 385–386
 factors influencing behavior, 378
 Goodman-Boller relationship, 384
 influence of mean stress, 383
 influence on properties, 375–377
 of high modulus fiber composites, 386–390
 of short fiber composites, 390–395
 schematic representation of, 373–374
 shear, 382–383
 S-N curve, 375
Fibers:
 advanced, 16–30
 aramid, 7, 26
 average stress on, 139
 boron, 27–28

INDEX 557

breaking of, 96, 100–101, 397
buckling of, 102, 103
carbon, 23–26
ceramic, 28
chopped, 10
critical length of, 136
crushing of, 107
debonding of, 399
end effects, 133
Fiber FP, 28
glass, 7, 16–23
  cake, 17
  composition, 18
  end, see strand
  production of, 17, 18
  properties, 18, 19
  roving, 21
  sizes, 17
  staple, 17, 18
  strand, 17
  surface treatment of, 18–20
  yield, 21
graphite, 7, 33–26
  see also Carbon-fibers
ineffective length of, 136
Kevlar, 26–27
load transfer length of, 136
man-made, 7
microbuckling of, 102
polyethylene, 28–30
properties of, 7, 8
pullout, 100–101
silicon carbide, 28
Spectra, 29
Fiber aspect ratio, 140
Fiber composites,
  applications of, 10–14
  properties of, 10
Fiber packing, 124
Fiber pullout, 396, 399–401
Fiber splitting, 106, 107
Fiber volume fraction
  minimum, 75, 146
  critical, 75, 146
Fiberglass, see glass fibers
Fiber-reinforced composites, 7–10
Filament winding, 49–51
  patterns, 50, 51
Fillers, 39
  inorganic, 6
Finite element analysis codes, 555, 556
Finish, 18
Flakes, 6
  mica, 6
Fracture mechanics, 335–355

Fracture mechanics concepts, 338–346
Fracture process in composites, 336–338
Fracture process in impact, 395–396
Fracture process zone, 397
Fracture surface work, 341
Fracture toughness
  measures of, 338
  of composite laminates, 346–349
Fundamental frequency, 303

Gel coat, 43
Generalized Hooke's law, 175
Generally orthotropic lamina, 164
Glass fibers, 7, 16–23
  cake, 17
  composition, 18
  chopped-strand mat, 22
  continuous-strand mat, 22
  coupling agent, 18
  E-, 19
  end, 17
  fabric, 23
  finish, 18
  forms of, 21–23
  mat, 22
  milled, 23
  production of, 17–18
  properties of, 8, 18, 19
  roving, 21
  S-, 19
  sizes, 17
    compatible, 18, 19
    temporary, 18
  staple, 17–18
  strand, 17
  surface treatment of, 15
  surfacing mat, 22
  veil, 22
  woven roving, 22
  yarn, 23
Goodman-Boller relationship, 383, 384
Graphite fibers, 23
  from PAN, 24
  precursor, 24
  tows, 26
Gel coating, 43

Haplin-Tsai equations
  for ribbon reinforced composites, 154
  for shear modulus, 93

  for short-fiber composites, 141–144
  for transverse modulus, 85
  for transverse transport properties, 115
Hand lay-up technique, 43
Hole size effect, 349
Homogeneity, 3
Hooke's law, 174
  for generally isotropic material, 174–177
  for generally orthotropic material, 174–177
  for isotropic material, 181–182
  for specially orthotropic material, 177–180
  for transversely isotropic material, 180–181
  generalized, 175
  in contracted notation, 179–180
Hybrid, see hybrid composites
Hybrid composites, 9, 397, 407–411
Hybrid laminates, 3, 5
Hybridization, 407
Hygrothermal forces, 268
Hygrothermal moments, 268
Hygrothermal stresses, 263–273
  Calculations, 264–273

IITRI test fixture, 450
Impact:
  energy absorbed, 396
  energy absorbing mechanisms, 396–401
  failure modes, 396–401
  initiation energy, 408
  low velocity, damage due to, 411–416
  propagation energy, 408
  hybrid composites, 407–411
  strength of short-fiber composites, 152
  Charpy, 475
  drop weight, 476
  instrumented Charpy, 477
  Izod, 475–476
Impact energy values for materials, 408
Impact properties,
  effect of materials variables on, 401–407
  effect of testing variables on, 401–407

558 INDEX

Impact properties (*Continued*)
  of unidirectional fiber-epoxy composites, 410
Ineffective length, 136
Interfacial bond, 79
Interfacial area, 1
Interfacial conditions, 79
Initiation energy, 405, 408
Intelligent structures, 503
Interlaminar fracture toughness,
  determination of, 471–475
Interlaminar shear strength,
  determination of, 324–335, 471
Interlaminar stresses, 471, 472
  approximate solutions for, 330–334
  concepts of, 324–326
  determination of, 326–328
  effect of stacking sequence on, 328–330
Invariant forms of
  compliance matrix, 194–196
  stiffness matrix, 194–195
Isotropic composite, 132
Isotropy, 3, 159
Izod tests, 475

J-curve, 471
J-integral, 345–346
  critical, 470
  determination of, 470–471
  energy interpretation, 470
Joints
  adhesively bonded, 355–360
    advantages of, 359
    configuration, 356–357
    design of, 355
    failure modes, 357
    stresses in, 358
  bonded-mechanically fastened, 361–362
  for composite structures, 355
  mechanically fastened, 360–361
    advantages of, 361
    disadvantages of, 361
    failure modes of, 36

K-calibration factor, 465
K_1-curve, *see* crack extension force curve
Kevlar fibers, 26–27
  chemistry of, 26

  properties of, 27
Knee of stress-strain curve, 250

Lamina, 63, 158, *see also* Orthotropic lamina
Laminate:
  analysis after initial failure, 247–262
  analysis of, 213–281
  analysis through computers, 272–277
  angle-ply, 228–229
  constitutive equations for, 221, 224–225
  cross-ply, 228–229
  curing stresses, 268
  definition of, 158
  description system, 225–226, *see also* laminate orientation code
  effective modulus, 251
  fracture mechanics of, 335–355
  hygrothermal forces in, 268
  hygrothermal moments in, 268
  hygrothermal stresses in, 263–272
  interlaminar stresses in, 324–335, 471
  load carrying capacity of, 255
  mechanical strains in, 266
  netting analysis, 280
  orientation code, 544–549
  primary modulus, 250
  quasi-isotropic, 229–230
  residual stresses, 268
  resultant force, 218
  resultant moment, 218
  secondary modulus, 250
  specially orthotropic, 228
  stacking sequence, 219, 328, 544–549
  stiffness matrices, 221
  strains in, 238
  strength analysis, 274–276
  stress analysis, 273–274
  stresses and strains in, determination of, 238–247
  stresses in, 238
  symmetric, 227–228
  thermal strain, 263
  thermal stresses, *see* Laminate, hygrothermal stresses

  unidirectional, 228–229
Laminated beams
  bending of, 315–318
  buckling of, 318–319
  free vibrations of, 319–320
  governing equations for, 314
Laminated plates
  bending of, 288
  buckling of, 295
  equilibrium equations for, 283–286
  free vibrations of, 301
  governing equations for, 283
  in terms of displacements, 286–288
Laminates, 8, 158
  bidirectional, 10
Laser shearography, 488
Law of mixtures, *see* Rule of mixtures
Load coefficients, 290
Load sharing, 71–73
Load transfer length, 136
Longitudinal direction, 63
Longitudinal stiffness:
  factors influencing, 76–80
  prediction of, 68–69
Longitudinal strength:
  factors influencing, 76–80
  prediction of, 75–76
Low velocity impact, 411–416
  damage due to, 411–416

Mass diffusion, 117–123
Major Poisson's ratio, 95
Mandrel, 50
MAPLE, 555, 556
Mat, 10, 22
  chopped-strand, 22–23
  continuous-strand, 22–23
  surfacing, 22–23
Material axes, 63
Mathcad, 520, 555, 556
MATLAB, 520, 555, 556
Matrix material, 2, 7, 30–41
  Bismaleimides, 38
  effect of temperature and moisture on, 422–426
  elevated temperature, degradation at, 426–429
  epoxy, 36–38
  metals, 39,41
  microcracking, 96
  phenolics, 38

plastics, 30–40
polyester, 34–35
polyimides, 38
polymers, 30–40
vinyl esters, 38
Matrix (mathematical):
  addition, 514
  column, 510
  definitions, 509
  determinant, 513
  diagonal, 511
  elements of, 509
  identity, 511
  inverse, 517
  multiplication, 516
  operations, 514–520
  orthogonal, 519
  principal diagonal of, 511
  row, 510
  skew symmetric, 511
  square, 510
  subtraction, 514
  symmetric, 511
  transformation of, 519
  transpose of, 510
  unit, 511
Matrix digestion method, 441
Matrix dissolution method, 441
Matrix ductility, 150
Maximum strain theory, 200–203
Maximum stress theory, 197–200
Maximum work theory, 203–205
Measure of fiber orientation, 148
Measurement of
  critical crack growth resistance, 464
  critical J-integral, 470
  critical strain-energy release rate, 463
  critical stress-intensity factor, 464
  density, 440
  diffusivity, 444–445
  flexure properties, 459
  impact properties, 475
  in-plane shear properties, 452
  interlaminar fracture toughness, 471
  interlaminar shear strength, 471
  measures of fracture toughness, 463
  mechanical properties, 445–481

  moisture absorption, 444–445
  moisture expansion coefficients, 444–445
  physical properties, 440–445
  properties in compression, 449
  properties in tension, 445
  stiffness and strength, see testing of composites
  thermal expansion coefficients, 442
  void volume fraction, 442
  volume fraction, 441
  weight fractions, 441
Microbuckling of fibers, 102, 103
  in extension mode, 102
  in shear mode, 102, 103
Micromechanics of transverse failure, 88
Microscopic failure initiation, 335
Milled fibers, 23
Minimum volume fraction:
  of short fiber composite, 146
  of unidirectional composite, 76
Minor Poisson's ratio, 95
Modulus:
  effective, 252
  longitudinal, 68, 160
  primary, 250, 375
  residual, 375
  secondary, 250, 375
  shear, 91–95, 160
  of short-fiber composites, 141–145
  transverse, 80–91, 141, 160
Moisture absorption, 114
Moisture expansion coefficients, 114
  longitudinal, 114
  transverse, 114
Mold release, 43
Molding compounds, 10, 53–55
  BMC, 42, 53
  bulk, 42, 53
  DMC, 53
  dough, 53
  prepregs, 42, 55
  sheet, 42, 53, 54
  SMC, 42, 53, 54

Nanocomposites, 7, 496–498
  Clay-reinforced, 7

  Nanotube-reinforced, 7
Nanotubes, 7, 496
  Single walled, 497
  Multi-walled, 497
Natural fibers, 499
  properties of, 500
Navier's approach
NDE, see nondestructive evaluation
Neutral axis, 460
Nodal lines, 304
Nondestructive evaluation, 481–488
Nondestructive evaluation techniques:
  acoustic emission, 483
  laser shearography, 488, 489
  thermography, 486
  ultrasonics, 481
  X-radiography, 485
Notch sensitivity, 346
Notched-bend tests, 464–465
Notched plate, 465
Notched-plate test, 472, 473

Orthotropic lamina, 161
  analysis of, 158
  balanced, 172
  engineering constants of, 160, 166
  generally, 161
  shear strength, importance of sign, 205–209
  specially, 161
  strength of, 196–209
  strength under biaxial stresses, 196–209
  stress-strain relations in arbitrary direction, 164–166
  transformation of engineering constants, 166–174
  variation of elastic constants, 170–172
Orthotropic materials, 158–160, see also Orthotropic lamina
  definition of, 158–160
  deformation behavior of, 159
  Hooke's law for, 160–174

PAN, 24
Particulate composites, 1, 5–7
Plasma spraying, 59

Plastics, *see* polymers
Platelets, 2, 496
Ply, 63
Point of instability, 466
Point-stress criterion, 349
Poisson's ratio:
  major, 95, 161
  minor, 95, 161
  prediction of, 95
  restriction on, 188, 189
Polyester, 34–36
  properties of, 35, 36
Polyethylene fibers, 28–30
  Dyneema, 29
  properties of, 29
  Spectra, 29
Polymerization, 42
Polymers, 30–40
  crystalline melt
    temperatures of, 32
  epoxy, 36–38
    properties of, 37
  glass transition temperatures
    of, 32, 33
  melting point of, 32
  network, 31
  polyester, 34–35
    properties of, 35, 36
  properties of, 31–34
    epoxy resin, 37
    phenolics, 38
    polyester resin, 34
    polyimide, 38
    thermoplastic resins, 40
  thermoplastic, 31, 38–39
    high temperature, 39
    properties of, 39, 40
  thermosets, 31
    temperature for
      processing of, 32
  thermosetting, 31
Preform, 10
Preimpregnated fibers, *see*
  prepregs
Premixes, 42
Prepregs, 9, 42, 55
Primary modulus, 375
Propagation energy, 405, 408
Properties of fiber composites,
  550–554
Properties of unidirectional
  composites, 123–124
Pultruded shapes, 51, 52

Quasi-isotropic laminate, 229–
  230

R-curve, *see* crack-growth
  resistance curve
Rail shear test, 458
Reinforcements, 2
  geometry of, 3
  orientation of, 3
Reinforcing material, 2
Residual strength, 376
Residual stresses, 79–80, 268
Resin transfer molding (RTM),
  49
  vacuum assisted (VARTM),
    49
Resultant forces, 218
Resultant moments, 218
Ribbon-reinforced composites,
  152–155
  Halpin-Tsai equations for,
    154
  in-plane transverse modulus
    of, 154
Rule of mixtures, 69
  for density, 65
  for longitudinal diffusivity,
    120
  for longitudinal modulus,
    69
  for Poisson's ratio, 96
  for stress, 68
  for transport properties, 114

Sandwich cross-beam, 459
Secondary cracks, 398
Secondary modulus, 250, 375
Self-consistent models, 84, 85
Self-similar crack growth, 337
SEN, *see* single edge notched
Series solution, 290
Shear coupling, 164
Shear deformation theory
  first-order, 306–311
  higher-order, 311–314
Shear failure, 107, 109
Shear-lag analysis, 134
Shear modulus, 91–95, 160
Shear strain:
  engineering, 180, 534–535
  tensorial, 180, 535
Shear strength, 197
Shearography, laser, 488
Short beam shear test, 472
Short-fiber composites, 132–
  157
  critical fiber length, 136
  critical fiber volume fraction
    of, 146, 147
  effect of matrix ductility on
    properties of, 150–152

examples of, 133
failure initiation of, 146
fatigue of, 390–395
impact strength of, 152
ineffective fiber length, 136
load transfer length, 136
matrix ductility, effect of,
  150–152
minimum fiber volume
  fraction of, 146, 147
modulus of, 140
  prediction of, 141
  randomly oriented, 143
randomly oriented, 143, 147
ribbon reinforced, 152–155
strength of, 140, 145–149
  prediction of, 145
strength of random fiber
  composite, 147, 148
stress distribution in, 137–
  139
theories of stress transfer
  for, 133–140
Single edge notched specimen,
  465
Sizes, 17, 18–20
  compatible, 18, 19
  temporary, 18
Smart structures, 503
S-N curve, 378
Software packages,
  commercially available,
  556
Sound deadener, 6
Specially orthotropic lamina,
  161
  stress-strain relations for,
    161–164
  under plane stress, 182
    constitutive equation for,
      192
    stiffness coefficients for,
      183
Specially orthotropic material,
  177
Specific stiffness, 7–8, 11
Specific strength, 7–8, 11
Speckle effect, 488
Spray-up, 44, 45
Stamping, 57
Staple fibers, 18
Static fatigue, 417
Static fatigue of fibers, 417
Static-rupture of fibers, 417
Stiffness:
  factors influencing, 76–80
  methods of predicting
    longitudinal, 68–69
  residual, 376
  transverse, 84–87

INDEX 561

Stiffness matrix, 183
  bending, 221
  coupling, 221
  engineering constants,
    relations with, 185–187
  extensional, 221
  for orthotropic materials,
    183
  invariant form of, 194
  inversion of, 238
  of fractured plies, 248
  synthesis of, 218–221
  transformation of, 189
Strain:
  analysis of, 532–535
  compatibility conditions,
    535
  determination of, 238
  engineering, 180, 535
  mechanical, 266
  tensorial, 180, 535
  transformation equation,
    534
Strain-displacement relations,
  533
Strain-energy release rate,
  339–341
  critical, 341
Strain magnification factor, 90
Strain-stress relations, see
    Stress-strain relations
Strength:
  factors influencing, 76–80
  longitudinal compressive,
    103–106, 197
  longitudinal tensile, 75–76,
    197
  notched, 351–355
  of orthotropic lamina, 196
  of quasi-isotropic laminate,
    257
  residual, 376
  of short-fiber composite,
    145–149
  transverse, 87–91, 197
  theories of, 197–205
Strength reduction factor, 90
Strengthening mechanisms, 3,
  5
Stress:
  analysis of, 536–540
  boundary conditions, 537,
    539
  curing, 268
  determination of, 238–247
  equilibrium equations, 539
  hygrothermal, 263–272
  interlaminar, 324–335
  residual, 79–80, 268

sign convention for, 205,
  537
symmetry of, 538
thermal, see, hygrothermal
  transformation, 539–540
Stress concentration factor, 88,
  90
Stress corrosion, 419
  features of, 416
  of glass fibers, 419–422
  of GRP, 419–422
Stress-intensity factor, 339,
  341–344
  critical, 464
  determination of, 464–470
  relation with strain energy
    release rate, 342
Stress-rupture characteristics,
  429–431
Stress-rupture of fibers, 418
Stress-strain relations, 160
  for anisotropic materials,
    175–176
  for generally orthotropic
    lamina, 164–166
  for isotropic materials, 540
  for specially orthotropic
    lamina, 161–164
  generalized, 175
  in terms of engineering
    constants, 163
Surface energy, 341
Surfacing mat, 44
Symmetric laminates, 227

Tensors:
  definition of, 527
  laws of transformation, 527
Tests,
  bend, four point, 459, 460
  bend, three point, 459, 460
  compression, 449
  coupon, $[\pm 45]_s$, 455
  coupon, off-axis, 456
  double cantilever beam, 472
  flexural, 459
  fracture toughness, 463
  impact, 475
    drop weight, 476
    Charpy, 476
    Izod, 476
  in-plane shear, 452
  Iosipescu shear, 453
  notched plate, 472
  picture frame, 458
  rail shear, 458
  sandwich cross beam, 459
  sandwich beam, 448

short beam shear, 472
tension, 445
torsion tube, 452
Testing of composites:
  bending, 459
  Charpy, 475–481
  compression, 449
  drop weight, 476
  flexural, 459
  four-point bending, 459
  fracture toughness, 463–475
  in-plane shear, 452–459
    Iosipescu, 453
    $[\pm 45]_s$ coupon, 455
    off-axis coupon, 456
    torsion tube, 452
  instrumented Charpy, 475–
    481
  Izod, 475–481
  notched bend, 465
  notched plate, 463
  off-axis shear, 457
  off-axis tension, 447
  picture frame, 458
  rail shear, 458
  sandwich crossbeam, 459
  short beam shear, 472
  tension, 445–449
  three point bending, 459
Thermal stresses
  concepts of, 263–264
Thermography, 486–488
  pulse-echo, 487
  through transmission, 487
Thickness shear, 472
Transverse shear
  deformations due to, 306–
    311
  effects of, 306
Transverse splitting, 102
Transversely isotropic, 63, 160
Theories of failure, see failure
    theories
Theories of stress transfer,
  133–140
Thermal conductivity, 115–
  117
Thermal expansion
  coefficients, 108–114
  longitudinal, 111
  transverse, 111
Thermal strains, 264
Thermal stress, see
    Hygrothermal stress
Thermoforming, 57
Thermoplastic polymers, 31
  properties of, 40
Total impact energy, 405
Transformation:
  matrix, 190

Transformation (*Continued*)
  of strain, 534
  of stress, 539
  relations, 190
Transport properties, 114
Transverse:
  isotropy, 63
  splitting, 102–106
  stiffness, 80–87
  strength, 87–91
Transverse direction, 63
Tsai-Hill theory, 203

Ultrasonics, 481–483
  pitch-catch method of, 481, 482
  pulse-echo method of, 481, 482
  through transmission method of, 482, 483
Unidirectional composites:
  anisotropy of thermal expansion, 112
  coefficient of thermal expansion, 108–114
  critical volume fraction, 75–76
  expansion coefficients, 108–114
  failure initiation, 74
  failure mechanism of, 74
  failure modes, 96–108
  longitudinal behavior of, 67
  longitudinal stiffness, 68–71
  longitudinal strength, 75–76
    factors influencing, 76
    statistical models for, 77
  mass diffusion in, 117–123
  model for, 68
  moisture in, 114
  minimum volume fraction, 75
  Poisson's ratio, 95
    major, 95
    minor, 95
    prediction of, 95–96
  properties of, typical, 113, 123, 124
  shear modulus of, 91–95
    prediction of, 91
    Haplin-Tsai equations for, 93
  thermal conductivities, 115–117
  thermal expansion, 108–114
  transport properties, 114
  transverse stiffness, 80–87
  transverse strength, 87–91
    prediction of, 90
    empirical approach for, 91
  transversely isotropic, 63
Unstable crack growth, 466

Variation of stresses in a laminate, 216–217
Veil, 23, 44
Voids, 67
  volume fraction, 67
Void content, 67
Volume fraction, 64, 65
  critical, 76, 146
  definition of, 65–66
  minimum, 75, 146

Weight fractions, 2, 64, 65
Whitney-Nuismer failure criteria, 349–355
  for notched composites, 349–355
Winding angle, optimal, 280
Wood-plastic composite, 501–502

X-radiography, 485–486
  penetrant-enhanced, 486